T0202889

Lecture Notes in Computer Science   14163

Founding Editors

Gerhard Goos
Juris Hartmanis

Editorial Board Members

The series Lecture Notes in Computer Science (LNCS), including its subseries Lecture Notes in Artificial Intelligence (LNAI) and Lecture Notes in Bioinformatics (LNBI), has established itself as a medium for the publication of new developments in computer science and information technology research, teaching, and education.

LNCS enjoys close cooperation with the computer science R & D community, the series counts many renowned academics among its volume editors and paper authors, and collaborates with prestigious societies. Its mission is to serve this international community by providing an invaluable service, mainly focused on the publication of conference and workshop proceedings and postproceedings. LNCS commenced publication in 1973.

Avi Arampatzis · Evangelos Kanoulas ·
Theodora Tsikrika · Stefanos Vrochidis ·
Anastasia Giachanou · Dan Li ·
Mohammad Aliannejadi · Michalis Vlachos ·
Guglielmo Faggioli · Nicola Ferro
Editors

Experimental IR Meets Multilinguality, Multimodality, and Interaction

14th International Conference of the CLEF Association, CLEF 2023
Thessaloniki, Greece, September 18–21, 2023
Proceedings

 Springer

Editors
Avi Arampatzis (iD)
Democritus University of Thrace
Xanthi, Greece

Evangelos Kanoulas (iD)
University of Amsterdam
Amsterdam, The Netherlands

Theodora Tsikrika (iD)
CERTH-ITI
Thessaloniki, Greece

Stefanos Vrochidis (iD)
CERTH-ITI
Thessaloniki, Greece

Anastasia Giachanou (iD)
Utrecht University
Utrecht, The Netherlands

Dan Li (iD)
Elsevier
Amsterdam, The Netherlands

Mohammad Aliannejadi (iD)
University of Amsterdam
Amsterdam, The Netherlands

Michalis Vlachos (iD)
University of Lausanne
Lausanne, Switzerland

Guglielmo Faggioli (iD)
University of Padua
Padova, Italy

Nicola Ferro (iD)
University of Padua
Padova, Italy

ISSN 0302-9743   ISSN 1611-3349 (electronic)
Lecture Notes in Computer Science
ISBN 978-3-031-42447-2   ISBN 978-3-031-42448-9 (eBook)
https://doi.org/10.1007/978-3-031-42448-9

This Springer imprint is published by the registered company Springer Nature Switzerland AG
The registered company address is: Gewerbestrasse 11, 6330 Cham, Switzerland

Paper in this product is recyclable.

Preface

Since 2000, *the Conference and Labs of the Evaluation Forum (CLEF)* has played a leading role in stimulating research and innovation in the domain of multimodal and multilingual information access. Initially founded as the *Cross-Language Evaluation Forum* and running in conjunction with the *European Conference on Digital Libraries* (ECDL/TPDL), CLEF became a standalone event in 2010 combining a peer-reviewed conference with a multi-track evaluation forum. The combination of the scientific program and the track-based evaluations at the CLEF conference creates a unique platform to explore information access from different perspectives, in any modality and language.

The CLEF conference has a clear focus on experimental information retrieval (IR) as seen in evaluation forums (like the CLEF Labs, TREC, NTCIR, FIRE, MediaEval, RomIP, TAC) with special attention to the challenges of multimodality, multilinguality and interactive search, ranging from unstructured to semi-structured and structured data. The CLEF conference invites submissions on new insights demonstrated by the use of innovative IR evaluation tasks or in the analysis of IR test collections and evaluation measures, as well as on concrete proposals to push the boundaries of the Cranfield/TREC/CLEF paradigm.

CLEF 2023[1] was organized by the Information Technologies Institute, Centre for Research and Technology Hellas (CERTH), Thessaloniki, Greece, from 18 to 21 September 2023. CLEF 2023 was the 14th year of the CLEF Conference and the 24th year of the CLEF initiative as a forum for IR Evaluation. The conference format remained the same as in past years and consisted of keynotes, contributed papers, lab sessions and poster sessions, including reports from other benchmarking initiatives from around the world. All sessions were organized in presence but also allowing for remote participation for those who were not able to attend physically.

CLEF 2023 continued the initiative introduced in the 2019 edition, during which the *European Conference for Information Retrieval (ECIR)* and CLEF joined forces: ECIR 2023[2] hosted a special session dedicated to CLEF Labs where lab organizers presented the major outcomes of their Labs and their plans for ongoing activities, followed by a poster session to favour discussion during the conference. This was reflected in the ECIR 2023 proceedings, where CLEF Lab activities and results were reported as short papers. The goal was not only to engage the ECIR community in CLEF activities but also to disseminate the research results achieved during CLEF evaluation cycles as submission of papers to ECIR.

The following scholars were invited to give keynote talks at CLEF 2023:*Barbara Plank* (Ludwig Maximilian University of Munich, Germany and IT University of Copenhagen, Denmark) and *Claudia Hauff* (Spotify, The Netherlands).

---

[1] https://clef2023.clef-initiative.eu/.

[2] https://ecir2023.org/.

CLEF 2023 received a total of 14 scientific submissions, of which a total of 11 papers (10 long & 1 short) were accepted. Each submission was reviewed by two program committee members, and the program chairs oversaw the reviewing and follow-up discussions. Several papers were a product of international collaboration. This year, researchers addressed the following important challenges in the community: authorship attribution, fake news detection and news tracking, noise-detection in automatically transferred relevance judgments, impact of online education on children's conversational search behavior, analysis of multi-modal social media content, knowledge graphs for sensitivity identification, a fusion of deep learning and logic rules for sentiment analysis, medical concept normalization and domain-specific information extraction.

Like in previous editions, since 2015, CLEF 2023 continued to invite CLEF lab organizers to nominate a "best of the labs" paper that was reviewed as a full paper submission to the CLEF 2023 conference, according to the same review criteria and PC. Seven full papers were accepted for this "best of the labs" section.

The conference integrated a series of workshops presenting the results of lab-based comparative evaluations. A total of 15 lab proposals were received and evaluated in peer review based on their innovation potential and the quality of the resources created. The 13 selected labs represented scientific challenges based on new datasets and real-world problems in multimodal and multilingual information access. These datasets provide unique opportunities for scientists to explore collections, to develop solutions for these problems, to receive feedback on the performance of their solutions and to discuss the challenges with peers at the workshops. In addition to these workshops, the labs reported results of their year-long activities in overview talks and lab sessions. Overview papers describing each of the labs are provided in this volume. The full details for each lab are contained in a separate publication, the Working Notes.[3]

The 13 labs running as part of CLEF 2023 comprised mainly labs that continued from previous editions at CLEF (BioASQ, CheckThat!, eRisk, iDPP, ImageCLEF, JOKER, LifeCLEF, PAN, SimpleText and Touché) and new pilot/workshop activities (DocILE, EXIST and LongEval). In the following we give a few details for each of the labs organized at CLEF 2023 (presented in alphabetical order):

BioASQ: Large-scale biomedical semantic indexing and question answering[4] aimed to push the research frontier towards systems that use the diverse and voluminous information available online to respond directly to the information needs of biomedical scientists. It offered the following tasks.

Task 1 - b: Biomedical Semantic Question Answering: benchmark datasets of biomedical questions, in English, along with gold standard (reference) answers constructed by a team of biomedical experts. The participants have to respond with relevant articles, and snippets from designated resources, as well as exact and "ideal" answers. *Task 2 - Synergy: Question Answering for developing problems*: biomedical experts pose unanswered questions for developing problems, such as COVID-19, receive the responses provided by the participating systems, and provide feedback, together with updated

---

[3] Aliannejadi, M., Faggioli, G., Ferro, N., and Vlachos, M. editors (2023). CLEF 2023 Working Notes. CEUR Workshop Proceedings (CEUR-WS.org), ISSN 1613-0073.

[4] http://www.bioasq.org/workshop2023.

questions in an iterative procedure that aims to facilitate the incremental understanding of developing problems in biomedicine and public health. *Task 3 - MedProcNER: Medical Procedure Text Mining and Indexing Shared Task*: focuses on the recognition and indexing of medical procedures in clinical documents in Spanish posing subtasks on (1) indexing medical documents with controlled terminologies, (2) automatic detection indexing textual evidence, i.e., medical procedure entity mentions in text, and (3) normalization of these medical procedure mentions to terminologies.

CheckThat!: Check-Worthiness, Subjectivity, Political Bias, Factuality, and Authority of News Articles and their Sources[5] aimed at producing technology to support the fight against misinformation and disinformation in social media, in political debates and in the news with a focus on check-worthiness, subjectivity, bias, factuality and authority of the claim. It offered the following tasks *Task 1 - Check-worthiness in textual and multimodal content*: determine whether an item, be it a text alone or a text plus an image, deserves the attention of a journalist to be fact-checked. *Task 2 - Subjectivity in News Articles*: assess whether a text snippet within a news article is subjective or objective. *Task 3 - Political Bias of News Articles and News Media*: identify the political leaning of an article or media source: left, centre or right.

Task 4 - Factuality of Reporting of News Media: determine the level of factuality of both a document and a medium. *Task 5 - Authority Finding in Twitter*: identify authorities that should be trusted to verify a contended claim expressed in an Arabic tweet.

DocILE: Document Information Localization and Extraction[6] ran the largest benchmark for the tasks of Key Information Localization and Extraction (KILE) and Line Item Recognition (LIR) from business documents such as invoices. It offered the following tasks. *Task 1 - Key Information Localization and Extraction (KILE)*: localize fields of each pre-defined category and read out their values. *Task 2 - Line Item Recognition (LIR)*: find all line items, e.g., a billed item in a table, and localize their corresponding fields in the document as in Task 1.

eRisk: Early Risk Prediction on the Internet[7] explored the evaluation methodology, effectiveness metrics and practical applications (particularly those related to health and safety) of early risk detection on the Internet. Early detection technologies can be employed in different areas, particularly those related to health and safety. For instance, early alerts could be sent when a predator starts interacting with a child for sexual purposes, or when a potential offender starts publishing antisocial threats on a blog, forum or social network. Our main goal is to pioneer a new interdisciplinary research area that would be potentially applicable to a wide variety of situations and to many different personal profiles. Examples include potential paedophiles, stalkers, individuals that could fall into the hands of criminal organisations, people with suicidal inclinations, or people susceptible to depression. It offered the following tasks. *Task 1 - Search for symptoms of depression*: the challenge consists of ranking sentences from a collection of user writings according to their relevance to a depression symptom. The participants will have to provide rankings for the 21 symptoms of depression from the BDI Questionnaire. A sentence will be deemed relevant to a BDI symptom when it conveys information about the

---

[5] http://checkthat.gitlab.io/.

[6] https://docile.rossum.ai/.

[7] https://erisk.irlab.org/.

user's state concerning the symptom. That is, it may be relevant even when it indicates that the user is OK with the symptom. *Task 2 - Early Detection of Signs of Pathological Gambling*: the challenge consists of sequentially processing pieces of evidence to detect early traces of pathological gambling (also known as compulsive gambling or disordered gambling), as soon as possible. The task is mainly concerned with evaluating Text Mining solutions and, thus, it concentrates on texts written in Social Media. *Task 3 - Measuring the severity of the signs of Eating Disorders*: the task consists of estimating the level of features associated with a diagnosis of eating disorders from a thread of user submissions. For each user, the participants will be given a history of postings and will have to fill a standard eating disorder questionnaire (based on the evidence found in the history of postings).

EXIST: sEXism Identification in Social neTworks[8] aimed to capture and categorize sexism, from explicit misogyny to more subtle behaviors, in social networks. Participants will be asked to classify tweets in English and Spanish according to the type of sexism they enclose and the intention of the persons that writes the tweets. It offered the following tasks. *Task 1 - Sexism Identification*: is a binary classification task. The systems have to decide whether or not a given tweet contains or describes sexist expressions or behaviors (i.e., it is sexist itself, describes a sexist situation or criticizes a sexist behavior). *Task 2 - Source Intention*: aims to categorize the sexist messages according to the intention of the author in one of the following categories: (i) direct sexist message, (ii) reported sexist message and (iii) judgemental message. *Task 3 - Sexism Categorization*: is a multiclass task that aims to categorize the sexist messages according to the type or types of sexism they contain (according to the categorization proposed by experts and that takes into account the different facets of women that are undermined): (i) ideological and inequality, (ii) stereotyping and dominance, (iii) objectification, (iv) sexual violence and (v) misogyny and non-sexual violence.

iDPP: Intelligent Disease Progression Prediction[9] aimed to design and develop an evaluation infrastructure for AI algorithms able to: (1) better describe the mechanism of the Amyotrophic Lateral Sclerosis (ALS) disease; (2) stratify patients according to their phenotype assessed throughout the disease evolution; and (3) predict ALS progression in a probabilistic, time dependent fashion. It offered the following tasks. *Task 1 – Predicting Risk of Disease Worsening (Multiple Sclerosis)*: focuses on ranking subjects based on the risk of worsening, setting the problem as a survival analysis task. More specifically the risk of worsening predicted by the algorithm should reflect how early a patient experiences the event "worsening". Worsening is defined based on the Expanded Disability Status Scale (EDSS), accordingly to clinical standards. *Task 2 – Predicting Probability of Worsening (Multiple Sclerosis)*: refines Task 1 by asking participants to explicitly assign a probability of worsening at different time windows (e.g., between years 4 and 6, 6 and 8, 8 and 10 etc.). *Task 3 – Impact of Exposure to Pollutants (Amyotrophic Lateral Sclerosis)*: evaluates proposals of different approaches to assess whether exposure to different pollutants is a useful variable to predict time to Percutaneous Endoscopic Gastrostomy (PEG), Non-Invasive Ventilation (NIV) and death in ALS patients.

---

[8] http://nlp.uned.es/exist2023/.

[9] https://brainteaser.health/open-evaluation-challenges/idpp-2023/.

ImageCLEF: Multimedia Retrieval[10] promoted the evaluation of technologies for annotation, indexing, classification and retrieval of multimodal data, with the objective of providing information access to large collections in various usage scenarios and domains. It offered the following tasks. *Task 1 - ImageCLEFmedical*: continues the tradition of bringing together several initiatives for medical applications fostering cross-exchanges, namely: medical concept detection and caption prediction, synthetic medical images generated with GANs, Visual Question Answering and generation, and doctor-patient conversation summarization. *Task 2 - ImageCLEF- aware*: the images available on social networks can be exploited in ways users are unaware of when initially shared, including situations that have serious consequences for the users' real lives. The task addresses the development of algorithms which raise the users' awareness about real-life impact of online image sharing. *Task 3 - ImageCLEFfusion*: despite the current advances in knowledge discovery, single learners do not produce satisfactory performances when dealing with complex data, such as class imbalance, high-dimensionality, concept drift, noise, multimodality, subjective annotations, etc. This task aims to fill this gap by exploiting novel and innovative late fusion techniques to produce a powerful learner based on the expertise of a pool of classifiers. *Task 4 - ImageCLEFrecommendation*: focuses on content recommendation for cultural heritage content in 15 broad themes that have been curated by experts in the Europeana Platform. Despite current advances, there is limited understanding of how well these perform and of how relevant they are for the final end-users.

JOKER: Automatic Wordplay Analysis[11] aimed to create reusable test collections for benchmarking and to explore new methods and evaluation metrics for the automatic processing of wordplay. It offered the following tasks. *Task 1 - Pun detection*: detection of puns in English, French and Spanish. *Task 2 - Pun interpretation*: interpretation of puns in English, French, and Spanish. *Task 3 - Pun translation*: translation of puns from English to French and Spanish.

LifeCLEF: Multimedia Retrieval in Nature[12] was dedicated to the large-scale evaluation of biodiversity identification and prediction methods based on artificial intelligence. It offered the following tasks. *Task 1 - BirdCLEF*: bird species recognition in audio soundscapes. *Task 2 - FungiCLEF*: fungi recognition from images and metadata. *Task 3 - GeoLifeCLEF*: remote-sensing-based prediction of species. *Task 4 - PlantCLEF*: global-scale plant identification from images. *Task 5 - SnakeCLEF*: snake species identification in medically important scenarios.

LongEval: Longitudinal Evaluation of Model Performance[13] focused on evaluating the temporal persistence of information retrieval systems and text classifiers. The goal is to develop temporal information retrieval systems and longitudinal text classifiers that survive through dynamic temporal text changes, introducing time as a new dimension for ranking models' performance. It offered the following tasks. *Task 1 - LongEval-Retrieval*: aims to propose a temporal information retrieval system which can handle changes over time. The proposed retrieval system should follow the temporal persistence

---

[10] https://www.imageclef.org/2023.
[11] http://joker-project.com/.
[12] http://www.lifeclef.org/.
[13] https://clef-longeval.github.io/.

of Web documents. This task will have 2 sub-tasks focusing on short-term and long-term persistence. *Task 2 - LongEval-Classification* aims to propose a temporal persistence classifier which can mitigate performance drop over short and long periods of time compared to a test set from the same time frame as training. This task will have 2 sub-tasks focusing on short-term and long-term persistence.

PAN: Digital Text Forensics and Stylometry[14] aimed to advance the state of the art and provide for an objective evaluation on newly developed benchmark datasets in those areas. It offered the following tasks. *Task 1 - Cross-Discourse Type Authorship Verification*: focuses on (cross-discourse type) authorship verification where both written (e.g., essays, emails) and oral language (e.g., interviews, speech transcriptions) are represented in the set of discourse types. *Task 2 - Profiling Cryptocurrency Influencers with Few-Shot Learning*: aims to profile cryptocurrency influencers in social media (Twitter) from a low-resource perspective. *Task 3 - Multi-Author Writing Style Analysis*: addresses multi-authored documents whose authorship cannot be easily determined by exploiting topic changes alone. *Task 4 - Trigger Detection*: addresses the task of assigning a single trigger warning label (violence) to narratives in a corpus of fanfiction.

SimpleText: Automatic Simplification of Scientific Texts[15] aimed to create a simplified summary of multiple scientific documents based on a popular science query which provides a user with an instant accessible overview on this specific topic. It offered the following tasks. *Task 1 - What is in, or out?*: selecting passages to include in a simplified summary. *Task 2 - What is unclear?*: difficult concept identification and explanation. *Task 3 - Rewrite this!*: rewriting scientific text.

Touché: Argument and Causal Retrieval[16] aimed to foster and support the development of technologies for argument and causal retrieval and analysis that includes argument quality estimation, stance detection, image retrieval and causal evidence retrieval. It offered the following tasks. *Task 1 - Argument Retrieval for Controversial Questions*: given a controversial topic and a collection of web documents, the task is to retrieve and rank documents by relevance to the topic, by argument quality, and to detect the document stance. *Task 2 - Evidence Retrieval for Causal Questions*: given a causality-related topic and a collection of web documents, the task is to retrieve and rank documents by relevance to the topic and detect each document's "causal" stance (i.e., whether a causal relationship from the topic's title holds). *Task 3 - Image Retrieval for Arguments*: given a controversial topic, the task is to retrieve images (from web pages) for each stance (pro/con) that show support for that stance. *Task 4 - Intra-Multilingual Multi-Target Stance Classification*: given a proposal on a socially important issue, its title and topic in different languages, the task is to classify whether a comment is in favor, against or neutral towards the proposal.

The success of CLEF 2023 would not have been possible without the huge effort of several people and organizations, including the CLEF Association[17], the Program Committee, the Lab Organizing Committee, the reviewers and the many students and volunteers who contributed.

---

[14] https://pan.webis.de/.

[15] http://simpletext-project.com/.

[16] https://touche.webis.de/.

[17] https://www.clef-initiative.eu/#association.

Finally, we thank the Information Technologies Institute (ITI) of the Centre for Research and Technology Hellas (CERTH) in Thessaloniki, Greece (with special mention to the Multimodal Data Fusion & Analytics Group of the Multimedia Knowledge and Social Media Analytics Laboratory) for their invaluable support. We also thank the Friends of SIGIR program for covering the registration fees for a number of student delegates.

July 2023

Avi Arampatzis
Evangelos Kanoulas
Theodora Tsikrika
Stefanos Vrochidis
Anastasia Giachanou
Dan Li
Mohammad Aliannejadi
Michalis Vlachos
Guglielmo Faggioli
Nicola Ferro

Organization

General Chairs

Avi Arampatzis   Democritus University of Thrace, Greece
Evangelos Kanoulas   University of Amsterdam, The Netherlands
Theodora Tsikrika   Information Technologies Institute, CERTH,
                                  Greece
Stefanos Vrochidis   Information Technologies Institute, CERTH,
                                  Greece

Program Chairs

Anastasia Giachanou   Utrecht University, The Netherlands
Dan Li   Elsevier, The Netherlands

Lab Chairs

Mohammad Aliannejadi   University of Amsterdam, The Netherlands
Michalis Vlachos   University of Lausanne, Switzerland

Lab Mentorship Chair

Jian-Yun Nie   University of Montreal, Canada

Proceedings Chairs

Guglielmo Faggioli   University of Padua, Italy
Nicola Ferro   University of Padua, Italy

CLEF Steering Committee

Steering Committee Chair

Nicola Ferro   University of Padua, Italy

Deputy Steering Committee Chair for the Conference

Paolo Rosso   Universitat Politècnica de València, Spain

Deputy Steering Committee Chair for the Evaluation Labs

Martin Braschler   Zurich University of Applied Sciences,
                                     Switzerland

Members

Alberto Barrón-Cedeño	University of Bologna, Italy
Khalid Choukri	Evaluations and Language resources Distribution Agency (ELDA), France
Fabio Crestani	Università della Svizzera italiana, Switzerland
Carsten Eickhoff	University of Tübingen, Germany
Norbert Fuhr	University of Duisburg-Essen, Germany
Lorraine Goeuriot	Universitè Grenoble Alpes, France
Julio Gonzalo	National Distance Education University (UNED), Spain
Donna Harman	National Institute for Standards and Technology, USA
Bogdan Ionescu	University "Politehnica" of Bucharest, Romania
Evangelos Kanoulas	University of Amsterdam, The Netherlands
Birger Larsen	University of Aalborg, Denmark
David E. Losada	Universidade de Santiago de Compostela, Spain
Mihai Lupu	Vienna University of Technology, Austria
Maria Maistro	University of Copenhagen, Denmark
Josiane Mothe	IRIT, Universitè de Toulouse, France
Henning Müller	University of Applied Sciences Western Switzerland (HES-SO), Switzerland
Jian-Yun Nie	Universitè de Montréal, Canada
Gabriella Pasi	University of Milano-Bicocca, Italy
Eric SanJuan	Avignon University, France
Giuseppe Santucci	Sapienza University of Rome, Italy

Laure Soulier	Sorbonne University, France
Theodora Tsikrika	Information Technologies Institute (ITI), Centre for Research and Technology Hellas (CERTH), Greece
Christa Womser-Hacker	University of Hildesheim, Germany

Past Members

Paul Clough	University of Sheffield, UK
Djoerd Hiemstra	Radboud University, The Netherlands
Jaana Kekäläinen	University of Tampere, Finland
Séamus Lawless	Trinity College Dublin, Ireland
Carol Peters	ISTI, National Council of Research (CNR), Italy (Steering Committee Chair 2000–2009)
Emanuele Pianta	Centre for the Evaluation of Language and Communication Technologies (CELCT), Italy
Maarten de Rijke	University of Amsterdam, The Netherlands
Jacques Savoy	University of Neuchâtel, Switzerland
Alan Smeaton	Dublin City University, Ireland

Supporters and Sponsors

Contents

Conference Papers

Inception Models for Fashion Image Captioning: An Extensive Study
on Multiple Datasets 3
   Mirko Del Moro, Serban Cristian Tudosie, Francesco Vannoni,
   Andrea Galassi, and Federico Ruggeri

The Best is Yet to Come: A Reproducible Analysis of CLEF eHealth TAR
Experiments 15
   Giorgio Maria Di Nunzio and Federica Vezzani

Predicting Retrieval Performance Changes in Evolving Evaluation
Environments 21
   Alaa El-Ebshihy, Tobias Fink, Gabriela Gonzalez-Saez,
   Petra Galuščáková, Florina Piroi, David Iommi, Lorraine Goeuriot,
   and Philippe Mulhem

When Sarcasm Hurts: Irony-Aware Models for Abusive Language
Detection 34
   Simona Frenda, Viviana Patti, and Paolo Rosso

Cem Mil Podcasts: A Spoken Portuguese Document Corpus
for Multi-modal, Multi-lingual and Multi-dialect Information Access
Research 48
   Ekaterina Garmash, Edgar Tanaka, Ann Clifton, Joana Correia,
   Sharmistha Jat, Winstead Zhu, Rosie Jones, and Jussi Karlgren

Using Authorship Embeddings to Understand Writing Style in Social Media ... 60
   Javier Huertas-Tato, Alejandro Martín, and David Camacho

Trend Detection in Crime-Related Time Series with Change Point
Detection Methods 72
   Apostolos Konstantinou, Despoina Chatzakou, Ourania Theodosiadou,
   Theodora Tsikrika, Stefanos Vrochidis, and Ioannis Kompatsiaris

DAVI: A Dataset for Automatic Variant Interpretation 85
   Francesca Longhin, Alessandro Guazzo, Enrico Longato, Nicola Ferro,
   and Barbara Di Camillo

qCLEF: A Proposal to Evaluate Quantum Annealing for Information
Retrieval and Recommender Systems   97
    Andrea Pasin, Maurizio Ferrari Dacrema, Paolo Cremonesi,
    and Nicola Ferro

Graph-Enriched Biomedical Entity Representation Transformer   109
    Andrey Sakhovskiy, Natalia Semenova, Artur Kadurin,
    and Elena Tutubalina

Supervised Machine-Generated Text Detectors: Family and Scale Matters   121
    Areg Mikael Sarvazyan, José Ángel González, Paolo Rosso,
    and Marc Franco-Salvador

Best of CLEF 2022 Labs

Cross-Lingual Candidate Retrieval and Re-ranking for Biomedical Entity
Linking   135
    Florian Borchert, Ignacio Llorca, and Matthieu-P. Schapranow

Humour Translation with Transformers   148
    Farhan Dhanani, Muhammad Rafi, and Muhammad Atif Tahir

Fight Against Misinformation on Social Media: Detecting
Attention-Worthy and Harmful Tweets and Verifiable and Check-Worthy
Claims   161
    Ahmet Bahadir Eyuboglu, Bahadir Altun, Mustafa Bora Arslan,
    Ekrem Sonmezer, and Mucahid Kutlu

A Re-labeling Approach Based on Approximate Nearest Neighbors
for Identifying Gambling Disorders in Social Media   174
    Hermenegildo Fabregat, Andres Duque, Lourdes Araujo,
    and Juan Martinez-Romo

Touché 2022 Best of Labs: Neural Image Retrieval for Argumentation   186
    Tobias Schreieder and Jan Braker

SimpleText Best of Labs in CLEF-2022: Simplify Text Generation
with Prompt Engineering   198
    Shih-Hung Wu and Hong-Yi Huang

Answer Retrieval for Math Questions Using Structural and Dense Retrieval   209
    Wei Zhong, Yuqing Xie, and Jimmy Lin

CLEF 2023 Lab Overviews

Overview of BioASQ 2023: The Eleventh BioASQ Challenge
on Large-Scale Biomedical Semantic Indexing and Question Answering 227
Anastasios Nentidis, Georgios Katsimpras, Anastasia Krithara,
Salvador Lima López, Eulália Farré-Maduell, Luis Gasco,
Martin Krallinger, and Georgios Paliouras

Overview of the CLEF–2023 CheckThat! Lab on Checkworthiness,
Subjectivity, Political Bias, Factuality, and Authority of News Articles
and Their Source 251
Alberto Barrón-Cedeño, Firoj Alam, Andrea Galassi,
Giovanni Da San Martino, Preslav Nakov, Tamer Elsayed,
Dilshod Azizov, Tommaso Caselli, Gullal S. Cheema, Fatima Haouari,
Maram Hasanain, Mucahid Kutlu, Chengkai Li, Federico Ruggeri,
Julia Maria Struß, and Wajdi Zaghouani

Overview of DocILE 2023: Document Information Localization
and Extraction 276
Štěpán Šimsa, Michal Uřičář, Milan Šulc, Yash Patel, Ahmed Hamdi,
Matěj Kocián, Matyáš Skalický, Jiří Matas, Antoine Doucet,
Mickaël Coustaty, and Dimosthenis Karatzas

Overview of eRisk 2023: Early Risk Prediction on the Internet 294
Javier Parapar, Patricia Martín-Rodilla, David E. Losada,
and Fabio Crestani

Overview of EXIST 2023 – Learning with Disagreement for Sexism
Identification and Characterization 316
Laura Plaza, Jorge Carrillo-de-Albornoz, Roser Morante,
Enrique Amigó, Julio Gonzalo, Damiano Spina, and Paolo Rosso

Intelligent Disease Progression Prediction: Overview of iDPP@CLEF 2023 ... 343
Guglielmo Faggioli, Alessandro Guazzo, Stefano Marchesin,
Laura Menotti, Isotta Trescato, Helena Aidos, Roberto Bergamaschi,
Giovanni Birolo, Paola Cavalla, Adriano Chiò, Arianna Dagliati,
Mamede de Carvalho, Giorgio Maria Di Nunzio, Piero Fariselli,
Jose Manuel García Dominguez, Marta Gromicho, Enrico Longato,
Sara C. Madeira, Umberto Manera, Gianmaria Silvello,
Eleonora Tavazzi, Erica Tavazzi, Martina Vettoretti,
Barbara Di Camillo, and Nicola Ferro

Overview of the ImageCLEF 2023: Multimedia Retrieval in Medical,
Social Media and Internet Applications 370

Bogdan Ionescu, Henning Müller, Ana-Maria Drăgulinescu,
Wen-Wai Yim, Asma Ben Abacha, Neal Snider, Griffin Adams,
Meliha Yetisgen, Johannes Rückert, Alba García Seco de Herrera,
Christoph M. Friedrich, Louise Bloch, Raphael Brüngel,
Ahmad Idrissi-Yaghir, Henning Schäfer, Steven A. Hicks,
Michael A. Riegler, Vajira Thambawita, Andrea M. Storås,
Pål Halvorsen, Nikolaos Papachrysos, Johanna Schöler, Debesh Jha,
Alexandra-Georgiana Andrei, Ioan Coman, Vassili Kovalev,
Ahmedkhan Radzhabov, Yuri Prokopchuk, Liviu-Daniel Ştefan,
Mihai-Gabriel Constantin, Mihai Dogariu, Jérôme Deshayes,
and Adrian Popescu

Overview of JOKER – CLEF-2023 Track on Automatic Wordplay Analysis ... 397
Liana Ermakova, Tristan Miller, Anne-Gwenn Bosser,
Victor Manuel Palma Preciado, Grigori Sidorov, and Adam Jatowt

Overview of LifeCLEF 2023: Evaluation of AI Models for the Identification
and Prediction of Birds, Plants, Snakes and Fungi 416
Alexis Joly, Christophe Botella, Lukáš Picek, Stefan Kahl,
Hervé Goëau, Benjamin Deneu, Diego Marcos, Joaquim Estopinan,
Cesar Leblanc, Théo Larcher, Rail Chamidullin, Milan Šulc,
Marek Hrúz, Maximilien Servajean, Hervé Glotin, Robert Planqué,
Willem-Pier Vellinga, Holger Klinck, Tom Denton, Ivan Eggel,
Pierre Bonnet, and Henning Müller

Overview of the CLEF-2023 LongEval Lab on Longitudinal Evaluation
of Model Performance 440
Rabab Alkhalifa, Iman Bilal, Hsuvas Borkakoty,
Jose Camacho-Collados, Romain Deveaud, Alaa El-Ebshihy,
Luis Espinosa-Anke, Gabriela Gonzalez-Saez, Petra Galuščáková,
Lorraine Goeuriot, Elena Kochkina, Maria Liakata, Daniel Loureiro,
Philippe Mulhem, Florina Piroi, Martin Popel, Christophe Servan,
Harish Tayyar Madabushi, and Arkaitz Zubiaga

Overview of PAN 2023: Authorship Verification, Multi-Author Writing
Style Analysis, Profiling Cryptocurrency Influencers, and Trigger
Detection: Condensed Lab Overview 459
Janek Bevendorff, Ian Borrego-Obrador, Mara Chinea-Ríos,
Marc Franco-Salvador, Maik Fröbe, Annina Heini, Krzysztof Kredens,
Maximilian Mayerl, Piotr Pęzik, Martin Potthast, Francisco Rangel,
Paolo Rosso, Efstathios Stamatatos, Benno Stein, Matti Wiegmann,
Magdalena Wolska, and Eva Zangerle

Overview of the CLEF 2023 SimpleText Lab: Automatic Simplification
of Scientific Texts 482
Liana Ermakova, Eric SanJuan, Stéphane Huet, Hosein Azarbonyad,
Olivier Augereau, and Jaap Kamps

Overview of Touché 2023: Argument and Causal Retrieval 507
Alexander Bondarenko, Maik Fröbe, Johannes Kiesel,
Ferdinand Schlatt, Valentin Barriere, Brian Ravenet, Léo Hemamou,
Simon Luck, Jan Heinrich Reimer, Benno Stein, Martin Potthast,
and Matthias Hagen

Author Index 531

Conference Papers

Inception Models for Fashion Image Captioning: An Extensive Study on Multiple Datasets

Mirko Del Moro, Serban Cristian Tudosie, Francesco Vannoni,
Andrea Galassi$^{(\boxtimes)}$ (ID), and Federico Ruggeri (ID)

Department of Computer Science and Engineering, University of Bologna,
Bologna, Italy
{a.galassi,federico.ruggeri6}@unibo.it

Abstract. Fashion e-commerce platforms are becoming increasingly popular. However, scanning, rendering, and captioning fashion items are still done mostly manually. In this work, we address the task of generating a textual description of a fashion item from an image portraying it. We carry out an extensive study with several neural architectures based on InceptionV3. We consider two existing fashion image captioning datasets, FACAD and InFashAI. We also curate a novel dataset, Fashion-Cap, that contains more than 290,000 images and 40,000 corresponding captions. In our analysis, we observe significant differences between the three datasets' captions, with Fashion-Cap having higher quality captions. To the best of our knowledge, this is the most extensive experimental study in fashion image captioning to date. Our experimental results show that our dataset is less challenging than FACAD but more than InFashAI, which confirms our insights, suggesting that it could be a valuable benchmark for this domain.

Keywords: Fashion · Dataset · Image Captioning · NLP

1 Introduction

In the last few years, the e-commerce fashion industry has witnessed significant growth. Major worldwide events like the recent COVID-19 pandemic defined a valuable playground for e-commerce sales platforms, whose growth has greatly exceeded even the most generous predictions. As a result, many fashion consumers are progressively adopting e-commerce platforms as their default shopping solution [24]. This phenomenon has led to the definition of e-commerce platforms that cover a wide variety of fashion items and services, which pose a challenge due to the great human effort that they require. Indeed, the definition of an autonomous pipeline for scanning, rendering, and captioning fashion items is still in its infancy, consequently most of the effort is still attributed

M. Del Moro, S. C. Tudosie and F. Vannoni—First Authors

to human workers. Current research mainly addresses the consumer perspective by defining adequate recommender systems [32]. However, a complete pipeline should contain other components designed to capture a consumer's attention and provide them with the necessary information in an effective way. For instance, captions should be short with minimal but relevant details, to be compatible with smartphone screens and voice-based searches.[1][2]

In this work, we discuss the definition of generative models for automatically defining captions for fashion items. Despite the growth of e-commerce fashion platforms, this problem is still scarcely addressed in the literature. To the best of our knowledge, only two datasets designed for fashion image captioning have been released so far: the FACAD dataset [30] and the InFashAI [12] project. We propose an extensive study on these datasets and release a novel one called Fashion-Cap, which we obtain by adapting an image generation dataset to the task of image captioning. We evaluate a well-known generative architecture for image captioning [29], experimenting with different configuration settings and variants to assess the task's difficulty. Compared to existing contributions, our method relies on the input fashion image and does not leverage additional domain knowledge like fashion attributes [30]. This design choice reflects the purpose of reducing human effort when defining fashion e-commerce platforms. Our contribution is twofold: (i) we release Fashion-Cap, a new dataset for the task of fashion image captioning, which is obtained by adapting and curating a dataset for image generation; (ii) we provide a reproducible and extensive study on three datasets for the fashion image captioning task using several encoder-decoder neural architectures. To the best of our knowledge, we are the first to propose a study on as many datasets in this domain. We make our code and data publicly available. [3]

2   Related Work

Model pre-training has become the default approach in the image captioning domain, especially for the encoder module [27,29], as well as in the image-text understanding domain for the vision and language multitask [2,4,15]. Yang et al. [30] were the first to propose large pre-trained models for image captioning by proposing the FAshion CAptioning Dataset (FACAD). In their study, the authors use an encoder-decoder neural architecture, as in [27], but they also integrate task-specific attribute embeddings trained via reinforcement learning. They rely on a set of fashion-related attributes extracted from the input image to regularize model training. More precisely, they introduce attribute-level semantic (ALS) and sentence-level semantic (SLS) rewards as metrics to improve the quality of generated image captions. In contrast, our proposed solution doesn't require the identification of domain-specific attributes to generate an image caption. Indeed, we speculate that acquiring domain knowledge can become a bot-

---

[1] https://content26.com/blog/product-description-word-counts-length-matters-2/.

[2] https://www.bigcommerce.com/blog/perfect-product-description-formula/.

[3] Publicly available repository: https://www.github.com/NoLogicPlease/Visionizer.

Table 1. Source datasets statistics.

Dataset	Images	Max Resolution	Categories	Captions	Avg. Caption Length	Poses	Task
FACAD [30]	993,000	1560 × 2392	78	130,000	21	multiple	I. Captioning
InFashAI+DeepFashion [12]	87,821	800 × 1070	n/a	87,821	9	single	I. Captioning
Fashion-Gen [21]	325,536	1360 × 1360	48	78,850	30	multiple	I. Generation

Table 2. Composition of datasets used in our study.

Dataset	Images	Train Images	Val Images	Test Images	Resolution	Images per Caption	Avg. Caption Length
Reduced-FACAD	55,021	44,016	5,502	5,503	299 × 299	1	17
Reduced-InFashAI	86,763	69.410	8.676	8.677	299 × 299	1	9
Fashion-Cap	290,441	232,352	29,044	29,045	299 × 299	up to 8	10

tleneck for defining efficient image captioning tools for the fashion industry. In particular, the absence of a standardized set of fashion attributes can lead to a time-consuming attribute identification annotation step.

Fashion image captioning has been taken into consideration also by Hacheme and Sayouti [12]. They implemented a model based on the *Show and tell* approach [27]: an encoder-decoder architecture in which the encoder is a Convolutional Neural Network (CNN), and the decoder is a Recurrent Neural Network (RNN). They initialized the encoder using a pre-trained ResNet152 [13] and used a Long Short-Term Memory (LSTM) as decoder. In their work, they jointly train their model on two datasets, one of Western-style items and one of African-style items, with the purpose of transferring knowledge between the two. With respect to their work, we add more recent techniques, namely Beam Search and Bahdanau attention [1]. The former is used to improve the decoder performance in the caption generation, while the latter is introduced to make the model more interpretable [28]. Another layer of controllability and interpretability could be added by using a framework for generating controllable and grounded captions through regions, as proposed by [6]. Lastly, differently from them, we do not rely on an index-based representation of words but employ Glove embeddings [19].

Beyond image captioning, artificial intelligence has been applied to the fashion domain for several other purposes, such as generating synthetic images from items description [21], assessing the similarity between two images of fashion items [8], recognizing items characteristics [17], and providing specialized and tailored recommendations [9,31]. Additional information can be found in the following surveys: [3,16] and [22].

3   Data

In this study, we consider three sources: the FACAD dataset [30], a collection presented in [12] containing two datasets (InFashAI and DeepFashion), and Fashion-Gen [21]. We select only a subset of the data available in these sources, according to the following principles: (i) the images must be publicly available; (ii) all the images related to the same item must have the same quality; (iii) the captions must be concise. In particular, we implement the last principle by measuring the average length of the captions across the three sources, which is 20 words, and filtering out any data with a longer caption. Table 1 provides a summary of the original sources, whereas Table 2 shows the datasets used in our study.

3.1 Reduced-FACAD

The FAshioning CAptioning Dataset (FACAD) [30] is a collection of 993,000 high-resolution fashion images. The dataset contains images of fashion items targeting different seasons, ages (kids and adults), and categories (clothing, shoes, bag, accessories, etc.). Each fashion item is collected from different angles (front, back, side, etc.). Figure 1a shows an example. FACAD is the first large dataset built specifically for the image captioning task in the fashion domain. In particular, the dataset contains 130K image captions, each one corresponding to a single clothing item represented in 6–7 images. The average length of the captions is 21 words and each of them contains a single sentence that often includes also information that can be considered subjective (e.g., "so-simple yet so-chic", "retro flair"). Concerning the image captioning task, FACAD is a challenging dataset since the images and captions were collected from the web through web scraping of fashion websites, and therefore there are cases where captions contain linguistic or format errors.

For each item, there is only one image with a proper background, object position, and image quality that properly represents the fashion item. The other ones, as shown in Fig. 1a, are less consistent and they contain noisy elements, e.g., the background. For this reason, we consider only such image for each item and ignore the remaining ones. Additionally, we filter out images with a corresponding caption of more than 20 words. Eventually, we obtain a dataset comprising 55,021 images with corresponding captions. We label this dataset subset as Reduced-FACAD hereafter.

3.2 Reduced-InFashAI

We consider the work of Hacheme and Sayouti [12] as the second source of data. They present a novel dataset, Inclusive Fashion AI (InFashAI), which contains 8,842 clothing images with corresponding captions targeting the African fashion culture. The images were collected from Afrikrea,[4] a well-known marketplace specializing in fashion items. They also use the DeepFashion dataset [17,34], which contains 78,979 images of Western culture items collected from Pinterest.[5] Instead of using the original captions, Hacheme and Sayouti constructed new ones through crowdsourcing, instructing a team of volunteers that followed a template-based approach such as: *The (man|woman|lady) is wearing (a|an)*

---

[4] https://www.afrikrea.com/.
[5] https://www.pinterest.com/.

(a) "A pearly button accents the stand collar that gives this so-simple, yet so-chic A-line dress its retro flair."

(b) "The lady is wearing an african gray long sleeved hoodie."

(c) "Long sleeve blazer in deep navy. Notched lapel collar. Padded shoulders. closure at front. Welt pocket at breast. Flap pockets at waist. Four-button cuffs. Two vents at back. Partial lining. Tonal stitching".

Fig. 1. Examples of fashion item images and corresponding caption in (a) FACAD, (b) InFashAI+DeepFashion, and (c) FashioGen, respectively.

*(western|african) *item description**. For this reason, image captions are relatively short, with an average length of 9 words. Figure 1b shows an example. Overall, the resulting dataset contains 87,821 images with corresponding captions. We consider the publicly available version of this dataset, which contains 86,763 images with corresponding captions. We denote this version as Reduced-InFashAI hereafter.

3.3   Fashion-Cap

The Fashion-Gen dataset [21] was originally proposed for the task of image generation. It contains 325,536 high-definition fashion images, but the publicly available version of the dataset only features images in 256 × 256 resolution. The items were photographed under consistent studio conditions, and the photos are paired with item captions provided by professional stylists. Similarly to FACAD, for each fashion item, multiple images taken from different angles were collected depending on the item category. Figure 1c shows an example. Overall, the dataset contains 78,850 image captions, whose average length is 30 words. This length is due to the fact that captions are articulated and verbose, usually spanning through multiple sentences. The first one typically describes the fashion item with the most relevant characteristics, while the following ones are shorter and contain minor details.

Starting from Fashion-Gen data, we curate a novel dataset for the task of image captioning. We consider the publicly available version of this dataset, which contains 293,018 image-captions pairs. Since all the images associated with a fashion item (and its caption) have the same quality and there are no relevant inconsistencies between them, we do not discard any of them, in contrast to what we have done with FACAD. However, to address the verbosity of the captions, we filter them by considering only those having 20 or fewer words to obtain concise textual descriptions comparable in length to the ones reported in Reduced-FACAD and Reduced-InFashAI datasets. Furthermore, we consider a text normalization preprocessing step based on regular expressions to remove impurities like excessive blank spaces and special characters. The resulting dataset contains 290,441 images paired with 42,172 unique captions, with an average length of 8 words. We denote the obtained dataset as Fashion-Cap hereafter.

4   Experimental Setting

4.1   Models

We experiment with several models based on a general encoder-decoder architecture. Each step of the captioning process generates a new token of the caption following this scheme:

1. An input image X is encoded by the encoder: $\tilde{X} = ENC(X)$;
2. The encoded image $\tilde{X}$ and the embedding of the $y_0 =$ <start> token for generation are concatenated and fed as input to the decoder;
3. The decoder generates the first token $y_1 = softmax\left(DEC([\tilde{X} \,||\, y_0], h_0)\right)$, where h_0 is the decoder initial hidden state;
4. The decoder iteratively generates the following caption tokens:

$$y_t = softmax\left(DEC([\tilde{X} \,||\, y_{t-1}], h_{t-1})\right)$$

The simplest model, which we address as Baseline, follows a popular encoder-decoder architecture for image captioning and is represented in Fig. 2 (top). This architecture was first introduced in [27] and is itself inspired by previous work on sequence-to-sequence translation [25]. The encoder is based on a pre-trained InceptionV3 architecture [26], a popular convolutional neural network for assisting in image analysis and object detection, followed by a single fully connected layer. The decoder comprises a recurrent layer and a stack of two fully connected layers for caption generation. The textual inputs are encoded through trainable embeddings of size 300. Differently from [27], to generate y_t, we use greedy search, which we denote as Max Search. Max Search concerns selecting the token with the highest probability as output at each generation step. We experiment with two variations of the Baseline that differ for the recurrent layer: one uses a GRU [5], the other one an LSTM [14]. This approach is similar to the one used in [12], except that we follow the original model of the decoder, while they replace it with a pre-trained ResNet152 [13].

We enhance the Baseline with more recent techniques, obtaining a model that we call Visionizer, as shown in Fig. 2 (bottom). Inspired by [29], we add an attention layer in the decoder, before the concatenation step. Specifically, we employ Bahdanau attention [1], using the hidden state of the recurrent layer as query element [10]. The introduction of this module is motivated by its many successes in Natural Language Processing and Computer Vision tasks, but also

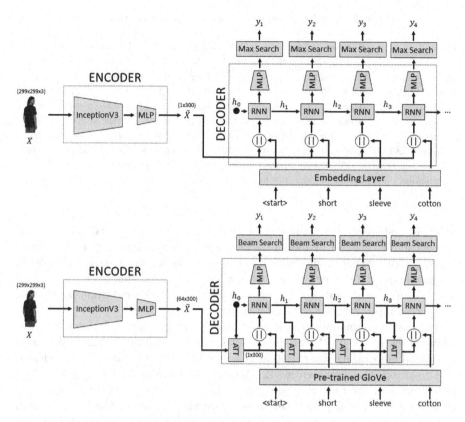

Fig. 2. The architecture of the Baseline approach (top) and Visionizer (bottom).

because it allows interpreting the output of the model [28]. In addition, we encode the textual input using 300-dimensional GloVe embeddings [19], but we keep the encoding layer trainable to also learn out of vocabulary terms (OOV) and fine-tunining the embeddings. As for Baseline, we experiment Visionizer with GRU and LSTM for the recurrent layer. Finally, we also add the possibility with Visionizer to generate captions through beam search. The Beam Search algorithm selects multiple tokens for a position in a given sequence based on conditional probability. Unlike the decoder with max search, on each step of the decoder, beam search keeps track of the top k most probable partial translations (hypotheses). The beam size parameter is used to determine how large is the space of hypothesis.

4.2 Setup

We split each described dataset into train (80%), validation (10%), and test (10%) splits (see Table 2). We train our models with Adam optimizer, using teacher forcing [11] as an additional regularization at training time. Teacher

Table 3. Model performance for fashion image captioning.

Model	Reduced-FACAD			Reduced-InFashAI			Fashion-Cap		
	BLEU	*CHRF*	*BERT*	*BLEU*	*CHRF*	*BERT*	*BLEU*	*CHRF*	*BERT*
Baseline (GRU)	0.056	0.105	0.846	0.849	0.822	0.977	0.402	0.395	0.903
Baseline (LSTM)	0.050	0.101	0.846	0.852	0.827	0.978	0.405	0.397	0.905
Visionizer (GRU)	0.086	0.123	0.848	0.897	**0.882**	0.984	0.509	0.483	0.923
-Beam Search	**0.142**	**0.157**	0.827	0.864	0.842	0.979	0.421	0.409	0.905
-Attention	0.097	0.141	0.789	0.847	0.831	0.967	0.412	0.399	0.895
Visionizer (LSTM)	0.087	0.121	**0.849**	**0.898**	0.880	**0.985**	**0.520**	**0.494**	**0.926**
-Beam Search	0.125	0.153	0.826	0.865	0.843	0.979	0.423	0.409	0.907
-Attention	0.083	0.112	0.788	0.848	0.820	0.972	0.391	0.388	0.894

forcing is a strategy for training recurrent neural networks that uses ground truth as input, instead of model output from a prior time step as an input. Training with teacher forcing allows to converge faster, but it leads to exposure bias problems at inference time, because of the unavailability of the ground truth. We fix the resolution of input images to 299×299 resolution to account for different input formats across datasets.

For what concerns hyper-parameters, we train for 2 epochs because the perplexity of the model on the validation set started to degenerate after. We chose the textual embedding size of 300 as suggested in [19]. The batch size is chosen as 64 to match the approach in [12]. We set the learning rate to 0.001 and we use 512 units in the fully connected layer. Lastly, we chose 2 for the beam size and 2 for the k beam parameter to evaluate the impact of the Beam Search using the minimum possible values. Model capacity is an important factor in deep learning and image captioning as shown in [23] and [15], thus suggesting a future study on the model size. Due to computational resource limitations, we did not perform an extensive hyper-parameter calibration search. We leave this as future work.

As evaluation metrics, we consider two syntactic-oriented metrics, namely BLEU [18], CHRF [20]. Additionally, we consider BERTScore [33], a recent metric that is based on neural networks and is semantic-oriented. More in detail, BERTScore computes a similarity score for each token in the candidate sentence with each token in the reference sentence, encoding them using BERT [7].

5   Results

Table 3 reports evaluation metrics regarding the image captioning task on the three discussed datasets. In particular, we evaluate each model when the recurrent layer is defined by a GRU and by an LSTM architecture. We also perform an ablation study on Visionizer by removing the Beam Search and the Attention module. Overall, all the models have similar behavior on the three datasets. Reduced-FACAD is clearly the most challenging one, and the best models achieve only a score of ~0.14 in BLEU and ~0.84 in BERT. On Fashion-Cap the best models reach about ~0.52 in BLEU and ~0.93 in BERT. Reduced-InFashAI is clearly the easier dataset: the best models obtain an almost perfect BERT score (~0.99) and a considerably high BLEU score (~0.90). This is probably due to the fact that the captions follow a template structure, and therefore the generation of many tokens (e.g., the first half of the caption) is quite easy. In all the considered cases, the CHRF score is similar to the BLEU one. We observe that the Visionizer models using the beam search outperform their Baseline counterparts in all datasets and metrics. In particular, the Visionizer with LSTM and beam search performs best across all datasets. The model achieves an improvement in the BLEU score over its baseline counterpart of ~3, ~5, and ~12 percentage points on, respectively, Reduced-FACAD, Reduced-InFashAI, and Fashion-Cap. We observe a similar improvement for the same model regarding the CHRF metric.

The alignment between the BLEU and CHRF metrics is expected as both metrics capture syntactic and lexical similarities. In contrast, we observe fewer improvements for BERTScore, possibly motivated by the limited length of the captions. This is particularly evident in Reduced-InFashAI, where captions are shorter and follow a template-based construction.

For what concerns the model without the beam search, we observe inconsistent results across datasets. In particular, the Visionizer model with beam search outperforms its counterpart in Fashion-Cap and Reduced-InFashAI. We observe this performance improvement in all the reported evaluation metrics. In contrast, removing the beam search leads to improved results in Reduced-FACAD. However, it is worth noting that model performance is notably lower compared to the other datasets. Indeed, FACAD is a challenging dataset containing noisy image captions. Therefore, syntactic-oriented metrics like BLEU and CHRF might favor noisy captions similar to the original ones. We speculate that this characteristic of the dataset is responsible for the observed experimental results.

6 Qualitative Analysis

We carry out a qualitative analysis of Visionizer results considering two cases for Reduced-FACAD and Fashion-Cap. For each test set, we analyze the image for which the Visionizer with Max Search obtained the best BLEU score and the one for which it obtained the worst score, to highlight the contribution of the Beam search.

Figure 3 shows examples from Reduced-FACAD dataset. In particular, in Fig. 3 (top), the Visionizer with Beam Search successfully captures part of the ground-truth caption concerning the 'soft and stretchy blend'. In contrast, Visionizer with Max Search fails at capturing these details, while we observe that the baseline model repeats this pattern with different adjectives. Concerning worst-generation performance cases, in Fig. 3 (bottom), we observe that all models fail at capturing the fashion details described in the ground-truth caption, which are particularly challenging since they involve domain-specific knowledge that may not be retrievable solely from the input image.

Ground Truth: an eye catching geometric pattern color jaunty sock knit from a stretchy cotton blend.
Baseline: a moisture wicking knit blend and a soft and stretchy combed cotton blend.
Visionizer Max Search: a fine linked toe.
Visionizer Beam Search: a soft and stretchy blend.

Ground Truth: a striking branch pattern cover a lustrous sport tee made from breathable egyptian cotton and tailored for a flattering fit.
Baseline: warm weather style.
Visionizer Max Search: a comfortably cut fit.
Visionizer Beam Search: a crisp spread collar.

Fig. 3. Examples of generated captions on Reduced-FACAD test set, chosen considering the best BLEU score (top) and worst BLEU score (bottom), with respect to Visionizer Max Search. We underline the main differences between the captions.

Ground Truth: short sleeve cotton jersey t shirt in white.
Baseline: oversize t shirt in white.
Visionizer Max Search: short sleeve t shirt in white.
Visionizer Beam Search: short sleeve cotton jersey t shirt in white.

Ground Truth: sleeveless virgin wool dress in black featuring tonal leather trim throughout.
Baseline: sleeveless coated cotton dress in black.
Visionizer Max Search: sleeveless ribbed and wool blend dress in black.
Visionizer Beam Search: sleeveless a line dress in black.

Fig. 4. Examples of generated captions on Fashion-Cap test set, chosen considering the best BLEU score (top) and worst BLEU score (bottom), with respect to Visionizer Max Search. We underline the main differences between the captions.

For what concerns Fashion-Cap, in Fig. 4 (top) we observe that Visionizer models recognize an additional characteristic of the item ("short sleeve") compared to the baseline model. Furthermore, the Visionizer model with Beam Search also correctly generates the term "cotton jersey", while its Max Search counterpart fails. In Fig. 4 (bottom), we observe that all the models perform similar errors (e.g., missing the second part of the ground-truth caption). However, it is worth noticing, that Visionizer with Beam Search is able to correctly recognize the material of the item ("wool").

7 Conclusions

We have presented an extensive study concerning fashion image captioning. We have provided background and motivation for the definition of efficient generative models oriented to online application scenarios in the fashion domain. We have released a novel dataset for this task, and we have experimentally assessed its difficulty and compared it to two existing ones. To the best of our knowledge, this is the first study that investigates this problem by covering multiple datasets. Our experiments suggest our dataset can be tackled with popular architectures for image captioning, obtaining satisfactory results. Nonetheless, it can be considered challenging and leaves room for future improvements with more advanced techniques. In future work, we want to integrate semantic-based metrics such as BERTscore during the training, as part of the loss function. Moreover, the use of *professor forcing* [11] regularization instead of *teacher forcing* would reduce the discrepancy between the inputs received by the networks at training and test time, potentially leading to a performance improvement.

Acknowledgments. This work was partially supported by the European Commission NextGeneration EU programme, PNRR-M4C2-Investimento 1.3, PE00000013-"FAIR" - Spoke 8

References

1. Bahdanau, D., Cho, K., Bengio, Y.: Neural machine translation by jointly learning to align and translate. In: ICLR (2015)
2. Chen, Y.-C., et al.: UNITER: UNiversal image-TExt representation learning. In: Vedaldi, A., Bischof, H., Brox, T., Frahm, J.-M. (eds.) ECCV 2020. LNCS, vol. 12375, pp. 104–120. Springer, Cham (2020). https://doi.org/10.1007/978-3-030-58577-8_7
3. Cheng, W.H., Song, S., Chen, C.Y., Hidayati, S.C., Liu, J.: Fashion meets computer vision: a survey. ACM Comput. Surv. **54**(4), 1–41 (2021)
4. Cho, J., Lei, J., Tan, H., Bansal, M.: Unifying vision-and-language tasks via text generation. In: ICML, pp. 1931–1942. PMLR (2021)
5. Cho, K., van Merrienboer, B., Bahdanau, D., Bengio, Y.: On the properties of neural machine translation: Encoder-decoder approaches. In: SSST@EMNLP, pp. 103–111 (2014)
6. Cornia, M., Baraldi, L., Cucchiara, R.: Show, control and tell: a framework for generating controllable and grounded captions. In: CVPR, pp. 8307–8316 (2019)
7. Devlin, J., Chang, M., Lee, K., Toutanova, K.: BERT: pre-training of deep bidirectional transformers for language understanding. In: NAACL-HLT (1), pp. 4171–4186 (2019)
8. Dong, J., et al.: Fine-grained fashion similarity prediction by attribute-specific embedding learning. IEEE Trans. Image Process. **30**, 8410–8425 (2021)
9. Dong, M., Zeng, X., Koehl, L., Zhang, J.: An interactive knowledge-based recommender system for fashion product design in the big data environment. Inf. Sci. **540**, 469–488 (2020)
10. Galassi, A., Lippi, M., Torroni, P.: Attention in natural language processing. IEEE Trans. Neural Netw. Learn. Syst. **32**(10), 4291–4308 (2021)
11. Goyal, A., Lamb, A., Zhang, Y., Zhang, S., Courville, A.C., Bengio, Y.: Professor forcing: a new algorithm for training recurrent networks. In: NIPS, pp. 4601–4609 (2016)

12. Hacheme, G., Sayouti, N.: Neural fashion image captioning : accounting for data diversity. CoRR abs/2106.12154 (2021)
13. He, K., Zhang, X., Ren, S., Sun, J.: Deep residual learning for image recognition. In: CVPR (2016)
14. Hochreiter, S., Schmidhuber, J.: Long short-term memory. Neural Comput. **9**(8), 1735–1780 (1997)
15. Hu, X., et al.: Scaling up vision-language pre-training for image captioning. CoRR abs/2111.12233 (2021)
16. Liu, S., Liu, L., Yan, S.: Fashion analysis: current techniques and future directions. IEEE Multimedia **21**(2), 72–79 (2014)
17. Liu, Z., Luo, P., Qiu, S., Wang, X., Tang, X.: Deepfashion: powering robust clothes recognition and retrieval with rich annotations. In: CVPR (2016)
18. Papineni, K., Roukos, S., Ward, T., Zhu, W.: Bleu: a method for automatic evaluation of machine translation. In: ACL, pp. 311–318 (2002)
19. Pennington, J., Socher, R., Manning, C.D.: Glove: global vectors for word representation. In: EMNLP, pp. 1532–1543 (2014)
20. Popovic, M.: chrF: character n-gram F-score for automatic MT evaluation. In: WMT@EMNLP, pp. 392–395 (2015)
21. Rostamzadeh, N., et al.: Fashion-gen: the generative fashion dataset and challenge. CoRR (2018)
22. Song, S., Mei, T.: When multimedia meets fashion. IEEE Multimedia **25**(3), 102–108 (2018)
23. Stefanini, M., Cornia, M., Baraldi, L., Cascianelli, S., Fiameni, G., Cucchiara, R.: From show to tell: a survey on deep learning-based image captioning. IEEE Trans. Pattern Anal. Mach. Intell. **45**(1), 539–559 (2023)
24. Sumarliah, E., Usmanova, K., Mousa, K., Indriya, I.: E-commerce in the fashion business: the roles of the Covid-19 situational factors, hedonic and utilitarian motives on consumers' intention to purchase online. Int. J. Fashion Des. Technol. Educ. **15**(2), 167–177 (2022)
25. Sutskever, I., Vinyals, O., Le, Q.V.: Sequence to sequence learning with neural networks. In: NIPS, pp. 3104–3112 (2014)
26. Szegedy, C., Vanhoucke, V., Ioffe, S., Shlens, J., Wojna, Z.: Rethinking the inception architecture for computer vision. In: CVPR, pp. 2818–2826 (2016)
27. Vinyals, O., Toshev, A., Bengio, S., Erhan, D.: Show and tell: a neural image caption generator. In: CVPR, pp. 3156–3164 (2015)
28. Wiegreffe, S., Pinter, Y.: Attention is not not explanation. In: EMNLP/IJCNLP (1), pp. 11–20. Association for Computational Linguistics (2019)
29. Xu, K., et al.: Show, attend and tell: neural image caption generation with visual attention. In: ICML, vol. 37, pp. 2048–2057 (2015)
30. Yang, X., et al.: Fashion captioning: towards generating accurate descriptions with semantic rewards. In: Vedaldi, A., Bischof, H., Brox, T., Frahm, J.-M. (eds.) ECCV 2020. LNCS, vol. 12358, pp. 1–17. Springer, Cham (2020). https://doi.org/10.1007/978-3-030-58601-0_1
31. Yin, R., Li, K., Lu, J., Zhang, G.: Enhancing fashion recommendation with visual compatibility relationship. In: WWW, pp. 3434–3440, New York, NY, USA (2019)
32. Zhang, S., Yao, L., Sun, A., Tay, Y.: Deep learning based recommender system: a survey and new perspectives. ACM Comput. Surv. 52(1), 5:1–5:38 (2019)
33. Zhang, T., Kishore, V., Wu, F., Weinberger, K.Q., Artzi, Y.: Bertscore: evaluating text generation with BERT. In: ICLR (2020)
34. Zhu, S., Fidler, S., Urtasun, R., Lin, D., Loy, C.C.: Be your own prada: Fashion synthesis with structural coherence. In: ICCV, pp. 1689–1697 (2017)

The Best is Yet to Come: A Reproducible Analysis of CLEF eHealth TAR Experiments

Giorgio Maria Di Nunzio[1]([✉]) [iD] and Federica Vezzani[2] [iD]

[1] Department of Information Engineering, University of Padua, Padua, Italy
giorgiomaria.dinunzio@unipd.it
[2] Department of Linguistic and Literary Studies, University of Padua, Padua, Italy
federica.vezzani@unipd.it

Abstract. The CLEF eHealth Technology Assisted Reviews (TAR) in Empirical Medicine Tasks focused on evaluating the effectiveness of various technology-assisted review systems in assisting healthcare professionals in retrieving relevant information from vast amounts of medical literature. It ran for three years, from 2017 until 2019, giving the opportunity to research groups to conduct experiments and share results on automatic methods to retrieve relevant studies with high precision and high recall. In this paper, we perform a reproducibility study of one of the top-performing systems of both the years 2018 and 2019 by rerunning the original code that was provided by the authors of the paper. The goals of this paper are 1) to document the pitfalls in the description of the code, 2) to reorganize the code using a better reproducibility approach (R markdown), 3) to propose some minor changes to the code that would improve the performances of the system.

Keywords: Technology Assisted Review Systems · Reproducibility · Systematic Reviews

1 Introduction

The CLEF eHealth Task[1] proposed a range of challenges and objectives—for example, how to deal with the large volume and diversity of medical data as well as the heterogeneity of text medical data, including variations in language, terminology, and document formats—to enhance the effectiveness of information retrieval in the field of empirical medicine. For three years, from 2017 to 2019 [10, 12,15], the CLEF eHealth TAR lab aimed to address these challenges to build high-recall retrieval systems to support the compilation of medical systematic reviews. Given the importance of the potential impact of the output of this task, ensuring the reproducibility of such experiments is a crucial objective,

---

[1] https://clefehealth.imag.fr/.

A. Arampatzis et al. (Eds.): CLEF 2023, LNCS 14163, pp. 15–20, 2023.
https://doi.org/10.1007/978-3-031-42448-9_2

since reproducibility allows for the identification of potential biases, errors, or limitations in the systems under review, leading to their refinement [2].

In this paper, we perform a reproducibility study of the experiments carried out in 2019 by one of the top-performing participants [5] who provided the original source code to reproduce the results. In particular, the goals of this paper are: 1) to review the source code and compare it with the original papers [3,8] in order to document the potential pitfalls in the code by means of a reproducibility pipeline proposed by [9], 2) to reorganize the code using a better reproducibility code template with R Markdown,[2] 3) to propose some changes to the code that would improve the performances of the system.

2   Background on CLEF 2019 eHealth TAR Lab

Systematic reviews are comprehensive and structured approaches to synthesizing evidence from multiple studies or sources to answer specific research questions. In the context of CLEF eHealth TAR lab, systematic reviews played a crucial role and served as the foundation for evaluating the effectiveness of technology-assisted review systems for evidence-based medicine in the eHealth domain [11]. The CLEF 2019 eHealth TAR lab, in particular, proposed two tasks: Task 1 focused on retrieving relevant studies from PubMed[3] without the use of a Boolean query, while Task 2 focused on the efficient and effective ranking of studies during the abstract and title screening phase of conducting a systematic review. The organizers of these tasks constructed a benchmark collection of 31 reviews published by Cochrane[4] together with the corresponding relevant and irrelevant articles found in PubMed by the original Boolean query. Task 2 – which is the focus of this paper – had two goals: 1) to produce an efficient ordering of the articles, such that all of the relevant abstracts are retrieved as early as possible, and 2) to identify a subset which contains all or as many of the relevant abstracts for the least effort. In particular, for each systematic review, or *topic* in a more traditional IR sense, that needs to be conducted, participants were provided with the following input data: the identifier of the topic, the title of the review written by Cochrane experts, the Boolean query manually constructed by Cochrane experts, the set of PubMed Document Identifiers (PMID's) returned by running the query in MEDLINE.[5] Three teams participated in Task 2 [1,3,14]. In this paper, we will focus on the results reported by [3] as they were shown to be the top-performing ones.

3   Original Experiments

This study follows the reproducibility pipeline proposed by [9] and stems from one of the Best of Labs papers of CLEF 2020 [5] that summarized the performance of the system presented the year before [3] at CLEF 2019 eHealth TAR.

---

[2] https://rmarkdown.rstudio.com/.
[3] https://pubmed.ncbi.nlm.nih.gov/.
[4] https://www.cochrane.org/.
[5] https://www.nlm.nih.gov/bsd/pmresources.html.

The main idea of that research was to analyze a stopping strategy approach (i.e., when the system has to stop presenting documents to the expert who is preparing the systematic review) based on a finite number of documents that the expert is willing to review. The research question in that case was: if the expert tells us that s/he can only assess 1,000 articles in total, is there a way to optimize this finite effort across different topics? The paper tackled the problem in two ways: given a set of T topics and N documents that the user is willing to read, one approach equally distributes N across topics in order to have N/T documents to read per topic; the other approach distributes the effort in order to have a number of reviews which is proportional to the size of the pool for that topic.

The paper presented a solution that mixes a part of "traditional" IR, with explicit relevance feedback, with a part of text classification. We report here, for the sake of clarity, the procedure followed in the original paper. In particular, given a systematic review s:

- set the number n_s of documents that the expert is willing to read for s;
- for the first half, $n_s/2$, use a BM25 retrieval model with explicit relevance feedback; in addition, use query expansion to add terms to the original query. The number of terms is proportional to the current feedback iteration (at iteration 1, add 1 term; at iteration 10, add 10 terms);
- for the second half of the documents, use a Naïve Bayes classifier trained on the first $n_s/2$ documents to classify the remaining documents still using the explicit relevance feedback as proposed by [4].

4   Reproducing the Experiments

The reproducible study starts from the source code provided by the authors of the paper on GitHub.[6] In the first part of this section (Sect. 4.1, 4.2, and 4.3), we will briefly discuss the organization of the original source code and the issues that we encountered in the experimental setup phase; in the second part (Sect. 4.4, 4.5), we will discuss the results and the proposed improvements.

4.1   Fixing the Code

The original code is organized into a set of scripts written in R. Despite the missing documentation, it was easy enough to reconstruct the order of execution of each part. The three main files (*baseline_bm25_2019.R*, *original_query.R*, and *query_sampling_2019.R*) are the ones that allow us to rebuild the experiments presented in [3,5]. Besides some minor fixes about file paths (the file *load_data.R* requires a clone of the GitHub repository of the original CLEF eHealth TAR dataset),[7] it was necessary to contact the authors to obtain the index files necessary to run the code since these files were not uploaded on GitHub for space reasons.

---

[6] https://github.com/gmdn/CLEF2019.
[7] https://github.com/CLEF-TAR/tar.

4.2   Understanding the Hyperparameters

The main hyperparameters of the model, described in Sect. 3, can be found at the beginning of the source code of the main files. However, there are some additional parameters that were not immediately understandable: for example, there are inline comments in the source code that refer to "the angular coefficient during the ranking of documents" or "the minimum level of precision during the classification phase". In order to get a picture of these settings, it was necessary to trace back the work of the authors to the seminal paper [6].

4.3   Additional Bugs Fixing

In order to understand the functioning of the code, we started running blocks of code with a smaller amount of data (fewer topics and documents with fewer feedback iterations). We discovered two additional issues about fitting linear regression models and computing the number of relevant documents. The former was a missing condition in the code that, in those particular cases where the amount of non-relevant documents in the top-k ranked documents is zero or one, produced an error in the computation of the linear regression line (used for deciding when to stop reviewing). The other was related to the wrong computation of the total number of relevant documents (in the file *evaluate.R*): this number was computed using the set of documents retrieved by the experiment rather than the relevance judgements. We believe that this issue was fixed by the authors at some point since, once we corrected the mistake, we obtained the same results reported in the original paper.

4.4   R Markdown for Better Reproducibility

Once everything was fixed and ready to be run, we rearranged the code in a more suitable way by using the literate programming approach [13] with R Markdown, where code and documentation are interwoven. This approach improves code readability and maintainability as the document becomes a self-contained narrative of the analysis, making it easier to understand and modify in the future.[8] After running the experiments with the correct settings of the hyperparameters, we were able to perfectly reproduce the results of the CLEF 2019 eHealth task presented in Table 1, Table 2 and Fig. 1 of [5]. For space reasons, we only report the summary of the results in Table 1.

4.5   Further Operations to Improve Results

Our last goal was to make some minor changes to improve the results of this state-of-the-art approach. We focused our effort in two directions: one suggested by [14] where an initial pseudo-document d_0 composed of the description of the

---

[8] https://github.com/gmdn/CLEF2023.

Table 1. The left side of the table reports the results of our experiments and show a perfect match with Table 2 of [5]. The four runs of the original paper are highlighted in *italics*. The right side of the table reports the new results when the number of terms added at each iteration is equal to 200 instead of 1 and the proportion of documents to assess is not purely proportional as suggested by the original paper.

run	original study reproduced			original study improved		
	recall@k	recall	doc shown	recall@k	recall	doc shown
equal-t1000	0.28	0.94	25,015	**0.30**	0.94	25,107
equal-t600	0.28	0.91	16,529	**0.30**	0.91	16,698
abs-hh-ratio	0.46	0.89	28,201	0.46	0.89	28,201
prop-t600	0.28	0.85	27,791	0.28	**0.90**	**14,860**
abs-th-ratio	0.43	0.83	26,708	0.43	0.83	26,708
bm25-t1000	0.18	0.79	23,241	0.18	0.79	23,241

topic, instead the title of the topic, is used to start the search for relevant documents; the other is a different use of the proportion of documents to show to the experts, especially in those cases where the set of documents to review is relatively small compared to the total available effort. In the first case, we operated on the hyperparameter that increases the number of terms to add at each round of relevance feedback. Instead of adding just one term, we added 200 terms (like a short abstract or description) whenever a new relevant document is found. In the second case, we devised a slightly more elaborated distribution of the effort where we 'save' some effort (the capacity to read documents) for the next topic, in case the topic we are reviewing has fewer documents compared to the threshold. For example, if the threshold is 600 documents and the current topic has 400 documents, we save 200 documents to review for the remaining topics. The results highlighted in bold in Table 1, show that with the first approach, we can improve the recall at k by 2-point percentage, while with the second approach, we can improve the recall and reduce the amount of work at the same time.

5   Conclusions

In this paper, we presented the problem of reproducibility as the means to ensure that scientific findings can be verified and validated by other researchers. In the context of the CLEF eHealth TAR lab, we proposed a reproducibility study in IR that replicated the pipeline suggested by [9] to identify potential sources of bias and errors and enable a critical evaluation of the methods and data used in the original study. Our ultimate goal was not only to check if the same results as the original paper could be achieved, but to promote literate programming as one of the cornerstones of the research process, enhancing transparency and reproducibility. We believe that the use of R markdown notebook in this work had a major impact on our findings as well as on the possibility to improve the original system. As future work, we plan to design an interactive TAR system where the experts can take or change decisions according to a visual inspection of the remaining documents to read as proposed by the original paper [7].

References

1. Alharbi, A., Stevenson, M.: Ranking studies for systematic reviews using query adaptation: university of sheffield's approach to CLEF ehealth 2019 task 2. In: Working Notes of CLEF (2019). https://ceur-ws.org/Vol-2380/paper_185.pdf
2. Breuer, T., et al.: How to measure the reproducibility of system-oriented IR experiments. In: Proceedings of SIGIR (2020). https://doi.org/10.1145/3397271.3401036
3. Di Nunzio, G.M.: A distributed effort approach for systematic reviews. IMS unipd at CLEF 2019 ehealth task 2. In: Working Notes of CLEF (2019). https://ceur-ws.org/Vol-2380/paper_205.pdf
4. Di Nunzio, G.M.: A study of an automatic stopping strategy for technologically assisted medical reviews. In: Proceedings of ECIR (2018). https://doi.org/10.1007/978-3-319-76941-7_61
5. Di Nunzio, G.M.: A study on a stopping strategy for systematic reviews based on a distributed effort approach. In: Arampatzis, A., et al. (eds.) CLEF 2020. LNCS, vol. 12260, pp. 112–123. Springer, Cham (2020). https://doi.org/10.1007/978-3-030-58219-7_10
6. Di Nunzio, G.: A new decision to take for cost-sensitive naïve bayes classifiers. Inf. Process. Manage. **50**(5), 653–674 (2014). https://doi.org/10.1016/j.ipm.2014.04.008
7. Di Nunzio, G.: An interactive analysis of the costs of technologically assisted medical reviews. In: 12th EAI International Conference on Pervasive Computing Technologies for Healthcare (2018). https://doi.org/10.4108/eai.20-4-2018.2276346
8. Di Nunzio, G.M., Ciuffreda, G., Vezzani, F.: Interactive sampling for systematic reviews. IMS unipd at CLEF 2018 ehealth task 2. In: Working Notes of CLEF 2018 (2018). https://ceur-ws.org/Vol-2125/paper_186.pdf
9. Di Nunzio, G., Minzoni, R.: A thorough reproducibility study on sentiment classification: methodology, experimental setting, results. Information **14**(2), 76 (2023). https://doi.org/10.3390/info14020076
10. Goeuriot, L., et al.: CLEF 2017 eHealth evaluation lab overview. In: Jones, G.J.F., et al. (eds.) CLEF 2017. LNCS, vol. 10456, pp. 291–303. Springer, Cham (2017). https://doi.org/10.1007/978-3-319-65813-1_26
11. Kanoulas, E., Li, D., Azzopardi, L., Spijker, R.: CLEF 2019 technology assisted reviews in empirical medicine overview. In: Working Notes of CLEF (2019). https://ceur-ws.org/Vol-2380/paper_250.pdf
12. Kelly, L., et al.: Overview of the CLEF eHealth evaluation lab 2019. In: Crestani, F., et al. (eds.) CLEF 2019. LNCS, vol. 11696, pp. 322–339. Springer, Cham (2019). https://doi.org/10.1007/978-3-030-28577-7_26
13. Knuth, D.E.: Literate programming. Comput. J. **27**(2), 97–111 (1984). https://doi.org/10.1093/comjnl/27.2.97
14. Li, D., Kanoulas, E.: Automatic thresholding by sampling documents and estimating recall. In: Working Notes of CLEF (2019). https://ceur-ws.org/Vol-2380/paper_187.pdf
15. Suominen, H., et al.: Overview of the CLEF ehealth evaluation lab 2018. In: Bellot, P., et al. (eds.) CLEF 2018. LNCS, vol. 11018, pp. 286–301. Springer, Cham (2018). https://doi.org/10.1007/978-3-319-98932-7_26

Predicting Retrieval Performance Changes in Evolving Evaluation Environments

Alaa El-Ebshihy[1,2,3]($\boxtimes$) (iD), Tobias Fink[1,2] (iD), Gabriela Gonzalez-Saez[4] (iD), Petra Galuščáková[4] (iD), Florina Piroi[1,2] (iD), David Iommi[1] (iD), Lorraine Goeuriot[4] (iD), and Philippe Mulhem[4] (iD)

[1] Research Studios Austria, Data Science Studio, Vienna, Austria
{alaa.el-ebshihy,tobias.fink,florina.piroi,david.iommi}@researchstudio.at
[2] Technische Universität Wien, Vienna, Austria
[3] Alexandria University, Alexandria, Egypt
[4] Univ. Grenoble Alpes, CNRS, Grenoble INP (Institute of Engineering Univ. Grenoble Alpes), LIG, Grenoble, France
{gabriela-nicole.gonzalez-saez,petra.galuscakova,lorraine.goeuriot,
philippe.mulhem}@univ-grenoble-alpes.fr

Abstract. Information retrieval (IR) systems evaluation aims at comparing IR systems either (1) one to another with respect to a single test collection, and (2) across multiple collections. In the first case, the evaluation environment (test collection and evaluation metrics) stays the same, while the environment changes, in the second case. Different evaluation environments may be seen, in fact, as evolutionary versions of some given evaluation environment. In this work, we propose a methodology to predict the statistically significant change in the performance of an IR system (i.e. result delta $\mathcal{R}\Delta$) by quantifying the differences between test collections (i.e. knowledge delta $\mathcal{K}\Delta$). In a first phase, we quantify differences between document collections (i.e. $\mathcal{K}_d\Delta$) in the test collections by means of TF-IDF and Language Models (LM) representations. We use the $\mathcal{K}_d\Delta$ to train SVM classification models to predict the significantly performance changes of various IR systems using evolving test collections derived from the Robust and TREC-COVID collections. We evaluate our approach against our previous $\mathcal{K}_d\Delta$ experiments.

Keywords: Evolving Test Collections · Performance Prediction · Knowledge Delta · Result Delta

1 Introduction

Traditional offline evaluation of Information Retrieval (IR) systems uses test collections composed of [15]: (1) a set of documents or passages, (2) a set of queries, or topics, and (3) a set of relevance judgments indicating which documents are relevant to each query. The components of a test collection together with (a

A. Arampatzis et al. (Eds.): CLEF 2023, LNCS 14163, pp. 21–33, 2023.
https://doi.org/10.1007/978-3-031-42448-9_3

set of) evaluation metrics to assess the efficiency of an IR system constitute an *Evaluation Environment* (EE) [5]. Any change of an EE's element affects an IR system's performance. In this paper, we systematically quantify differences between representations of text document collections and analyse how these differences impact on the changes in IR system performance. In this research, we start from the situation where a document collection is constantly evolving in terms of number of documents, with some documents being added and some removed. An EE at the timestamp t can thus be derived from the EE at the previous timestamp t-1. In the following, we refer to this setup as an *Evolving Evaluation Environment* (EvEE). This behaviour is frequent in real world information retrieval scenarios, such as web search. To continuously evaluate IR systems over different EEs, including EvEEs, we have previously introduced the notion of *Results Delta* ($\mathcal{R}\Delta$) [5], which aims to quantify the differences in performance of an IR system used on different EEs wrt. a single metric.

As a simplification, for the purpose of this work, we only consider variations in the set of documents in an EE, keeping the topic set, the set of answers to the topics (the qrels) and the metrics static. We also use the same IR system for the different EEs, allowing us to focus on the effects of the differences in document representations.

We aim to understand and to quantify variations between the components of two different EEs by means of defining *Knowledge Delta* ($\mathcal{K}\Delta$) and observing its impact on the $\mathcal{R}\Delta$. In our view, $\mathcal{K}\Delta$ for two EEs is a combination of a document (collection) representation delta, $\mathcal{K}_d\Delta$, and a query representations delta, $\mathcal{K}_q\Delta$, both defined as difference functions between pairs of text representations. In this work, we focus on the $\mathcal{K}_d\Delta$ definition and its relation with $\mathcal{R}\Delta$.

In our introductory study on the relation between $\mathcal{K}_d\Delta$ and $\mathcal{R}\Delta$ [6] we identified $\mathcal{K}_d\Delta$ using Query Performance Prediction (QPP) features [7], where the analysis showed average correlation between $\mathcal{K}_d\Delta$ and $\mathcal{R}\Delta$. In this work, we:

1. use TF-IDF and Language Models (LM) as representations of documents sets, and refine the previous $\mathcal{K}_d\Delta$ formulation.
2. build SVM-based prediction models, which utilizes $\mathcal{K}_d\Delta$ to predict if there is a statistical significant change in $\mathcal{R}\Delta$.
3. examine our model using training data derived from the Robust [18] and the TREC-COVID [17] test collections to simulate EvEE.
4. compare this approach to define $\mathcal{K}_d\Delta$ with a QPP $\mathcal{K}_d\Delta$ baseline [6].

2   Related Work

The Evolving EEs is a setup most dominant in web search scenarios, where documents, queries, and also the notion of what is considered to be relevant are changing along the time. Companies operating Web search engines have access to large numbers of users and their search logs, allowing them to use various online evaluation methods to improve the engines' output [11].

The core of our work is predicting changes between different EE. Our work is thus well related with Query Performance Prediction (QPP) [8]. However, there are three major difference between QPP and our work: (1) QPP focuses on predicting performance for a given query, while we focus on changes of the document sets collections, (2) QPP typically does not consider the collection to be evolving, and (3) QPP focuses on a prediction of the exact performance of the system, while we focus on predicting the changes in the performance for subsequent EEs. However, we also use concepts from QPP also in this work to establish a baseline for our proposed methods. The effect of the document collection on the predictivity of the experiments was for example studied by [13] who experiment with different types of collection features, such as length of the documents and content type. Similar question was studied by [16], who split the collections into sub-collections based on different features, and measure the impact on retrieval quality. ANOVA analysis [3] is then looking at different components of the EE, including the queries and documents and analysing their effect on the performance. However, in all these papers, there is no notion of evolvement of the collections, which is a crucial assumption in our work.

3   Approach and Formalizations

Recall that we want to understand the influence $\mathcal{K}_d\Delta$ has on the $\mathcal{R}\Delta$. We formulate this in terms of a binary classification problem, where we predict whether $\mathcal{R}\Delta$ is statistically significantly changed given $\mathcal{K}_d\Delta$ on an evolving evaluation environment. For this purpose, we: (1) simulate an EvEE using classical test collection (Sect. 3.1), (2) give three definitions for $\mathcal{K}_d\Delta$, using different text representations (Sect. 3.2), (3) define what a significant change in $\mathcal{R}\Delta$ change is ($\mathcal{S}_e\Delta$ in Sect. 3.3), and (4) train a prediction model to predict significant $\mathcal{R}\Delta$ changes for some input $\mathcal{K}_d\Delta$. The model is classification-based and we see it as a useful tool to understand the impact of $\mathcal{K}_d\Delta$ on $\mathcal{R}\Delta$. The results of using this model are presented in Sect. 4.

3.1   Simulating an Evolving Evaluation Environment

For our experiments, we need to simulate an *Evolving Evaluation Environment* (EvEE) that approximates the evolution of an evaluation environment (EE) in a controlled manner. The EvEE is built by creating shards of a classical test collection, TC [2][1], that contains timestamped documents. We use the timestamps to assign documents to shards according to their temporal order, and to set the size of document overlap between shards. A shard tc_i of a test collection, TC, is constructed, then, as follows:

$$tc_i = \{d_k \in TC | ix \leq k < ix + s\} \tag{1}$$

where k is a timestamp, d_k is document with the k timestamp, x is the number of documents that are exchanged from one shard to another, and s is the size (number of documents) of a shard. In our experiments, x is set to 10% of s and i ranges from 0 to (including) 40.

---

[1] Recall that, in our work, a test collection, TC together with a set of appropriate metrics form an Evaluation Environment, EE.

3.2   Knowledge Delta

To understand how changes in document collections reflect in the retrieval results of IR systems using them, we design a delta function that quantifies the difference between pairs of document collections. More specifically, in the definition of our delta function, $\mathcal{K}_d\Delta$, we utilize three types of text representations for test collections: (1) Query Performance Prediction features (QPP) (2) TF-IDF scores and (3) Language Models (LM).

$\mathcal{K}_d\Delta^{qpp}$ **Representation:** Previous work used two popular features that are often used in query performance prediction [7], to study their correlation with $\mathcal{R}\Delta$ [6]: Averaged Term Weight Variability ($avVAR$) [20] and Averaged Collection Query Similarity ($avSCQ$) [20]. For these representations, we define $\mathcal{K}_d\Delta$ as a two dimensional vector of the difference between these feature values, $avVAR$ and $avSCQ$, for two given test collections, tc_i and tc_j:

$$\mathcal{K}_d\Delta^{qpp}(tc_i, tc_j) = [(avVAR^{tc_i} - avVAR^{tc_j}), (avSCQ^{tc_i} - avSCQ^{tc_j})] \qquad (2)$$

$\mathcal{K}_d\Delta^{tfidf}$ **Representation:** For this representation we compute a document-term matrix of TF-IDF values (Fig. 1)[2], where we previously applied tokenization, stopword removal, lemmatization, and stemming. Aiming to understand the contribution of words in characterizing document collections, we calculate, then, box-plot statistics of the TF-IDF values over one document collection, namely the lower quartile ($Q1$), the median ($Q2$), the upper quartile ($Q3$), the lower whisker boundary (L) and the upper whisker boundary (U)[3]. From here, for each word in the vocabulary, we compute a three dimensional vector (or embedding) by counting the number of documents for which the TF-IDF value of the word occurs in intervals $I_L = [L, Q1]$, $I_M = [Q1, Q3]$, and $I_U = [Q3, U]$. These embeddings are, then, used to construct a further matrix as a representation of the document collection, where the columns correspond to each term in the vocabulary, and has three rows, $docs_{I_L}$, $docs_{I_M}$ and $docs_{I_U}$ corresponding to each interval (right-side table in Fig. 1).

For a pair of test collections, (tc_i, tc_j), we define $\mathcal{K}_d\Delta$ to be the weighted sum of the differences between the corresponding rows in the test collection representation (right-most table in Fig. 1)[4]:

$$\mathcal{K}_d\Delta^{tfidf}(tc_i, tc_j) = \beta_1(docs_{I_L}^{tc_i} - docs_{I_L}^{tc_j}) + \beta_2(docs_{I_M}^{tc_i} - docs_{I_M}^{tc_j}) + \beta_3(docs_{I_U}^{tc_i} - docs_{I_U}^{tc_j}) \quad (3)$$

where β_1, β_2 and β_3 are parameters to control the influence of each row in $\mathcal{K}_d\Delta$ calculation. The motivation behind this $\mathcal{K}_d\Delta$ definition is derived from the idea that a term's TF-IDF set of values are indicative of its usefulness in discriminating documents in a collection. We generalize this to a document collection, deriving comparable scores for sets of documents by taking the document count of each word in the vocabulary for different TF-IDF ranges.

---

[2] In order for the test collections to be comparable, we consider as our vocabulary all tokens across all test collections.

[3] where: $L = Q1 - 1.5 * (Q3 - Q1)$ and $U = Q3 + 1.5 * (Q3 - Q1)$.

[4] Where we apply a min-max normalization to the entries of these rows.

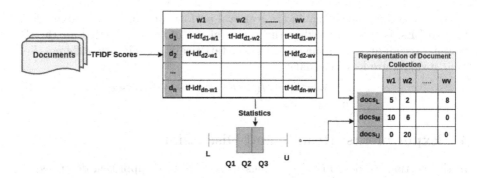

Fig. 1. The process of TF-IDF document collection representation.

$\mathcal{K}_d \Delta^{lm}$ *Representation:* The third representation choice we present in this paper is by means of Language Models (LMs). Having a LM trained on a corpus, we calculate the difference between the model and some text sequence $W = (w_1, w_2, ..., w_N)$ of length N using the perplexity measure, ([10], Chap. 4), PP, defined as an exponential of the cross-entropy H for the text sequence W.

$$PP(W; LM) = 10^{H(W;LM)} \tag{4}$$

$$H(W; LM) = -\frac{1}{N} \sum_{i=1}^{N} \log P_{LM}(w_i|w_{<i}) \tag{5}$$

where $P_{LM}(w_i|w_{<i})$ is the probability that w_i follows $w_{<i}$ as by the LM model.

To compute $\mathcal{K}_d \Delta^{lm}$ between two test collections, tc_i and tc_j, we first compute a Language Model for each of them, where we pre-process the text by splitting it into sentences, lowercasing, removing special characters, punctuation, and numbers. For each tc we compute a 4-gram LM model with modified Kneser-Ney smoothing [9]. Then, we compute a sequence of cross-entropies for the following cases: a) $H(tc_i; LM_i)$ and $H(tc_j; LM_j)$, the cross entropy between a model and a data it was trained on; b) $H(tc_j; LM_i)$ and $H(tc_i; LM_j)$ which represent the delta between a model of a collection and the collection it should be compared with; c) $H(tc_i; LM_i) - H(tc_j; LM_i)$ and $H(tc_j; LM_j) - H(tc_i; LM_j)$, the difference between the previous two cross entropy values.

During experimentation, we investigate the impact of a), b), and c) as features by weighting these value pairs with the parameters β_1, β_2, and β_3 respectively.

3.3   Result Delta

We define $\mathcal{R}\Delta^5$ as the absolute difference in the IR system performance measured for two EEs as:

$$\mathcal{R}\Delta = M(S_1, EE_1) - M(S_1, EE_2) \tag{6}$$

where M is the IR metric computed for the system S_i evaluated on the test collection of the evaluation environment EE_j.

---

[5] In our previous work, we defined different types of $\mathcal{R}\Delta$, in this paper $\mathcal{R}\Delta$ coincides with $\mathcal{R}_e\Delta$ in [5].

We state that the change in the retrieval performance of an IR system, $\mathcal{R}\Delta$ for two EEs, is significant when the means of the metrics computed for the IR system over all queries of the two EEs are statistically significantly different, i.e. their p-values are less than 0.05:

$$S_e\Delta = \begin{cases} 1, & \text{if } p(t\text{-}test(M(S_1, EE_1), M(S_1, EE_2))) \leq 0.05 \\ 0, & \text{otherwise} \end{cases} \tag{7}$$

4   Experiments, Results and Discussion

In this section we describe the settings used to test the approach described in Sect. 3 and the results of our experiments. We give information on the data sets, the classifier and its parameters, and on the experiments and the results.

4.1   Dataset Preparation

We use the Robust [18] and the TREC-COVID [17] test collections to create two sets of controlled evolving evaluation environments. The Robust test collection contains 250 queries and 528,155 documents from four news corpora, timestamped by the news' publication day. TREC-COVID[6] contains 50 queries and 191,160 documents–scientific papers about Covid-19 from the CORD-19 dataset [19]–each timestamped with its publication date.

From each of the two test collection (Robust and TREC-COVID), we create EvEEs by applying our approach described in Sect. 3.1, keeping the same topics for all EEs. For each EE, we pick 8 classical IR systems: BM25, PL2, Dirichlet language model, and DLH [1], all in two versions: with and without blind relevance feedback. The systems were implemented in Pyterrier with default parameters [12]. We evaluate each EE using seven performance metrics: Precision at 10 (P@10), RPrecision (RPrec), binary preference (bpref), Mean Average Precision (MAP), normalized Discounted Cumulative Gain (nDCG), normalized Discounted Cumulative Gain at 10 (nDCG@10), and Reciprocal Rank (RR). The metrics are computed using Pyterrier's implementation.

4.2   Classification Process

Since we are dealing with a binary classification problem with numerical features, we chose the Support Vector Machine (SVM) method to build the classification models. To build the training corpus, for a given EvEE, we assume that differences between consecutive EEs are not large and we compute $\mathcal{K}_d\Delta$ and $S_e\Delta$ between non-consecutive pairs of EEs (i.e. EE_1 vs. EE_3, EE_1 vs. EE_4, etc.). We use SVC scikit-learn[7] (sklearn) Python implementation with the default parameters to build the classification models. To insure the stability of the trained models, we train and test the SVM model using k-fold cross validation with $k = 5$. We report the classification performance on the testing data using Accuracy, Precision, Recall, and F1-score.

---

[6] https://ir.nist.gov/covidSubmit/data.html.
[7] https://scikit-learn.org/stable/.

Table 1. Description of β_i selection criteria for $\mathcal{K}_d\Delta$ computation. **Id** is used to identify the different sets of weight settings used in the experiments.

Description	Id	β_1	β_2	β_3
One feature on, the others are ignored.	W_1	0	0	1
	W_2	0	1	0
	W_3	1	0	0
Two features on (equally weighted), the third one off	W_4	0	0.5	0.5
	W_5	0.5	0.5	0
	W_6	0.5	0	0.5
Use all feature, with equal weight	W_7	0.333	0.333	0.333
Use all features, giving one more weight than the others	W_8	0.660	0.167	0.167
	W_9	0.167	0.660	0.167
	W_{10}	0.167	0.167	0.660

4.3   Experiments

We built and evaluated SVM models for each IR system and evaluation metric (Sect. 4.1) (i.e. a total of $8 * 7 = 56$ experiments per $\mathcal{K}_d\Delta$). In our presentation of the result, we identify each classification model, $\mathcal{C}$, model as a function of two parameters: (1) S, the IR system and (2) $\mathcal{S}_e\Delta^m$, where m is the evaluation metric used to compute the $\mathcal{S}_e\Delta$ for system S between EE pairs. For example $\mathcal{C}(BM25, \mathcal{S}_e\Delta^{MAP})$, is the classifier model build to predict the significant change of the MAP scores for BM25 across EEs. We experiment with the following different $\mathcal{K}_d\Delta$ inputs (i.e. feature vectors), shown in Sect. 3.2:

1. $\mathcal{K}_d\Delta^{qpp}$: a baseline $\mathcal{K}_d\Delta$.
2. $\mathcal{K}_d\Delta^{tfidf}$: computed with different β_i settings in Table 1.
3. $\mathcal{K}_d\Delta^{lm}$: computed with different β_i settings in Table 1.
4. $\mathcal{K}_d\Delta^{tfidf+lm}$: computed by combining both of $\mathcal{K}_d\Delta^{tfidf}$ and $\mathcal{K}_d\Delta^{lm}$ into one by selecting the best performing β_i combination of each.

4.4   Results

In this section we present a selection of our experimental results[8] That is, we show results for the BM25 model used with the Robust EvEEs, since: (1) we notice similar behaviour for the eight evaluated IR systems and (2) we have not observed changes for models trained using the TREC-COVID EvEE in terms of classification performance. We use the F1-score of the testing samples to report on classification models $\mathcal{C}(BM25, \mathcal{S}_e\Delta^m)$ performance and selecting, only, three of them to make the results readable. The class distribution of selected classification models is shown in Table 2.

---

[8] The full set of results for the 56 classifiers can be found here: https://owncloud. tuwien.ac.at/index.php/s/opUP9QlFEUHlfsx.

Table 2. $\mathcal{S}_e\Delta$ significance classes for 3 classifiers trained on top of BM25

Class label ($\mathcal{S}_e\Delta$)	Classifier		
	$\mathcal{C}(BM25,\mathcal{S}_e\Delta^{P@10})$	$\mathcal{C}(BM25,\mathcal{S}_e\Delta^{MAP})$	$\mathcal{C}(BM25,\mathcal{S}_e\Delta^{nDCG})$
0	173	198	215
1	427	402	385

$\mathcal{K}_d\Delta^{tfidf}$: Table 3 and Fig. 2 show the performance of applying different values of β (Table 1) to calculate the $\mathcal{K}_d\Delta^{tfidf}$ (Eq. 3). We notice drop (**bolded** values in Table 3) in the F1-Score values when β_1 contributes with greater than or equal values compared to β_2 and β_3 (i.e. W_3, W_5, W_6, W_7, W_8). In order to verify that this drop is not by chance, we divided the results for each classification model from different β_{is} setting into two sets: (1) $Set_1 = \{W_3, W_5, W_6, W_7, W_8\}$ and (2) $Set_2 = \{W_1, W_2, W_4, W_9, W_{10}\}$ and computed the significance difference in F1-Score using t-$test$ between the two sets which resulted in an average $p\_value = 0.054$. We conclude that the F1-Score using β values from Set_2 is significantly higher than that of values from Set_1. We interpret this result that words which have high TF-IDF values, within the documents set of an EE, can discriminate an EE and indicate the performance change of a system trained on documents from different EE.

We have done similar analyses for the results on the TREC-COVID EvEE experiments. The average p_value of the F1-Score differences is 0.28 and so we cannot hypothesis that the results from Set_2 is higher than that of Set_1. We assume that this result is due to the fact that TREC-COVID EvEE belongs to the same domain, so there is no significant change the vocabulary between EEs compared to the Robust EvEE which is a collection of news articles from different domains.

Table 3. The F1-score measure of selected classifiers to predict $\mathcal{S}_e\Delta$ for the BM25 system using the $\mathcal{K}_d\Delta^{tfidf}$ on Robust EvEE.

Id	$\mathcal{C}(BM25,\mathcal{S}_e\Delta^{P@10})$	$\mathcal{C}(BM25,\mathcal{S}_e\Delta^{MAP})$	$\mathcal{C}(BM25,\mathcal{S}_e\Delta^{nDCG})$
W_1	0.855 ± 0.139	0.828 ± 0.128	0.720 ± 0.279
W_2	0.902 ± 0.105	0.857 ± 0.111	0.772 ± 0.249
W_3	$\mathbf{0.643 \pm 0.11}$	$\mathbf{0.583 \pm 0.000}$	$\mathbf{0.593 \pm 0.003}$
W_4	0.867 ± 0.125	0.848 ± 0.105	0.760 ± 0.205
W_5	0.868 ± 0.233	0.838 ± 0.078	$\mathbf{0.643 \pm 0.159}$
W_6	0.793 ± 0.366	0.805 ± 0.081	$\mathbf{0.570 \pm 0.140}$
W_7	0.863 ± 0.195	0.835 ± 0.083	0.732 ± 0.318
W_8	0.745 ± 0.212	$\mathbf{0.662 \pm 0.043}$	$\mathbf{0.583 \pm 0.070}$
W_9	0.907 ± 0.101	0.882 ± 0.082	0.782 ± 0.303
W_{10}	0.845 ± 0.206	0.838 ± 0.088	0.712 ± 0.282

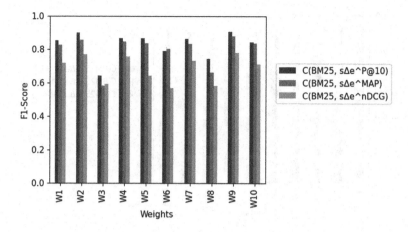

Fig. 2. The graph corresponding to the values in Table 3.

$\mathcal{K}_{\mathbf{d}} \mathbf{\Delta}^{\mathrm{lm}}$: Table 4 and Fig. 3 show the performance for different β value sets (Table 1) to calculate the $\mathcal{K}_{\mathbf{d}} \mathbf{\Delta}^{\mathrm{lm}}$ (Sect. 3.2). We were assuming that the classification models should not work when β_1 value is low (W_3 and W_8), since the features that are obtained by comparing two different EEs contribute with no or low weight during the training. In the case of W_3 and W_8 using the $\mathcal{C}(BM25, \mathcal{S}_e \Delta^{P@10})$ data set, the models achieve the same F1-score one would achieve if one would label every sample as 1, meaning that the models learn nothing. Moreover, this leads us to the converse conclusion that each metric where the results for W_3 and W_8 are not significantly worse (e.g. $\mathcal{C}(BM25, \mathcal{S}_e \Delta^{nDCG})$) does not yield a data set fit for modeling the difference between EEs.

Table 4. The F1-score for selected classifiers predicting $\mathcal{S}_e \Delta$ for the BM25 system using $\mathcal{K}_{\mathbf{d}} \mathbf{\Delta}^{\mathrm{lm}}$ on Robust EvEE.

Id	$\mathcal{C}(BM25, \mathcal{S}_e \Delta^{P@10})$	$\mathcal{C}(BM25, \mathcal{S}_e \Delta^{map})$	$\mathcal{C}(BM25, \mathcal{S}_e \Delta^{nDCG})$
W_1	0.927 ± 0.105	0.843 ± 0.055	0.738 ± 0.262
W_2	0.923 ± 0.123	0.858 ± 0.053	0.732 ± 0.254
W_3	0.833 ± 0.209	0.768 ± 0.174	0.767 ± 0.216
W_4	0.933 ± 0.113	0.840 ± 0.062	0.735 ± 0.246
W_5	0.913 ± 0.119	0.832 ± 0.065	0.745 ± 0.240
W_6	0.923 ± 0.108	0.822 ± 0.081	0.743 ± 0.238
W_7	0.928 ± 0.116	0.828 ± 0.077	0.738 ± 0.237
W_8	0.860 ± 0.192	0.815 ± 0.083	0.752 ± 0.227
W_9	0.922 ± 0.127	0.853 ± 0.054	0.740 ± 0.249
W_{10}	0.920 ± 0.121	0.832 ± 0.070	0.742 ± 0.247

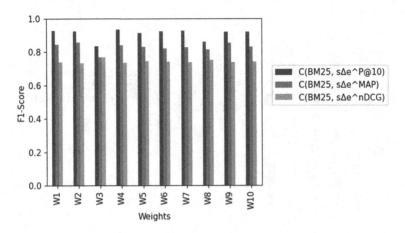

Fig. 3. The graph corresponding to the values in Table 4.

$\mathcal{K}_{d}\mathbf{\Delta}^{qpp}$: We take $\mathcal{K}_{d}\mathbf{\Delta}^{qpp}$ (Eq. 2) [6], to be a baseline estimate of the performance. As there is not enough space to present the complete comparison, we limit ourselves to the following models to compare with $\mathcal{K}_{d}\mathbf{\Delta}^{qpp}$; (1) $\mathcal{K}_{d}\mathbf{\Delta}^{tfidf}$: selected the models trained on W_4 (i.e. $\beta_1 = 0$ and $\beta_2 = \beta_3$) since β_1 cause the performance to drop significantly, (2) $\mathcal{K}_{d}\mathbf{\Delta}^{lm}$: selected the models trained on W_7 (i.e. $\beta_1 = \beta_2 = \beta_3$) since we do not obtain significance difference in performance using different variations of β_{is}, and (3) $\mathcal{K}_{d}\mathbf{\Delta}^{tfidf+lm}$: trained a model by combining $\mathcal{K}_{d}\mathbf{\Delta}^{tfidf}$ and $\mathcal{K}_{d}\mathbf{\Delta}^{lm}$ into a single $\mathcal{K}_{d}\Delta$ feature vector.

Table 5 and Fig. 4 show the comparison of performance between different $\mathcal{K}_{d}\Delta$ and the baseline. The F1-Score of each of $\mathcal{K}_{d}\mathbf{\Delta}^{tfidf}$, $\mathcal{K}_{d}\mathbf{\Delta}^{lm}$ and combining both is higher than that of $\mathcal{K}_{d}\mathbf{\Delta}^{qpp}$. From this result we assume that, identifying variations between document representations across EE helps in predicting the change of the system performance for EvEE. However, combing both of $\mathcal{K}_{d}\mathbf{\Delta}^{tfidf}$ and $\mathcal{K}_{d}\mathbf{\Delta}^{lm}$ do not result in change in the prediction of the performance change. This shows that the features from $\mathcal{K}_{d}\mathbf{\Delta}^{tfidf}$ dominates that of $\mathcal{K}_{d}\mathbf{\Delta}^{lm}$ which can be clarified more with feature selection techniques.

Table 5. Comparison of F1-Scores between proposed $\mathcal{K}_{d}\Delta$ (i.e. $\mathcal{K}_{d}\mathbf{\Delta}^{tfidf}$, $\mathcal{K}_{d}\mathbf{\Delta}^{lm}$, and $\mathcal{K}_{d}\mathbf{\Delta}^{tfidf+lm}$) and baseline $\mathcal{K}_{d}\mathbf{\Delta}^{qpp}$.

	$\mathcal{C}(BM25, \mathcal{S}_e\Delta^{P@10})$	$\mathcal{C}(BM25, \mathcal{S}_e\Delta^{MAP})$	$\mathcal{C}(BM25, \mathcal{S}_e\Delta^{nDCG})$
$\mathcal{K}_{d}\mathbf{\Delta}^{qpp}$	0.588 ± 0.090	0.778 ± 0.161	0.582 ± 0.074
$\mathcal{K}_{d}\mathbf{\Delta}^{tfidf}$	0.882 ± 0.143	0.85 ± 0.178	0.818 ± 0.165
$\mathcal{K}_{d}\mathbf{\Delta}^{lm}$	0.933 ± 0.105	0.84 ± 0.055	0.735 ± 0.194
$\mathcal{K}_{d}\mathbf{\Delta}^{tfidf+lm}$	0.882 ± 0.143	0.853 ± 0.168	0.838 ± 0.127

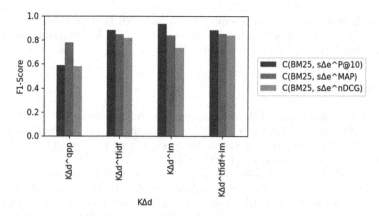

Fig. 4. Comparison of F1-Scores between proposed $\mathcal{K}_d\Delta$ (i.e. $\mathcal{K}_d\Delta^{\mathrm{tfidf}}$, $\mathcal{K}_d\Delta^{\mathrm{lm}}$, and $\mathcal{K}_d\Delta^{\mathrm{tfidf+lm}}$) and baseline $\mathcal{K}_d\Delta^{\mathrm{qpp}}$(Table 5).

5   Summary and Future Work

Aiming to understand the impact of the changes of elements of EE ($\mathcal{K}\Delta$), in EvEE, on the IR systems performance changes ($\mathcal{R}\Delta$), we present two approaches to compute differences between document sets of EvEE ($\mathcal{K}_d\Delta$) using text representations: TF-IDF and LM. We present a SVM-based classification model to predict if there is a significance change in $\mathcal{R}\Delta$ given $\mathcal{K}_d\Delta$. We experimented on two EvEEs which were built by using the timestamps from the Robust and TREC-COVID test collections. We trained classification models to predict the performance change of different IR systems and IR evaluation metrics. We evaluated the models performance using $\mathcal{K}_d\Delta^{\mathrm{tfidf}}$, $\mathcal{K}_d\Delta^{\mathrm{lm}}$ and a combined $\mathcal{K}_d\Delta^{\mathrm{tfidf+lm}}$ against a baseline ($\mathcal{K}_d\Delta^{\mathrm{qpp}}$). The results are based on simple representations and settings, however they are motivating as a first step to have an IR evaluation framework in which we can compare EEs by injecting differences between different IR systems also between test collections from different domains.

As future work, we plan to apply our proposed $\mathcal{K}_d\Delta$ on the LongEval [4] test collection, which is a benchmark for continuous IR sytem evaluation. In addition, we plan to look at other text representations to extend the $\mathcal{K}_d\Delta$ definitions. We plan to utilize keyword-based methods in combination with knowledge graphs and topic detection methods to define $\mathcal{K}_d\Delta$. We plan to explore $\mathcal{K}_d\Delta$ definitions for the use with neural network-based models by utilizing the text representations for different layers of a neural network model (e.g. modelling $\mathcal{K}_d\Delta$ in terms of the knowledge gained at different layers of BERT model [14]). Moreover, as we now simplify the problem as a binary classification problem to detect statistically significant $\mathcal{R}\Delta$ changes, we will extend this aspect of our work by looking at other methods of predicting $\mathcal{R}\Delta$.

Acknowledgement. This work is supported by the ANR Kodicare bi-lateral project, grant ANR-19-CE23-0029 of the French Agence Nationale de la Recherche, and by the Austrian Science Fund (FWF), grant I-4471-N.

References

1. Amati, G.: Frequentist and Bayesian approach to information retrieval. In: Lalmas, M., MacFarlane, A., Rüger, S., Tombros, A., Tsikrika, T., Yavlinsky, A. (eds.) ECIR 2006. Lecture Notes in Computer Science, vol. 3936, pp. 13–24. Springer, Berlin (2006). https://doi.org/10.1007/11735106_3
2. Ferro, N., Kim, Y., Sanderson, M.: Using collection shards to study retrieval performance effect sizes. ACM Trans. Inf. Syst. (TOIS) **37**(3), 1–40 (2019)
3. Ferro, N., Silvello, G.: Towards an anatomy of IR system component performances. J. Assoc. Inf. Sci. Technol. **69**, 187–200 (2018). https://doi.org/10.1002/asi.23910
4. Galuščáková, P., et al.: Longeval-retrieval: French-english dynamic test collection for continuous web search evaluation. arXiv preprint arXiv:2303.03229 (2023)
5. González-Sáez, G.N., Mulhem, P., Goeuriot, L.: Towards the evaluation of information retrieval systems on evolving datasets with pivot systems. In: Candan, K.S., et al. (eds.) CLEF 2021. Lecture Notes in Computer Science, vol. 12880, pp. 91–102. Springer International Publishing, Cham (2021). https://doi.org/10.1007/978-3-030-85251-1_8
6. González-Sáez, G., et al.: Towards result delta prediction based on knowledge deltas for continuous IR evaluation. In: Faggioli, G., Ferro, N., Mothe, J., Raiber, F. (eds.) The QPP++ 2023: Query Performance Prediction and Its Evaluation in New Tasks Workshop (QPP++), pp. 20–24, no. 3366 in CEUR Workshop Proceedings, Aachen (2023). http://ceur-ws.org/Vol-3366/#paper-04
7. Hauff, C.: Predicting the effectiveness of queries and retrieval systems. In: SIGIR Forum, vol. 44, p. 88 (2010)
8. He, B., Ounis, I.: Query performance prediction. Inf. Syst. **31**(7), 585–594 (2006) https://doi.org/10.1016/j.is.2005.11.003, https://www.sciencedirect.com/science/article/pii/S0306437905000955. (1) SPIRE 2004 (2) Multimedia Databases
9. Heafield, K., Pouzyrevsky, I., Clark, J.H., Koehn, P.: Scalable modified Kneser-Ney language model estimation. In: Proceedings of the 51st Annual Meeting of the Association for Computational Linguistics (Volume 2: Short Papers), pp. 690–696 (2013)
10. Jurafsky, D., Martin, J.H.: Speech and Language Processing, 2Nd edn. Prentice-Hall Inc, Upper Saddle River (2009)
11. Kanoulas, E.: A short survey on online and offline methods for search quality evaluation. In: Russian Summer School on Information Retrieval (2015)
12. Macdonald, C., Tonellotto, N.: Declarative experimentation in information retrieval using PyTerrier. In: Proceedings of the 2020 ACM SIGIR on International Conference on Theory of Information Retrieval, pp. 161–168 (2020)
13. Rashidi, L., Zobel, J., Moffat, A.: Evaluating the predictivity of IR experiments. In: Proceedings of the 44th International ACM SIGIR Conference on Research and Development in Information Retrieval, SIGIR 2021, pp. 1667–1671. Association for Computing Machinery, New York, NY, USA (2021). https://doi.org/10.1145/3404835.3463040
14. Rogers, A., Kovaleva, O., Rumshisky, A.: A Primer in BERTology: what We know about how BERT works. Trans. Assoc. Comput. Linguist. **8**, 842–866 (2021). https://doi.org/10.1162/tacl_a_00349

15. Sanderson, M.: Test collection based evaluation of information retrieval systems. Now Publishers Inc (2010)
16. Sanderson, M., Turpin, A., Zhang, Y., Scholer, F.: Differences in effectiveness across sub-collections. In: Proceedings of the 21st ACM International Conference on Information and Knowledge Management, CIKM 2012, pp. 1965–1969. Association for Computing Machinery, New York, NY, USA (2012). https://doi.org/10.1145/2396761.2398553
17. Voorhees, E., et al.: TREC-COVID: constructing a pandemic information retrieval test collection. In: ACM SIGIR Forum, vol. 54, no. 1, pp. 1–12. ACM New York (2021)
18. Voorhees, E.M.: The TREC 2005 robust track. In: ACM SIGIR Forum, vol. 40, pp. 41–48. ACM, New York (2006)
19. Wang, L.L., et al.: Cord-19: The covid-19 open research dataset. ArXiv (2020)
20. Zhao, Y., Scholer, F., Tsegay, Y.: Effective pre-retrieval query performance prediction using similarity and variability evidence. In: Macdonald, C., Ounis, I., Plachouras, V., Ruthven, I., White, R.W. (eds.) ECIR 2008. LNCS, vol. 4956, pp. 52–64. Springer, Heidelberg (2008). https://doi.org/10.1007/978-3-540-78646-7_8

When Sarcasm Hurts: Irony-Aware Models for Abusive Language Detection

Simona Frenda[1,2][(✉)][iD], Viviana Patti[1][iD], and Paolo Rosso[3][iD]

[1] University of Turin, Turin, Italy
{simona.frenda,viviana.patti}@unito.it
[2] aequa-tech srl, Turin, Italy
[3] Universitat Politècnica de València, Valencia, Spain
prosso@dsic.upv.es
http://www.di.unito.it/ , https://www.prhlt.upv.es/

Abstract. Linguistic literature on irony discusses sarcasm as a form of irony characterized by its biting nature and the intention to mock a victim. This particular trait makes sarcasm apt to convey hate speech and not only humour. Previous works on abusive language stressed the need to address ironic language to lead the system to recognize correctly hate speech, especially in spontaneous texts, like tweets [13]. In this context, our main hypothesis is that information about the presence of sarcasm could help to improve the detection of hateful messages, especially when they are camouflaged as sarcastic. To corroborate this hypothesis: i) we perform analysis on HASPEEDE20_EXT, an Italian corpus of tweets about the integration of cultural minorities in Italy, ii) we carry out computational experiments injecting the knowledge of sarcasm in a system of hate speech detection, and iii) we adopt strategies of validation in terms of performance and significance of the obtained results. Results confirm our hypothesis and overcome the state of the art.

Keywords: Abusive Language Detection · Irony Detection · Multi-Task Learning

1 Introduction

Nowadays, abusive language detection turns into a task of growing interest in Natural Language Processing (NLP) due to the recent necessity of monitoring the pervasive hostile user-generated contents that intimidate or incite violence, targeting many vulnerable groups especially in social platforms. In the typology delineated by [29], and adopted also in [23], this kind of content generally defined as *hate speech*[1] is gathered under the umbrella term of *abusive language*.

---

[1] One of the most complete definitions is provided by [25]: a content is considered hateful on the basis of its *action* and its *target*. The action is the illocutionary act of the utterance aimed to spread or justify hate, incite violence, or threat people's freedom, dignity, and safety. The target must be a protected group or an individual belonging to such a group, attacked for his/her individual characteristics.

A. Arampatzis et al. (Eds.): CLEF 2023, LNCS 14163, pp. 34–47, 2023.
https://doi.org/10.1007/978-3-031-42448-9_4

The authors of [29] classify the sub-tasks of abusive language on the basis of the "type of target" and the "degree" to which it is explicit. The former furnishes an interesting distinction between *individual* or entity targeted by cyberbullying and trolling, and the generalized *other* or group with certain ethnicity or protected characteristics targeted by racism, homophobia, or misogyny. The latter implies the linguistic and semiotic definitions of *denotation* (literal meaning) and *connotation* (sociocultural associations or assumptions). Therefore, on the one hand abusive language could be unambiguous and explicit, on the other one it implies some connotations that are difficult to interpret as abusive for the lack of profanities, the use of rhetorical elements (i.e., sarcasm) that recall contextual knowledge (Example 1) and negative stereotypes (Example 2).

(1) *Signore, hanno tutti diritto a una vita dignitosa, ma mettete un migrante sulla mia strada e io saró Salvini. (Matteo 15, 83)*[2]
(2) *Un piatto di pasta e chiediamogli scusa per non essere anche noi musulmani. Magari così diventano nostri amichetti e non ci uccidono più.*[3]

The fact that specific linguistic styles, such as humour, could make *hate speech* implicit, is emphasized by various scholars [12,15,20,31]. Analysing various benchmark datasets in English, [32] identified specific subtypes of implicit abuse: stereotypes, perpetrators, comparisons, dehumanization, euphemistic constructions, call-for-action, multimodal abuse, and all the phenomena that require world knowledge and inferences such as jokes, sarcasm and rhetorical questions. Some of these subtypes have been identified by scholars [4,33] as problematic challenges in abusive language detection, demonstrating that only their explicit manifestations are identified by current classifiers (supervised and unsupervised).

In this context, we concentrated on the role of ironic language in hateful contents online, focusing in particular on a specific form of irony, that is sarcasm. Differently from other ironic jokes, sarcasm aims to mock, scorn or ridicule a victim [17], and for this reason, is apt to disguise hurtful messages, lowering the tones without losing the hurtfulness of the message. Funny messages, moreover, are more likely to be accepted and shared by the community, making the abuse viral [25]. Therefore, our hypothesis is that information about the presence of sarcasm could help to improve the detection of hateful messages, even when they are camouflaged as sarcastic. To verify it, in this work, we aim to answer the following research question:

RQ *Could the awareness of the presence of sarcasm increase the performance of abusive language detection systems?*

To give the system an overall perception of irony that co-occur with abusive language, we designed a neural architecture based on multi-task learning that makes the system of hate speech detection more sensitive even to indirect abuses. In the past, some studies already proposed to train the models in different related

---

[2] Sir, everybody has the right to a dignified life, but if you put a migrant in my way, I will be Salvini. (Matthew 15, 83).
[3] A plate of pasta and let's apologize for not being Muslims too. Maybe then they become our friends and won't kill us anymore.

tasks [6,16]. Multi-task learning, indeed, gives systems more evidences to evaluate whether a feature is relevant or not, especially, in the cases where the data have different characteristics and thus various labels. In this work, we present:

- the analysis of HASPEEDE20_EXT [13], the extended version of the HaSpeeDe benchmark corpus composed of hateful Italian tweets[4] [24] that encodes also the presence of irony and sarcasm. To our knowledge, this is the first corpus that contains labels of hate speech and sarcasm for the same texts, allowing us to address our study;
- the design of a model for detecting hate speech in tweets that is aware of ironic language;
- a deep analysis, in terms of performance and significance, of the obtained results that contributes to confirm our hypothesis.

The proposed approach is validated in different settings of evaluation and compared with the actual state of the art for abusive language detection in Italian.

2   Related Work

In various surveys and positional papers on abusive language detection [12, 26,32], sarcasm is considered one of the figures of speech that make implicit hate speech along with euphemism, rhetorical questions, litotes, and absence of explicit accusations, negative evaluations or insults. These elements tend to elude the offensiveness of the text, making its recognition hard, especially for machines [19,21]. Differently from the other figures of speech, sarcasm is used to create jokes and amusing people; and taking into account its stinging purposes, in their definition of hate speech, [12] stressed the attention on linguistic styles and humour used to convey abuses: "Hate speech is language that attacks or diminishes, that incites violence or hate against groups [...] and it can occur with different linguistic styles, even in subtle forms or when humour is used".

The harshness of some jokes is a real problem. Look for example at the ethnic jokes. Some scholars consider the ethnic joke not serious [7], whereas others underline the importance of the context and, thus, the relation between stereotypes-based jokes with the social exclusion of the group targeted [15]. Indeed, this type of jokes can reinforce negative stereotypes and foster, especially online, the spread of hateful discourse leading to serious consequences. In [31] the seriousness of humour relies principally on the *rhetoric* of jokes (when are related to hostility, exclusion, or hierarchies) and on the hurtfulness of their *content*. That could have psychological repercussions because this kind of jokes is experienced by the target like harassment [3,10].

In linguistic literature, some scholars stress the 'muting the meaning' hypothesis that considers ironic language as a device to mute the negative meaning [9], whereas others underline its characteristic of increasing the negativity of messages. In particular, [22] proposed a pragmatic analysis of ironic insults and ironic compliments and how they are perceived by society. Ironic insults are perceived as more polite, whereas ironic compliments as more mocking and sarcastic: speakers tend to criticize someone lowering the social cost of doing so, and ironic language seems appropriate to cover the scorn.

---

[4] http://www.di.unito.it/~tutreeb/haspeede-evalita20/index.html.

The studies on abusive language detection and on its sub-tasks are various. The efforts of the NLP community focused especially on the detection of specific manifestations of abusive language [28,30,34], and on the exploration of its different characteristics, such as expressions of dehumanization, dominance or sexual harassment especially in misogynistic and sexist behaviours [11], the type of target (individual or communities) [1], and the topic (like in HaSpeeDe3[5] organized at EVALITA 2023[6]).

However, the attention to its implicit forms is less popular. To our knowledge, the second edition of HaSpeeDe at EVALITA 2020 [24] was the first shared task that has encouraged the investigation of hate speech and stereotypes as orthogonal axes that could occur in the same text. Other studies, like [33], focused on the study of implicit hate speech looking at the vocabulary used in the texts to convey abusive language. While, to approach directly the implicitness of hate speech, [4] designed a new schema of annotation extending the existing OLID/OffensEval dataset. In particular, they showed: the annotation of implicit forms of hate speech required contextual information, the recognition of implicit abuse is more challenging than the explicit one, and that the most used English datasets about hate speech and offensiveness contain especially explicit abusive language, encouraging a reflection on how the datasets are created.

The interest in implicit abusive language is recent, and the computational studies that use the information on implicit forms of abuses to improve the detection of hate speech are few. To our knowledge, only metaphorical and stereotyped information have been exploited for abusive language detection [16,18]. Finally, [13] focused on discovering the characteristics of implicit and explicit hate speech and stereotypes in the different textual genres (tweets and news headlines) present in the corpus released by the organizers of HaSpeeDe 2020. Especially the training set of this corpus (composed only of tweets) has been extended by [13] and called HASPEEDE20_EXT. Their analyses showed that hate speech is strongly associated with the presence of sarcasm in tweets, while the expression of stereotypes is characterized by patterns (such as semantic incongruity) typical also of ironic language. For the computational analysis, they designed different classifiers that detect hate speech and stereotypes based on the simultaneous learning of more tasks. These classifiers have also been informed with specific linguistic features to explore the patterns of hate speech and stereotype in tweets and news headlines.

Differently from our current work, [13] described exploratory analysis oriented to provide solid basis for the development of systems well-informed with linguistic information to detect correctly hate speech and stereotypes. And on the basis of these previous findings, we investigate deeply the important role played by humour in hate speech, especially in spontaneous texts, such as tweets, and propose a system of hate speech detection aware of ironic language.

---

[5] http://www.di.unito.it/~tutreeb/haspeede-evalita23/index.html.
[6] https://www.evalita.it/campaigns/evalita-2023/.

3   Methodology

The methodology used in this work is oriented to prove the hypothesis that the awareness of the presence of sarcasm could help the system to detect hate speech even when it is masked as sarcastic. To this purpose:
1) Firstly, we examined the compresence of ironic language when the text is annotated as hate speech, observing the contingency tables and percentages of compresence. This corpus-based analysis helps us to corroborate the basis of our hypothesis in the scenario of the discussion online about immigration, represented by HASPEEDE20_EXT.
2) Secondly, we designed a system of hate speech detection that is aware of ironic language using the multi-task approach (Model-MTL). This system combines the knowledge coming from Italian language models (LMs) with the information on the presence of irony in the text. In order to validate the contribution of multi-task learning, we also created fine-tuned models (Model-FT) from the three considered LMs. All these models have been tested on the test set of tweets released by the organizers of HaSpeeDe 2020 (Test_TW) to compare the obtained results with the best systems of the competition, the ones obtained by [13] and a new strong baseline called *Baseline_Avg_LMs*.
3) Thirdly, we analysed the confusion matrix, comparing the predictions of Model-MTL with the predictions obtained with *Baseline_Avg_LMs*; and we performed a statistical experiment based on the bootstrap sampling significance test, to know how significant the injection of the knowledge of ironic language is for hate speech detection respect to the awareness of other phenomena like stereotypes.

System Design. The design of the irony aware system for hate speech detection (Model-MTL) is based on the idea of combining general knowledge coming from LMs and linguistic information about irony and sarcasm, obtained with the injection of ironic language recognition within a multi-task learning framework. The choice of employing a multi-task learning (MTL) based model is motivated by the strong correlation observed especially between the presence of hate speech and sarcasm in tweets in [13], and, at computational level, by the advantages derived from the use of MTL techniques such as the *hard parameter sharing*. Firstly, this technique gives the system more evidences to evaluate whether a feature is relevant or not, focusing strictly on the most relevant ones for each task. Then, the hard parameter sharing allows a better generalization for each task [2]: learning simultaneously more tasks means to find a representation that is appropriate for learning all the tasks, reducing consequently the overfitting on the original task. Model-MTL is built fine-tuning three Italian LMs (AlBERTo, UmBERTo, and GilBERTo)[7] to solve two tasks: hate speech and irony/sarcasm detection. For tuning the models on these tasks, we took into account only the CLS token of the BERT-based model. Indeed, in accordance with [8], the purpose of this token is to contain the information useful for the classification task at the

---

[7] These three language models are trained on different genres of texts in Italian and available on the Hugging Face platform: https://huggingface.co/models.

end of the forwarding process. Then a classifier can just take this CLS token as input to classify the whole text. Moreover, we added a dropout layer (with a value of 0.1) and double linear output layers, one for each task, to get the class-related probabilities employing a Sigmoid function. To understand the real contribution of the simultaneous learning of correlated tasks, we also modelled a fine-tuned classifier of hate speech for each language model (Model-FT). The only difference with the network designed for Model-MTL, is a unique final linear layer for hate speech classification. Moreover, we created a new baseline (*Baseline_Avg_LMs*) that takes into account the mean of the probabilities obtained employing every language model for each text.

Corpus. HaSpeeDe is a shared task organized in different campaigns of EVALITA: in 2018, 2020 and 2023. The three editions propose as main task the detection of hate speech in Italian texts collected from different social media platforms. In particular, in its second edition in 2020, the organizers of HaSpeeDe [24] collected texts from Twitter to create the training set and an in-domain test set, and headlines from newspapers to create an out-domain test set. All the collected data report as topic the immigration in Italy and in particular the integration of immigrants, Roma community and Muslims. The subtasks proposed in the second edition of HaSpeeDe are: 1) the recognition of hate speech, 2) the recognition of stereotypes, and 3) the identification of nominal utterances expressing hate. In this work, we focus exclusively on hate speech detection exploiting the extended version of the training set provided by [13] (the Train_TW_ext in HASPEEDE20_EXT) and the test set of tweets (Test_TW) released for the competition. A summary of the amount of tweets and the distribution of labels used in this work is reported in Table 1. The choice of data is based on the results of the exploratory analysis in [13], and on the fact that, especially in case of negative and hateful opinions, [25] noticed that the users tend to be less explicit in their claims in order to limit their exposure using expedients like irony. Sarcasm in the annotation of [13][8] is a type of irony, sharper than other ironic languages, and with the aim to scorn a victim [3]. Therefore, in this dataset, if a tweet is ironic it could also be sarcastic. Some examples are reported in Table 2.

Table 1. Distribution of labels in the collection of tweets.

set	hs	non-hs	iro	non-iro	sarc	iro non-sarc	total
Train_TW_ext	3,035	5,226	1,806	6,455	1,111	695	8,261
Test_TW	622	641	361	902	239	122	1,263

---

[8] The schema of annotation of ironic language is inherited by [5] who annotated the IronITA corpus of tweets for the first time, to our knowledge, as ironic and sarcastic.

Table 2. Some examples from our dataset of tweets.

hs	iro	sarc	text
1	1	0	*"Anziché far venire gli immigrati diamo il Reddito di Cittadinanza e gli italiani incominceranno a trombare come ricci..." (Massimo Baroni, deputato M5S)* → "Instead of letting immigrants come, we give the Citizenship Income (an economic support to combat poverty, inequality and social exclusion) and the Italians will begin to f**k like hedgehogs..." (Massimo Baroni, M5S deputy)
1	1	1	*Per l'ONU la capotreno sarebbe colpevole di di 'razzismo' e 'intolleranza' verso un'immigrata.. Come si è permessa di chiedere il biglietto ad un nigeriana?? Insomma, noi italiani non sappiamo proprio... URL* → According to the UN, the conductor would be guilty of 'racism' and 'intolerance' towards an immigrant.. How did she dare to ask a Nigerian for a ticket? In short, we Italians just don't know ... URL
0	1	1	*@USER @USER @USER Accettiamo scommesse sul tipo di 'lavoro' che sta andando a fare il rom in... URL* → @USER @USER @USER We accept bets on the type of 'work' the roma guy is going to do in... URL

4   Irony in Hateful Contents

In this section, we present: a corpus-based analysis to corroborate our hypothesis, the computational experiments to validate the proposed approach, and a deep analysis of results of our approach in terms of performance and significance.

Corpus-Based Analysis. Looking at the contingency tables about the presence of hate speech and ironic language in the tweets of Train_TW_ext, we notice that the frequencies of compresence of hate speech (hs) and sarcasm are higher than the compresence of hate speech with other forms of irony (Table 3).

Taking into account the definition of sarcasm, for obtaining the contingency values between hate speech and sarcasm, we selected all the tweets that are annotated as ironic (i.e., 1,806 tweets). For obtaining the contingency values between hate speech and non-sarcastic forms of irony, we selected all the tweets that could be annotated as ironic but are not sarcastic (i.e., 7,150 tweets). Comparing them, about the 69% of tweets labelled as hate speech are also annotated as sarcastic, differently from the amount of hate speech containing non-sarcastic irony (7%). That justifies the positive and significative association already seen in [13] between hate speech and sarcasm (Yule's Q = 0.24).

Computational Experiments. To test this association also at computational level, we employed the Model-MTL described in Sect. 3 using as correlated task irony (iro) and sarcasm (sarc) detection, and compared its performances with

Table 3. Contingency tables.

	non-sarcastic irony				sarcasm	
	0	1			0	1
hs 0	4,020	518		hs 0	518	688
hs 1	2,427	185		hs 1	185	415
total		7,150		total		1,806

the Model-FT of each considered LMs, and the strong baseline model *Baseline_Avg_LMs* that takes into account the mean of all the probabilities obtained by each Model-FT for each instance. For testing and comparison, we employed the same evaluation measures used in HaSpeeDe 2020: $f1$-macro as average of the $f1$ of each class. Table 4 reports the obtained results.

Table 4. Results of hate speech detection obtained on Test_TW.

	Model-FT	Model-MTL (iro)	Model-MTL (sarc)
AlBERTo	0.741	0.753	0.765
UmBERTo	0.790	0.780	**0.816**
GilBERTo	0.762	0.756	0.778
Baseline_Avg_LMs	*0.800*	*0.792*	*0.795*

Looking at the performance of each language model, we notice that UmBERTo in general performs better than AlBERTo and GilBERTo, and that the awareness of sarcasm reports optimal results regardless the used language model. Moreover, the used baseline is in all the cases hard to overcome, except when UmBERTo is fine-tuned to solve hate speech and sarcasm detection.

Finally, we compared in Table 5 these results with the best ranked systems and baselines models of the HaSpeeDe 2020 shared task, and the results of the best models presented in [13] whose systems of hate speech detection are informed adding linguistic features (LingFeat) and knowledge about stereotypes detection (stereo). Observing Table 5, it is clear that the addition of linguistic information enriched with external features or knowledge about stereotype plays an important role in making the system more sensible (0.809 and 0.808). TheNorth employed the UmBERTo pre-trained model in both submitted runs, juxtaposing its fine-tuning (*TheNorth_1*) with the additional learning of the stereotype detection task in *TheNorth_2*[9]. This second run achieved the best score (0.809) in hate speech detection in the competition. However, the awareness about sarcasm proves to be another important element to take into account

---

[9] In particular, they used a linear layer with a softmax on top of the CLS token, applying a novel technique of layer-wise learning rate. That is the main difference with our approach.

in the detection of abusive messages. Indeed, the model trained also on sarcasm detection achieves the best score (0.816).

Table 5. Results in hate speech detection in the Ranking of HaSpeeDe 2020.

		$f1\_Tw$
UmBERTo-MTL (sarc)		**0.816**
TheNorth_2	[16]	0.809
ItBERT-MTL (stereo)+LingFeat	[13]	0.808
Baseline_Avg_LMs		*0.800*
TheNorth_1	[16]	0.790
CHILab[a]	[14]	0.789
UmBERTo-MTL (iro)		0.780
Baseline_SVC		*0.721*
Baseline_MFC		*0.337*

[a] [14] experimented transformer encoders in the first run creating specifically two transformer/convolution blocks for each input (texts and Part-of-Speech or PoS tags) averaged through max pooling and processed finally by a dropout and dense layer to obtain the predictions.

Table 6. Values of confusion matrix for hate speech detection in tweets.

	FP (%)	FN (%)	TP (%)	TN (%)
UmBERTo-MTL (sarc)	24	**12**	**88**	76
Baseline_Avg_LMs	**20**	21	79	**80**

Analysis of Results. In order to understand the advantage to inject sarcasm knowledge in the system of hate speech detection, we carried out the analysis of the True Positive cases (TP), True Negatives (TN), False Positives (FP) and False Negatives (FN) obtained with Model-MTL, comparing it with *Baseline_Avg_LMs*. Table 6 reports the percentages of cases. The system based on MTL of hate speech and sarcasm detection shows an improvement of its accuracy in detecting hateful messages. This improvement on retrieving TPs could be convenient in a real word application. Another interesting finding that emerged from a manual analysis is that making aware the system of sarcasm, even tweets containing other figures of speech have been correctly classified:

(3) *Quindi se un italiano muore in ospedale in mezzo alle formiche è 'episodio' mentre se un nigeriano muore per una circoncisione si richiede la sanitá gratuita per gli immigrati. Roba da guerra civile e di sommosse fino ai bastioni di Orione.*[10]

---

[10] So if an Italian dies in the hospital in the midst of ants it is an 'episode' while if a Nigerian dies of a circumcision, free healthcare is required for immigrants. Stuff from civil war and riots up to the ramparts of Orion.

Considered this performance, we carried out also a statistical experiment to understand how significant our approach is. Taking into account the fact that we cannot meet the assumptions of perfect metrics and unbiased dataset, we followed the suggestions of [27], reporting the significance results across the used dataset with regard to all available metrics (i.e., precision, recall and $f1$-score). As thresholds for the p-value, we adopted the typical cut-off at 0.05. In particular, for this experiment we used the library *boostsa*[11] that gives us the possibility to run a bootstrap sampling significance test, observing how fair are the obtained significance values with respect to different sizes of the unbalanced dataset. To examine, particularly, the contribution of the knowledge of sarcasm respect to the stereotypes in hate speech detection, we propose to investigate the significance of *TheNorth_2* (Fig. 1(a)) and *UmBERTo-MTL (sarc)* (Fig. 1(b)) comparing them with the baseline model *Baseline_Avg_LMs* that is representative of general knowledge transferred from language models. As parameters, we used a sample size that ranges from 0.05 to 0.5 of the total size of the test set, and 1000 iterations for computing the bootstrap sampling.

Observing the curves of levels of significance of the considered models in Fig. 1, the system aware of sarcasm (b), differently from TheNorth's (a), reports significant p-value lower than typical 0.05 (0,038 for $f1$-score, 0.012 for precision and 0.022 for recall). Moreover, increasing the size of samples we can notice that the p-values tend to decrease proving that with a bigger test set, despite unbalanced, the model could perform optimally.

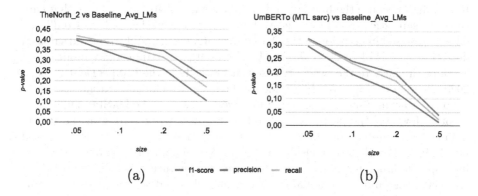

(a)   f1-score   precision   recall   (b)

Fig. 1. Levels of significance of models for hate speech detection in tweets.

5 Conclusions

Linguistic and pragmatic studies suggest sarcasm as the kind of irony that is perceived as aggressive, and therefore suitable to convey abusive language, especially in spontaneous texts about delicate issues such as the integration of other

---

[11] https://github.com/fornaciari/boostsa#readme.

cultures in Italy. Taking into account the statistical findings in tweets performed by [13], in this work, we aimed at answering the RQ: *Could the awareness of the presence of sarcasm increase the performance of abusive language detection systems?*

Approaching the issue as a learning problem, we trained the system of abusive language detection also on irony and sarcasm identification. The reached performance shows that a system, even when it is able to generalize better because of the use of pre-trained language model (i.e., *Baseline_Avg_LMs*), needs to be aware mainly of sarcasm to understand the sarcastic messages that are abusive towards cultural minorities. From the obtained results, we notice that the awareness of sarcasm helps the system of hate speech detection to retrieve especially positive examples, and in a real world-context, it could be convenient. Indeed, sarcasm is the type of irony that could reinforce the negative message and, at the same time, lessen the hurtful tones, hindering the detection of abusive text. Moreover, the knowledge about sarcasm leads the system to recognize also other figurative languages that typically make hate speech implicit. However, although the results about the simultaneous learning of stereotypes is not significant like sarcasm in hate speech detection, the obtained scores in Table 5 suggest that hate speech is characterized sometimes also of stereotypes. Therefore, as future work, we want to experiment with a more complex system that could combine the different knowledges and specific linguistic patterns that orient the identification of more implicit manifestation of hate speech.

A limitation of this work is the language. In this work, we focused only on Italian data because the fine-grained annotation about ironic language that captures the sarcastic form of irony is available, to our knowledge, only in IronITA, a dataset released in occasion of the shared task organized about irony and sarcasm detection at EVALITA 2018 [5]. The IronITA dataset contains data about immigrations similarly to the tweets released successively in the HaSpeeDe dataset in 2020. Therefore, [13] were able to merge the two datasets, and extend the annotation of irony and sarcasm to the rest of tweets, creating the HASPEEDE20_EXT corpus. However, the proposed methodology proves to be strong in the Italian immigration context, and supports also the social and linguistic studies proposed by [3,10,15,31]. Finally, considering the lack of linguistic resources encoding hate speech and sarcasm in other languages, in the future, we plan to perform cross-language and cross-domain experiments taking advantage of the available multilingual language models.

Acknowledgments. The work of S. Frenda and V. Patti was partially funded by the Multilingual Perspective-Aware NLU Project in partnership with Amazon Alexa. The work of the P. Rosso was done in the framework of the FairTransNLP-Stereotypes research project on Fairness and Transparency for equitable NLP applications in social media: Identifying stereotypes and prejudices and developing equitable systems (Grant PID2021-124361OB-C31 funded by MCIN/AEI/10.130 39/501100011033 and by ERDF, EU A way of making Europe).

References

1. Basile, V., et al.: SemEval-2019 task 5: multilingual detection of hate speech against immigrants and women in Twitter. In: Proceedings of the 13th International Workshop on Semantic Evaluation, pp. 54–63 (2019)
2. Baxter, J.: A Bayesian/information theoretic model of learning to learn via multiple task sampling. Mach. Learn. **28**(1), 7–39 (1997)
3. Bowes, A., Katz, A.: When sarcasm stings. Discourse Process. Multi. J. **48**(4), 215–236 (2011)
4. Caselli, T., Basile, V., Mitrović, J., Kartoziya, I., Granitzer, M.: I feel offended, don't be abusive! implicit/explicit messages in offensive and abusive language. In: Proceedings of the 12th Language Resources and Evaluation Conference, pp. 6193–6202 (2020)
5. Cignarella, A.T., Frenda, S., Basile, V., Bosco, C., Patti, V., Rosso, P.: Overview of the EVALITA 2018 task on irony detection in Italian tweets (IronITA). In: Proceedings of the Sixth Evaluation Campaign of Natural Language Processing and Speech Tools for Italian (EVALITA 2018) Co-located with the Fifth CLiC-it, vol. 2263, pp. 1–6 (2018)
6. Cimino, A., De Mattei, L., Dell'Orletta, F.: Multi-task learning in deep neural networks at EVALITA 2018. In: Proceedings of the Sixth Evaluation Campaign of Natural Language Processing and Speech Tools for Italian (EVALITA 2018) Co-located with the Fifth CLiC-it, vol. 2263 (2018)
7. Davies, C.: Jokes and Targets. Indiana University Press, Bloomington (2011)
8. Devlin, J., Chang, M., Lee, K., Toutanova, K.: BERT: pre-training of deep bidirectional transformers for language understanding. In: Proceedings of the 2019 Conference of the North American Chapter of the Association for Computational Linguistics: Human Language Technologies (NAACL-HLT 2019), pp. 4171–4186 (2019)
9. Dews, S., Winner, E.: Muting the meaning a social function of irony. Metaphor Symbolic Act. **10**(1), 3–19 (1995)
10. Douglass, S., Mirpuri, S., English, D., Yip, T.: 'They were just making jokes': ethnic/racial teasing and discrimination among adolescents. Cultur. Divers. Ethnic Minor. Psychol. **22**(1), 69–82 (2016)
11. Fersini, E., Rosso, P., Anzovino, M.: Overview of the task on automatic misogyny identification at IberEval 2018. In: Proceedings of the Third Workshop on Evaluation of Human Language Technologies for Iberian Languages (IberEval 2018) co-located with 34th Conference of SEPLN, vol. 2150, pp. 214–228 (2018)
12. Fortuna, P., Nunes, S.: A survey on automatic detection of hate speech in text. ACM Comput. Surv. **51**(4), 85:1–85:30 (2018)
13. Frenda, S., Patti, V., Rosso, P.: Killing me softly: creative and cognitive aspects of implicitness in abusive language online. Nat. Lang. Eng. 1–22 (2022)
14. Gambino, G., Pirrone, R.: CHILab @ HaSpeeDe 2: enhancing hate speech detection with part-of-speech tagging. In: Proceedings of the Seventh Evaluation Campaign of Natural Language Processing and Speech Tools for Italian (EVALITA 2020), vol. 2765 (2020)
15. Kuipers, G., Van der Ent, B.: The seriousness of ethnic jokes: ethnic humor and social change in the Netherlands, 1995–2012. Humor **29**(4), 605–633 (2016)
16. Lavergne, E., Saini, R., Kovács, G., Murphy, K.: TheNorth @ HaSpeeDe 2: BERT-based language model fine-tuning for Italian hate speech detection. In: Proceedings of the Seventh Evaluation Campaign of Natural Language Processing and Speech Tools for Italian (EVALITA 2020), vol. 2765, pp. 142–147 (2020)

17. Lee, C.J., Katz, A.N.: The differential role of ridicule in sarcasm and irony. Metaphor. Symb. **13**(1), 1–15 (1998)
18. Lemmens, J., Markov, I., Daelemans, W.: Improving hate speech type and target detection with hateful metaphor features. In: Proceedings of the Fourth Workshop on NLP for Internet Freedom: Censorship, Disinformation, and Propaganda, pp. 7–16 (2021)
19. MacAvaney, S., Yao, H.R., Yang, E., Russell, K., Goharian, N., Frieder, O.: Hate speech detection: challenges and solutions. PLoS ONE **14**(8), 1–16 (2019)
20. Merlo, L.I., Chulvi, B., Ortega, R., Rosso, P.: When humour hurts: linguistic features to foster explainability. Procesamiento Leng. Nat. (SEPLN) **70**, 85–98 (2023)
21. Nobata, C., Tetreault, J., Thomas, A., Mehdad, Y., Chang, Y.: Abusive language detection in online user content. In: Proceedings of the 25th International Conference on World Wide Web (WWW 2016), pp. 145–153 (2016)
22. Pexman, P.M., Olineck, K.M.: Does sarcasm always sting? Investigating the impact of ironic insults and ironic compliments. Discourse Process. **33**(3), 199–217 (2002)
23. Poletto, F., Basile, V., Sanguinetti, M., Bosco, C., Patti, V.: Resources and benchmark corpora for hate speech detection: a systematic review. Lang. Resour. Eval. **55**, 477–523 (2021)
24. Sanguinetti, M., et al.: Haspeede 2 @ EVALITA2020: overview of the EVALITA 2020 hate speech detection task. In: Proceedings of the Seventh Evaluation Campaign of Natural Language Processing and Speech Tools for Italian (EVALITA 2020), vol. 2765 (2020)
25. Sanguinetti, M., Poletto, F., Bosco, C., Patti, V., Stranisci, M.: An Italian twitter corpus of hate speech against immigrants. In: Proceedings of the Eleventh International Conference on Language Resources and Evaluation (LREC 2018) (2018)
26. Schmidt, A., Wiegand, M.: A survey on hate speech detection using natural language processing. In: Proceedings of the Fifth International Workshop on Natural Language Processing for Social Media, pp. 1–10 (2017)
27. Søgaard, A., Johannsen, A., Plank, B., Hovy, D., Martínez Alonso, H.: What's in a p-value in NLP? In: Proceedings of the Eighteenth Conference on Computational Natural Language Learning, pp. 1–10 (2014)
28. Taulé, M., Ariza, A., Nofre, M., Amigó, E., Rosso, P.: Overview of DETOXIS at IberLEF 2021: DEtection of TOXicity in comments in Spanish. Procesamiento Leng. Nat. **67**, 209–221 (2021)
29. Waseem, Z., Davidson, T., Warmsley, D., Weber, I.: Understanding abuse: a typology of abusive language detection subtasks. In: Proceedings of the First Workshop on Abusive Language Online, pp. 78–84 (2017)
30. Waseem, Z., Hovy, D.: Hateful symbols or hateful people? Predictive features for hate speech detection on twitter. In: Proceedings of the NAACL Student Research Workshop, pp. 88–93 (2016)
31. Weaver, S.: A rhetorical discourse analysis of online anti-Muslim and anti-Semitic jokes. Ethn. Racial Stud. **36**(3), 483–499 (2013)
32. Wiegand, M., Ruppenhofer, J., Eder, E.: Implicitly abusive language - what does it actually look like and why are we not getting there? In: Proceedings of the 2021 Conference of the North American Chapter of the Association for Computational Linguistics: Human Language Technologies, pp. 576–587 (2021)
33. Wiegand, M., Ruppenhofer, J., Kleinbauer, T.: Detection of abusive language: the problem of biased datasets. In: Proceedings of the 2019 Conference of the North American Chapter of the Association for Computational Linguistics: Human Language Technologies, vol. 1, pp. 602–608 (2019)

34. Zampieri, M., Malmasi, S., Nakov, P., Rosenthal, S., Farra, N., Kumar, R.: SemEval-2019 task 6: identifying and categorizing offensive language in social media (OffensEval). In: Proceedings of the 13th International Workshop on Semantic Evaluation, pp. 75–86 (2019)

Cem Mil Podcasts: A Spoken Portuguese Document Corpus for Multi-modal, Multi-lingual and Multi-dialect Information Access Research

Ekaterina Garmash[1(✉)], Edgar Tanaka[1], Ann Clifton[1], Joana Correia[1],
Sharmistha Jat[1], Winstead Zhu[1], Rosie Jones[1], and Jussi Karlgren[2]

[1] Spotify, Stockholm, Sweden
katyag@spotify.com
[2] Silo AI, Helsinki, Finland

Abstract. In this paper we describe the Portuguese-language podcast dataset we have released for academic research purposes. We give an overview of how the data was sampled, descriptive statistics over the collection, as well as information about the distribution over Brazilian and Portuguese dialects.

We give results from experiments on multi-lingual summarization, showing that summarizing podcast transcripts can be performed well by a system supporting both English and Portuguese. We also show experiments on Portuguese podcast genre classification using text metadata. Combining this collection with previously released English-language collection opens up the potential for multi-modal, multi-lingual and multi-dialect podcast information access research.

Keywords: Dataset · Podcast · Spoken audio · Speech retrieval · Multi-modal · Summarization

1   Introduction

Podcasts, a new and emergent spoken mass communication medium, come in many formats and levels of formality. Podcasts are typically produced as topically or stylistically consistent *shows* that consist of *episodes*, which are published serially with some regularity over time. Podcast consumption has been growing rapidly [23] and podcasts have in the past years become a topic of interest for research in speech and language technology, linguistics, information access technology, and media studies.

Podcast shows can be educational, journalistic, or fictional; formal or informal; conversational or monologic; and vary over type, style, form, and topic. Podcast material appears to differ in several aspects from other types of recorded speech or textual material [12]. This breadth of variation and contrast with other collections of human language motivates using podcasts for research both to develop infrastructure and tools for podcast distribution and consumption

A. Arampatzis et al. (Eds.): CLEF 2023, LNCS 14163, pp. 48–59, 2023.
https://doi.org/10.1007/978-3-031-42448-9_5

as well as to broaden the scope of the general study of human communicative behavior.

Academic research on podcasts requires availability of open-source and representative datasets. The currently largest available collection is Spotify's English language podcast dataset [5] which differs from other collections of English language data in that it is orders of magnitude larger than previous collections of spoken language and contains a rich variety of genres, subject matter, speaking styles, and structural formats.

Podcasts are available in a wide variety of languages. For example, the Anchor podcast creation app is available in thirty-five languages [1] with many more podcasters creating podcasts in other languages. However, the Spotify English language podcast dataset is, as are most available datasets, composed entirely of English-language material. Research on English alone risks results to be biased towards linguistic specifics of one language and cultural arenas where English mostly is produced and used.

To address this source of bias in podcast research, we have compiled a complementary dataset of Portuguese-language material, with comparable size and breadth using the same general methodology as in [5]. This dataset contains metadata similar to what is provided together with the Spotify English-language dataset, which can be used as proxy labels for supervised learning tasks, such as creator-provided textual episode descriptions for the summarization task and names of show publishers for authorship attribution tasks.

The Portuguese language is the sixth most-spoken language in the world, with 250 million native speakers, in many cultural areas, and with 24 million more L2 speakers [24], and the Lusophone markets are, taken together in the top ten of the world by GDP [9]. This motivates a growing interest in working on language technology for Portuguese: as shown not least by the recent release of a large transformer model based on written Portuguese [21].

In order to facilitate research on spoken Portuguese in general, and podcasts in Portuguese more specifically, we now make available a podcast dataset consisting of 123,054 podcast episodes in Portuguese from 16,131 shows, encompassing more than 76,000 h of speech audio.

We know of no previous large-scale study of Portuguese language podcasts. There have been smaller-scale studies of Portuguese podcasts [2]. Morais et al. [19] surveyed 566 Brazilian podcast listeners, and found that they listen to podcasts across a range of topic areas, and that they like podcasts to convey information that is complementary to the information found in other media formats, similarly to what has been found in other linguistic and cultural areas.

2   Dataset Construction

To construct the Portuguese dataset we followed a procedure patterned on the approach used Clifton et al. [5]. to build the English-language podcast dataset.

From a fairly comprehensive list of Portuguese podcasts we selected data based on the following filters:

- The language of the show as given by the podcast creator in the show meta-data specification must be Portuguese (pt-BR or pt-PT).
- The language of the episode description must be identified as Portuguese using the langid Python package [17].
- We only selected episodes published between September 9, 2019 and March 31, 2022.
- The episode must have more than 50% of speech over its duration. A propri-etary speech detection algorithm was used here to filter out podcasts which contain mostly music, white noise or ambient sounds, rather than speech.

From the list of filtered candidates, episodes were randomly selected to obtain just over 150 000 individual items.

The next step was to transcribe this set of episodes using Azure's speech-to-text service[1]. One of the parameters of this service was the target language-variant which we set to either pt-PT (Portuguese from Portugal) or pt-BR (Portuguese from Brazil) according to the following metadata in this order of precedence: creator-provided language code given in the *show* metadata is either 'pt-PT', the creator-provided language code given in the *episode* metadata, the *show's* country of origin is 'pt' or 'br'. If no metadata is set to either pt-PT or pt-BR, we fall back to 'pt-BR' because the number of podcast creators in Brazil is larger.

Despite sending the entire pool of approximately 150 000 episodes to tran-scription, some of them failed to be transcribed. In the end, 114,387 episodes were transcribed using 'pt-BR' and 8,667 were transcribed using 'pt-PT' as tar-get language. Examples of each are shown in Fig. 1. Manual inspection reveals that some classification errors between 'pt-PT' and 'pt-BR' remain: words such as *"legal"* (Brazilian for "cool") or the Brazilian-only pronoun "cê" appear in the pt-PT set.

2.1   Dataset Schema

For each episode, we provide the audio file, the transcription of this audio file and the associated metadata. The following metadata is provided:

- *show_uri*: URI for the show
- *show_name*: Name of the show (e.g. "Hoje no TecMundo Podcast").
- *show_description*: Description of the show provided by podcast creator (e.g. "O Hoje no TecMundo é o tradicional programa diário do TecMundo no YouTube...")
- *publisher*: Publisher of the show (e.g. Hoje no TecMundo - Podcast).
- *language*: Language of the show in in BCP 47 format (e.g. pt-BR).
- *rss_link*: URL of the show's RSS feed (e.g. https://anchor.fm/s/11c4550c/podcast/rss).
- *episode_uri*: URI for the episode

---

[1] https://docs.microsoft.com/en-us/azure/cognitive-services/speech-service/index-speech-to-text.

pt-PT	*Olá **és** curioso sobre o que se passa no mundo, gosta de saber o que afetou diversosecossistemas e como podes ajudá-los. E esta ao **sítio** certo, não podcast e como escola São Pedro do Sul, **podes** encontrar informação sobre o ambiente, novidades relacionadas com o nosso planeta e muitas curiosidades, incluindo as diversas medidas que podes aplicar para ajudares a proteger a natureza. **Serás** capaz de aprofundar os teus conhecimentos sobre diversos temas, desde alterações climáticas até ao ruído. A **equipa** de 2 ou 3 vezes por mês que vai contar tudo é composta pelo Miguel Almeida, Rodrigo Cardoso. Tiago Rocha e Filipe Correia. Para não perder nada, subscreve. Já o nosso podcast na tua plataforma preferida e não te esqueças após cada episódio, Podes sempre visitar o nosso website para saberes mais.*
pt-BR	*Olá **você** que é nosso ouvinte do podcast de arte saúde o fiba hoje estaremos nossa segunda entrevista e contaremos com a presença de uma convidada mais do que especial? Ela estéfane psicóloga e arte terapeuta. Oi eu sou estefani eu sou psicóloga formada pela universidade de Passo Fundo com especialização em arteterapia também pela universidade de Passo Fundo e recentemente eu encontrei uma ponte entre a psicologia e a arte terapia através de uma especialização em psicologia clínica e um. Indo pela fam acne Porto Alegre eu tô muito feliz com o convite da área de saúde para falar um pouquinho sobre esse assunto que deixa o meu coração tão quentinho que é arte terapia é sempre ...*

Fig. 1. Example transcripts for the two target language varieties of Portuguese. The pt-PT example was extracted from the show "EESPS: Podcast sobre o ambiente". The pt-BR was extracted from the show "Arte e Saúde - UFBA".

- *episode_name*: Name of the episode. (e.g."Hoje no TecMundo 17/01/2020 - Preço do Galaxy Fold no Brasil, imagens do Huawei P40 Pro").
- *episode_description*: Description of the episode (e.g. "No programa de hoje, falamos do preço caríssimo do Galaxy Fold no Brasil, a Google ...").
- *duration*: duration of the episode in minutes (e.g. 9.113833333333334).
- *show_filename_prefix*: Filename path for the show
- *episode_filename_prefix*: Filename of the episode file
- *show_category*: The genre of the show extracted from the *itunes:category* tag in the show's RSS feed.

2.2   Access to the Dataset

The Portuguese language podcast dataset is available for non-commercial research purposes in the same way and under the same agreement as the English language podcast dataset. The English language podcast dataset has been most notably used in shared tasks on segment retrieval and summarization in TREC 2020 [10] and TREC 2021 [13] but is also currently used for many other research purposes such as document segmentation or dialogue modelling and there are several annotations and enrichments such as a search index, human assessments, and precomputed audio features available for the English language section. We expect that this extension will broaden the scope of research and lower the threshold to apply methods to more than one language. We welcome contributions to further enrich the dataset through annotations of various kinds.

To request the dataset, please go to https://podcastsdataset.byspotify.com/ and follow the instructions.

3 Descriptive Statistics and Comparison to English Podcast Dataset

In Table 1, we give some descriptive statistics for the Portuguese dataset and compare them to the English-language dataset released by [5] to demonstrate that they are of comparable size (in terms of number of episodes) and quality. We note however that the English set turns out to have a slightly larger diversity of shows: the show-to-episode ratio for English is 17%, while it is 13% for Portuguese. This is an expected consequence of there being more podcast shows in English than in Portuguese, and thus selecting approximately the same number of episodes will yield slightly more episodes per show for Portuguese than for English. The distribution of episode durations also differ: the Portuguese episodes tend to be longer on average and the distribution is skewed to the right. As a consequence, the average number of words per episodes is also higher for Portuguese. This observation entails, in particular, that development of machine learning models of podcast understanding may be more challenging for Portuguese, since its input size will be larger. To illustrate this we provide a case study on episode summarization in (Sect. 5).

Given that the general dataset construction procedure is similar for both datasets, and that the analyzed samples are of comparable sizes, the detected differences in the distributions of various features support our original claim that linguistically diverse data is necessary to avoid biased conclusions in podcast research.

Table 1. Descriptive statistics for the Portuguese Language Podcast Dataset, with corresponding data for the Spotify English Language Podcast Dataset given as comparison.

	English	Portuguese
Number of episodes	105 360	123 054
Number of shows	18 376	16 131
Average episode duration (minutes)	33.8	37.3
- 25%	13.6	10.9
- 50%	31.6	31.2
- 75%	50.4	55.0
- max	305	695
Average number of words per transcript	5 726	9 539
- 25%	2 036	2 203
- 50%	5 204	6 746
- 75%	8 672	13 693
- max	43 504	205 163

4   Podcast Genre Prediction Case Study

Besides being multi-modal, our dataset comes with rich metadata annotation. We demonstrate the usefulness of metadata by benchmarking the task of podcast genre prediction. Podcast genres are essential when understanding user taste in order to recommend related listening experiences. Intrinsically, the genre is a characterization of the podcast's content as a whole and could therefore be inferred based on the raw podcast data (audio or text). However, we show in our experiments that the metadata we provide in the dataset could be sufficient for the task, which is beneficial from a practical perspective. Specifically, we run genre prediction experiments where input is restricted to episode names and a short episode description (summary), which is often provided by creators and which does not need to be generated separately.

4.1   Genre Prediction Experiment Setup

As target labels, we use creator-provided genre labels located in the *show_category* column of the metadata provided. The taxonomy consists of 19 genre labels. In Table 2, we can see the distribution of episodes per genre. The top 5 genres account for 69% of all episodes. We also note that *Business, Education, Sports,* and *Comedy* are all within the top 5 genres. This is similar to what was reported for Spotify's English Language dataset [5]. Please note that the number of episodes in the genre prediction task is a subset of the total number of episodes in the dataset. We provide the train and test split of the dataset used in the prediction task.

We frame genre prediction as a classification task. Predictions are made per episode. As input, we use episode names and episode descriptions, individually and combined (see Results for an ablation study). We use a multi-class support vector machine for the classification task using *sklearn*'s [4] *SVC* class implementation with default hyperparameter settings. The episode-name and episode-description text inputs are pre-processed using *bert-base-multi-lingual-uncased* [6] to generate 768-sized embeddings. Train and test splits are created using an 80:20 split, making sure that the split is partitioned by show URIs in order to avoid any information leakage between test and train splits (since episodes from the same show have a common genre).

We consider three experimental conditions: (1) using episode name input only; (2) using episode description input only; (3) using both episode name and description as input. We report precision, recall, and F1 of test prediction results by genre, as well as aggregate accuracy and macro & weighted averages of these metrics over all the genres.

4.2   Genre Prediction Experiment Results

All the results are summarized in Table 3. Firstly, we see that for the top genres (that we identified in Table 2), the F1 scores are substantially above 0.5, which

Table 2. Portuguese Podcasts dataset: number of episodes per genre

S.No.	Genre	Number of Episodes in Portuguese Dataset
1	Business	26 915
2	Education	23 541
3	Sports	13 422
4	Comedy	11 089
5	Arts	9 799
6	TV & Film	5 445
7	Science	5 371
8	Music	3 916
9	Technology	3 186
10	Society & Culture	3 172
11	Kids & Family	3 026
12	Leisure	2 647
13	Health & Fitness	2 337
14	History	2 213
15	Fiction	2 017
16	True Crime	1 748
17	News	1 097
18	Religion & Spirituality	1 033
19	Government	531
Total		122 5273

indicates that given enough data, just the name and the description are sufficient to identify a podcast genre correctly. Second, as intuitively expected, the description of episodes contributes more to prediction quality. However, combining them both yields the best result. As for the rest of the genres, the prediction quality is much lower, which could be explained by the low data support; *News* e.g. has only 1097 episodes compared to 26,915 in the *Business* genre.

5 Episode Summarization Case Study

In this section we present a case study of one of the possible machine learning applications of our dataset: episode summarization. Automatic Text Summarization is the task of taking a source document as input and producing a much shorter version of it while preserving the most important pieces of information [7]. One of the differences between podcast episode summarization and the many other domains where summarization is applied is the extensive length of the podcast input document, which poses a challenge since currently most neural

Table 3. Results of the genre prediction classification experiments: precision, recall, and F1 are reported by genre and as averages across genres. The top most popular genres (according to Table 2) are marked in boldface. We observe that the popular genres have an F1 score above 0.5, showing that metadata alone can detect the genre of a given episode. Although episode description features dominate the Genre prediction accuracy, adding the episode name features helps improve the genre prediction metrics in most genres.

Genre	Name			Description			Name & Description		
	Prec	Rec	F1	Prec	Rec	F1	Prec	Rec	F1
Arts	0.31	0.3	0.31	0.34	0.38	0.36	0.42	0.47	0.44
Business	0.44	0.68	0.54	0.72	0.63	0.57	0.75	0.65	
Comedy	0.38	0.32	0.35	0.54	0.59	0.56	0.53	0.57	0.55
Education	0.48	0.66	0.56	0.5	0.72	0.59	0.52	0.73	0.61
Fiction	0.08	0.03	0.04	0.27	0.03	0.05	0.16	0.07	0.1
Government	0.0	0.0	0.0	0.0	0.0	0.0	0.0	0.0	0.0
Health and Fitness	0.0	0.0	0.0	0.0	0.0	0.0	0.0	0.0	0.0
History	0.22	0.05	0.08	0.36	0.16	0.22	0.39	0.11	0.17
Kids and Family	0.33	0.08	0.13	0.47	0.07	0.12	0.51	0.09	0.15
Leisure	0.07	0.06	0.07	0.23	0.14	0.18	0.21	0.13	0.16
Music	0.45	0.3	0.36	0.64	0.48	0.55	0.65	0.55	0.6
News	0.0	0.0	0.0	0.0	0.0	0.0	0.0	0.0	0.0
Religion and Spirituality	0.0	0.0	0.0	0.67	0.03	0.05	0.75	0.02	0.04
Science	0.39	0.08	0.13	0.41	0.11	0.17	0.47	0.13	0.21
Society and Culture	0.0	0.0	0.0	0.76	0.04	0.08	0.52	0.02	0.04
Sports	0.52	0.67	0.58	0.81	0.79	0.8	0.76	0.83	0.79
TV and Film	0.38	0.27	0.32	0.49	0.42	0.46	0.5	0.4	0.45
Technology	0.14	0.02	0.03	0.46	0.09	0.15	0.52	0.09	0.16
True Crime	0.37	0.10	0.16	0.51	0.34	0.41	0.48	0.32	0.38
macro average	0.23	0.18	0.18	0.4	0.25	0.27	0.4	0.26	0.27
weighted average	0.39	0.44	0.39	0.52	0.53	0.49	0.53	0.55	0.51
Accuracy	0.44			0.53			0.55		

network-based approaches are limited in terms of input size. Moreover, as we saw in Table 1, Portuguese episodes tend to be longer than the English ones, which suggests a research question: is episode summarization more challenging for Portuguese? We address this question by conducting a series of machine learning experiments to train a transcript-based summarization model, and compare the resulting quality for English (based on [5]) and Portuguese (based on the dataset from this paper).

5.1 Data Preparation for Summarization

As input to the summarization model, we use the automatically generated episode transcripts described in Sect. 2. We treat creator-provided episode descriptions as the summaries and train the model to generate them.

We further clean the data using the following filters:

- We remove episodes with repeated descriptions (any description used in more than one episode). We applied a TF-IDF vectorization of the descriptions which were compared to each other using the cosine distance. Any data points with too similar descriptions (threshold 95%) were filtered out.
- We remove episodes where the episode description is too similar to the show description (threshold 95%).
- We remove any email addresses or URLs from episode descriptions as we did not want our trained models to hallucinate such information in the generated summaries.
- We remove episodes where the creator descriptions are either too long or too short with the boundary conditions set to between 10 and 1300 characters.
- We remove *boilerplate* content from episode descriptions. Briefly speaking, boilerplate is any extraneous content which does not describe the episode in natural language text. Common cases of boilerplate in podcasts are advertisements and promotional content for social media [20]. We train a sentence-level binary classifier for that, based on a small manually annotated dataset and fine-tune a pretrained language model (*bert-base-cased* for English and *bert-base-multi-lingual-cased* for Portuguese).

After applying the filters above, we split the remaining data into 3 parts: train (90%), dev (5%) and test (5%). The split was a per-show partitioning of the data. Tables 4 contains statistics of the resulting splits.

Table 4. Experimental data for summarization: split size by number of episodes.

	ratio	EN	PT
train	90%	80 895	90 859
dev	5%	4 503	5 073
test	5%	4 511	5 058

5.2 Models for Summarizatiom

For baselines, we follow [5,10,11,13] and use the **first minute** transcript and **TextRank** [18], a graph-based model which can be used as an unsupervised method to extract both keywords or key sentences. Both baselines are extractive summarization methods and do not require any training.

We run machine learning experiments with **MBART** [22], a multi-lingual version of BART [14]. We chose to use the MBART-50 [16] model because it has been pre-trained in 50 languages (including Portuguese and English) and also because it is an encoder-decoder model, i.e. capable of generating text. Additionally, we replace MBART's original full attention mechanism with the one of Longformer [3] – we refer to this model variation as **LongMBART**. LongMBART's linear attention allows to increase the input size limit from 512 tokens to 4096 tokens. Our hypothesis is that passing more information (i.e. more transcript text) to the model would lead to higher scores.

For both MBART and LongMBART, we consider three experimental conditions: (1) unfinetuned base model; (2) finetuned on the language of the test set ("monolingual fine-tuning"); (3) finetuned on both the English and Portuguese training set ("bilingual fine-tuning"). The fine-tuning was set to early stop once the ROUGE-2 [15] score didn't improve after 3 validation checkpoints.

5.3   Results for Summarization

We report ROUGE-1 (unigram overlap), ROUGE-2 (bigram overlap), ROUGE-L (longest matching sequence of words) [15] scores on the test set in Table 5. First, we see that both languages obtain the highest scores for the same types of model (fine-tuned MBART). Second, we see no substantial difference between the best scores of English and Portuguese, contrary to our hypothesis that Portuguese should be more challenging given its larger input size. Finally, we see that using LongMBART as base model does not result in better summarization.

To sum up, having two comparable podcast datasets for different languages with intrinsically different data distribution allowed us to test a series of hypotheses: effect of linguistic and cultural specifics on the quality of machine learning-based summarization model, effect of size of input, effect of multi-lingual fine-tuning.

Table 5. ROUGE-1, ROUGE-2 and ROUGE-L F1 scores for test set of 4511 English-language podcast episodes and test set of 5073 Portuguese-language podcast episodes. In bold, the top two highest ROUGE scores.

	EN			PT		
	R1	R2	RL	R1	R2	RL
First Minute baseline	0.17	0.03	0.15	0.17	0.03	0.14
TextRank Top 5 sentences	0.14	0.02	0.12	0.17	0.01	0.1
MBART unfinetuned	0.16	0.03	0.14	0.16	0.03	0.13
MBART finetuned monoling	**0.19**	**0.06**	**0.17**	**0.19**	**0.05**	**0.16**
MBART finetuned biling	**0.19**	0.05	**0.17**	0.18	0.05	**0.16**
LongMBART unfinetuned	0.16	0.03	0.14	0.11	0.01	0.1
LongMBART finetuned monoling	0.03	0.0	0.03	0.18	**0.05**	0.16
LongMBART finetuned biling	0.18	0.05	0.16	0.18	0.05	0.15

6 Conclusions

In this paper we presented a new dataset of Portuguese language podcasts, containing audio and transcript data, as well as rich metadata annotation, following the methodology used to put together the English language podcast dataset [5].

Having a dataset in a language other than English, developed under the same methodology, allows for a more comprehensive and unbiased research of the podcast domain. From the point of view of machine learning research, linguistically diverse podcast data allows to study the effect of various input characteristics on the quality of the final model. In particular, in a case study presented in this paper, we show how our Portuguese dataset is used to train a podcast summarization model, and to compare it to an English summarization model, and thus test the effect of input language and size of input on the quality of the final model. Moreover, having data in multiple languages opens up possibility of research into multi-lingual machine learning models, widely adopted in other domains of natural language processing [8,22,25]. Finally, metadata features which we supply in the dataset allow to train lightweight prediction models that do not need to take the full raw input. In another case study presented in this paper, we demonstrate how metadata can be used for genre prediction where we only used episode name and short description text as input.

Beyond the case studies presented in this paper, the new dataset can be used for many more experiments into multi-lingual and multi-modal (text and audio) machine learning models of podcasts. From a methodological point of view, we have demonstrated the reproducibility of the dataset construction procedure and plan to extend it to further languages.

References

1. Anchor: Anchor web: Now localized for more creators around the world (2022), https://blog.anchor.fm/updates/anchor-web-localization. Accessed Sept 2022
2. Antunes, M.J., Salaverría, R.: Examining independent podcasts in Portuguese iTunes. In: Stephanidis, C., Antona, M. (eds.) HCII 2020. CCIS, vol. 1226, pp. 149–153. Springer, Cham (2020). https://doi.org/10.1007/978-3-030-50732-9_20
3. Beltagy, I., Peters, M.E., Cohan, A.: Longformer: The long-document transformer (2020). https://arxiv.org/abs/2004.05150
4. Buitinck, L., et al.: API design for machine learning software: experiences from the scikit-learn project. In: ECML PKDD Workshop: Languages for Data Mining and Machine Learning (2013)
5. Clifton, A., et al.: 100,000 podcasts: a spoken english document corpus. In: Proceedings of the 28th International Conference on Computational Linguistics (COLING). International Committee on Computational Linguistics (2020). https://podcastsdataset.byspotify.com/
6. Devlin, J., Chang, M.W., Lee, K., Toutanova, K.: BERT: pre-training of deep bidirectional transformers for language understanding. In: Proceedings of the 2019 Conference of the North American Chapter of the Association for Computational Linguistics: Human Language Technologies, Volume 1 (Long and

Short Papers). Association for Computational Linguistics, Minneapolis, Minnesota (2019). https://doi.org/10.18653/v1/N19-1423, https://aclanthology.org/N19-1423

7. El-Kassas, W.S., Salama, C.R., Rafea, A.A., Mohamed, H.K.: Automatic text summarization: a comprehensive survey. Expert Syst. Appl. **165**, 113679 (2021). https://doi.org/10.1016/j.eswa.2020.113679, https://www.sciencedirect.com/science/article/pii/S0957417420305030

8. Fan, A., et al.: Beyond English-centric multilingual machine translation (2020). https://arxiv.org/abs/2010.11125

9. International Monetary Fund: World economic outlook database (2023). https://www.imf.org/en/Publications/WEO/weo-database/2023/April/weo-report. Accessed May 2023

10. Jones, R., et al.: TREC 2020 podcasts track overview. In: Voorhees, E.M., Ellis, A. (eds.) NIST Special Publication 1266: The Twenty-Ninth Text REtrieval Conference Proceedings (TREC 2020). NIST, Gaithersburg (2021)

11. Karlbom, H.: Abstractive Summarization of Podcast Transcriptions. Master's thesis, Uppsala University (2021)

12. Karlgren, J.: Lexical variation in English language podcasts, editorial media, and social media. North Eur. J. Lang. Technol. **8** (2022)

13. Karlgren, J., et al.: TREC 2021 podcasts track overview. In: Voorhees, E.M., Ellis, A. (eds.) NIST Special Publication 335: The Thirtieth Text REtrieval Conference Proceedings (TREC 2021). NIST, Gaithersburg (2022)

14. Lewis, M., et al.: BART: denoising sequence-to-sequence pre-training for natural language generation, translation, and comprehension (2019). https://arxiv.org/abs/1910.13461

15. Lin, C.Y.: ROUGE: a package for automatic evaluation of summaries. In: Text Summarization Branches Out. Association for Computational Linguistics (2004). https://aclanthology.org/W04-1013

16. Liu, Y., et al.: Multilingual denoising pre-training for neural machine translation (2020). https://arxiv.org/abs/2001.08210

17. Lui, M., Baldwin, T.: langid.py: an off-the-shelf language identification tool. In: Proceedings of the ACL 2012 System Demonstrations. Association for Computational Linguistics (2012). https://pypi.org/project/langid/

18. Mihalcea, R., Tarau, P.: Textrank: bringing order into text. In: Proceedings of the 2004 conference on Empirical Methods in Natural Language Processing (2004)

19. Morais, R., Giacomelli, F., Grafolin, T., Rocha, F.: Audience transformations and new audio experiences: an analysis of the trends and consumption habits of podcasts by Brazilian listeners. J. Audience Reception Stud. **18**(1) (2021)

20. Reddy, S., Yu, Y., Pappu, A., Sivaraman, A., Rezapour, R., Jones, R.: Detecting Extraneous Content in Podcasts (2021). https://arxiv.org/abs/2103.02585

21. Rodrigues, J., et al.: Advancing neural encoding of Portuguese with transformer albertina pt-* (2023). https://arxiv.org/abs/2305.06721

22. Tang, Y., et al.: Multilingual translation with extensible multilingual pretraining and finetuning (2020). https://arxiv.org/abs/2008.00401

23. Whitner, G.: The meteoric rise of podcasting (2020). https://musicoomph.com/podcast-statistics

24. Wikipedia: Portuguese language (2022). https://en.wikipedia.org/wiki/Portuguese_language. Accessed Sept 2022

25. Xue, L., et al.: mT5: a massively multilingual pre-trained text-to-text transformer (2021). https://arxiv.org/abs/2010.11934

Using Authorship Embeddings to Understand Writing Style in Social Media

Javier Huertas-Tato [ID], Alejandro Martín[(✉)] [ID], and David Camacho [ID]

Department of Computer Systems Engineering, Universidad Politécnica de Madrid, Madrid, Spain
{javier.huertas.tato,alejandro.martin,david.camacho}@upm.es
https://aida.etsisi.upm.es

Abstract. With the escalation of misinformation and malicious behavior issues on social media platforms, traditional detection-based measures often fail to address the problem in time. The use of multiple accounts or the continuous creation of new accounts makes it difficult to re-detect the presence of a user who, for example, has disseminated false information. In this paper, we present a novel approach to understanding and characterizing authorship in social media using a model called $PART_{SCL}$, an improvement of the previous PART model. $PART_{SCL}$ generates "authorship embeddings", numerical representations of an author's writing style, allowing for more accurate and earlier detection of malicious behavior. Our main contributions include the $PART_{SCL}$ model itself, a new pre-training approach for authorship attribution, and the application of our model on different datasets. These advances help bridge the gap between popular Natural Language Processing techniques such as Transformers and feature engineering, providing a robust tool for the ongoing fight against online misbehavior and misinformation.

Keywords: authorship embedding · transformer · writing style

1 Introduction

In recent years, social media has become a hotbed for misinformation and harmful behavior. Predominantly, the focus has been on the malevolent content circulating across these platforms, a reflection of the rampant problems that have emerged within these digital societies. The conventional approaches to tackling these issues are largely detection-oriented, identifying and removing problematic content after it has been posted. However, this reactive methodology has its shortcomings; whilst uploading disinformation or hate speech requires minimal effort and often escapes immediate notice, the identity of the author remains constant.

As such, the identification and characterization of authors based on the content they produce is a crucial element in the fight against malicious conduct.

A. Arampatzis et al. (Eds.): CLEF 2023, LNCS 14163, pp. 60–71, 2023.
https://doi.org/10.1007/978-3-031-42448-9_6

With this premise in mind, advanced Natural Language Processing (NLP) techniques such as Transformers provides us with the necessary tools to tackle this problem. However, despite their popularity and potential, these techniques still fall short of the precision and comprehensiveness offered by feature engineering. In this paper, we propose an approach aimed at bridging this gap.

We present $PART_{SCL}$, an enhanced version of the PART model [6], capable of translating writing characteristics into a numerical representation, or an 'authorship embedding'. This novel approach allows us to numerically capture the unique style of an author, offering a powerful tool for author identification and characterization.

Our contributions extend beyond the model itself. Notably, the pre-training approach we use for authorship attribution marks a groundbreaking direction in the field. Additionally, we provide an evaluation of our model's applications in different state-of-the-art datasets, to illustrate its effectiveness. This paper not only introduces a new model but also uncovers fresh perspectives in comprehending writing style in social media, paving the way for proactive measures against disinformation and harmful behaviors online.

The rest of the paper is organized as follows: Section 2 presents a summary of the state of the art, necessary to understand the background of this proposal such as the concept of authorship with transformers, authorship embeddings and supervised contrastive pretraining. In Sect. 3 we will describe the data processing required to train the architecture, the key part of this research. Section 4 includes the experimentation to finally show a series of conclusions in Sect. 5.

2   Related Work

Authorship identification has traditionally been explored through the use of specific features extracted from the text such as n-grams. These are carefully designed attributes drawn from one or several training sets that, for a considerable time, have dominated Machine Learning (ML) competitions concerning this specific problem. However, there is a growing trend in the use of transformer-based models for authorship detection, due to their capabilities to deal with small nuances, context and semantic. While these models have shown promising results, they often require specific fine-tuning for attribution tasks. The question then arises: Can we verify authorship merely with a threshold? The practicability of learning representations, as part of our proposed solution, could be a viable answer to this question.

The domain of methods that focus on learning representations, though potent, is rather scarce. Existing models such as Style-BERT employ contrastive learning to distinguish between writing styles. While contrastive learning has significantly advanced, there always exists room for further improvement. For instance, current methods tend to rely on datasets with similar styles, as seen with the PART model, which can potentially limit the model's generalizability. In particular, PART relies on constrastive learning that focus on authorship embeddings instead of semantics. It compares pairs of sentences or documents written by the same author to determine the proprietary of a text, using cosine similarity. Moreover, the loss functions used, such as the triplet loss in Style-BERT, can be sub-optimal when it comes to handling diverse and complex data.

Notably, InfoNCE, a commonly employed loss function, performs adequately but its potential for generalization can be boosted. We propose the adoption of the "Supervised Contrastive Loss (SCL)" in our model to overcome these limitations. The integration of SCL in our $PART_{SCL}$ model is a significant stride towards achieving superior performance in authorship attribution tasks. This approach promises a more nuanced understanding of authorship in social media, thus better equipping us to address the challenges of misinformation and harmful content.

2.1   Authorship with Transformers

Transformers have been already studied as an instrument to improve previous results in authorship verification tasks. Manolache et al. [13] analysed the effectiveness of BERT-like transformers for authorship verification using vast availability of data online and large-scale authorship verification datasets. The authors showed that these models outperformed manual feature extraction methods, underlining the importance of analyzing dataset properties.

Other researchers focus on cross-domain situations considering text's topic or genre [2]. The authors augment a successful authorship verification technique based on a multi-headed neural network language model with pre-trained language models. They conduct experiments on a controlled corpus encompassing various text genres and maintain specific control over topic and genre. The results show promising outcomes for the proposed approach. A key finding from their research is the pivotal role of an appropriate normalization corpus in cross-domain attribution. They emphasize the need to consider stylistic properties over topic or genre in the authorship attribution process.

DeepStyle is another proposal that follows this research line [5]. The authors aim to tackle the limitations in existing methods, such as the inability to handle individual social media posts effectively, limited use of various feature types in representing users' writing styles, and a lack of explainability due to the "black-box" nature of deep learning models. It extracts salient features from social media posts and uses deep neural networks with a triplet loss as the objective function to learn post embeddings.

2.2   Towards Universal Authorship Embeddings

On previous works [6,7] we established the feasibility of encoding authorship features into an embedding vector. For short, authorship embeddings are meant to contain information about writing style, not content. Features such as rhythm, flow, punctuation or registry are shared by the same author and thus can be encoded. The model PART with its pretrain contrastive objective achieved fair results, but further research has shown that better stylistic understanding can be achieved, thus we present here $PART_{SCL}$.

There are three notable differences with our previous pre-training. First, PART used InfoNCE [14] which does not process more than two positive examples at once. We already required the author label to pair texts, thus we can simultaneously compare several texts from the same author to achieve better performance. In this new version, we apply a supervised contrastive objective such as SupCon [11] to enable more robust embeddings. Another novelty of $PART_{SCL}$ is that we remove the LSTM layer in favour of a pure-transformer approach, which results in cheaper inference later. Finally, the data used for pre-training has been increased with new large datasets; improving generalization of the model.

2.3   Supervised Contrastive Pretraining

A summary of the used method is presented in Fig. 1. Let $A = \{A_1, A_2, ..., A_n\}$ be the set of authors with each author containing an authored document set such as $A_i = \{d_1, d_2, ..., d_m\}$. Each document of an author is considered a view for the purpose of SupCon, thus picking k authors and l random documents from these authors results in a multi-viewed batch of authored documents. We avoid repeating the same author twice within the batch, so it has l positive documents for each author i with indices $P(i)$, and $(k - 1) * l$ negative documents with indices $A(i)$. For each document we want to generate an embedding such as $e_j = encoder(d_j)$.

We use the SupCon loss term as presented in Eq. 1:

$$\mathcal{L}_{SCL}(e) = \sum_{i \in I} \frac{-1}{|P(i)|} \sum_{p \in P(i)} \log \frac{\exp(e_i \cdot e_p / \tau)}{\sum_{a \in A(i)} \exp(e_i \cdot e_a / \tau)} \qquad (1)$$

where $|P(i)|$ is the cardinality and τ is a trainable temperature parameter.

With this loss term similar positive examples are rewarded while dissimilar negative examples are also rewarded; related documents are grouped together and will present greater cosine similarity with same-author text than any other text. This way we force the transformer to learn authorship representations, which requires to contain defining common features of style to minimize loss.

Inference Example on Attribution: Performing inference requires a document set K from the known authors, and a document set from anonymous sources U. These two sets of documents are processed through $PART_{SCL}$, which outputs an embedding for each document. Each unknown document is matched to the author embedding with highest cosine similarity (or $max(S_c(K, U_j))$).

We further explore how to use the model in the experimentation, applying our methods to attribution and fine-tuning a minimal model for verification.

2.4   Network Architecture and Hyper-parameters

An encoder is required for $e_j = encoder(d_j)$ thus we use a transformer encoder-only model. Warm-starting a model is more efficient time-wise for pre-training,

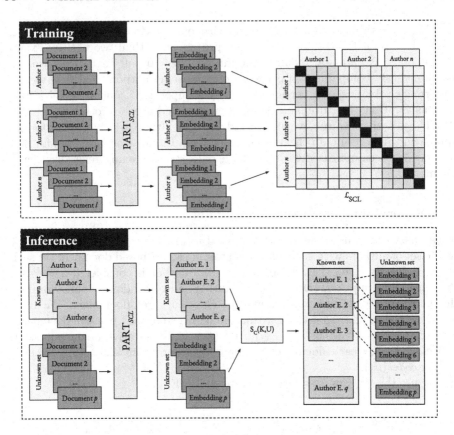

Fig. 1. Summary of PART$_{SCL}$ training and inference.

thus we use the public RoBERTa [12] large checkpoint[1] The RoBERTa transformer outputs a weighted average of word embeddings, to form a sentence embedding representative of the input content, this sentence embedding is processed by a final dense layer with dimension d containing the final authorship embedding, with no activation whatsoever. Hyper-parameters used in training are shown in Table 1.

3   Data

We aim to learn a high variety of textual stylistic features, therefore data requires to match this variety to recognize new, unseen authors successfully. Other key to this technique is how to handle the data, over-representation of a single style can be harmful but some datasets have limited amount of authors.

---

[1] RoBERTa large huggingface checkpoint: https://huggingface.co/roberta-large.

Table 1. Hyper-parameters used for pretraining the authorship model.

	Training hyperparameters
Batch size	$b = 16384$, $l = 1024$, $k = 16$
Max sequence length	512
Base model	RoBERTa large
Training steps	3000
Schedule	Warmup with linear decay
Warmup steps	180
Optimizer	AdamW
Learning Rate	$1e-2$
Weight decay	$1e-4$

3.1 Data Handling

Constructing a batch is performed by sampling l texts from n different authors. Generalization of the network is higher as the batch size $b = l \cdot n$, but in-batch representation affects the capacity of the model as reviewed in previous work, therefore each batch is balanced equally ensuring fair representation for all datasets. Authors from underrepresented datasets are upsampled to match larger datasets. Sampling texts from an author is performed at random, this mitigates the upsampling overrepresentation of some authors in the dataset. We do not consider documents as datapoints instead we consider each author to be a point in data, therefore the number of texts per authors does not lead to imbalance.

Texts written by the same author may be longer than the maximum determined length, this is common on datasets containing books, therefore all texts are split in equal sized chunks without overlap. Only authors with more than 16 texts are introduced to the training dataset. If a text has been written by someone anonymous or by several authors, the text is not included in the dataset either, independently of the number of texts found.

Additionally some datasets have received special consideration and preprocessing described as follows:

Standardized Gutenberg [4]: The Gutenberg Project is an open repository of literary works from authors whose works have entered the public domain. The writing styles range from novels, to poetry and essays although it presents imbalance in the representation. Works are divided into books, each book contains identifying information within the first and last text chunks, which are removed from the dataset. After preprocessing, we have extracted 1270 authors with, on average, 542 documents. There is a total of $6.89 \cdot 10^5$ documents in this dataset.

Blog Authorship [15]: The blog profiling dataset was originally constructed to study the age, gender and other personal features from textual content. We use it for attribution with our methods. It contains text ranging from personal opinion to short stories or fan-fiction. No additional preprocessing was required for this dataset. We found 4837 eligible authors, with 60 documents per author, for a total of $2.9 \cdot 10^5$.

Twitter Users [3]: Using the geolocation dataset we retrieve 10000 tweets (or all, whichever is lower) from all users in the original dataset. We clean each twit removing user information and links, replaced by special tokens $< h >$ and $< u >$ for hyperlinks and usernames respectively. To make training more efficient we join tweets with a triple linespace to fill up the maximum 512 token length. This dataset contains stylistically varied tweets, including emoji usage, slang, irony, shorthand and other informal registry text. At training time, there were 52465 accounts, with 57 documents available on average. A total of $3.02 \cdot 10^6$ documents are retrieved from this dataset.

Reddit TLDR [17]: The Reddit TLDR dataset contains texts from reddit users posted in different forums explaining several topics and ranging from informal divulgation to gossip. The train dataset has 14548 authors with 36 documents on average with a total of $2.9 \cdot 10^5$ documents.

Full Dataset: In total, we balance the datasets so each represents 1/4th of the authors. The final dataset contains 73466 unique authors with $4.5 \cdot 10^6$ texts to train with.

4   Experimentation

For the experimentation, we focus on the PAN competition, series of scientific events and shared tasks on digital text forensics and stylometry. We extract experimental results with two different approaches. First we explore the PAN authorship attribution common tasks, to test the zero-shot capabilities of the model. The second section explores the PAN style change detection, to showcase the capabilities of fine-tuning the $PART_{SCL}$ model.

On all experiments we focus on comparing our new method $PART_{SCL}$ against PART and RoBERTa. PART is the previous version of the model without the described upgrades, on the other hand RoBERTa is the backbone of PART and the warm startup weights of $PART_{SCL}$. If available we report scores from the competition.

PAN Attribution: For attribution we use the described method in Sect. 2.3, we first split each candidate document in chunks of 512; compute the embeddings of candidates and unknown documents; perform the pairwise cosine similarity and assign each unknown document to its most similar candidate. As in training, as the model is not capable of naming unknown authors, the only change we make is removing the unknown texts from the challenge. We explore all available challenges on this category [1, 8–10].

PAN Style Change: For verification we design a simple siamese topology to classify whether two texts belong to the same author or not. We extract both embeddings from the frozen models and concatenate each embedding along with their

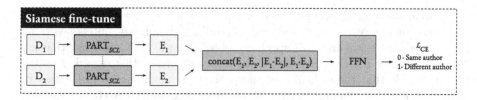

Fig. 2. Siamese architecture for fine tuning.

absolute difference and their product, which is classified by a dense layer and a output neuron (a feed forward network). The siamese architecture is presented in Fig. 2.

4.1 Authorship Attribution

Results for each competition are shown in Tables 2 and 3. On Table 2 the global results are explored, computed as the average among related problems in the competition. On average, the new technique dominates earlier versions and baselines, specially in PAN19. PAN19 is particularly close to the training domain due to it containing fanfiction which has style overlap with the blog dataset and the literary works of the gutenberg corpus. In practice, the PAN19 dataset achieves very high accuracy in english outclassing the best reported method in the competition by 4.5% fl-score.

Table 2. Zero-shot results for PAN-CLEF attribution tasks averaged. *Best* column indicates first place for the competition. The best scores for zero-shot methods marked in bold.

	Accuracy				F1-score			
	PART$_{SCL}$	PART	RoBERTa	Best	PART$_{SCL}$	PART	RoBERTa	Best
PAN11 - Large	**0.399**	0.375	0.287	-	**0.281**	0.269	0.188	*0.321*
PAN12	**0.873**	0.841	0.692	*0.884*	**0.854**	0.827	0.635	*0.859*
PAN18	**0.695**	0.692	0.564	-	**0.621**	0.594	0.431	*0.762*
PAN19	**0.809**	0.617	0.516	-	**0.710**	0.546	0.445	*0.665*

It is to be noted that he PAN11 challenge is derived from the Enron Mail dataset [16] which, as shown in previous work, was troubling for PART; our new methodology and training results in much higher accuracy and fl-scores. Despite the improvement we still score 2nd in the competition, off by 4% from the best method.

Finally, PAN 12 results in close scores to the best model, averaging to 85.4%, behind by 0.5%. PAN18 results are behind the best model in the cometition by 14%. Summarizing, PART$_{SCL}$ outclasses previous zero-shot approximations to attribution on average, but still lags behind when not properly tuned to the specific domain. However, when the domain is tangentially close to our training styles, metrics improve by a large margin.

A breakdown of the problems is shown in Table 3. Here we see how the model $PART_{SCL}$ compares to previous work. $PART_{SCL}$ frequently outclasses other embeddings but there are some specific problems in the PAN challenge that are better solved by PART. There is no apparent pattern in this regard. Another observation is that the difficulty of the problem is not indicative of higher or lower accuracy, as exemplified by PAN18 - 10 authors and PAN18 - 5 authors. The first should be harder to solve but achieves the highest accuracy and F1-score, while the second has worse metrics despite being significantly easier. This phenomenon repeats on the PAN19 challenge too.

Across challenges we observe the underperforming of a vanilla RoBERTa large model, lagging behind both pretrained authorship models at almost every available problem. In summary, we observe that $PART_{SCL}$ also works at a finer level and is generally better on most problems, including without specific training, making it very a very powerful feature extractor.

Table 3. Zero-shot results for all PAN-CLEF attribution tasks. Accuracy and F1-Score are reported, best results reported in bold for each metric.

	Accuracy			F1-Score		
	$PART_{SCL}$	PART	RoBERTa	$PART_{SCL}$	PART	RoBERTa
PAN11 - *Large*	**0.3985**	0.3754	0.2869	**0.2809**	0.2618	0.1882
PAN12 - *A*	**0.8333**	**0.8333**	0.5000	**0.8222**	**0.8222**	0.4127
PAN12 - *B*	0.8333	**1.0000**	0.6667	0.8222	**1.0000**	0.6556
PAN12 - *C*	**1.0000**	**1.0000**	**1.0000**	**1.0000**	**1.0000**	1.0000
PAN12 - *D*	**1.0000**	**1.0000**	0.6250	**1.0000**	**1.0000**	0.5417
PAN12 - *I*	**0.8571**	0.5714	0.6429	**0.8095**	0.5357	0.5357
PAN12 - *J*	**0.7143**	0.6429	**0.7143**	**0.6667**	0.5714	**0.6667**
PAN18 - *20 authors*	**0.6582**	0.6203	0.4684	**0.6566**	0.5549	0.3359
PAN18 - *15 authors*	0.5946	**0.6351**	0.4730	**0.5788**	0.5220	0.3186
PAN18 - *10 authors*	**0.9000**	0.8250	0.6250	**0.7557**	0.7544	0.5617
PAN18 - *5 authors*	0.6250	**0.6875**	0.6875	0.4914	**0.5455**	0.5073
PAN19 - *r = 100%*	**0.9252**	0.8504	0.7350	**0.8071**	0.7166	0.5730
PAN19 - *r = 80%*	**0.6327**	0.4898	0.4694	**0.5263**	0.3986	0.4036
PAN19 - *r = 60%*	**0.6742**	0.5379	0.5000	**0.6564**	0.4714	0.4665
PAN19 - *r = 40%*	**0.9079**	0.4868	0.1645	**0.7877**	0.6035	0.2753
PAN19 - *r = 20%*	**0.8788**	0.7197	0.7121	**0.7723**	0.5420	0.5067

4.2 Authorship Verification

Table 4 shows the results when comparing PART, our new $PART_{SCL}$ approach, Roberta and the best result in different PAN author style change detection competitions. According to the proposed approach, the results of PART, $PART_{SCL}$

and Roberta are results after constructing a Siamese architecture (4) in which only a linear classification layer with 2 outputs is trained. The rest of the layers of the architecture are not trained. In case of the PART 2023 competition as of the date of this article, there are still no published results of the competition.

As can be seen, PART$_{SCL}$ improves PART in all cases and Roberta with very noticeable differences in the results. In comparison with methods with the best results obtained in the competition, our method is very competitive, since despite not being a complete architecture trained on the data, it is always close to the best results obtained by researchers using training with all the data. We believe that, based on these results, the capabilities of $PART_{SCL}$ to perform zero-shot classification tasks have great potential.

Table 4. Results for 2020–2023 PAN-CLEF competition on writing style change detection for PART, $PART_{SCL}$, Roberta and the best result obtained in the conference. It must be noted that PART, $PART_{SCL}$ and Roberta columns show results where only one final classification layer is trained, while the rest of layers are frozen. The best result from the competition consider the best method, which usually includes feature engineering or training a whole architecture. F1-score is reported for each experiment.

Dataset	PART	$PART_{SCL}$	Roberta	Best result
PAN 2023 Dataset 1	86,69%	92,24%	80,47%	-
PAN 2023 Dataset 2	74,39%	75,62%	41,33%	-
PAN 2023 Dataset 3	62,21%	63,84%	36,15%	-
PAN 2022 Task 2 Dataset 1	79,51%	83,70%	79,51%	70,7%
PAN 2022 Task 2 Dataset 2	69,32%	73,64%	48,02%	70,7%
PAN 2022 Task 2 Dataset 3	62,27%	63,12%	40,13%	70,7%
PAN 2021 (task 2)	69,42%	70,43%	40,13%	75,1%
PAN 2020 Task 2 narrow	84,12%	87,93%	82,71%	85,67%
PAN 2020 Task 2 wide	82,38%	85,32%	75,41%	85,67%

5   Conclusions

In this paper we have presented PART$_{SCL}$, a system for authorship attribution and the generation of authorship embeddings containing relevant information about the author, numerically encoded. We have performed experiments on benchmark challenges to demonstrate the performance of the system even at zero-shot or as a frozen feature extractor. It presents competitive behaviour in both authorship attribution and style change detection, being on-par with dominant approaches such as n-grams or transformer ensembles.

The system shows clear capacity to discern authorship at domains it has not been trained for, shown with the authorship attribution at zero shot, and capable of successful adaption to tasks different to exactly attribution, as shown by the style change detection tasks.

Further research is required in the capabilities of the system concerning applications to other tasks suck as authorship profiling, obfuscation and verification. We foresee useful adaptions of the system to these tasks but doing so is a non-trivial issue. Improvements on studied tasks could be made, for instance, attribution has been performed at zero-shot, a short fine-tuning of the model to the target domain could surely boost results considering the training dataset of these challenges is never used. On the other hand the verification uses the model as a frozen encoder, where $PART_{SCL}$ could have been fully fine-tuned too.

The system could be improved with further data but datasets with authorship information are scarce and typically very small (such as the PAN datasets). Better generalization could be achieved with higher batch sizes, either tweaking the number of authors per batch or the number of positive samples per author; higher batch sizes present technical challenges beyond the scope of this work. New techniques have to be developed to recognize unknown authors in the available documents.

Acknowledgments. This work is part of the project PCI2022-134990-2 (MARTINI) of the CHISTERA IV Cofund 2021 program, funded by MCIN/AEI/10.13039/ 501100011033 and by the "European Union NextGenerationEU/PRTR", by the Spanish Ministry of Science and Innovation under FightDIS (PID2020-117263GB-100) grant and MCIN/AEI/10.13039/501100011033/ and European Union NextGeneration EU/PRTR for XAI-Disinfodemics (PLEC2021-007681) grant, by European Comission under IBERIFIER - Iberian Digital Media Research and Fact-Checking Hub (2020-EU-IA-0252), and by "Convenio Plurianual with the Universidad Politécnica de Madrid in the actuation line of *Programa de Excelencia para el Profesorado Universitario*". This publication is also part of the I+D+i project PLEC2021-007681, financed by MCIN/AEI/10.13039/501100011033/ and the European Union NextGeneration/PRTR.

References

1. Argamon, S., Juola, P.: Overview of the international authorship identification competition at pan-2011. In: CLEF (Notebook Papers/Labs/Workshop) (2011)
2. Barlas, G., Stamatatos, E.: Cross-domain authorship attribution using pre-trained language models. In: Maglogiannis, I., Iliadis, L., Pimenidis, E. (eds.) AIAI 2020. IAICT, vol. 583, pp. 255–266. Springer, Cham (2020). https://doi.org/10.1007/ 978-3-030-49161-1_22
3. Cheng, Z., Caverlee, J., Lee, K.: You are where you tweet: a content-based approach to geo-locating twitter users. In: Proceedings of the 19th ACM international conference on Information and knowledge management, pp. 759–768 (2010)
4. Gerlach, M., Font-Clos, F.: A standardized project Gutenberg corpus for statistical analysis of natural language and quantitative linguistics. Entropy **22**(1), 126 (2020)
5. Hu, Z., Lee, R.K.-W., Wang, L., Lim, E., Dai, B.: DeepStyle: user style embedding for authorship attribution of short texts. In: Wang, X., Zhang, R., Lee, Y.-K., Sun, L., Moon, Y.-S. (eds.) APWeb-WAIM 2020. LNCS, vol. 12318, pp. 221–229. Springer, Cham (2020). https://doi.org/10.1007/978-3-030-60290-1_17

6. Huertas-Tato, J., Huertas-Garcia, A., Martin, A., Camacho, D.: PART: Pre-trained Authorship Representation Transformer. arXiv (2022). https://doi.org/10.48550/arXiv.2209.15373

7. Huertas-Tato, J., Martin, A., Huertas-Garcia, A., Camacho, D.: Generating authorship embeddings with transformers. In: 2022 International Joint Conference on Neural Networks (IJCNN), pp. 1–8. IEEE (2022)

8. Juola, P.: An overview of the traditional authorship attribution subtask. In: CLEF (Online Working Notes/Labs/Workshop), vol. 1178, p. 1 (2012)

9. Kestemont, M., Stamatatos, E., Manjavacas, E., Daelemans, W., Potthast, M., Stein, B.: Overview of the cross-domain authorship attribution task at {PAN} 2019. In: Working Notes of CLEF 2019-Conference and Labs of the Evaluation Forum, Lugano, Switzerland, September 9–12, 2019, pp. 1–15 (2019)

10. Kestemont, M., et al.: Overview of the author identification task at pan-2018: cross-domain authorship attribution and style change detection. In: Working Notes Papers of the CLEF 2018 Evaluation Labs. Avignon, France, September 10–14, 2018/Cappellato, Linda [edit.] et al, pp. 1–25 (2018)

11. Khosla, P., et al.: Supervised Contrastive Learning. arXiv (2020). 10.48550/arXiv.2004.11362

12. Liu, Y., et al.: Roberta: a robustly optimized bert pretraining approach. arXiv preprint arXiv:1907.11692 (2019)

13. Manolache, A., Brad, F., Burceanu, E., Barbalau, A., Ionescu, R., Popescu, M.: Transferring Bert-like transformers' knowledge for authorship verification. arXiv preprint arXiv:2112.05125 (2021)

14. Oord, A.V.D., Li, Y., Vinyals, O.: Representation learning with contrastive predictive coding. arXiv preprint arXiv:1807.03748 (2018)

15. Schler, J., Koppel, M., Argamon, S., Pennebaker, J.W.: Effects of age and gender on blogging. In: AAAI Spring symposium: Computational Approaches to Analyzing Weblogs, vol. 6, pp. 199–205 (2006)

16. Shetty, J., Adibi, J.: The Enron email dataset database schema and brief statistical report. Information sciences institute technical report, University of Southern California, vol. 4, no. 1, pp. 120–128 (2004)

17. V"olske, M., Potthast, M., Syed, S., Stein, B.: TL;DR: mining reddit to learn automatic summarization. In: Proceedings of the Workshop on New Frontiers in Summarization, pp. 59–63. Association for Computational Linguistics, Copenhagen, Denmark (2017). https://doi.org/10.18653/v1/W17-4508, https://www.aclweb.org/anthology/W17-4508

Trend Detection in Crime-Related Time Series with Change Point Detection Methods

Apostolos Konstantinou$^{(\boxtimes)}$, Despoina Chatzakou, Ourania Theodosiadou, Theodora Tsikrika, Stefanos Vrochidis, and Ioannis Kompatsiaris

Information Technologies Institute, Centre for Research and Technology Hellas, Thessaloniki, Greece
{konstantinou,dchatzakou,raniatheo,theodora.tsikrika, stefanos,ikom}@iti.gr

Abstract. Time series analysis can be an asset in the hands of the authorities, as it can enable the understanding and monitoring of trends of criminal activities. In this work, a variety of methods is exploited to detect significant points of change in crime-related time series that may indicate the occurrence of events that require attention. In particular, change point analysis is applied in relevant time series, both offline (retrospective change detection when all data is available) and online (detection of changes as soon as they occur). The focus is on the Crimes in Boston and London Police Records datasets, examining how change point detection can benefit relevant authorities in understanding crime trends to better allocate and manage resources. The experimental results allow us to gain valuable insights, including the observation of seasonal patterns in some cases, with corresponding crimes peaking at specific times, the somewhat different change points identified by online and offline methods, and the observation that domain knowledge is desired for better method selection and parameters configuration.

Keywords: Trend detection · Change point detection · Crime-related time series

1 Introduction

With the emergence of online platforms (such as social media, blogs, forums, etc.) and the Internet as a whole, new opportunities for delinquent and criminal activities have arisen, ranging from hacking and financial frauds [8] to even terrorism-related activities, including propaganda spreading, recruitment, and training, as well as hate spreading towards specific social groups [3]. Thus far, significant effort has been placed into developing a wide range of tools to tackle criminal activities from different perspectives, including real time detection of online terrorism-related content [1], crime hotspots detection through spatio-temporal analysis [27], and linkage of online identities to criminal investigations [18].

Further to traditional data mining methods, time series have been effectively applied to a wide range of tasks, including the development of methods to detect

A. Arampatzis et al. (Eds.): CLEF 2023, LNCS 14163, pp. 72–84, 2023.
https://doi.org/10.1007/978-3-031-42448-9_7

and predict criminal activities (e.g., [6]). In this context, change point detection methods have been considered and employed for the detection of significant changes in time series, with the accurate and early detection of change points being a pivotal point for drawing valuable insights. Change point detection has many important applications in several areas, including, but not limited to, the financial sector [20], network traffic analysis [25], and climatology [26].

When it comes to fighting crime and terrorism, little effort has been placed thus far to detect critical points of change. An example is the application of the Cumulative Sum change point detection method for the identification of statistically significant changes in Houston's daily crime totals during Hurricane Harvey [5]. In the same vein, a framework that builds on top of a nonparametric multivariate change point detection algorithm has been proposed to detect statistically significant change points in terrorism-related time series [35]. In both cases, the focus has been on analyzing time series data in an *offline* manner, where change detection is applied retrospectively when all data is available.

Analyzing trends as well as identifying changes as soon as they occur in real-time (i.e., in an *online* manner) could be particularly valuable for the authorities, as it could enable a more effective response and allocation of resources in order to mitigate serious incidents. To this end, this work aims to investigate the effectiveness of both online and offline change point detection methods towards identifying critical changes in crime-related time series; to the best of our knowledge, there is no other work in the literature that performs such analyses on crime-related data in an online setting. Overall, to enable the effective evaluation of the most popular online and offline change point detection methods, first, a wide range of ground-truth datasets from different domains are examined, ultimately leading to the development of a framework that allows for the identification of trends and significant change points in an effective manner. The applicability of the proposed framework in the crime-related domain is demonstrated on two popular relevant datasets, namely the 'Crimes in Boston' and the 'London Police Records' datasets; these datasets do not though have an associated ground truth and thus an insightful qualitative analysis is performed.

2   Related Work

Change point detection methods are typically divided into online and offline [2]. Although offline methods are characterized by higher accuracy, one of their main features is that they need access to the entire time series, which makes them inapplicable in real time scenarios. Contrary, online methods process data in real time, thus being suitable for crime detection in real world applications.

Offline Change Point Detection Methods (supervised and unsupervised). Supervised methods include Decision Trees [29], Bayesian Networks [16], Hidden Markov Models [17], and Gaussian Mixture Models [12]. A key drawback of such methods is their need for large amount of annotated data for training, while most real world data is sparsely annotated or not annotated at all. Training can be done on artificial data, but such models usually do not generalize well.

On the contrary, unsupervised methods do not require any kind of annotations and include likelihood ratio methods, probabilistic, kernel-based, and graph-based approaches, as well as clustering methods [2]. The first attempts at unsupervised change point detection have been with the Cumulative Sum Control Chart (CUSUM), which allows for step detection in time series [28]. In the same direction, one of the most commonly used methods is the Binary Segmentation [31], which is characterized by low complexity and operates in a sequential manner. Pruned Exact Linear Time (PELT) [22] is another high-performance offline algorithm that can also be used on multivariate signals. PELT is both computationally efficient and versatile, and in many cases outperforms Binary Segmentation making it one of the top change point detection algorithms. Finally, the Prophet forecasting and change point detection tool [34] implements several models and selects the most appropriate for the data at hand.

Online Change Point Detection Methods. Online (real time) methods run concurrently with the activity being monitored (e.g. crime rate), processing data, one point at a time, as it becomes available. Such a point could be the temperature at a location [2], crime rate in a area [32], or the effect of an outside factor (e.g. changes on cannabis regulation laws [24]) on crime rates.

Many common algorithms for online change point detection are often variations of their offline counterparts. CUSUM is a typical example with several variations for online detection (e.g. [30,37]). Another commonly used approach is the Bayesian Online Change Point Detection method [11], which allows for effective detection of long-term changes in online setups. Moreover, Change Finder [33] is an online learning framework based on a probabilistic model that enables the detection of outliers and change points in streaming time series data. Online methods often perform worse compared to their offline counterparts, since they require data from both before and after a data point to effectively determine whether it constitutes a change point, with different methods requiring different amounts of such data; their performance is thus a trade-off between the amount of data after the point considered and the time criticality of the task at hand.

3 Methodology

This section briefly overviews the methods considered in this work to ultimately enable effective detection of changes in crime-related time series.

3.1 Offline Change Point Detection (CPD) Methods

Binary Segmentation [31] is characterized by low complexity and uses a recursive approach: first a change point in the complete input signal is detected, then the series is split into two parts around this change point, and the operation is repeated in each part. The process stops when a specified number of change points is detected; in case the number is unknown, a penalty parameter is given.

Pruned Exact Linear Time (PELT) [22] relies on a pruning rule and detects change points by minimizing a cost function over their possible numbers and locations. In particular, it combines optimal partitioning and pruning, and achieves efficient computational cost, while maintaining high accuracy, thanks to the pruning rule that discards many indexes under the assumption that they can never be minima in terms of the minimization performed at each iteration.

Cumulative Sum (CUSUM) [15] requires a set of parameters to be calculated first in order to condition the change detection, namely the mean, the standard deviation, the shift of interest (which is the smallest deviation we wish to detect), the allowance parameter K, and the decision parameter H that determines whether a change has occurred or not. Various values of H have been used; e.g., H was set to 100 in [15], while values between 1 and 40 were explored in [10].

Segment Neighborhood [4] searches the entire segmentation space by first defining a maximum number of change points, denoted as Q. By computing a cost function for all possible segments, then all segmentations with change point between 0 and Q are considered. Due to the exhaustive search performed, an important drawback of this method is the significant computational cost.

Prophet [34] first determines a large number of possible change points at which the rate is allowed to change. It then places a sparse priority on the magnitudes of rate changes (equivalent to L1 regularization) to limit the number of the possible change points to use. By default, 25 potential change points are specified that are uniformly placed in the first 80% of the time series.

3.2 Online Change Point Detection (CPD) Methods

Bayesian Online Change Point Detection (BOCPD) [11] focuses on generating an accurate distribution of the next datum in the series given only the previously observed data. Central point to the algorithm is the time since the last checkpoint, i.e. the run length. The algorithm assumes that the points in the observed time series can be partitioned into non-overlapping segments.

BOCPD with model selection (BOCPDMS) [23] extends BOCPD by introducing multiple models and a method for online model selection; it aggregates over all models and prunes run lengths, keeping the most probable ones per model.

SWAB. Two common approaches to CPD are top-down and bottom-up; top-down approaches (e.g., Binary Segmentation) start with the entire time series and recursively segment it until a halting condition is met, while in the bottom-up, the process starts with the maximum number of segments and merges them using a cost function, until a stopping threshold. SWAB [21] combines a sliding window with bottom-up segmentation, achieving good results in online setups.

This section presented various online and offline CPD methods, with each one approaching the CPD problem in a different way; e.g., Binary segmentation follows a recursive approach, while PELT builds on a pruning rule. While some methods tend to perform better overall (e.g. [36]), variations in their performance can be observed depending on the data and the choice of initial parameters (if any). This is also evident in our experiments (Sect. 5) indicating that domain knowledge is also needed for choosing the best method in each case.

4   Datasets

To evaluate the different change point detection methods described in Sect. 3, two types of datasets are considered: (i) a collection of 42 datasets for which the ground truth labels are known and thus a direct comparison among different methods is feasible; and (ii) two crime-related datasets for which there are no ground-truth labels and thus a more qualitative analysis takes place.

4.1   TCPD Benchmark and Dataset Collection

A benchmarking framework consisting of a collection of methods and time series data (also referred to as TCPD) has been proposed that allows testing and comparing CPD algorithms [9]. Overall, it consists of 37 annotated datasets and 5 artificial control datasets, including e.g. the daily closure price of Apple Inc. stock, the price of bitcoin, and the GDP of several countries. All these datasets have been annotated by one or more field experts in time series analysis.

4.2   Crimes in Boston Dataset

The 'Crimes in Boston' dataset [7] contains crimes reported to the Boston Police Department from June 15, 2015 to September 29, 2019. For each reported crime, additional information is available, e.g. its incident number, the date and time of the crime, and the location. In total, there are 576 different crime types. To allow for a more coarse-grained analysis, we manually grouped the different types in eight general categories, e.g. crimes related to 'Theft', 'Robbery', and 'Burglary' are grouped into the *Theft* category. The crime categories along with their frequency are: Person-related: $125,747$ (18.79%), Assault: $124,750$ (18.64%), Theft: $119,788$ (17.90%), Fraud: $108,983$ (16.29%), Traffic: $58,844$ (8.79%), Narcotics: $23,928$ (3.58%), Misc: $22,450$ (3.35%), and Other: $84,702$ (12.66%).

Figure 1 depicts the time series for each category, with each point corresponding to the number of crimes committed in each of the 52 months in the dataset. Through visual inspection one can identify underlying patterns that may be of interest and assess potential critical points of change. For instance, we observe that for 'Theft' a periodicity appears that could be useful for police authorities to make a better allocation of resources with the aim of dealing with this type of crime as best as possible. Moreover, for 'Assault', there appear to be four points (at around the 10, 15, 35 and 50 points on the x-axis) of change that may require further study to draw useful conclusions. On the other hand, for the 'Person'

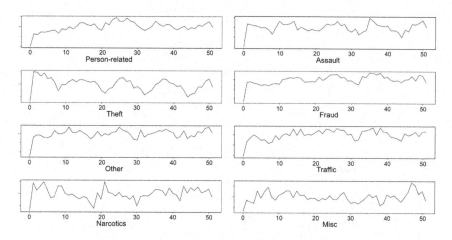

Fig. 1. Boston crime series: count of crimes per category in each month.

Table 1. Crimes in Boston dataset: correlation matrix.

	Assault	Fraud	Misc	Narcotics	Other	Person-related	Theft	Traffic
Assault	1	0.690	0.454	0.335	0.729	0.584	0.688	0.645
Fraud	0.690	1	0.322	0.279	0.754	0.822	0.317	0.856
Misc	0.454	0.322	1	0.180	0.401	0.310	0.278	0.195
Narcotics	0.335	0.279	0.180	1	0.383	0.038	0.166	0.135
Other	0.729	0.754	0.401	0.383	1	0.649	0.400	0.603
Person-related	0.584	0.822	0.310	0.038	0.649	1	0.202	0.820
Theft	0.688	0.317	0.278	0.166	0.400	0.202	1	0.264
Traffic	0.645	0.856	0.195	0.135	0.603	0.820	0.264	1

and 'Fraud' crimes there do not seem to be any obvious points of interest, which suggests that there is a relatively stable pattern regarding these crimes.

To determine if there is any connection and co-occurrence between crimes, we also examined the correlation matrix of each crime type against every other, using the Pearson's correlation coefficient [14]. Table 1 indicates that there is for instance high correlation between 'Traffic' and 'Fraud' (0.856), and 'Fraud' and 'Assault' (0.690), but also between 'Person-related' crimes and 'Fraud' (0.822). The correlations that emerge from such an analysis can be useful, as they can be important piece of information on how and whether it makes sense to deal with not just one crime at a time, but a set of crimes in a more effective way.

4.3   London Police Records Dataset

The 'London Police Records' [13] dataset consists of a list of crimes committed in the area of London from June 2014 to May 2017. Overall, it consists of 14 crime types: 1. Vehicle crime: $262,309$ (8.90%), 2. Violence and sexual offences:

78   A. Konstantinou et al.

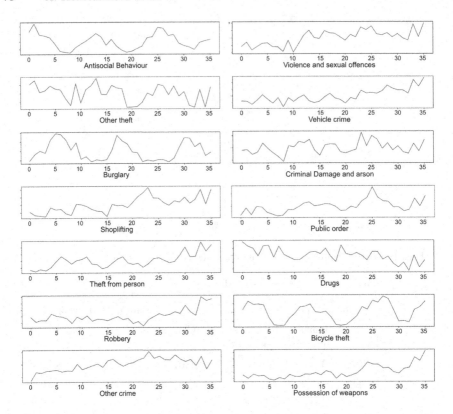

Fig. 2. London crime series: count of crimes per category in each month.

596, 107 (20.23%), 3. Antisocial Behavior: 708, 264 (24.04%), 4. Bicycle theft: 54, 649 (1.85%), 5. Other theft: 333, 817 (11.33%), 6. Theft from the person: 109, 168 (3.71%), 7. Other crime: 29, 208 (0.99%), 8. Drugs: 106, 836 (3.63%), 9. Burglary: 213, 125 (7.23%), 10. Public order: 130, 653 (4.43%), 11. Shoplifting: 135, 780 (4.61%), 12. Criminal damage and arson: 184, 772 (6.27%), 13. Robbery: 68, 920 (2.34%), and 14. Possession of weapons: 12, 871 (0.44%).

From Fig. 2 we observe that 'Burglary' and 'Bicycle theft' crimes seem to have quite distinguishable change points, while periodicity is also observed in both cases; e.g. for 'Burglary', a seasonality is observed, with crime rates peaking each year during the months of November (5, 17, 29), December (6, 18, 30), and January (7, 19, 31). 'Bicycle theft' is also observed in the summer period during June (0, 12, 24), July (1, 13, 25), and August (2, 14, 26), while 'Antisocial behavior' seems to peak every July (1, 13, 25). Change points that may be of interest and could receive more attention can also be seen in the crimes of 'Violence and sexual offences', 'Criminal damage and arson', and 'Shoplifting'. For the rest of the crimes, no particularly obvious change points seem to appear.

Finally, similarly to before, we also estimated the correlation matrix as presented in Table 2. Overall, a high correlation is observed between Antisocial

Table 2. London Police Records dataset: correlation matrix.

	1	2	3	4	5	6	7	8	9	10	11	12	13	14
1	1	0.70	0.15	0.37	0.09	0.73	0.58	−0.46	0.04	0.59	0.71	0.49	0.84	0.84
2	0.70	1	0.23	0.31	0.25	0.46	0.79	−0.22	−0.17	0.82	0.71	0.68	0.46	0.71
3	0.15	0.23	1	0.89	0.64	−0.29	0.01	0.29	−0.57	0.53	−0.07	0.50	0.08	0.29
4	0.37	0.31	0.89	1	0.55	−0.13	0.17	0.11	−0.60	0.62	0.06	0.54	0.24	0.45
5	0.09	0.25	0.64	0.55	1	−0.09	−0.06	0.34	−0.05	0.37	−0.13	0.59	0.10	0.01
6	0.73	0.46	−0.29	−0.13	−0.09	1	0.35	−0.61	0.36	0.32	0.61	0.25	0.83	0.59
7	0.58	0.79	0.01	0.17	−0.06	0.35	1	−0.33	−0.20	0.68	0.74	0.51	0.22	0.54
8	−0.46	−0.22	0.29	0.11	0.34	−0.61	−0.33	1	−0.05	−0.18	−0.42	−0.04	−0.40	−0.35
9	0.04	−0.17	−0.57	−0.60	−0.05	0.36	−0.20	−0.05	1	−0.43	−0.06	−0.24	0.23	−0.25
10	0.59	0.82	0.53	0.62	0.37	0.32	0.68	−0.18	−0.43	1	0.54	0.75	0.37	0.70
11	0.71	0.71	−0.07	0.06	−0.13	0.61	0.74	−0.42	−0.06	0.54	1	0.56	0.47	0.66
12	0.49	0.68	0.50	0.54	0.59	0.25	0.51	−0.04	−0.24	0.75	0.56	1	0.28	0.43
13	0.84	0.46	0.08	0.24	0.10	0.83	0.22	−0.40	0.23	0.37	0.47	0.28	1	0.74
14	0.84	0.71	0.29	0.45	0.01	0.59	0.54	−0.35	−0.25	0.70	0.66	0.43	0.74	1

behavior and Bicycle theft (0.889), Sexual offences and Vehicle crimes (0.703), Violence and Public order (0.825), as well as Violence and Shoplifting (0, 705).

5   Experimental Results

This section first presents the results obtained on the TCPD dataset collection. Then, the best performing (offline and online) methods are employed to conduct a qualitative analysis on the two crime-related datasets (i.e. 'Crimes in Boston' and 'London Police Records'), for which no ground truth labels are available.

5.1   Experimental Results on the TCPD Dataset Collection

In the TCPD framework (see Sect. 4.1) various CPD methods have been implemented [9]. To gain a good understanding of the methods most commonly considered in the literature (Sect. 3), we tested and compared them against each other (using the settings defined in the aforementioned framework) based on two metrics: (i) the cover metric, which is based on the Jaccard Index (also known as Intersection over union); and (ii) the F_1 metric that treats change points detection as a classification problem [9].

Table 3 presents the corresponding results, through averaging the scores obtained across the 42 datasets presented in Sect. 4. Overall, TCPD consists of both univariate and multivariate datasets[1], but as in our case of interest (i.e. crime rate monitoring per observation period) the focus is on univariate data analysis, we proceed only with the univariate ones. As for the offline methods,

---

[1] Multivariate data analysis involves more than two dependent variables to result in an outcome, compared to univariate where only one variable at a time is considered.

Table 3. Experimental results on TCPD dataset collection.

Method	Cover score	F1 score
Offline CPD methods		
Binary Segmentation	**0.672**	**0.698**
CUSUM	0.526	0.572
Segment Neighborhoods	0.642	0.635
PELT	0.652	0.674
Prophet	0.522	0.47
Online CPD methods		
Bayesian online change point detection (BOCPD)	**0.594**	**0.662**
BOCPD with model selection (BOCPDMS)	0.590	0.495
SWAB	0.543	0.487

the best performance is achieved by Binary Segmentation with 0.672 cover and 0.698 F1 scores, followed by PELT achieving 0.652 cover and 0.674 F1 score.

For online methods, the best performance is obtained with BOCPD, a quite popular approach, followed by BOCPDMS, which trains multiple models choosing the appropriate model each time. Although one would expect BOCPDMS to perform better compared to BOCPD, it was observed that in some datasets (e.g. with small time series length) the performance was particularly poor, consequently leading to an overall reduced performance. Moreover, the BOCPDMS is a variant of the Bayesian method, targeting mainly multivariate data; when evaluating BOCPDMS on TCPD's multivariate datasets, better performance was observed compared to BOCPD (0.496 vs. 0.455 in the cover score).

In the following sections, a qualitative analysis on the crime-related datasets is conducted. Due to space limits for each case (online and offline), we proceed with one method per case. In particular, focusing on the offline methods, although Binary Segmentation achieved better performance, we proceed with PELT (the second best performing method), as according to the literature PELT leads in most cases to a more accurate detection of change points [36]. In relation to the online methods, BOCPD is used for the remaining analyses.

5.2   Experimental Results on the Crime-Related Datasets

Crimes in Boston Dataset. Focusing illustratively (due to space limits) on the 'Assault' crime, Fig. 3 depicts the corresponding time series in addition to the identified change points (depicted by dashed lines) as detected with PELT (offline) and BOCPD (online). Overall, we sampled the time series with various sample rates (i.e. grouping of crime rates by hour, day, week, and month), and here we indicatively present the results at a monthly rate. Based on the illustrated results, different change points are identified with the offline vs. online methods. This could be attributed to the fact that online methods process the

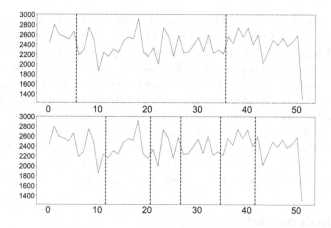

Fig. 3. Crimes in Boston (assault crimes): PELT (top) and BOCPD (bottom).

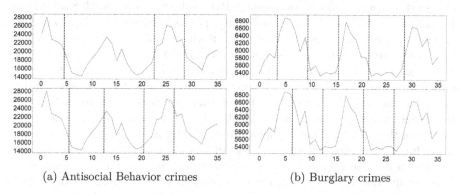

(a) Antisocial Behavior crimes   (b) Burglary crimes

Fig. 4. London Police Records: PELT (top) and BOCPD (bottom).

data as it arrives as opposed to offline methods that consider the entire dataset at once, thus affecting the change points detection process. Moreover, although PELT is known for its efficacy, it relies heavily on the choice of the initial parameters, meaning that for instance a small *penalty* (hyperparameter) would make the method too sensitive and thus leading to too many points being predicted. Overall, the depicted results indicate that there is no method that fits all, and therefore targeted configuration (based on domain knowledge) is required to best describe the data at hand.

London Police Records. Similarly to above, here we indicatively focus on the 'Antisocial behavior' (Fig. 4a), and 'Burglary' (Fig. 4b) crimes. As the figures show, there are in some cases seasonal patterns (e.g. crimes related to antisocial behavior tend to be more intense in the winter months) indicating that the respective crimes peak at certain times during the year.

In contrast to the Crimes in Boston dataset, the PELT and BOCPD methods here make very similar predictions. This could be attributed to the difference in the length of the datasets (Boston covers 52 months, while London 36 months). Based on how PELT works, keeping the *penalty* hyperparameter constant, the shorter the length, the more the number of change points detected. For a more effective detection of change points, the value of *penalty* should be set taking also into account the size of the time series; there are works focusing on the appropriate selection of this value (e.g. [19]). Finally, in the case of the 'Antisocial behavior' and 'Burglary' crimes (London) stronger fluctuations are observed in the time series compared to the 'Assault' crimes (Boston), e.g. 'Antisocial behavior' has 3493.87 standard deviation, while 'Assault' has 268.34. As BOCPD is based on detecting changes in variance, this could explain why it performs better on the London dataset and predicts more similar change points to PELT.

6   Conclusions and Future Work

This work examined offline and online CPD methods to enable effective understanding and detection of trends and change points in time series. The focus was on crime-related time series, having first performed a fairly extensive analysis of data coming from other domains, but characterized by ground truth labels. The analysis conducted indicates that CPD methods can be a valuable tool for police authorities as they will be able to better understand the trend on topics of interest so that they can then proceed with better management of resources. Through time series analysis, patterns can be identified (such as seasonality), while at the same time the detected change points can be pivotal points for decision-making. In the future, we intend to conduct a similar study to additional crime-related datasets, while also deep neural network-based approaches will be examined to allow for an even better detection of changes in crime-related time series data.

Acknowledgment. This project has received funding from the European Union's H2020 research and innovation programme as part of the STARLIGHT (GA No 101021797) project.

References

1. Abrar, M.F., Arefin, M.S., Hossain, M.S.: A framework for analyzing real-time tweets to detect terrorist activities. In: ECCE, pp. 1–6. IEEE (2019)
2. Aminikhanghahi, S., Cook, D.J.: A survey of methods for time series change point detection. Knowl. Inf. Syst. **51**(2), 339–367 (2017)
3. Asongu, S.A., Orim, S.M.I., Nting, R.T.: Terrorism and social media: global evidence. J. Glob. Inf. Technol. **22**(3), 208–228 (2019)
4. Auger, I.E., Lawrence, C.E.: Algorithms for the optimal identification of segment neighborhoods. Bull. Math. Biol. **51**(1), 39–54 (1989)
5. Augusto, D.: Change-point analysis of Houston crime during hurricane Harvey. J. Appl. Secur. Research **16**(1), 1–18 (2021)
6. Borowik, G., Wawrzyniak, Z.M., Cichosz, P.: Time series analysis for crime forecasting. In: ICSEng, pp. 1–10. IEEE (2018)

7. Boston, A.: Crimes in boston (2023). https://kaggle.com/datasets/AnalyzeBoston/crimes-in-boston
8. Burden, K., Palmer, C.: Internet crime: cyber crime-a new breed of criminal? Comput. Law Secur. Rev. **19**(3), 222–227 (2003)
9. Van den Burg, G.J., Williams, C.K.: An evaluation of change point detection algorithms. arXiv preprint arXiv:2003.06222 (2020)
10. Burkatovskaya, Y., Kabanova, T., Khaustov, P.: Choice of the parameters of the CUSUM algorithms for parameter estimation in the Markov modulated poisson process. In: ITSMSSM, pp. 456–462. Atlantis Press (2016)
11. Chowdhury, M.F.R., Selouani, S.A., O'Shaughnessy, D.: Bayesian on-line spectral change point detection: a soft computing approach for on-line ASR. Int. J. Speech Technol. **15**, 5–23 (2012)
12. Cleland, I., et al.: Evaluation of prompted annotation of activity data recorded from a smart phone. Sensors **14**(9), 15861–15879 (2014)
13. Dane, S.: London police records (2023). https://kaggle.com/datasets/sohier/london-police-records
14. Derrick, T.R., Bates, B.T., Dufek, J.S.: Evaluation of time-series data sets using the Pearson product-moment correlation coefficient. MSSE **26**(7), 919–928 (1994)
15. Granjon, P.: The CUSUM algorithm-a small review (2013)
16. Grzegorczyk, M., Husmeier, D.: Non-homogeneous dynamic Bayesian networks for continuous data. Mach. Learn. **83**, 355–419 (2011)
17. Han, M., Vinh, L.T., Lee, Y.K., Lee, S.: Comprehensive context recognizer based on multimodal sensors in a smartphone. Sensors **12**(9), 12588–12605 (2012)
18. Han, X., Wang, L., Cui, C., Ma, J., Zhang, S.: Linking multiple online identities in criminal investigations: a spectral co-clustering framework. IEEE Trans. Inf. Forensics Secur. **12**(9), 2242–2255 (2017)
19. Haynes, K., Eckley, I.A., Fearnhead, P.: Computationally efficient changepoint detection for a range of penalties. J. Comput. Graph. Stat. **26**(1), 134–143 (2017)
20. Jeon, S.Y., Ryou, H.S., Kim, Y., Oh, K.J.: Using change-point detection to identify structural changes in stock market: application to Russell 2000. Quant. Bio-Sci. **39**(1), 61–69 (2020)
21. Keogh, E., Chu, S., Hart, D., Pazzani, M.: An online algorithm for segmenting time series. In: International Conference on Data Mining, pp. 289–296. IEEE (2001)
22. Killick, R., Fearnhead, P., Eckley, I.A.: Optimal detection of changepoints with a linear computational cost. JASA **107**(500), 1590–1598 (2012)
23. Knoblauch, J., Damoulas, T.: Spatio-temporal Bayesian on-line changepoint detection with model selection. In: International Conference on Machine Learning, pp. 2718–2727. PMLR (2018)
24. Lu, R., et al.: The cannabis effect on crime: time-series analysis of crime in Colorado and Washington state. Justice Q. **38**(4), 565–595 (2021)
25. Lung-Yut-Fong, A., Lévy-Leduc, C., Cappé, O.: Distributed detection/localization of change-points in high-dimensional network traffic data. Stat. Comput. **22**(2), 485–496 (2012)
26. Maidstone, R.: Efficient Analysis of Complex Changepoint Problems. Lancaster University (United Kingdom) (2016)
27. Malleson, N., Andresen, M.A.: Spatio-temporal crime hotspots and the ambient population. Crime Sci. **4**, 1–8 (2015)
28. Page, E.S.: Continuous inspection schemes. Biometrika **41**(1/2), 100–115 (1954)
29. Reddy, S., Mun, M., Burke, J., Estrin, D., Hansen, M., Srivastava, M.: Using mobile phones to determine transportation modes. TOSN **6**(2), 1–27 (2010)

30. Romano, G., Eckley, I.A., Fearnhead, P., Rigaill, G.: Fast online changepoint detection via functional pruning CUSUM statistics. J. Mach. Learn. Res. **24**, 1–36 (2023)
31. Scott, A.J., Knott, M.: A cluster analysis method for grouping means in the analysis of variance. Biometrics **30**, 507–512 (1974)
32. Shamsuddin, N.H.M., Ali, N.A., Alwee, R.: An overview on crime prediction methods. In: ICT-ISPC, pp. 1–5. IEEE (2017)
33. Takeuchi, J.I., Yamanishi, K.: A unifying framework for detecting outliers and change points from time series. TKDE **18**(4), 482–492 (2006)
34. Taylor, S.J., Letham, B.: Forecasting at scale. Am. Stat. **72**(1), 37–45 (2018)
35. Theodosiadou, O., et al.: Change point detection in terrorism-related online content using deep learning derived indicators. Information **12**(7), 274 (2021)
36. Wambui, G.D., Waititu, G.A., Wanjoya, A.: The power of the pruned exact linear time (pelt) test in multiple changepoint detection. AJTAS **4**(6), 581 (2015)
37. Wei, S., Xie, Y.: Online kernel CUSUM for change-point detection. arXiv preprint arXiv:2211.15070 (2022)

DAVI: A Dataset for Automatic Variant Interpretation

Francesca Longhin[1] , Alessandro Guazzo[1] , Enrico Longato[1] , Nicola Ferro[1] ,
and Barbara Di Camillo[1,2](✉)

[1] Department of Information Engineering, University of Padova, 35131 Padova, Italy
barbara.dicamillo@unipd.it
[2] Department of Comparative Biomedicine and Food Science, University of Padova,
35020 Legnaro, PD, Italy

Abstract. The analysis of an individual's genetic material may uncover genetic variants, which can be classified as disease-causing (pathogenic) or benign. Identifying pathogenic variants among millions of variants relies on the research of evidence in support of or against variant pathogenicity, a process regulated by the American College of Molecular Genetics (ACMG) guidelines, which leverages data from the scientific literature. Despite recent improvements towards automation, searching shreds of evidence for pathogenicity in the literature still requires manual curation, a time-consuming process, due to the ever-growing number of published papers.

In this work, we built DAVI (Dataset for Automatic Variant Interpretation), a reliable, manually curated dataset comprising articles both containing (positive) and not containing (negative) evidence activating two opposing ACGM criteria, namely PS3 and BS3, for a pool of 41 variants. Moreover, we demonstrated that DAVI can be used to train a predictive model that automatically identifies positive *(variant, article)* associations.

DAVI contains 311 *(variant, article)* pairs: 154 positive and 157 negative associations. We used three different text representation models combined with a logistic regression to efficiently identify positive associations, with an F1-score of 0.84. The model's performance constitutes a clear proof of concept for automatic PS3/BS3 evidence identification. DAVI represents a useful resource to train further models.

Keywords: Clinical Genetics · Variant Interpretation · Natural Language Processing

1 Introduction

Deoxyribonucleic acid, more commonly known as DNA, is a complex molecule that stores the genetic information needed for the development and functioning of an organism. The DNA molecule is contained in each of an organism's cells, which are the basic biological building blocks that provide structure to its tissues. The DNA

© The Author(s), under exclusive license to Springer Nature Switzerland AG 2023
A. Arampatzis et al. (Eds.): CLEF 2023, LNCS 14163, pp. 85–96, 2023.
https://doi.org/10.1007/978-3-031-42448-9_8

is composed of a series of four different smaller molecules, called nucleotides: adenine ("A"), thymine ("T"), guanine ("G"), and cytosine ("C"). In 2003, with the completion of the Human Genome Project [1], the first human genome, i.e. the four-letter sequence encoding a person's DNA, was determined through a laboratory technique called sequencing. Thereafter, sequencing technologies have become more and more sophisticated and widely accessible, enabling the resolution of thousands of genomes and the detection of small differences among them, known as genetic *variants*. Variants can be inherited from a parent or occur during a person's lifetime. Identification of genetic variants, which consists in assessing the variants' positions in the genome and affected nucleotides, is crucial as variants are not only responsible for differences in appearance among individuals of the same species, but also associated to their health status. For example, some variants are located within genes, which are chunks of nucleotides in the genome carrying instructions for the synthesis of proteins, complex molecules that play many critical roles (signalling, structural support, nutrients storage) in the organism. Alterations in gene sequences can result in the production of inactive proteins, increasing an individual's susceptibility to a certain disease (*pathogenic variants*), or they can have no impact on the function of the gene/protein (*benign variants*).

Recently, sequencing technologies have been increasingly used for personalised healthcare, as the identification of a person's genetic variants and the assessment of their benignity/pathogenicity allow clinicians to provide suitable therapies to patients [2]. However, a correct variant benignity/pathogenicity assessment, a process also known as variant interpretation, does not rely only on information about variant position, affected gene, and affected protein, but it requires the clinician to perform a complete *variant annotation*, gathering all relevant evidence about the nature and the effect of the variant from biological databases and the scientific literature [3]. To be meaningful, variant annotation should follow the recommended guidelines defined by the American College of Medical Genetics and Genomics (ACMG) in 2015 [4]. These guidelines contain 28 criteria, each identified by an *evidence code* representing evidence in support of variant benignity or pathogenicity. Some criteria are applied to a specific variant based on the evidence contained in databases (variant frequencies in healthy reference populations, prediction scores based on the probability of damaging protein structure, etc.), while others require information contained in the literature (results of experimental tests carried out on the variant, disease-association studies, etc.). An example of two opposing ACMG criteria which are applied based on information contained in the literature are PS3 and BS3. These criteria are alternatively assigned to a specific variant when an experimental test, in which the variant is injected in the DNA of an animal (in vivo) or cell culture (in vitro), proved that the variant has a damaging or null effect on protein function, respectively.

The task of mining for evidence of a variant's benignity/pathogenicity from the literature, a process known as *manual curation*, is extremely complex and time-consuming. It requires highly qualified curators who scan the continuously growing biomedical literature in the quest for the required evidence. This typically happens by querying a literature search engine with "Variant_Name AND Gene_Name", where Variant_Name is a variant's identification code and Gene_Name is the symbol of the gene where the variant is located. Then, curators have to proceed by reading all retrieved articles, looking for relevant information in figures, tables, sentences that contain the variant's identification code, and those nearby (e.g. typically only the previous and next sentences) [5]. When curators find a relevant article, they assign the specific ACMG criterion to the *(variant, article)* pair.

Considering that the number of biomedical publications that contain genetic variants grows day by day and that the research community uses multiple forms to refer to genetic variants (variant synonyms), it is increasingly difficult to have enough expert curators to read all available publications and to find all relevant information about each discovered variant [6]. As a result, currently, there is the lack of a complete and constantly updated database containing *(variant, article)* associations curated following the ACMG guidelines for each variant and for each criterion. The only resource of this kind is ClinGen [7]: it contains expert-curated assertions regarding variant pathogenicity, as well as supporting evidence summaries, and it is used for consultation in clinical decision-making. However, the number of variants annotated in ClinGen is very limited and the information used for variant interpretation is partial: indeed, most of the time, ClinGen curators make their statements on variant pathogenicity when they think they have collected enough evidence from a restricted number of analysed publications, possibly missing lots of useful information contained in the remaining overlooked papers.

Given the abovementioned considerations, there is a need for a tool able to automatically identify the evidence needed for an ACMG-compliant variant interpretation, which could be easily applied to any variant at any time. Indeed, several tools have been proposed to automatically collect variant annotations [8, 9], but none of them performs a comprehensive screening of the extensive and ever-growing literature, nor automatically extract the information needed for the activation of ACMG criteria.

The first aim of this work is to build a high quality manually curated dataset of articles that either activate, for a specific variant, one of two opposing ACMG criteria, namely PS3 and BS3, or that activate neither. This dataset, named DAVI (Dataset for Automatic Variant Interpretation), will be available on Zenodo. Besides being a useful resource by itself, will be the basis for developing automatic methods for variant annotation. To the best of our knowledge, this is the first type of such dataset available for research. The second aim is to perform a preliminary exploratory analysis of DAVI via the development of an automatic, machine-learning-based predictive model to identify *(variant, article)* pairs where either PS3 or BS3 are activated. This can be thought as a first step before a second classification step to distinguish between articles that activate PS3 vs. those which activate BS3.

The paper is organized as follows: Sect. 2 describes the methodology used to create DAVI; Sect. 3 describes the implementation of a predictive model, trained on DAVI, that automatically performs identification of *(variant, article)* associations where either PS3 or BS3 are activated; finally, Sect. 4 draws some conclusions and outlooks for future work.

2 Dataset Construction

In the following, we call *positive articles* those articles which activate either the PS3 or the BS3 criterion, while we call *negative articles* those activating neither of them. Typically, in positive articles, the result of the experimental test is summarised in one or more sentences (*positive sentences*), which trigger the activation of the PS3 or BS3 criteria. Positive sentences contain the functional comparison between two analysed models, in vivo or in vitro, one carrying the variant (mutant) and the other carrying the non-mutated sequence of DNA (wild-type). All other sentences activate neither PS3 nor BS3 (*negative sentences*).

2.1 Article Retrieval

To build DAVI, we started from downloading the ClinGen Evidence Repository, whose rows contain information about 4980 curated genetic variants, distributed across 88 genes of interest. Each curated variant is reported with its Human Genome Variation Society (HGVS) [10] standard nomenclature. According to HGVS, variants (e.g., NM_000277.2(PAH): c.472C > T (p.Arg158Trp)) are unambiguously described at the DNA level through an accepted reference DNA sequence (e.g., NM_000277.2), also called transcript, which is located in a gene (e.g., PAH); the position of the variant (e.g., 472), calculated with respect to that specific transcript; and replaced and replacing nucleotides (e.g., 472C > T means cytosine becomes thymine in position 472). In addition, variants can be described at the protein level, specifying the position of the variant in the amino acids sequence (e.g., 158), replaced and replacing amino acids (e.g., Arg158Trp means arginine becomes tryptophan in position 158).

Each variant is associated to a list of ACMG evidence codes assigned by ClinGen manual curators on the basis of information contained in databases (e.g., ACMG criterion applied: PM2, source of evidence: ExAC [11]) or one or more scientific papers, identified by PubMed [12] identification codes (PMIDs) (e.g., ACMG criterion applied: PS3, source of evidence: PMID:24401910).

We focused only on variant curations where either PS3 or BS3 evidence codes were assigned, given that the evidence needed for these assignments is often contained in articles' plain texts (most of times, the manual curator does not need to study tables and figures, but only textual information). Therefore, we filtered the ClinGen Evidence Repository for variant curations where either PS3 or BS3 evidence codes were assigned (*ClinGen variants*) and we extracted their corresponding articles' PMIDs (*ClinGen articles*). As we wanted to analyse the articles' full-text, we converted the PMIDs, which only refer to articles' abstracts, to PubMed Central [13] identification codes (PMCIDs). *ClinGen articles* with no PMCID were ignored. In this way, we obtained a set of variants for which ClinGen experts' manual curation produced at least one positive (*variant, ClinGen article*) association. Then, we applied our own manual curation to *ClinGen variants* in order to assess ClinGen's completeness in reporting positive (*variant, article*) associations; and to find negative (*variant, article*) associations for training a classifier to perform automatic positive evidence identification. We chose EuropePMC[1] as our reference literature search engine, for it has a very convenient R interface, provided by the package europepmc [14]. Specifically, the user can define a query through the function `epmc_search` and obtain PMCIDs of retrieved articles. As we want to mimic the same procedure followed by manual curators, our queries were structured as "Variant_Name AND Gene_Name AND Keywords".

---

[1] https://europepmc.org/.

`Variant_Name` is the variant identifier. Given the variety of formats commonly used in publications to refer to genetic variants [6], we used, for each *ClinGen variant*, five different queries in which `Variant_Name` was respectively represented by:

  i) the nucleotide change in HGVS format (e.g., 1A > G);
 ii) the nucleotide change in a non-HGVS format (e.g., A1G);
iii) the amino acid change in an HGVS format (e.g., Met1Val);
 iv) the amino acid change in a non-HGVS format (e.g., M1V);
  v) the RefSeq [15] Identification (rsID) code (e.g., rs786204467).

While i), ii), iii) and iv) can be derived from the *ClinGen variant*'s HGVS nomenclature reported in the ClinGen Evidence Repository, v) was obtained using the VEP [8] REST API. `Gene_name` is the gene identifier. It is needed together with the `Variant_Name` to ensure we are referring to the correct variant: two distinct articles might contain information about variants with the same variant identifier, but found on two different genes. `Keywords` is a set of 113 words extracted from two resources: Mastermind [16], which is a commercial search engine that allows paid users to rank retrieved articles according to ACMG relevance through criteria-specific keywords, and *ClinGen articles*. In particular, Mastermind contained 68 keywords for PS3/BS3, while the other 45 keywords were words recurrently found in *ClinGen articles*, which are known to be positive for PS3/BS3.

Furthermore, we refined the query syntax by adding the following flags:

• BODY: query terms were searched within the body of full-text articles. Sections such as "References" and "Acknowledgements" were not considered.
• OPEN_ACCESS: search results were limited to articles that are Open Access in EuropePMC. This was needed to access their full text.
• PUB_TYPE: filter by publication type. Only journal articles were considered.

These customised queries produced five lists of retrieved PMCIDs for each *ClinGen variant*, a list for each `Variant_Name` synonym. As some *ClinGen articles* are not Open Access in EuropePMC, some *ClinGen variant* queries did not retrieve any *ClinGen article* and thus they were discarded in the current analysis.

2.2 Manual Curation

Manual curation, i.e. manual variant annotation, was needed to distinguish between positive *(variant, article)* associations (assigned to the label 1), i.e., articles that contain at least one positive sentence activating either PS3 or BS3 evidence codes for a certain variant, and negative *(variant, article)* associations (assigned to the label 0), which do not contain any positive sentence.

For each article, we selected for manual curation only *target sentences*, i.e., sentences containing `Variant_Name` as used in all the five queries related to the same *ClinGen variant*, concatenated to the ones immediately adjacent (the previous and next sentences). In this way, we considered only textual information, easily interpretable for an automatic algorithm (tables and figures are excluded, as they would require additional, specialised modules). Using the R package tidypmc [17], we downloaded the articles' XML code given their PMCIDs and then we performed *target sentences* extraction. For each *ClinGen variant*, we read all *target sentences* extracted from its *ClinGen articles* and, given the burden of human curation in terms of time, from a random subset of articles not included in *ClinGen articles* but retrieved by our queries. The number of articles to be curated R for each *ClinGen variant* was chosen considering the total number T of articles retrieved with all of its five queries as follows.

- If $T \leq 30$, then $R = T$.
- If $30 < T < 50$, then $R = 30$.
- If $T \geq 50$, then $R = 50$.

In this way, we included in DAVI a number of curated articles for each *ClinGen variant* that was representative of its presence in the EuropePMC database. We applied the same reasoning to choose the number r_i of articles to be curated for each of the five queries related to the same *ClinGen variant*. Considering the number t_i of articles retrieved with each of the five queries related to the same *ClinGen variant*, r_i was calculated as follows for $i = 1, ..., 5$:

$$r_i = \frac{t_i}{T} \times R \tag{1}$$

We performed manual curation considering the following rules and ensuring consistencies with the Genomic Variant Analysis & Clinical Interpretation [6] procedure. We assigned to each *(variant, article)* association the negative label (0) if none of *target sentences*, extracted from the considered article, contained sufficient information for assigning PS3 or BS3 (i.e., all *target sentences* were negative), regardless of the content of tables or figures (which might have contained information for assigning PS3 or BS3, but whose automated analysis was out-of-scope for this work). Instead, the positive label (1) was assigned to *(variant, article)* associations for which at least one *target sentence* contained information for assigning PS3 or BS3 (i.e., at least one *target sentence* was positive).

3 Automatic Variant Annotation

3.1 Pre-processing

We trained the automatic classification model on DAVI according to a by-sentence perspective, where we considered each *target sentence* as an independent entry. For performance evaluation only, we considered a by-article perspective, distinguishing between positive and negative *(variant, article)* associations according to the classification of each of their extracted *target sentences* (association is positive if the article contains at least one positive *target sentence*). We pre-processed the *target sentences* included in DAVI according to the following typical steps [18].

- English stop word removal, using the stop list provided by the package nltk [19]. We excluded the word "not", which is a relevant word in the context of PS3 or BS3 assignment, and we handled negation by concatenating it to the following word.
- Stemming using the snowball stemming algorithm implemented by the package nltk.

- Removal of words with an absolute frequency less than the 90th percentile of the absolute frequency distribution of words in the vocabulary.
- Exclusion of sentences consisting of less than 3 words

We split the pre-processed dataset into a training set, a test set and a validation set (70%, 15%, 15%), making sure that proportions of positive and negative *(variant, article)* associations and *target sentences* were similar (within a tolerance of $\varepsilon = 0.01$) in the 3 subsets. Finally, given that we were considering the classification of *(variant, article)* pairs, but we needed to construct a single dataset comprising all *target sentences*, we had to deal with the presence of duplicated *target sentences*. As duplicated *target sentences* could cause over-fitting (identical *target sentences* with concordant labels) or bias (identical *target sentences* with discordant labels), we removed one copy, if concordant, or both, if discordant, of such sentences from the training and validation sets. This reasoning was not applied to the test-set, as it was used for performance evaluation only: predicted labels were correctly computed considering *target sentences* extracted from articles in *(variant, article)* pairs.

3.2 Model Construction

We applied three different text representation schemes, implemented through the python package scikit-learn [20], to transform *target sentences* in the pre-processed DAVI into sequences of numbers.

- Binary Bag Of Words (BBOW) [21], in which each word was represented by 1, if the word is present in the *target sentence*, and 0 otherwise.
- Bag Of Words (BOW), in which each word was represented by its frequency in the *target sentence*.
- Term-frequency Inverse Document-Frequency (TF-IDF), in which each word was represented by its frequency in the *target sentence* weighted by how often it appeared in all *target sentences*.

We performed a preliminary exploratory analysis on automatic PS3/BS3 evidence identification using a logistic regressor (LR) trained on the three versions of DAVI. For this model, we considered a L2 regularisation loss-function with a single hyperparameter, the inverse of the regularisation strength C. For each version of the dataset (BBOW, BOW, TF-IDF), we performed hyperparameter optimisation considering only the training set, using a 5-fold cross validation [22] and a random search approach [23] accounting for 10000 values of C, randomly sampled from a log uniform distribution ranging from 10^{-4} to 10^2. We selected the best hyperparameter as the one that led to the minimum average binary cross-entropy across the 5 folds.

To transform the model from a ranker into a classifier, useable in practice for automatic PS3/BS3 evidence identification, we implemented a thresholding approach by identifying one probability threshold (th) to discriminate between positive (1, if predicted probability $p \geq$ th) and negative (0, if $p <$ th) predictions on *target sentences*. We selected the optimal threshold by using each probability value predicted for *target sentences* in the validation set as a threshold and choosing the one associated to the maximum geometric mean between true positive and true negative rate in the validation set itself.

3.3 Performance Measures

In the by-sentence perspective, we evaluated the discrimination performance of the model via five measures: area under the receiver operating characteristic (AUROC) and area under the precision-recall curve (AUPRC) [24] for the continuous probability output; as well as precision, recall, and F1-score after applying the aforementioned thresholding approach.

In the by-article perspective, we did not consider AUROC and AUPRC as predicted labels were assigned as the logical OR of by-sentence outputs after thresholding and, hence, were Boolean in nature.

4 Results

4.1 Manual Curation Results

DAVI is organised into 6 columns, containing, for each *(variant, article)* pair, the variant's HGVS standard nomenclature, variant's HGVS nomenclature used in query, the article's PMCID, the label assigned to the article, a *target sentence* extracted from the article and, the label assigned to that *target sentence*. Table 1 shows an example of a DAVI entry.

Table 1. Example of an entry in DAVI

HGVS standard nomenclature	HGVS nomenclature used in query	Article PMCID	Article Label	*Target Sentence*	*Target Sentence* Label
NM_021133.4 (RNASEL): c.793G > T (p. Glu265Ter)	G793T	PMC2361943	0	All sequence variations [...]. [...], we discovered one protein-truncating variant, nt g793t, [...]. This point mutation [...]	0

Overall, DAVI contains the results of manual curation for 41 *ClinGen variants*, yielding 1239 *target sentences* extracted from 311 *(variant, article)* pairs, namely 44 *(variant, ClinGen article)* pairs and 267 *(variant, non-ClinGen article)* pairs. Table 2 provides a comparison of the labels assigned to *target sentences* and *(variant, article)* pairs in *ClinGen articles* and non-*ClinGen articles*.

Table 2. Comparison of the assigned labels in *ClinGen articles* and non-*ClinGen articles*

		Total	Positive	Negative
ClinGen articles	*Target sentences*	388	219	169
	(Variant, article) pairs	44	37	7
Non-*ClinGen articles*	*Target sentences*	851	378	473
	(Variant, article) pairs	267	117	150

DAVI contained almost the same amount of positive and negative *target sentences*, i.e., respectively 597 and 642, and positive and negative *(variant, article)* pairs, i.e., respectively 154 and 157. The number of *target sentences* extracted per *ClinGen article* was three times greater than the number of *target sentences* extracted per non-*ClinGen article*. Moreover, even though we assumed *(variant, ClinGen article)* associations to be positive, 7 *(variant, ClinGen article)* pairs were re-classified as negative after manual curation (see Sect. 2.2).

4.2 Pre-processing Results

Initially, the vocabulary of *target sentences* contained 7745 words. Following the approach described in Sect. 3.1, we removed 733 stop words and 6309 words as they had an absolute frequency below the 90th percentile of the absolute frequency distribution of words in the vocabulary. We removed 2 negative *target sentences*, as they comprised less than 3 words. Lastly, we removed 66 and 22 duplicated concordant *target sentences* from training-set and validation-set, respectively, whereas no duplicated discordant *target sentences* were found. Thus, the pre-processed DAVI finally contained 1149 *target sentences*. Table 3 provides a comparison of the assigned labels of *target sentences* and *(variant, article)* pairs in the training, test, and validation sets after pre-processing.

Table 3. Comparison of assigned labels in the pre-processed training, test, and validation sets

		Total	Positive	Negative
Training-set	*Target sentences*	644	320	324
	(Variant, article) pairs	196	99	97
Test-set	*Target sentences*	302	144	158
	(Variant, article) pairs	52	25	27
Validation-set	*Target sentences*	203	99	104
	(Variant, article) pairs	48	24	24

4.3 Classification Results

This section reports the performance of the models constructed following the approach described in Sect. 3.2 and using the measures introduced in Sect. 3.3. Results of hyperparameter C optimization on the three versions of the training set (BBOW, BOW, TF-IDF), minimizing the score (binary cross entropy) across the 5-folds are reported in Table 4.

Table 4. Hyperparameter C optimization on the BBOW, BOW and TF-IDF versions of DAVI

Text representation scheme	Hyperparameter (C)	Best Score
BBOW	0.103	−0.541
BOW	0.055	−0.549
TF-IDF	2.236	−0.557

The combination of TF-IDF text representation model and hyperparameter $C = 2.236$ led to the lowest value of binary cross entropy. The performance metrics obtained in the by-sentence and by-article perspectives, using BBOW+LR, BOW+LR, and TF-IDF+LR, are shown in Table 5.

Table 5. Classification results according to by-sentence and by-article perspectives, using BBOW+LR, BOW+LR and TF-IDF+LR

Perspective	Model	AUROC	AUPRC	TP	TN	FP	FN	Precision	Recall	F1-score
By-sentence	BBOW+LR	0.805	0.767	132	81	77	12	0.631	0.917	0.748
	BOW+LR	0.815	0.771	124	108	50	20	0.713	0.861	0.780
	TF-IDF+LR	0.819	0.796	121	99	59	23	0.672	0.840	0.747
By-article	BBOW+LR	-	-	24	15	12	1	0.667	0.960	0.787
	BOW+LR	-	-	23	19	8	2	0.742	0.920	0.821
	TF-IDF+LR	-	-	24	19	8	1	0.750	0.960	0.842

In the by-sentence perspective, the TF-IDF+LR model performed better, yielding an AUROC and a AUPRC of 0.819 and 0.796, respectively. However, the BBOW+LR model showed a higher recall and, overall, the BOW+LR model had a higher F1-score. In the by-article perspective, the best performing model was TF-IDF+LR. As the number of false negatives was lower than the one of false positives, recall was higher than precision. This result suggests that correctly identifying positive sentences and articles was slightly more challenging than correctly identifying negative cases. While not directly comparable, performance was overall better in the by-article setting than in the by-sentence one, which was expected as it is easier to obtain a correct classification looking at multiple *target sentences* for each *(variant, article)* pair rather than classifying each *target sentence* independently.

5 Discussion and Future Work

The main aim of this work was to build a high quality and manually-curated dataset that associates each variant to its PS3/BS3-activating articles (positive associations) and non PS3/BS3-activating articles (negative associations), as such resource is critical for clinical decision-making and it is currently missing. The second aim was to use such dataset to train a predictive model that efficiently performs automatic positive associations identification.

We built DAVI, a manually-curated dataset comprising 1239 sentences related to 311 *(variant, article)* associations. In order to guarantee a sufficient number of positive associations, we included in DAVI 44 *(variant, ClinGen article)* pairs and, to consider a more representative sample of the entire corpus of articles retrieved when querying for a specific variant, 267 *(variant, non-ClinGen article)* pairs. As expected, most *(variant, ClinGen articles)* pairs were positive, but 7 were reclassified as negative, on the basis of textual information only. Overall, about half of the extracted sentences contained sufficient evidence for activating PS3 or BS3 evidence codes, and same for the *(variant, article)* pairs. A positivity offset is given by the fact that we forcedly included in DAVI an elevated number of positive sentences extracted from few *ClinGen articles*, but, generally, positive sentences and *(variant, article)* associations are respectively fewer than negative ones (378 vs. 473 sentences extracted from 117 vs. 150 *(variant, non-ClinGen articles)*). However, we found a significant number of positive examples in *(variant, non-ClinGen articles)* pairs (117 out of 267), suggesting that the manually curated information contained in ClinGen is incomplete and/or not updated frequently enough. As ClinGen has been recognised by the Food and Drug Administration as a source of valid scientific evidence for support in clinical decisions, it should be always up-to-date, containing all new evidence about all discovered variants.

ACMG criteria and, specifically, PS3 and BS3, can activate for a *(variant, article)* in relation to multiple specific diseases or through the use of different types of experimental texts, sometimes even at the same time. Therefore, it is crucial to provide to the clinician all available positive evidence, even if this implies higher costs for manual curation. In order to reduce these costs, automatic models could be integrated in the curation pipeline. As an exploratory analysis on the feasibility of this approach, we tested the discrimination performances of three predictive models, trained on DAVI, for the automatic identification of positive *(variant, article)* associations. Performance was good both in the by-sentence and by-article perspective, with F1-scores well above 0.70 and 0.80 respectively. This result suggests that reliable tools could be developed in support of manual curation, efficiently enriching biological databases with all the information needed for a complete and correct variant interpretation.

Future developments include the further distinction of (PS3 or BS3)-positive examples into PS3-positive vs. BS3-positive examples. Moreover, the solid manual curation procedure described in this work may be applied to variants which are not included in ClinGen and expanded to the evaluation of other evidence codes among the 28 covered by ACMG guidelines. Lastly, we may focus on the development of more complex architectures for text representation and classification, including deep learning approaches.

References

1. Collins, F.S., Fink, L.: The human genome project. Alcohol Health Res. World **19**(3), 190–195 (1995)
2. Morash, M., Mitchell, H., Beltran, H., Elemento, O., Pathak, J.: The role of next-generation sequencing in precision medicine: a review of outcomes in oncology. J. Personalized Med. **8**(3), 30 (2018)
3. Amendola, L.M., et al.: Performance of ACMG-AMP variant-interpretation guidelines among nine laboratories in the clinical sequencing exploratory research consortium. Am. J. Hum. Genet. **98**(6), 1067–1076 (2016)
4. Richards, S., et al.: Standards and guidelines for the interpretation of sequence variants: a joint consensus recommendation of the American college of medical genetics and genomics and the association for molecular pathology. Genet. Med. **17**(5), 405–424 (2015)
5. GVACI Course 2022. https://gvaci.genomes.in/home. Accessed 29 Dec 2022
6. Lee, K., Wei, C.-H., Lu, Z.: Recent advances of automated methods for searching and extracting genomic variant information from biomedical literature. Brief Bioinform **22**(3), bbaa142 (2020)
7. Welcome to ClinGen. https://www.clinicalgenome.org/. Accessed 29 Dec 2022
8. McLaren, W., et al.: The ensembl variant effect predictor. Genome Biol **17**, 1–14 (2016)
9. Wang, K., Li, M., Hakonarson, H.: ANNOVAR: functional annotation of genetic variants from high-throughput sequencing data. Nucleic Acids Res **38**(16), e164 (2010)
10. Den Dunnen, J.T., et al.: HGVS Recommendations for the description of sequence variants: 2016 update. Hum. Mutat. **37**(6), 564–569 (2016)
11. Karczewski, K.J., et al.: The ExAC browser: displaying reference data information from over 60 000 exomes. Nucleic Acids Res. **45**, D840-D 845 (2017)
12. PubMed. https://pubmed.ncbi.nlm.nih.gov/. Accessed 03 Jan 2023
13. Home - PMC – NCBI. https://www.ncbi.nlm.nih.gov/pmc/. Accessed 03 Jan 2023
14. Levchenko, M., et al.: Europe PMC in 2017. Nucleic Acids Res. **46**, D1254–D1260 (2017)
15. RefSeq: NCBI Reference Sequence Database. https://www.ncbi.nlm.nih.gov/refseq/. Accessed 03 May 2023
16. Chunn, L.M., et al.: Mastermind: a comprehensive genomic association search engine for empirical evidence curation and genetic variant interpretation. Front Genet **11**, 577152 (2020)
17. Stubben, C.: tidypmc: Parse Full Text XML Documents from PubMed Central. (2019)
18. Kathuria, A., Gupta, A., Singla, R.K.: A review of tools and techniques for preprocessing of textual data. In: Singh, V., Asari, V.K., Kumar, S., Patel, R.B. (eds.) Computational Methods and Data Engineering. AISC, vol. 1227, pp. 407–422. Springer, Singapore (2021). https://doi.org/10.1007/978-981-15-6876-3_31
19. Bird, S., Klein, E., Loper, E.: Natural language processing with Python: analyzing text with the natural language toolkit. O'Reilly Media, Inc., (2009)
20. Pedregosa, F., et al.: Scikit-learn: machine learning in python. J. Mach. Learn. Res. **12**, 2825–2830 (2011)
21. Qader, W. A., Ameen, M. M., Ahmed, B. I.: An overview of bag of words; importance, implementation, applications, and challenges. In: International Engineering Conference (IEC) 2019, pp. 200–204, (2019)
22. Berrar, D.: Cross-Validation (2018)
23. Bergstra, J., Bengio, Y.: Random search for hyper-parameter optimization. J. Mach. Learn. Res **13**, 281–305 (2012)
24. Keilwagen, I.G., Grau, J.: Area under precision-recall curves for weighted and unweighted data. PLoS ONE **9**(3), e92209 (2014)

qCLEF: A Proposal to Evaluate Quantum Annealing for Information Retrieval and Recommender Systems

Andrea Pasin[1]([✉]), Maurizio Ferrari Dacrema[2,3], Paolo Cremonesi[2], and Nicola Ferro[1]

[1] University of Padua, Padua, Italy
pasinandre@dei.unipd.it, nicola.ferro@unipd.it
[2] Politecnico di Milano, Milan, Italy
{maurizio.ferrari,paolo.cremonesi}@polimi.it
[3] ICSC, Milan, Italy

Abstract. Quantum Computing (QC) has been a focus of research for many researchers over the last few years. As a result of technological development, QC resources are also becoming available and usable to solve practical problems in the Information Retrieval (IR) and Recommender Systems (RS) fields. Nowadays IR and RS need to perform complex operations on very large datasets. In this scenario, it could be possible to increase the performance of these systems both in terms of efficiency and effectiveness by employing QC and, especially, Quantum Annealing (QA). The goal of this work is to design a Lab composed of different Shared Tasks that aims to:

- compare the performance of QA approaches with respect to their counterparts using traditional hardware;
- identify new ways of formulating problems so that they can be solved with quantum annealers;
- allow researchers from to different fields (e.g., Information Retrieval, Operations Research...) to work together and learn more about QA technologies.

This Lab uses the QC resources provided by CINECA, one of the most important computing centers worldwide, thanks to an already met agreement. In addition, we also show a possible implementation of the required infrastructure which uses Docker containers and the Kubernetes orchestrator to ensure scalability, fault tolerance and that can be deployed on the cloud.

1 Introduction

Information Retrieval (IR) and Recommender Systems (RS) play a fundamental role in providing access to and retrieving relevant resources to address our information needs. To this end, they face ever increasing amounts of data and rely

© The Author(s), under exclusive license to Springer Nature Switzerland AG 2023
A. Arampatzis et al. (Eds.): CLEF 2023, LNCS 14163, pp. 97–108, 2023.
https://doi.org/10.1007/978-3-031-42448-9_9

on more and more computational demanding approaches. For example, search engines have to deal with the estimated 50 billions indexed pages[1] of the Web.

In this challenging scenario, Quantum Computing (QC) can be employed to improve the performance of IR and RS methods, thanks to the development and implementation of more and more powerful QC devices, now able to tackle realistic problems. Although QC has been applied to many mathematical problems with applications in several domains, limited work has been done specifically for the IR and RS fields [6,9,11]. In particular, we focus on Quantum Annealing (QA), which exploits a special-purpose device able to rapidly find an optimal solution to optimization problems by leveraging quantum-mechanical effects. Therefore, the goal of this work is to better understand if QA can be used to improve the efficiency and effectiveness of IR and RS systems. In particular, the contribution of this work is the design of an evaluation lab, called Quantum CLEF (qCLEF), aimed at:

- evaluating the efficiency and effectiveness of QA with respect to traditional approaches;
- identifying new ways for formulating IR and RS algorithms and methods, so that they can be solved with QA;
- growing a research community around this new field in order to promote a wider adoption of QC technologies for IR and RS.

qCLEF will consist of different tasks, each one specifically focused on a computationally-intensive problem related to IR and RS that is solvable with quantum annealers, namely:

1. **Feature selection**: identify the subset of the most relevant features that can be used to optimize a learning model.
2. **Clustering**: group items based only on their characteristics and similarities.
3. **Boosting**: find the optimal subset of *weak* predictors that can be combined together to form a *strong* predictor which performs better according to the considered dataset.

To run QA algorithms developed by qCLEF participants, we will use the QA resources provided by CINECA[2], one of the most important computing centers worldwide, located in Italy. Since participants cannot have direct access to quantum annealers, we will design and develop a dedicated infrastructure to sandbox participants' systems and execute them on the CINECA resources.

The paper is organized as follows: Sect. 2 discusses related works; Sect. 3 presents the tasks which will constitute the qCLEF lab while Sect. 4 introduces the design and implementation of the infrastructure for the lab; Sect. 5 shows a practical example of how to solve the feature selection problem, using the developed infrastructure; finally, Sect. 6 draws some conclusions and outlooks some future work.

---

[1] https://www.worldwidewebsize.com/.

[2] https://www.cineca.it/en.

2   Related Works

What Is Quantum Annealing. QA is a QC paradigm that is based on special-purpose devices (quantum annealers) able to tackle optimization problems with a certain structure, such as the famous Travelling Salesman Problem (TSP). The basic idea of a quantum annealer is to represent a problem as the energy of a physical system and then leverage quantum-mechanical phenomena, i.e. superposition and entanglement, to let the system find a state of minimal energy, which corresponds to the solution of the original problem. This can be seen in Fig. 1, where we consider an example of the energy landscape obtained by two entangled qubits after the annealing process.

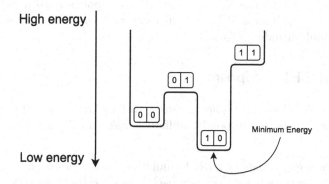

Fig. 1. Example of the Quantum Annealing process with two qubits. The optimal solution is **10**, representing the lowest point in the energy landscape.

In order to use a quantum annealer, you need to formulate the optimization problem as a minimization one using the Quadratic Unconstrained Binary Optimization (QUBO) formulation, which is defined as follows:

$$\min \quad y = x^T Q x$$

where x is a vector of binary decision variables and Q is a matrix of constant values representing the problem we wish to solve. Note that the QUBO formulation is very general and can be used to represent many interesting problems [7]. Once the problem has been formulated as QUBO, a further step called *minor embedding* is required to map the general mathematical formulation into the physical quantum annealer hardware, accounting for the limited number of qubits and the physical connections between them. This step can be done automatically, relying on some heuristic methods. Generally, a QUBO problem can be solved by a quantum annealer in a few *milliseconds*.

Applications of Quantum Annealing. QA can have practical applications in several fields thanks to its ability to tackle integer optimization problems which are *NP-Hard*. These problems can be found in different areas such as IR, RS, banking, finance, chemistry, drug development, and many others.

Quantum annealers have been previously applied to tackle IR and RS tasks such as feature selection [9], showing the feasibility of the task and promising improved efficiency and effectiveness. Indeed, as the technology matures, these devices have the potential to offer significant speedups for *NP-Complete* and *NP-Hard* problems that are difficult to tackle on traditional hardware.

QA has also been applied for Machine Learning (ML) tasks. For example, Willsch et al. [16] proposes a formulation of kernel-based Support Vector Machine (SVM) on a D-Wave 2000Q Quantum Annealer, while Delilbasic et al. [4] proposes a quantum multiclass SVM formulation aiming to reduce the execution time as the training set size increases. Other works explore the application of QA to clustering; for example, Zaiou et al. [18] applies it to a balanced K-means method which showed better efficiency and effectiveness, according to the Davies-Bouldin Index (DBI).

3   The qCLEF Proposal

In this section, we describe 3 different problems that can be solved with a quantum annealer and that correspond to different tasks in qCLEF. Each task has 2 main goals:

- find one or more possible QUBO formulations of the problem;
- evaluate the quantum annealer approach compared to a corresponding traditional approach to assess both its efficiency and its effectiveness.

In general, we expect that quantum annealers can solve problems in a shorter amount of time compared to traditional approaches obtaining results that are similar, or even better, in terms of effectiveness.

The evaluation of efficiency and effectiveness is further discussed in Sect. 3.5. Moreover, effectiveness will be measured according to different evaluation measures specific to each task.

3.1   Task 1 - Quantum Feature Selection

This task focuses on formulating the well-known *NP-Hard* feature selection problem in such a way that it can be solved with a quantum annealer, similarly to what has already been done in previous works [6, 9].

Feature selection is a widespread problem for both IR and RS which requires to identify a subset of the available features with certain characteristics (e.g., the most informative, less noisy etc.) to train a learning model. This problem is very impacting, since many of IR and RS systems involve the optimization of learning models, and reducing the dimensionality of the input data can improve their performance.

If the input data has n features, we can enumerate all the possible sets of input data having a fixed number k of features, thus obtaining $\binom{n}{k}$ possible subsets. Therefore to obtain the best subset of k features we should train our learning model on all the possible $\binom{n}{k}$ subsets of features, which is infeasible even for small datasets. There are nowadays heuristics to find good solutions in a short amount of time, but they do not guarantee to find the optimal one.

Therefore, in this task, we aim to understand if QA can be applied to solve this problem more efficiently and effectively. Feature selection fits very well the QUBO formulation, in which there is one variable x per feature and its value indicates whether it should be selected or not. The challenge lies in designing the objective function, i.e., matrix Q.

We have identified some possible datasets such as MQ2007 or MQ2008 [13] and The Movies Dataset[3] which have already been used in previous works [6, 9], LETOR4.0 and MSLR-WEB30K [14]. These datasets contain pre-computed features and the objective is to select a subset of these features to train a learning model, such as LambdaMART [3] or a content-based RS, in order to achieve best performance according to metrics such as nDCG@10.

3.2   Task 2 - Quantum Clustering

This task focuses on the formulation of the clustering problem in such a way that it can be solved with a quantum annealer. Clustering is a relevant problem for IR and RS and it involves grouping the items together according to their characteristics. In this way, "similar" items fall in the same group while different items will belong to different groups. Clustering can be helpful for organizing large collections, helping users to explore a collection and providing similar search results to a given query. Furthermore, it can be helpful to divide users according to their interests or build user models with the cluster centroids [17] speeding up the runtime of the system or its effectiveness for users with limited data.

There are different clustering problem formulations, such as centroid-based Clustering or Hierarchical Clustering. In this task, we focus on centroid-based clustering, since each document can be seen as a vector in the space and it is possible to cluster points based on their distances, which can be interpreted as a dissimilarity function: the more distant two vectors are, the more different the corresponding documents are likely to be. A similar reasoning can be applied in the case of features corresponding to users.

In this context, k-means clustering has a formal definition as an optimization problem but is known to be an *NP-Hard* problem. The Lloyd's algorithm is usually employed to return an approximation of the optimal solution. However, Lloyd's algorithm does not guarantee to return the optimal clustering solution, even though it provides some computational guarantees [8]. In addition, the number of iterations needed to compute the final clusters can still be exponential.

Clustering fits very well with a QUBO formulation and various methods have already been proposed [1,2,15]. Most of these methods use variables x to indicate in which cluster should the data point be put, hence the number of points in the space is the main limitation. There are ways to overcome this issue, such

---

[3] https://www.kaggle.com/datasets/rounakbanik/the-movies-dataset.

as by applying a weighted-centroid approach, which results in an approximate solution but allows to use quantum annealers for large datasets.

For this task, we have identified as a possible dataset the MSMARCO dataset [10]. In addition, since the high number of documents in MSMARCO could be an issue, we have identified a smaller dataset such as 20 Newsgroups[4]. From the considered dataset we will produce embeddings using powerful models such as BERT [5]. The cluster quality will be measured with user queries that undergo the same embedding process. These queries will match only the most representative embeddings of the found clusters, avoiding having to compute the similarity between the whole collection. For the recommendation task, we will generate user and item embeddings using state-of-the-art collaborative recommendation algorithms such as graph neural networks, on datasets Yelp and Amazon-Books. The cluster quality will be measured based on whether the centroids can be used to improve the efficiency and effectiveness of the user modeling similarly to what done in [17]. In this case the cluster quality will be measured according to the Silhouette coefficient and P@10.

3.3   Task 3 - Quantum Boosting

This task focuses on the formulation of the boosting problem for a quantum annealer. This is the most challenging task in our proposal.

Boosting is another problem that finds wide application in IR and RS. It involves identifying the best subset of *weak* predictors that can be combined together to form a *strong* predictor which performs better. A possible application of boosting is LambdaMART [3], which is a combination of LambdaRank and Multiple Additive Regression Trees (MART). It uses gradient boosted decision trees with a cost function derived from LambdaRank to order documents.

Similarly to feature selection and clustering, also boosting is a combinatorial problem that cannot be solved easily. In fact, it would require to try $\binom{n}{k}$ possible subsets of weak classifiers to find the optimal one, where n is the total number of classifiers and k is the desired number of classifiers to employ. QA can provide a boost in terms of both efficiency and effectiveness, allowing to retrieve the optimal solution in few microseconds if the size of the problem is small enough to fit the quantum annealer.

Also in this case, we consider as a viable dataset the LETOR4.0 dataset [14]. The aim here is to build a strong predictor which performs the best according to the dataset itself and evaluation measures such as nDCG@10.

3.4   Additional Challenges

When using quantum annealers to solve optimization problems, identifying an appropriate QUBO formulation is only part of the challenge. State-of-the-art quantum annealers nowadays have thousands of qubits (e.g., the D-Wave Advantage has ~ 5000 qubits) and more powerful devices are planned to become available in the near future.

---

[4] http://qwone.com/~jason/20Newsgroups/.

One crucial limitation of currently available quantum annealers is that each qubit is physically connected only to a limited number of other qubits (15–20) in a graph of a certain topology.

The process of *minor embedding* transforms the QUBO formulation in an equivalent one that fits in the particular topology of the quantum annealer. This process may require to use multiple physical qubits to represent a single problem variable, therefore even if the quantum annealer has ~ 5000, qubits in practice one can fit in its topology only problems with at most hundreds of variables.

Furthermore, if the problem does not fit on the device, hybrid traditional-quantum methods exist to split the problem in smaller ones that can be solved on the quantum annealer and then combine the results. This is usually done in a general way independently on the specific problem, thus not exploiting its possible structure and properties.

One possible further challenge consists in finding a better ways to split a problem in sub-problems exploiting its structure, as well as developing new problem formulations that account for the limited connectivity of the quantum annealer.

3.5   Evaluation of Quantum Annealing

Evaluating QA approaches is not straightforward. Using a quantum annealer requires several stages:

Formulation: compute the QUBO matrix Q;

Embedding: generate the *minor embedding* of the QUBO for the quantum annealer hardware;

Data Transfer: transfer the problem and the embedding on the global network to the datacenter that hosts the quantum annealer;

Annealing: run the quantum annealer itself. This is an inherently stochastic process, therefore it is usually run a large number of times (hundreds).

Considering effectiveness, one must account for the fact that there are at least two layers of stochasticity, the embedding phase and the annealing phase. First, in the embedding phase heuristic methods transform the QUBO formulation of the problem in an equivalent problem that accounts for the limited connectivity of the physical qubits. This process includes some randomization steps and therefore may result in different embeddings for the same problem. Different embeddings will create different physical systems that are, in principle, equivalent but in practice may affect the final result.

Second, the annealing phase is a highly stochastic process and operates by *sampling* a low-energy solution, therefore depending on the problem one may require a large number of samples to obtain a good solution with sufficient probability. Usually one selects the best solution found, but this may result in experiments with high variance. Due to this, statistical evaluation measures are essential to account for the inherent stochastic behaviour of the quantum device.

Considering efficiency, while the annealing phase in which the quantum annealer is actually used may last in the range of *tens of milliseconds*, transferring the problem on the global network will introduce a delay of seconds and

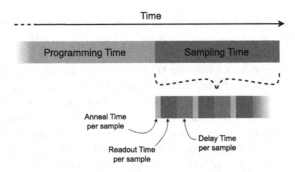

Fig. 2. The quantum annealer access time split in several steps.

generating the minor embedding may require minutes for particularly large problems. Furthermore, the total quantum annealer runtime can be split in several phases, see Fig. 2: first the device needs to be programmed for the specific problem, then the quantum-mechanical annealing process is run and lastly the result is read. The annealing process is extremely fast, requiring in the range of 20 μs, but because the device is inherently stochastic the annealing process is repeated multiple times. Clearly it is unfair to evaluate the efficiency based on a single annealing step, it is instead necessary to consider the time requirements of all the steps involved.

4   Implementation of the Infrastructure

Since participants cannot have direct access to the quantum annealers and we want the measurements to be as fair and reproducible as possible, we provide here a possible design and implementation of the infrastructure required to carry out the Lab. This infrastructure has been designed following the principles of scalability, availability, security and fault-tolerance. As depicted in Fig. 3, our infrastructure is composed of several components which have specific purposes:

- **Workspace**: each team has its own workspace which is accessible through the browser by providing the correct credentials. The workspace has a preconfigured git repository that is fundamental for reproducibility reasons.
  There is a custom library installed in the workspace which allows the communication with the dispatcher to submit problems to the actual quantum annealer.
- **Dispatcher**: it manages and keeps track of all the submissions done by the teams. It also holds the secret API Key that is used to submit problems to the quantum annealer. In this way, participants will never know what is the actual secret Key used.
  The dispatcher is accessible only from inside the system so that attackers cannot reach it.

– **Web Application**: it is the main source of information to the external users about the ongoing tasks. Moreover, it allows teams to view their quotas and some statistics through a dashboard.
Also organizers have their own dashboard through which it is possible to manage teams and tasks.

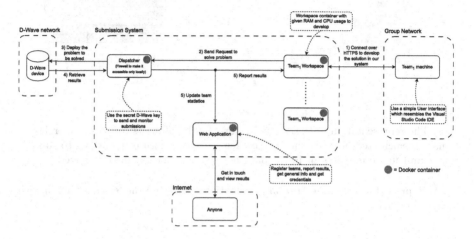

Fig. 3. High-level representation of the infrastructure.

We employ technologies such as Docker containers and Kubernetes in order to make the system scalable and fault-Tolerant. In fact, the system can be deployed on cloud making use of different physical machines to handle several teams working together. We ensure to apply the correct security measures to handle possible vulnerabilities such as SQL-Injection and Cross-Site Scripting. Note that our infrastructure plays for QA a role similar to other infrastructures, such as TIRA [12], for more general evaluation purposes.

5   Feature Selection in Practice

In this section we will show how the feature selection problem can be solved with a quantum annealer using our infrastructure. This is based on an example taken from the D-Wave tutorials[5].

The task is to identify the subset of the most relevant k features that can be used to predict the survival of Titanic passengers, using the D-Wave quantum annealer. The problem is formulated as a QUBO, where the matrix Q contains the *Conditional Mutual Information* of the features associated to the row and column, and the *survival* feature. This approach is called MIQUBO and in Fig. 4a it is possible to see the Mutual Information (MI) values considering each feature

---

[5] https://github.com/dwave-examples/feature-selection-notebook.

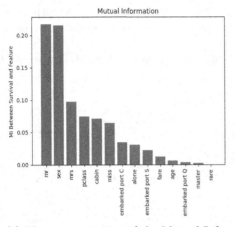
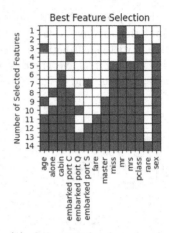
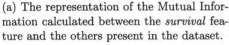

(a) The representation of the Mutual Information calculated between the *survival* feature and the others present in the dataset.

(b) Subsets of selected relevant features according to subset sizes highlighted in red.

Fig. 4. Representations of the Mutual Information values and the features chosen with the QA approach.

and the *survival* feature. The method further requires to define the number of features to select, k, by introducing penalties to the QUBO model so that solutions with a different number of selected features are penalized. In total the dataset contains 15 features and 1045 rows representing passengers. We can solve the problem directly on the quantum annealer without applying hybrid approaches by calling the D-Wave APIs. Considering different values of $k \in [1, 14]$ and applying the opportune penalties, it is possible to identify which are the most relevant features that can be used to establish the survival of a passenger. This is seen in Fig. 4b, where the subsets of selected relevant features for each k are highlighted in red.

Using a number of samples $n_{samples} = 100$ and a number of features $k = 4$, we report here in Table 1 the times required for some of the steps:

Table 1. Timings to solve the considered problem on the QPU for $k = 4$ features.

Access time	Sampling time	Programming time	Anneal time per sample
28615.97 μs	12856.0 μs	15759.97 μs	20.0 μs

Once the most relevant features for each k are identified, we can train a Tree Classifier and evaluate its classification accuracy. In order to do this, the data is split selecting 90% for training and the remaining 10% for testing. Table 2 reports the results in terms of Accuracy for $k = \{2, 3, 4, 5, 6\}$ for the Tree Classifier trained respectively on the most relevant subset of features found with the

Table 2. Comparison of the Accuracy of the Tree Classifier trained on different subsets of features.

	$k = 2$	$k = 3$	$k = 4$	$k = 5$	$k = 6$
Selected Features	0.621	0.766	0.828	0.786	0.793
Highest MI features	0.621	0.621	0.786	0.793	0.793

quantum annealer and on the subset of features that had the highest MI values with respect to the *survival* feature (e.g., for $k = 3$ *mr*, *sex*, *mrs* have the highest corresponding MI values as in Fig. 4a).

It is possible to see that the Accuracy measured on the subsets obtained through the feature selection process is always similar or even better than the Accuracy measured on the subsets having the highest MI values.

6 Conclusions and Future Work

In this paper we have proposed qCLEF, a new lab composed of 3 different tasks that aims at evaluating the performances of QA applied to IR and RS. These tasks represent some practical problems that are often faced by these systems. We have also discussed about the potential benefits that QA can bring to the IR and RS fields and we have highlighted how the evaluation of both efficiency and effectiveness should be performed. Finally, we have proposed an infrastructure that has been designed and implemented to satisfy both participants and organizers' needs.

qCLEF can represent a starting point for many researchers worldwide to know more about these new cutting-edge technologies that will likely have a big impact on the future of several research fields. Through this lab it will be also possible to assess whether QA can be employed to improve the current state-of-the-art approaches, hopefully delivering new performing solutions using quantum annealers.

References

1. Arthur, D., Date, P.: Balanced k-means clustering on an adiabatic quantum computer. Quantum Inf. Process. **20**(9), 294 (2021). https://doi.org/10.1007/s11128-021-03240-8
2. Bauckhage, C., Piatkowski, N., Sifa, R., Hecker, D., Wrobel, S.: A QUBO formulation of the k-medoids problem. In: "Lernen, Wissen, Daten, Analysen", Berlin, Germany, CEUR Workshop Proceedings, vol. 2454, pp. 54–63, CEUR-WS.org (2019). https://ceur-ws.org/Vol-2454/paper_39.pdf
3. Burges, C.J.C.: From RankNet to LambdaRank to LambdaMART: an overview. Technical report, Microsoft Research, MSR-TR-2010-82 (2010)
4. Delilbasic, A., Saux, B.L., Riedel, M., Michielsen, K., Cavallaro, G.: A single-step multiclass SVM based on quantum annealing for remote sensing data classification. arXiv preprint arXiv:2303.11705 (2023)

5. Devlin, J., Chang, M.W., Lee, K., Toutanova, K.: BERT: pre-training of deep bidirectional transformers for language understanding. arXiv preprint arXiv:1810.04805 (2018)

6. Ferrari Dacrema, M., Moroni, F., Nembrini, R., Ferro, N., Faggioli, G., Cremonesi, P.: Towards feature selection for ranking and classification exploiting quantum Annealers. In: Proceedings of the 45th Annual International ACM SIGIR Conference on Research and Development in Information Retrieval (SIGIR 2022), pp. 2814–2824, ACM Press, New York, USA (2022)

7. Glover, F.W., Kochenberger, G.A., Du, Y.: Quantum bridge analytics I: a tutorial on formulating and using QUBO models. 4OR **17**(4), 335–371 (2019). https://doi.org/10.1007/s10288-019-00424-y

8. Lu, Y., Zhou, H.H.: Statistical and computational guarantees of Lloyd's algorithm and its variants (2016)

9. Nembrini, R., Ferrari Dacrema, M., Cremonesi, P.: Feature selection for recommender systems with quantum computing. Entropy **23**(8), 970 (2021)

10. Nguyen, T., et al.: MS MARCO: a human generated machine reading comprehension dataset. Choice **2640**, 660 (2016)

11. Pilato, G., Vella, F.: A survey on quantum computing for recommendation systems. Information **14**(1), 20 (2023). https://doi.org/10.3390/info14010020

12. Potthast, M., Gollub, T., Wiegmann, M., Stein, B.: TIRA integrated research architecture. In: Information Retrieval Evaluation in a Changing World. TIRS, vol. 41, pp. 123–160. Springer, Cham (2019). https://doi.org/10.1007/978-3-030-22948-1_5

13. Qin, T., Liu, T.Y.: Introducing LETOR 4.0 datasets. arXiv preprint arXiv:1306.2597 (2013a)

14. Qin, T., Liu, T.Y.: Introducing LETOR 4.0 datasets. arXiv.org. Information Retrieval (cs.IR) arXiv:1306.2597 (2013b)

15. Ushijima-Mwesigwa, H., Negre, C.F.A., Mniszewski, S.M.: Graph partitioning using quantum annealing on the D-wave system. CoRR abs/1705.03082 (2017). https://arxiv.org/abs/1705.03082

16. Willsch, D., Willsch, M., De Raedt, H., Michielsen, K.: Support vector machines on the D-wave quantum Annealer. Comput. Phys. Commun. **248**, 107006 (2020)

17. Wu, Y., Cao, Q., Shen, H., Tao, S., Cheng, X.: INMO: a model-agnostic and scalable module for inductive collaborative filtering. In: SIGIR 2022: The 45th International ACM SIGIR Conference on Research and Development in Information Retrieval, Madrid, Spain, pp. 91–101, ACM (2022). https://doi.org/10.1145/3477495.3532000

18. Zaiou, A., Bennani, Y., Matei, B., Hibti, M.: Balanced K-means using quantum annealing. In: 2021 IEEE Symposium Series on Computational Intelligence (SSCI), pp. 1–7 (2021). https://doi.org/10.1109/SSCI50451.2021.9659997

Graph-Enriched Biomedical Entity Representation Transformer

Andrey Sakhovskiy[1,3]($\boxtimes$) (iD), Natalia Semenova[1,2] (iD), Artur Kadurin[2] (iD),
and Elena Tutubalina[2,3] (iD)

[1] Sber AI, Moscow, Russia
andrey.sakhovskiy@gmail.com
[2] Artificial Intelligence Research Institute, Moscow, Russia
[3] Kazan (Volga Region) Federal University, Kazan, Russia

Abstract. Infusing external domain-specific knowledge about diverse biomedical concepts and relationships into language models (LMs) advances their ability to handle specialised in-domain tasks like medical concept normalization (MCN). However, existing biomedical LMs are primarily trained with contrastive learning using synonymous concept names from a terminology (e.g., UMLS) as positive anchors, while accurate aggregation of the features of graph nodes and neighbors remains a challenge. In this paper, we present Graph-Enriched Biomedical Entity Representation Transformer (GEBERT) which captures graph structural data from the UMLS via graph neural networks and contrastive learning. In GEBERT, we enrich the entity representations by introducing an additional graph-based node-level contrastive objective. To enable mutual knowledge sharing among the textual and the structural modalities, we minimize the contrastive objective between a concept's node representation and its textual embedding obtained via LM. We explore several state-of-the-art convolutional graph architectures, namely GraphSAGE and GAT, to learn relational information from local node neighborhood. After task-specific supervision, GEBERT achieves state-of-the-art results on five MCN datasets in English.

Keywords: Natural language processing · Biomedical entity representations · Knowledge representation · Graph neural network · Entity linking

1 Introduction

Biomedical entity representation finds application in numerous biomedical tasks, such as knowledge discovery, information extraction, and search [5,9,15,21,29, 31]. Nonetheless, identifying specific biomedical concepts like diseases, symptoms, and drugs in free-form text can be problematic because their names, abbreviations, and spelling inconsistencies are highly variable. Moreover, a single biomedical concept can appear in numerous nonstandard forms. This challenge

© The Author(s), under exclusive license to Springer Nature Switzerland AG 2023
A. Arampatzis et al. (Eds.): CLEF 2023, LNCS 14163, pp. 109–120, 2023.
https://doi.org/10.1007/978-3-031-42448-9_10

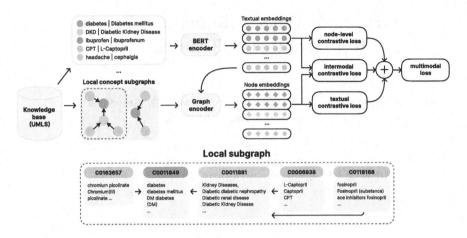

Fig. 1. GEBERT model's architecture overview. Our model consists of two encoders for text and graph data. Graph encoder uses textual embeddings from BERT as an additional input. The loss function is a weighted sum of three terms: textual and node-level contrastive losses, and intermodal contrastive loss to match representations between different encoders.

can be addressed by medical concept normalization (MCN; also called medical concept linking) which is the task where entity mentions are mapped against a large set of medical concept names and their concept unique identifiers (CUIs) from a knowledge base (KB). In addition to a high variation of mentions, the biomedical domain is characterized by extensive KBs such as the Unified Medical Language System (UMLS) [3].

Early models for MCN [19,26] commonly used classification type losses that are often trained on narrow benchmarks and lead to significant performance degradation on other domains and structurally different texts. Modern approaches usually employ similarity between *embeddings* (distributed representations) of entity mentions and concepts constructed by language models (LMs) and a BERT [7]-like ranking architecture [30,32,38]. However, the problem of learning meaningful and robust entity representations still poses a challenge for LMs.

Biomedical knowledge has been injected into neural networks by metric learning and contrastive learning [17,22,24,27,37]. Naturally, knowledge from KBs is typically represented as triples (head, relation, tail); head and tail terms of the same and different concepts serve as positive and negative pairs (e.g., *diabetic nephropathy* is a synonym to *diabetic kidney disease* and differs from *diabetes mellitus*, as shown in Fig. 1). In addition to representation learning with textual triples [17,22,27,37] proposed to use term-relation-term similarity inspired by semantic matching methods like TransE [4] and DistMult [36]. However, these structural approaches are unable efficiently use textual node features.

In this paper, we present **G**raph-**E**nriched **B**iomedical **E**ntity **R**epresentation **T**ransformer (GEBERT), which uses contrastive learning and graph neural networks to capture graph structural data from a KB. As shown in Fig. 1, the GEBERT architecture consists of three losses: (i) a textual contrastive loss that learns on synonymous concept names; (ii) a node-level contrastive loss that learns to produce concept embeddings that are independent of the surface form choice; (iii) an intermodal contrastive loss that allows the information exchange between the textual and graph encoder. The source code and pre-trained models are freely available[1].

2 Related Work

MCN is typically formulated as a classification or ranking problem with a wide range of features, including syntactic, morphological parsing, dictionaries of medical concepts and their synonyms, and distances between formal concept names and raw entity mentions in terms of sparse/dense representations [1,6,33].

Classification approaches [19,26] are typically trained on labeled datasets with mentions linked to a small set of target concepts, while existing biomedical KBs, such as the UMLS, have millions of concepts. Ranking models use training pairs of positive and negative terms from a dictionary to determine how similar entity mention and concept names are. [24] trained a triplet network to rank the candidate concept names based on their similarity with a disease mention. Convolutional and pooling layers based on *word* embeddings were chosen as the encoder. [30] proposed the BioSyn model, which maximizes the likelihood that all synonym representations are present in the top 20 candidates. As a similarity function, BioSyn combines the sparse and dense scores with a scalar weight. To encode the morphological information of given strings, sparse scores are computed on character-level TF-IDF representations. Dense scores are defined by the similarity between CLS tokens of a single vector of input in BioBERT [14]. [22] proposed a DILBERT model which optimizes the relative similarity of mentions and concept names from a terminology via triplet loss. Different negative sampling strategies were applied to DILBERT models including random sampling and re-sampling using concept names from concepts' parents (parent-child or broader-narrower relationships). However, both DILBERT and BioSyn were trained on a dataset in English with a narrow subsample of concepts from a specific terminology.

There are few attempts to inject external domain-specific knowledge (e.g., UMLS) into pre-trained language models (LMs) in order to learn entity representations [17,18,20,27,37]. [27] presented an encoding framework with context, concept, and synonym-based objectives. Synonym-based objective enforces similar representations between synonymous names, while concept-based objective pulls the name's representations closer to its concept's centroid. This model was trained on 29 million PubMed abstracts annotated with UMLS concepts of diseases and chemicals. However, ranking on these embeddings shows worse results than models with dictionaries and features on three sets in English. Umls-BERT [20], a bert-like LM, integrates the domain knowledge from UMLS during

---

[1] https://github.com/Andoree/GEBERT.

the pre-training process via a novel knowledge augmentation strategy. Recently, a self-alignment pretraining (SAP) [17] procedure for learning on synonymous term pairs from the UMLS has been proposed. The authors of the procedure released a BERT-based SapBERT model that is pre-trained on English synonyms from UMLS. SapBERT pre-trained on UMLS outperformed on MCN task several domain-specific LMs such as BioBERT [14], SciBERT [2], and UmlsBERT [20].

The SAP procedure makes no use of the UMLS graph's structure that describes the relations between concepts. To address the limitation, a relation-aware language model named CODER was proposed [37]. The authors infused the relational knowledge from the UMLS graph into the original SAP procedure by introducing a relational loss in addition to synonym-based contrastive loss. The main difference compared to SapBERT is that CODER simultaneously learns from synonyms and related concepts.

3   Background and Architecture

Let V denote a set of all concepts present in a knowledge base. Knowledge graphs, such as UMLS, usually store relational information in the form of relation triplets (h, r, t) where h and t are concepts from V and r is a relation type. In this work, we omit the relation types and view the UMLS graph as an oriented unlabelled graph $G = G(V, \mathcal{E})$, where $\mathcal{E}$ is the set of oriented edges with relation types dropped. For each concept $c \in V$, UMLS presents a set of k synonymous terms $S_c = \{s_1^c, s_2^c, \ldots, s_k^c\}$. For each term from S_c, the UMLS stores the label of the language it came from. Let s denote an arbitrary textual term which, in other words, is a concept name. The goal of the biomedical entity linking task is to predict a concept $c \in V$ that s belongs to.

3.1   Self-alignment Pretraining

A reasonable and straightforward way to learn an informative representation space of biomedical entities is to represent textual knowledge from KG in the form of positive and negative term pairs and optimize some contrastive learning loss function.

In this work, we adopt the self-alignment pretraining (SAP) procedure [17]. To enrich the training procedure with harder negative samples, SAP employs online hard mining for valid triplets [10,23]. During SAP, the model is encouraged to produce similar representations for all terms that represent the same concept (share the same CUI). At each pretraining step, we sample a batch B that consists of N positive samples $(c, s_i^c, s_j^c) \in V \times S_c \times S_c$. Given B, SAP constructs all possible term triplets (s^p, s^a, s^n) such that $p = a$ and $n \neq a$. s^a is called an anchor term; s^p is a positive term for s^a (i.e., s^p and s^a are synonymous terms representing the same concept $a = p$); s^n is a negative term for s^a (i.e., s^n and s^a represent non-matching concepts). Each triple produces a positive pair (s^a, s^p) and a negative pair (s^a, s^n). To keep only the most informative triples, we use online hard mining for valid triplets with respect to the following constraint:

$$\|f_{enc}(s^a) - f_{enc}(s^p)\| < \|f_{enc}(s^a) - f_{enc}(s^n)\| + \lambda$$

where f_{enc} is a BERT-based textual encoder, $\|\cdot\|$ is the normalized L_2-norm, and λ is a pre-defined mining margin. Thus, the mining procedure discards all the triplets such that the distance from an anchor to its negative sample is greater than the distance to its positive sample by more than λ. Let $\mathcal{P}$ and $\mathcal{N}$ denote the sets of all positive and negative term pairs, respectively. The SAP procedure utilizes the Multi-Similarity (MS) loss [35] to learn from $\mathcal{P}$ and $\mathcal{N}$.

$$\mathcal{L}_{sap} = \frac{1}{|B|} \sum_{i=1}^{|B|} \left(\frac{1}{\alpha} \log \left(1 + \sum_{n \in \mathcal{N}_i} e^{\alpha(S_{in}-\epsilon)} \right) + \frac{1}{\beta} \log \left(1 + \sum_{p \in \mathcal{P}_i} e^{-\beta(S_{ip}-\epsilon)} \right) \right),$$

where α, β, and ϵ are the parameters of MS-loss. $\mathcal{P}_i$ and $\mathcal{N}_i$ are the sets of positive and negative samples for the anchor concept i.

3.2   Graph Neural Networks

Message Passing Framework. A common way to learn structured knowledge from graph is to iteratively update the representation of node v by passing and aggregating messages from local node neighborhood $N(v)$ using a graph neural network. Message Passing Neural Networks (MPNN) [11] framework that describes an update of node representation $h_v^{(l)}$ at the $(l+1)$-th MPNN layer as the composition of a message function f_m and an update function f_u:

$$h_v^{(l+1)} = f_u(h_v^{(l)}, \sum_{u \in N(v)} f_m(h_v^{(l)}, h_u^{(l)}))$$

where $N(v)$ is the set of neighboring nodes of node v. As the number of neighbors can significantly vary across different nodes and result in excessive computational complexity, we use a uniformly drawn fixed-size subset of neighbors instead of the full node neighborhood as proposed by [13]. The choice of f_m and f_u functions is the key difference between various GNN models that fall under the MPNN framework. In GraphSAGE [13], a common and rather simple implementation of MPNN framework, an element-wise operator (e.g., max- or mean-pooling) is used as an f_m to aggregate the vectors of neighbor nodes $N(v)$ into a single vector. The aggregated representation is further concatenated with the original representation and passed to a linear layer W^{l+1} with a non-linear activation function σ. In this work, we use the GraphSAGE implementation with mean-pooling aggregation:

$$h_v^{(l+1)} = \sigma(W^l \cdot [h_v^{(l)} \parallel MEAN(N(v))])$$

where $MEAN$ is the mean-pooling operator, $[\cdot \parallel \cdot]$ is the concatenation of two vectors. The simplicity of GraphSAGE prevents a context-aware message passing since the mean-pooling treats all nodes from $N(v)$ with equal weights.

It means that graphSAGE is not able to weigh neighborhoods with respect to their relevance to the target node.

Graph attention network (GAT) [34] addresses the limitation by introducing the self-attention over neighboring nodes and learning the aggregated neighborhood representation as the weighted sum of neighboring nodes representations. Given two node representations $h_u^{(l-1)}$ and $h_v^{(l-1)}$, the l-th GAT layer computes the relevance of node u for the target node v as the normalized attention score $\alpha_{uv}^{(l)}$:

$$e_{uv}^{(l)} = a^T \cdot LeakyReLU(W^{(l)} \cdot [h_u^{(l-1)} \parallel h_v^{(l-1)}])$$

$$\alpha_{uv}^{(l)} = \frac{exp(e_{uv}^{(l)})}{\sum_{w \in N(v)} exp(e_{wv}^{(l)})}$$

With the attention scores obtained, the aggregated neighborhood representation is computed as a weighted sum of neighboring nodes embeddings.

3.3 GEBERT

Textual Loss. In GEBERT, we adopt and extend the pretraining procedure described in Sect. 3.1. At each training step, we begin by sampling a batch B of random positive samples. Each positive sample is a triplet $t = (c, s_i^c, s_j^c) \in V \times S_c \times S_c$ which consists of concept (node) identifier and two synonymous concept names. For each t, we randomly sample a set of concept node's neighbors (concept's neighborhood) $N(c)$ using the graph G. Next, we produce a textual embedding for each term present in B using a textual encoder f_{enc} and calculate the textual loss $\mathcal{L}_{sap}$ using the representations of concept names from the batch B.

Node-Level Loss. We define the batch B's subgraph $G_B = G(V_B, \mathcal{E}_B)$ as the union of concept nodes from batch B and all nodes and edges from the concept's neighborhood $N(c), c \in B$. Our goal is to enrich the embedding space of the textual encoder f_{enc} with the structural knowledge stored in G_B while keeping the embeddings of terms representing the same concept close to each other by cosine distance. As shown in Fig. 1, for each positive pair, textual encoder f_{enc} produces two textual embeddings: for the first and the second terms of the pair, respectively. These embeddings are passed to a graph encoder for node initialization.

Let $H_1 \in R^{|B| \times d}$ and $H_2 \in R^{|B| \times d}$ denote the matrices of d-dimensional textual embeddings of the first and the second terms of positive pairs from B, respectively. To obtain two graph-enriched representations g_c^1 and g_c^2 of the node (concept) c, we stack multiple MPNN layers to aggregate the structural information from the node's neighborhood $N(c)$ using the H_1 and H_2 as the initial representations of nodes V_B. Next, we collect all positive node samples (c, g_c^1, g_c^2) and pass them to the SAP procedure to obtain a node-level contrastive loss $\mathcal{L}_{node}$. Thus, the major difference between $\mathcal{L}_{sap}$ and $\mathcal{L}_{node}$ is that the latter operates on the graph-aware node (concept) embeddings rather than textual embeddings of concept terms.

Intermodal Loss. Let (c, g_c^1, g_c^2) and $(c, f_{enc}(s_i^c), f_{enc}(s_j^c))$ denote a term-level and node-level positive samples of concept c, respectively. We construct two intermodal positive samples $(c, g_c^1, f_{enc}(s_j^c))$ and $(c, f_{enc}(s_i^c), g_c^2)$ each containing a node-level and a term-level representation of c. To allow a mutual knowledge exchange between the textual encoder f_{enc} and a graph encoder, we collect all intermodal positive pairs and once again apply the SAP procedure to optimize the intermodal contrastive MS-loss $\mathcal{L}_{int}$, that minimizes the distance between textual and node representations of the same concept and pushes away the representations of non-matching concepts.

$$\mathcal{L}_{GEBERT} = \mathcal{L}_{sap} + \lambda_{node}\mathcal{L}_{node} + \lambda_{int}\mathcal{L}_{int}, \tag{1}$$

where λ_{node} and λ_{int} are the pre-selected weights of $\mathcal{L}_{node}$ and $\mathcal{L}_{int}$.

4   Experimental Evaluation

We initialized GEBERT with PubMedBERT[2] [12]. The model was trained on an English UMLS graph for 1 epoch with a learning rate of $2 \cdot 10^{-5}$. We set $\lambda_{node} = \lambda_{int} = 0.1$ and the maximum size of node neighborhood to 3. As a graph encoder, we use 3 consecutive layers of either GraphSAGE or GAT.

We implemented two versions of GEBERT that differ in graph encoder architecture: (i) GraphSAGE-GEBERT and (ii) GAT-GEBERT. To train our implementations of GEBERT, we use the UMLS 2020AB release which contains approximately 4.4 million concepts and 15.9 million unique concept names from 215 source vocabularies. We remove all concept names that originate from non-English source vocabularies and remove all duplicated edges. We follow the batching strategy proposed by the authors of SapBERT [17]: to ensure each batch includes a sufficient number of positive pairs, we pre-compute synonym pairs with common CUIs. If a concept produces more than 50 positive pairs, we randomly sample 50 of them.

Data. To evaluate our models, we use 5 datasets: (i) NCBI [8], (ii) BC5CDR-D [16], (iii) BC5CDR-D [16], (iv) TAC2017ADR [28], (v) BC2GN [25]. Due to overlap between official train/test sets, we follow [32] and use the presented *refined* test sets. For details on preprocessing and sets, please refer to [32]. We have used the publicly available code provided by the authors at https://github.com/insilicomedicine/Fair-Evaluation-BERT.

The NCBI Disease Corpus [8] is a collection of 793 abstracts from PubMed, which include mentions of diseases and their corresponding concepts. [16] introduces a task for the extraction of chemical-disease relations (CDR) from 1500

---

[2] huggingface.co/BiomedNLP-PubMedBERT-base-uncased-abstract-fulltext.

Table 1. Evaluation of models on academic evaluation datasets (*refined* test sets).

Model	NCBI		BC5CDR D		BC5CDR C		TAC ADR		BC2GN	
	@1	@5	@1	@5	@1	@5	@1	@5	@1	@5
Zero-shot evaluation										
enSapBERT	**71.57**	**84.31**	73.67	84.32	85.88	91.29	**82.58**	**90.93**	**87.72**	92.18
enCODER	69.12	**84.31**	73.36	**85.54**	84.24	90.82	79.15	88.41	84.47	90.96
GraphSAGE-GEBERT	70.59	82.84	73.97	84.02	**86.12**	**91.76**	81.54	90.61	86.19	93.10
GAT-GEBERT	70.59	83.33	**74.58**	85.39	85.41	**91.76**	82.12	89.90	87.31	**92.79**
Evaluation after fine-tuning										
enSapBERT	75.49	**84.80**	74.89	84.02	86.12	**93.41**	86.20	91.26	88.83	93.30
enCODER	73.53	82.84	75.34	**85.24**	86.35	92.71	84.72	91.00	88.32	92.49
GraphSAGE-GEBERT	**76.47**	**84.80**	**75.80**	84.93	**87.53**	93.18	**86.33**	**91.97**	**88.93**	93.50
GAT-GEBERT	73.04	**84.80**	75.49	84.78	87.06	92.71	85.82	91.32	88.63	**93.60**

Table 2. Error analysis examples of mentions, predicted and golden concept names from GraphSAGE-GEBERT on TAC ADR refined test set.

Mention	Predicted concept	Golden concept
clinical deterioration	clinical worsening	general physical health deterioration
mean change in heart rate 1 2 beats per minute	mean heart rate higher by an average of 1 to 2 bpm	heart rate abnormal
increased number of lashes	increased lacrimation	growth of eyelashes
body temperature dysregulation	body temperature fluctuation	temperature regulation disorder
emerging suicidality	suicidality	suicidal intention
homicidal threats	homicidal attempt	homicidal ideation

PubMed abstracts, with annotations for both chemicals and diseases. BioCreative II GN (BC2GN) [25] contains human gene and gene product mentions in PubMed abstracts for gene normalization (GN). TAC 2017 ADR challenge [28] focuses on extracting adverse drug reactions (ADRs) from product labels, such as prescribing information or package inserts.

Experimental Setup. We evaluate the proposed models in two settings: (i) zero-shot evaluation and (ii) evaluation with fine-tuning.

For zero-shot evaluation, we employ a ranking approach [32] that is built on the embeddings of mentions and potential concepts. Each entity mention and concept name is first passed through a model that produces their embeddings and then through an average pooling layer that yields a fixed-sized vector. The inference task is then reduced to finding the closest concept name representation to entity mention representation in a common embedding space, where the Euclidean distance can be used as the metric. Nearest concept names are chosen as top-k concepts for entities.

For the evaluation with fine-tuning, we utilize BioSyn [30], a model that iteratively updates candidates by applying synonym marginalization. The model utilizes two distinct similarity functions designed to capture both morphological and semantic information. The sparse representations are obtained with TF-IDF and dense representations are obtained using a BERT-based model. We adopt the default BioSyn hyper-parameters [30]. For each dataset, we trained BioSyn for 20 epochs, following [32].

We evaluate the models in the IR scenario, where the goal is to find top-k concepts for every entity mention in a dictionary of concept names and their identifiers. Following previous works [17,18,27,30,32,37], we use the top-k accuracy as the evaluation metric: Acc@k = 1 if the correct UMLS concept unique identifier is retrieved at the rank $\leq k$, otherwise Acc@k = 0.

Compared Representations. We compare the following representations:

- *enSapBERT*: a BERT-based metric learning framework that generates hard triplets based on the UMLS for pre-training [17]. The model is adopted from huggingface.co/cambridgeltl/SapBERT-from-PubMedBERT-fulltext.
- *enCODER*: a contrastive learning model inspired by semantic matching methods that uses both synonyms and relations from the UMLS [37]. We have used the model provided at huggingface.co/GanjinZero/coder_eng.

4.1   Results

Table 1 shows the Acc@1 and Acc@5 metrics for five datasets. In zero-shot evaluation, basic enSapBERT outperformed CODER and GEBERT on 3 of 5 datasets in terms of Acc@1. On disease and chemical mentions from BC5CDR, the best models are GraphSAGE-GEBERT and GAT-GEBERT with a slight improvement over enSapBERT. An interesting finding is that enCODER is the worst performing model on all five datasets in terms of Acc@1 despite the fact it inherited one of two its training objectives from enSapBERT. The situation changes after the fine-tuning: our GraphSAGE-GEBERT model becomes a leader on all five academic datasets with an insignificant improvement against enSapBERT on TAC ADR and BC2GN (0.13% and 0.1%, respectively) and a notable improvement on NCBI, BC5CDR Disease, and BC5CDR Chemical (0.98%, 0.91%, and 1.41%, respectively). On average, GraphSAGE-GEBERT outperformed enSapBERT and enCODER by 0.71% and 1.36% Acc@1, respectively. enCODER remains the worst-performing on 3 of 5 datasets. Thus, having a decent performance in zero-shot setting, our proposed GraphSAGE-GEBERT shows superior performance in the biomedical domain after in-domain fine-tuning.

Discussion and Error analysis. We looked through erroneous predictions of the fine-tuned GraphSAGE-GEBERT model on the refined test set of the TAC 2017 ADR corpus. Some examples of the model's errors are presented in Table 2. After the error analysis, we can draw the following key observations. First, in many cases, the model predicts a concept that is in some relation (e.g., hyponymic or hypernymic) with the true concept. For example, the model marks a mention related to heart rate change as the partial case of it – a heart rate decrease. Second, as can be seen from the examples, the normalization problem with a rich vocabulary poses a great challenge by providing a plethora of distinct but semantically related concepts (such as 'homicidal attempt' and 'homicidal ideation'). Thus, in many cases, a true concept and the wrongly predicted one are connected by some relation in the UMLS. We believe that a proper utilization of this relational knowledge is the key to the improvement of normalization quality. Presumably, neither GEBERT nor enCODER fully reveal the power of relational knowledge stored in the UMLS graph. More tricky and effective methods to encode structural knowledge from graphs into LMs are yet to be explored.

5   Conclusion

In this work, we have presented a new model called GEBERT which allows a mutual knowledge exchange between the textual encoder and a graph encoder. We pre-trained two GEBERT models with different state-of-the-art GNN encoders on an English UMLS graph which contains 4M concepts (nodes), 15M textual concept names, and 38.8M relationships (edges). The experimental results on five benchmark datasets in English demonstrate that after task-specific fine-tuning GEBERT outperforms existing state-of-the-art concept normalization models. We consider the following two directions for future work. First, we plan to adopt the proposed model for multilingual pre-training. Second, we plan to infuse relation types at the node neighborhood aggregation stage.

Acknowledgments. The work has been supported by the Russian Science Foundation grant # 23-11-00358.

References

1. Aronson, A.R.: Effective mapping of biomedical text to the UMLS Metathesaurus: the MetaMap program. In: Proceedings of the AMIA Symposium, p. 17 (2001)
2. Beltagy, I., Lo, K., Cohan, A.: SciBERT: a pretrained language model for scientific text. In: Proceedings of the 2019 Conference on Empirical Methods in Natural Language Processing and the 9th International Joint Conference on Natural Language Processing (EMNLP-IJCNLP), pp. 3615–3620 (2019)
3. Bodenreider, O.: The unified medical language system (UMLS): integrating biomedical terminology. Nucleic Acids Res. **32**(suppl_1), D267–D270 (2004)
4. Bordes, A., Usunier, N., Garcia-Duran, A., Weston, J., Yakhnenko, O.: Translating embeddings for modeling multi-relational data. In: Advances in Neural Information Processing Systems, vol. 26 (2013)

5. Chen, H., Chen, W., Liu, C., Zhang, L., Su, J., Zhou, X.: Relational network for knowledge discovery through heterogeneous biomedical and clinical features. Sci. Rep. **6**(1), 29915 (2016)
6. Dermouche, M., Looten, V., Flicoteaux, R., Chevret, S., Velcin, J., Taright, N.: ECSTRA-INSERM@ CLEF eHealth2016-task 2: ICD10 code extraction from death certificates. In: CLEF (2016)
7. Devlin, J., Chang, M.W., Lee, K., Toutanova, K.: BERT: pre-training of deep bidirectional transformers for language understanding. In: Proceedings of the 2019 Conference of the North American Chapter of the Association for Computational Linguistics: Human Language Technologies, pp. 4171–4186 (2019)
8. Doğan, R.I., Leaman, R., Lu, Z.: NCBI disease corpus: a resource for disease name recognition and concept normalization. J. Biomed. Inform. **47**, 1–10 (2014)
9. Fiorini, N., et al.: Best match: new relevance search for PubMed. PLoS Biol. **16**(8), e2005343 (2018)
10. Gillick, D., Kulkarni, S., Lansing, L., Presta, A., Baldridge, J., Ie, E., Garcia-Olano, D.: Learning dense representations for entity retrieval. In: Proceedings of the 23rd Conference on Computational Natural Language Learning, pp. 528–537 (2019)
11. Gilmer, J., Schoenholz, S.S., Riley, P.F., Vinyals, O., Dahl, G.E.: Neural message passing for quantum chemistry. In: International Conference on Machine Learning, pp. 1263–1272. PMLR (2017)
12. Gu, Y., et al.: Domain-specific language model pretraining for biomedical natural language processing. ACM Trans. Comput. Healthcare **3**(1), 1–23 (2021)
13. Hamilton, W., Ying, Z., Leskovec, J.: Inductive representation learning on large graphs. In: Advances in Neural Information Processing Systems, vol. 30 (2017)
14. Lee, J., et al.: BioBERT: pre-trained biomedical language representation model for biomedical text mining. Bioinformatics **36**, 1234–1240 (2019)
15. Lee, S., et al.: Best: next-generation biomedical entity search tool for knowledge discovery from biomedical literature. PLoS ONE **11**(10), e0164680 (2016)
16. Li, J., et al.: BioCreative V CDR task corpus: a resource for chemical disease relation extraction. Database **2016** (2016)
17. Liu, F., Shareghi, E., Meng, Z., Basaldella, M., Collier, N.: Self-alignment pretraining for biomedical entity representations. In: Proceedings of the 2021 Conference of the North American Chapter of the Association for Computational Linguistics: Human Language Technologies, pp. 4228–4238 (2021)
18. Liu, F., Vulić, I., Korhonen, A., Collier, N.: Learning domain-specialised representations for cross-lingual biomedical entity linking. In: Proceedings of the 59th Annual Meeting of the Association for Computational Linguistics and the 11th International Joint Conference on Natural Language Processing, pp. 565–574 (2021)
19. Lou, Y., Qian, T., Li, F., Zhou, J., Ji, D., Cheng, M.: Investigating of disease name normalization using neural network and pre-training. IEEE Access **8**, 85729–85739 (2020)
20. Michalopoulos, G., Wang, Y., Kaka, H., Chen, H., Wong, A.: UmlsBERT: clinical domain knowledge augmentation of contextual embeddings using the unified medical language system metathesaurus. In: Proceedings of the 2021 Conference of the North American Chapter of the Association for Computational Linguistics: Human Language Technologies, pp. 1744–1753 (2021)
21. Miftahutdinov, Z., Alimova, I., Tutubalina, E.: On biomedical named entity recognition: experiments in interlingual transfer for clinical and social media texts. In: Jose, J.M., et al. (eds.) ECIR 2020. LNCS, vol. 12036, pp. 281–288. Springer, Cham (2020). https://doi.org/10.1007/978-3-030-45442-5_35

22. Miftahutdinov, Z., Kadurin, A., Kudrin, R., Tutubalina, E.: Medical concept normalization in clinical trials with drug and disease representation learning. Bioinformatics **37**(21), 3856–3864 (2021)
23. Mikolov, T., Sutskever, I., Chen, K., Corrado, G.S., Dean, J.: Distributed representations of words and phrases and their compositionality. In: Advances in Neural Information Processing Systems, pp. 3111–3119 (2013)
24. Mondal, I., et al.: Medical entity linking using triplet network, pp. 95–100 (2019)
25. Morgan, A.A., et al.: Overview of biocreative ii gene normalization. Genome Biol. **9**(S2), S3 (2008)
26. Niu, J., Yang, Y., Zhang, S., Sun, Z., Zhang, W.: Multi-task character-level attentional networks for medical concept normalization. Neural Process. Lett. **49**, 1239–1256 (2019)
27. Phan, M.C., Sun, A., Tay, Y.: Robust representation learning of biomedical names. In: Proceedings of the 57th Annual Meeting of the Association for Computational Linguistics, pp. 3275–3285 (2019)
28. Roberts, K., Demner-Fushman, D., Tonning, J.M.: Overview of the TAC 2017 adverse reaction extraction from drug labels track. In: TAC (2017)
29. Soni, S., Roberts, K.: An evaluation of two commercial deep learning-based information retrieval systems for COVID-19 literature. J. Am. Med. Inform. Assoc. **28**(1), 132–137 (2021)
30. Sung, M., Jeon, H., Lee, J., Kang, J.: Biomedical entity representations with synonym marginalization. In: Proceedings of the 58th Annual Meeting of the Association for Computational Linguistics, pp. 3641–3650 (2020)
31. Sutton, R.T., Pincock, D., Baumgart, D.C., Sadowski, D.C., Fedorak, R.N., Kroeker, K.I.: An overview of clinical decision support systems: benefits, risks, and strategies for success. NPJ Digit. Med. **3**(1), 17 (2020)
32. Tutubalina, E., Kadurin, A., Miftahutdinov, Z.: Fair evaluation in concept normalization: a large-scale comparative analysis for BERT-based models. In: Proceedings of the 28th International Conference on Computational Linguistics, pp. 6710–6716 (2020)
33. Van Mulligen, E., Afzal, Z., Akhondi, S.A., Vo, D., Kors, J.A.: Erasmus MC at CLEF eHealth 2016: concept recognition and coding in French texts. In: CLEF (2016)
34. Veličković, P., Cucurull, G., Casanova, A., Romero, A., Liò, P., Bengio, Y.: Graph attention networks. In: International Conference on Learning Representations (2018). https://openreview.net/forum?id=rJXMpikCZ. accepted as poster
35. Wang, X., Han, X., Huang, W., Dong, D., Scott, M.R.: Multi-similarity loss with general pair weighting for deep metric learning. In: Proceedings of the IEEE/CVF Conference on Computer Vision and Pattern Recognition, pp. 5022–5030 (2019)
36. Yang, B., Yih, S.W.T., He, X., Gao, J., Deng, L.: Embedding entities and relations for learning and inference in knowledge bases. In: Proceedings of the International Conference on Learning Representations (ICLR) 2015 (2015)
37. Yuan, Z., Zhao, Z., Sun, H., Li, J., Wang, F., Yu, S.: CODER: knowledge-infused cross-lingual medical term embedding for term normalization. J. Biomed. Inform. **126**, 103983 (2022)
38. Zhu, M., Celikkaya, B., Bhatia, P., Reddy, C.K.: LATTE: latent type modeling for biomedical entity linking. In: Proceedings of the AAAI Conference on Artificial Intelligence, vol. 34, pp. 9757–9764 (2020)

Supervised Machine-Generated Text Detectors: Family and Scale Matters

Areg Mikael Sarvazyan[1]([✉]) [iD], José Ángel González[1] [iD], Paolo Rosso[2] [iD], and Marc Franco-Salvador[1] [iD]

[1] Symanto Research, Valencia, Spain
{areg.sarvazyan,jose.gonzalez,marc.franco}@symanto.com
[2] Universitat Politècnica de València, Valencia, Spain
prosso@dsic.upv.es
https://www.symanto.com, https://www.upv.es

Abstract. This work studies the generalization capabilities of supervised Machine-Generated Text (MGT) detectors across model families and parameter scales of text generation models. In addition, we explore the feasibility of identifying the family and scale of the generator behind an MGT, instead of attributing the text to a particular language model. We leverage the AuTexTification corpus, comprised of multi-domain multilingual human-authored and machine-generated text, and fine-tune various monolingual and multilingual supervised detectors for Spanish and English. The results suggest that supervised MGT detectors generalize well across scales but are limited in cross-family generalization. Contrariwise, we observe that MGT family attribution is practical and effective, while scale attribution has some limitations. Code and results are available here.

Keywords: Machine-Generated Text Detection · Model Attribution · Generalization · Robustness

1 Introduction

The strong language capabilities of current Large Language Models (LLMs) such as GPT [2,18], BLOOM [23], and LLaMA [26] are motivating a large-scale adoption in the workflows of businesses and individuals. Tasks like creative writing, coding, or information seeking through search services, are nowadays aided with LLMs to reduce human effort. The impact of LLMs on society is not negligible, and it has been estimated that the adoption of LLMs could affect to at least 10% of the tasks performed by 80% of the workforce in some countries [4].

These LLMs have the potential to be used in cutting-edge applications. However, they could also be leveraged for malicious intents, e.g., spreading propaganda or disinformation by generating human-like fake news, opinions, or scientific papers, posing a threat to the reputation of companies, academic institutions and individuals [12]. Since these technologies are used by millions,[1] in

---

[1] https://tinyurl.com/reuters-chatgpt.

A. Arampatzis et al. (Eds.): CLEF 2023, LNCS 14163, pp. 121–132, 2023.
https://doi.org/10.1007/978-3-031-42448-9_11

the foreseeable future, we will need to decide how to deal with these malicious applications.

A promising approach to ensure a responsible use of LLMs consists on detecting Machine-Generated Text (MGT) and applying content moderation techniques on top. In this line of research, there has been a recent surge of models [16], services [17], and watermarking techniques [13], aimed towards detecting or assisting to detect MGT. This approach has been explored in specific scenarios, including detecting fake news [30], bots in online environments [25], and MGT in technical research [20]. However, from the legal, security, and forensic points of view, only detecting whether a text has been automatically generated is not enough to identify who and what model, or family of models, is behind that text. In that sense, model attribution [27] can be employed to attribute a text to a specific LLM or a family of LLMs, yielding more insights into the actor behind the malicious MGT.

Most approaches to detect MGT and perform model attribution can be roughly divided into: (i) zero-shot detectors based on text statistics [16,30], and (ii) supervised detectors trained on human-authored and MGT texts [10,15,27]. Zero-shot methods usually assume access to the model(s) and are not always feasible, while supervised detectors do not make that assumption but are typically limited to specific domains and text-generation models [11]. The generalization capabilities of supervised detectors are still unexplored, and their high specialization could potentially limit their applicability.

In this work, we study the generalization of Transformer-based [28] supervised MGT detectors, across text generation model families and parameter scales. We also explore the feasibility of model family and parameter scale attribution, where the family and scale of the MGT generator must be identified instead of attributing an MGT to a specific language model, thus reducing the space of possible outcomes. To do this, we group models that share the same underlying architecture and are trained with the same data in the same manner, referring to them as *families*.[2] Similarly, we use *scale* to refer to models of different families but with similar number of parameters. We perform experiments both in English and Spanish, with monolingual and multilingual detectors: BLOOM [23], DeBERTa [9], MarIA [5], and XLM-RoBERTa [3], leveraging the AuTexTification corpus [22] which includes multi-domain and multilingual human-authored and machine-generated text, the latter generated by BLOOM and GPT models at different parameter scales. Under that setting, we propose the following research questions:

- **RQ1**: How well does a supervised MGT detector, trained on a particular family or scale of models, generalize to other families or scales?
- **RQ2**: Can a model of a particular family detect text generated by other models of the same family better than other MGT detectors?
- **RQ3**: Is it feasible to perform family or scale attribution?

---

[2] For instance, BLOOM refers to the family consisting of BLOOM-1b7, BLOOM-3b, BLOOM-7b1, etc.

2   Related Work

Recent LLMs are showing impressive text generation capabilities obtained by means of self-supervised pre-training on large-scale datasets, and, more recently, through instruction tuning to align to human preferences [18]. Large efforts have been dedicated to evaluate these LLMs in a consistent and comprehensible manner [14], where they showcase outstanding capabilities in most NLP tasks. Several of the best-performing LLMs are publicly available through different endpoints and under various licenses. Some are permissively licensed and open-sourced [23], some are restricted for research-only purposes [26], and others are made available through black-box APIs [2,18]. While there are other publicly-available but less powerful LLMs [19], we expect their use to gradually decrease in favor of better-performing LLMs, especially in applications aimed to deceive readers with human-looking MGT. Hence, given the current paradigm shift and in anticipation of future shifts, the generalization of MGT detectors is of key relevance [11].

MGT detection has shown remarkable results under assumptions of identical domain distribution and access to the generative model. However, in cross-domain settings or with certain writing styles, few works showed that the performance of specific detectors plummets [1]. This suggests a lack of generalization when these assumptions are broken. Generally, most works striving to detect MGT and performing model attribution fall under the following perspectives.

Machine-Aided Detection: in this paradigm, a human detector of MGT is assisted with statistical methods that capture generation artifacts. Since humans are good at noticing incoherence or factual errors in text and automatic methods are good detecting statistical abnormalities of token distributions, machine-aided detection strives to leverage the best of both worlds. The most prominent example is GLTR [7], a suite of statistical tools that improve humans' detection rate of text generated by GPT-2 [19] from 54% to 72% without any training. Yet, it requires a significant amount of human effort to be useful in practical scenarios, e.g., preventing massive campaigns of disinformation.

Zero-Shot: these approaches usually work under the white-box assumption, where a defender has access to the text-generation model that generated an MGT. Then, the same model is used to detect texts generated by itself or similar models, focusing on log-probabilities of the generated tokens. Two prominent examples of zero-shot detectors are presented in [16] and [24]. In [24], a baseline based on thresholding the sum of log-probabilities was found to detect MGT from a GPT-2 model with 85% accuracy. Likewise, DetectGPT [16] improves upon a zero-shot baseline for detecting fake news using log-probability ratios of text and perturbed samples. Zero-shot approaches are practical as they require no human intervention or training data. Nevertheless, their generalization capabilities to new generators are limited due to the white-box assumption, which severely constrains their application.

Supervised: these detectors are trained in a supervised fashion using datasets consisting of human-authored and machine-generated texts. Most are fine-tuned Transformer-based [28] language models such as RoBERTa [20,24,27], BERT

[10,15] or GROVER+LogisticRegression [30], with results usually higher than 90 macro F_1 under in-domain and in-model scenarios. Supervised detectors require diverse high-quality datasets that encompass various domains, text-generation models, generation hyper-parameters, and writing styles. However, the generalization capabilities of supervised detectors to new scenarios is still unexplored, and few works studied it tangentially for very specific detectors [1]. Depending on how well supervised detectors generalize, building these datasets could be impractical. To our knowledge, this is the first work that studies the generalization of Transformer-based supervised detectors across model families and parameter scales.

Watermarking: instead of aiming to detect MGT, these techniques are designed to distinguish MGT from human-authored text by modifying the generator's decoding strategy. Thus, the MGT includes a signature that makes it easily identifiable as MGT for automatic detectors. A notable example is watermarking by randomly ranking logit scores [13]. Another interesting approach [8] is to use a multi-task learning framework, where the model learns a set of backdoors pre-defined by its owner. However, watermarking could incentivize LLMs to generate lower quality text in an effort to satisfy watermark rules. Moreover, it requires enforcement, and malicious users could simply avoid using watermarked LLMs. Lastly, recent efforts have shown that watermarking can be beaten via paraphrasing MGT with another non-watermarked LLM [21].

3   Experiments

We carry out a set of experiments (i) to analyse the generalization of MGT detectors across families and scales of text generation models, and (ii) to explore the feasibility of MGT attribution to model families and parameter scales.

We frame the experiments to study the generalization of MGT detectors as binary classification tasks. For cross-family generalization, we train detectors with human text and MGT from a single family, then we evaluate them on detecting human text and MGT from different families, one family at a time. Following the same methodology, we also study MGT detectors' generalization across parameter scales.

We approach model family and parameter scale attribution as classification tasks between families or scales. We exclude human-authored text to consider the scenario where attribution is applied after a text has been identified as MGT. This way, we separately identify both the family and the scale of the MGT generator.

3.1   Experimental Set-Up

We use the AuTexTification Shared Task [22] datasets from both *Subtask 1: MGT detection* and *Subtask 2: Model Attribution*. The dataset statistics are presented in Table 1. These corpora include English and Spanish labeled human-authored and machine-generated text in five domains: tweets, reviews, how-to articles, news, and legal documents. The MGTs were obtained with two model families in various parameter scales: GPT-3 (babbage, curie and davinci) and BLOOM (BLOOM-1b7, -3b, and -7b), using nucleus sampling as a decoding strategy. For a more detailed description of the data we refer to the AuTexTification shared task overview [22].

We leverage data from both subtasks, and fine-tune Transformer-based classifiers, to answer our research questions. For MGT detection generalization across families, we use the MGT and labels from Subtask 2 by grouping them to obtain a training and test split per family. We add human text from Subtask 1 to these splits, matching the amount of MGT, thus obtaining our final training and test splits for each family. This way, we ensure that all the domains are in the training and test splits, and our data is balanced with respect to domains, classes (generated and human) and generators within families (or scales) in both splits. The same procedure is carried out to obtain per-scale training and test splits for scale-wise generalization. To study the attribution of MGT into families or scales we only use the Subtask 2 texts, grouping by families or scales, respectively.

Table 1. AuTexTification data statistics.

	Language	Split	**Human**			**Generated**		
Subtask 1: *MGT* *Detection*	English	Train	17,046			16,799		
		Test	10,642			11,190		
	Spanish	Train	15,787			16,275		
		Test	11,209			8,920		
			BLOOM			**GPT**		
	Language	Split	1b7	3b	7b	babbage	curie	davinci
Subtask 2: *Model* *Attribution*	English	Train	3,562	3,648	3,687	3,870	3,822	3,827
		Test	887	875	952	924	979	988
	Spanish	Train	3,422	3,514	3,575	3,788	3,770	3,866
		Test	870	867	878	946	1,004	917

For training our classifiers at each experiment, we employ three models: a language specific model (DeBERTa [9] for English and MarIA [5] for Spanish) a multilingual model (XLM-RoBERTa [3]), and a small model from a family of generators used to compile the AuTexTification dataset (BLOOM-560M [23]).[3] We fine-tune these models, with a randomly initialized classification head, in FP16 for 5 epochs using a linearly decaying learning rate schedule starting at 5e-5. Finally, we evaluate the models using class-wise and macro F_1 scores. All the experiments have been conducted using the HuggingFace ecosystem [29]. Code and results with additional metrics are available.[4]

---

[3] For the sake of fairness, our generalization experiments exclude the BLOOM models originally used to create the corpora. Likewise, we exclude GPT models given their limited transparency in the offered fine-tuning methodologies which could lead to unfair comparisons against the chosen classifiers.

[4] Due to space constraints, additional experiments, results and source code can be found at: https://github.com/symanto-research/supervised-mgt-family-scale.

3.2 Generalization in MGT Detection

Across Families. To study cross-family generalization, we split the generated text into two groups: GPT and BLOOM. We train MGT detectors with human-authored text and text from one family, then evaluating on both families separately. The results are presented in Tables 2 and 3 for English and Spanish, respectively.

In both languages we observe how all MGT detectors perform much better when tested on the same family, reaching differences of 29 macro F_1 with respect to cross-family evaluation when training with BLOOM in Spanish. Overall, **detectors do not generalize well to other families**.

In English, the language-specific detector (DeBERTa) outperforms the multilingual detectors in most scenarios. This also holds in Spanish for in-family evaluation, whereas in a cross-family setting the language-specific model, MarIA, lags behind XLM-R, with BLOOM-560 again having the worst performance, meaning that **language-specific detectors are generally preferable**.

In both languages BLOOM-560 obtains lower F_1 scores in the generated class than DeBERTa and XLM-R when trained with GPT and evaluated with BLOOM. Differences in terms of F_1 scores regarding the other detectors are generally large, with the largest difference being of 13 points in English and 18 in Spanish. More research is needed to determine whether family-specific detectors generalize well to their own families. Nonetheless, from this experiment we conclude that **BLOOM-560 does not generalize well to its family**.

In cross-family settings, and independently of the language, **most detectors obtain higher F_1 scores on the human class than in the generated one**, reaching 22 points of difference. This may be because the generated class contains MGT of different quality levels from the same family, while human text quality is consistently similar.

Finally, **the training family of generators matters**: cross-family generalization depends on the training family of an MGT detector. For example, when training with MGT from BLOOM and evaluating on GPT in English, all detectors obtain worse results than in the opposite generalization direction. Interestingly, this behaviour is reversed in Spanish, where detectors trained with BLOOM and evaluated on GPT perform better. Thus, **one must carefully choose the training families** when building datasets to train supervised detectors in order to generalize well to other model families. Besides, **this choice may be different for different languages**.

Across Parameter Scales. Similarly to the cross-family experiment, we train MGT detectors with human-authored text and text from one parameter scale of models. In this case, given the selection of models used to compile the AuTexTification datasets, we opt for three groups: **1b**, comprised of BLOOM-1b7 and babbage; **7b**, consisting of BLOOM-7b1 and curie; and **175b** which only includes davinci. This last group is only comprised of GPT models given their popularity and the lack of APIs that provide access to BLOOM-175b or other LLMs with similar parameter scales. We carry out in-scale and cross-scale evaluation in English (Table 4) and Spanish (Table 5).

Table 2. F_1 scores of the detectors for the generated (GEN) and human (HUM) classes when trained and evaluated on BLOOM and GPT model families (English). Best results in bold.

Train	Detector	BLOOM			GPT		
		GEN	HUM	Mean	GEN	HUM	Mean
BLOOM	BLOOM-560	93.70	93.92	93.81	59.32	75.81	67.57
	DeBERTa	**95.21**	**94.79**	**95.00**	76.19	80.66	78.43
	XLM-R	93.13	92.14	92.63	**79.26**	**80.86**	**80.06**
GPT	BLOOM-560	72.17	79.82	75.99	89.61	89.78	89.69
	DeBERTa	**85.61**	**85.05**	**85.33**	**89.94**	**87.82**	**88.88**
	XLM-R	82.40	82.04	82.22	89.52	87.22	88.37

Table 3. F_1 scores of the detectors for the generated (GEN) and human (HUM) classes when trained and evaluated on BLOOM and GPT model families (Spanish). Best results in bold.

Train	Detector	BLOOM			GPT		
		GEN	HUM	Mean	GEN	HUM	mean
BLOOM	BLOOM-560	88.05	87.78	87.91	65.03	73.52	69.28
	MarIA	**96.25**	**96.29**	**96.27**	58.95	75.91	67.43
	XLM-R	91.74	90.32	91.03	**73.93**	**76.29**	**75.11**
GPT	BLOOM-560	52.68	73.91	63.30	90.69	91.12	90.91
	MarIA	56.91	75.64	66.27	**94.97**	**94.98**	**94.98**
	XLM-R	**70.58**	**76.76**	**73.67**	91.14	89.50	90.32

We observe that most cross-scale evaluations result in +80 macro F_1, meaning that in general, **MGT detectors generalize well to other scales**, sometimes performing better than their in-scale counterparts as is the case of DeBERTa when evaluated in the 7b scale after training with MGT from 1b-scaled models. However, in some particular cases, we find **bad generalization from very large scales to small ones**. For example, when training on MGT from the 175b scale, the cross-scale performance is lower than in other scenarios, which can be due to this scale only including MGT generated by GPT and not of the largest BLOOM model. Interestingly, when we analyze this behaviour from the text readability and complexity viewpoint[6] (see Table 6), we observe that the readability of generated texts incorrectly classified as human is generally similar to that of correctly classified human texts: they are both easier to read. This is also in line with the training instances, where the generated texts have a mean readability score of 72.13 in contrast to a 77.09 in human-authored texts. In addition, the average number of *difficult words*[5] is greater in the human class (see our additional results[6]). This signal is captured by some models: when

---

[5] Difficult words according to: https://github.com/textstat/textstat

Table 4. F_1 scores of the detectors for the generated (GEN) and human (HUM) classes when trained and evaluated on 1b, 7b and 175 parameter scales (English). Best results in bold.

Train	Detector	1b			7b			175b		
		GEN	HUM	Mean	GEN	HUM	Mean	GEN	HUM	Mean
1b	BLOOM-560	89.69	90.04	89.89	85.22	86.45	85.84	76.37	83.43	79.90
	DeBERTa	**93.46**	**92.88**	**93.17**	**91.84**	**91.04**	**91.44**	89.90	**91.45**	90.67
	XLM-R	89.29	86.96	88.13	87.87	84.67	86.27	**91.12**	90.86	**90.99**
7b	BLOOM-560	87.49	**88.25**	**87.87**	86.02	**86.72**	**86.37**	79.16	84.75	81.96
	DeBERTa	**88.71**	85.99	87.35	**87.20**	83.14	85.17	**92.38**	**92.03**	**92.20**
	XLM-R	86.92	82.89	84.91	85.30	79.59	82.45	90.02	88.87	89.44
175b	BLOOM-560	56.14	74.47	65.30	64.47	77.36	70.92	91.52	**91.97**	91.75
	DeBERTa	69.77	75.51	72.64	**81.36**	**81.86**	**81.61**	**92.64**	91.48	**92.06**
	XLM-R	**73.31**	**75.67**	**74.49**	**81.36**	80.61	80.99	90.50	88.45	89.48

testing in the 1b and 175b scales, there are in average over 5% more difficult words in the predicted human class. Note that, when evaluating in the 175b scale in a cross-scale scenario, the detectors obtain reasonably good scores. In fact, it is possible to detect texts generated by davinci with $+90\,F_1$ using a training set comprised of text generated by models of 1b parameters. As in our previous study, this shows that one must **carefully choose the LLMs' scale** when building datasets to train supervised detectors in order to generalize well to other model scales.

Similarly to the cross-family experiment, we observe that **language-specific detectors perform better than multilingual ones**. However, in contrast to the previous experiment, for cross-scale generalization we find that models are typically **not biased towards the human class** given that the F_1 scores for each class are similar in most cases; in fact the opposite is sometimes true, especially when training with MGT from the 7b scale.

Finally, we find that **BLOOM-560 does not generalize well when trained with the 175b scale** in both English and Spanish, obtaining macro F_1 scores of as much as 11 points lower than the best detector when evaluated on MGTs from the 7b scale. This could be due to BLOOM-560 being trained with MGT that is very different to the distribution it had originally learned. Additionally, in English it obtains low results when generalizing to the 175b scale.

3.3   Family and Scale Attribution

Family Attribution. We study the family attribution problem in English and Spanish by fine-tuning Transformer-based language models to classify MGT into two classes, BLOOM and GPT. The results are presented in Table 7, where we observe that attributors slightly favor GPT texts, obtaining better F_1 scores compared to the BLOOM class. Additionally, BLOOM-560 **does not perform**

Table 5. F_1 scores of the detectors for the generated (GEN) and human (HUM) classes when trained and evaluated on 1b, 7b and 175 parameter scales (Spanish). Best results in bold.

		1b			7b			175b		
Train	Detector	GEN	HUM	Mean	GEN	HUM	Mean	GEN	HUM	Mean
1b	BLOOM-560	90.57	90.09	90.33	86.76	86.77	86.72	86.58	88.98	87.78
	MarIA	**94.13**	**94.25**	**94.19**	**90.90**	**91.54**	**91.22**	83.33	87.50	85.42
	XLM-R	87.85	84.35	86.10	86.67	82.62	84.64	**91.58**	**91.18**	**91.38**
7b	BLOOM-560	88.03	88.35	88.19	87.54	87.75	87.65	88.48	90.41	89.44
	MarIA	**91.75**	**92.00**	**91.88**	**92.52**	**92.54**	**92.53**	93.43	94.20	**93.82**
	XLM-R	85.61	80.24	82.92	84.64	78.37	81.51	90.16	88.69	89.43
175b	BLOOM-560	51.85	73.16	62.50	55.37	74.22	64.80	93.27	93.64	93.45
	MarIA	53.77	74.23	64.00	64.16	77.27	70.71	**96.29**	**96.30**	**96.29**
	XLM-R	**73.45**	**75.17**	**74.31**	**79.97**	**78.88**	**79.42**	90.74	88.80	89.77

Table 6. Mean Flesch Reading Ease Scores [6] for the predictions of XLM-R trained using the 175b data in English.

		1b		7b		175b	
Train	True Labels	GEN	HUM	GEN	HUM	GEN	HUM
175b	GEN	75.16	78.31	75.05	83.27	71.75	78.24
	HUM	77.90	77.96	78.06	76.81	78.99	77.25

Table 7. F_1 scores of attributors of MGT in English and Spanish in the BLOOM or GPT families. Best results in bold.

	English				Spanish		
Attributor	BLOOM	GPT	Mean	Attributor	BLOOM	GPT	Mean
BLOOM-560	90.55	91.23	90.89	BLOOM-560	91.25	92.46	91.86
DeBERTa	**94.09**	**94.51**	**94.30**	MarIA	94.77	95.25	95.01
XLM-R	93.97	93.97	93.97	XLM-R	**95.10**	**95.48**	**95.29**

on par with other attributors, especially when attributing text to its own family. It does not find a bias towards MGT of its own family for attribution, which follows from what was observed in previous experiments. Moreover, **language-specific attributors are not necessarily better**, seeing as XLM-R performs on par with DeBERTa and MarIA. Given the observed +90 macro F_1 scores, we conclude that **MGT can be feasibly attributed to model families**, thus reducing the space of possible outcomes.

Scale Attribution. We study cross-scale generalization in MGT attributors in English and Spanish by fine-tuning attributors to classify MGT into two classes:

Table 8. F_1 scores of attributors of MGT in English and Spanish in the 1b or 7b scales. Best results in bold.

| Attributor | English | | | Attributor | Spanish | | |
	1b	7b	Mean		1b	7b	Mean
BLOOM-560	56.47	60.59	58.53	BLOOM-560	59.90	57.56	58.73
DeBERTa	**67.15**	**69.93**	**68.54**	MarIA	**70.42**	**72.40**	**71.41**
XLM-R	65.23	0.00	32.61	XLM-R	65.87	0.00	32.93

1b, comprised of MGT from BLOOM-1b7 and babbage, and 7b which contains MGT generated by curie and BLOOM-7b1. We exclude BLOOM-3b and GPT davinci since, in our experiment data, they cannot be paired with other models from the other family in their respective parameter scales. The results are shown in Table 8, where we observe lower scores in comparison to the family attribution experiment. In this experiment, XLM-R always predict the 1b scale. We hypothesize that this could be either due to overfitting or some aspect that degenerates the training dynamics, such as the random seed or learning rate scheduler. The best attributor is the language-specific MarIA, which reaches 71 macro F_1. MarIA obtains a similar F_1 score both for both 1b and 7b scales, suggesting that either (i) generators of 1b and 7b scales generate text of similar quality, or (ii) they include MGT from different model families, meaning that within each scale class the texts do not have many underlying similarities. Thus, we conclude that, while not being as feasible as family attribution, **scale attribution is promising and has potential for high performance with further developments.**

4   Conclusions

We have studied cross-family and cross-scale generalization for MGT detectors, as well as the feasibility of family and scale attribution. In the former case, we have observed how MTG detectors do not generalize well to other families but they generalize to scales, that language-specificity instead of multilinguality should be favored, and that the choice of training family or scale significantly affects MTG detection generalization capabilities. Additionally, we found that the BLOOM-560, part of the family of generators, does not generalize well to its family, nor does it generalize when trained on MGT from GPT alone. In the latter, we discovered that while scale attribution requires further research, family attribution can be carried out with very good results, in which case language-specific attributors need not be favored.

Acknowledgements. The work from Symanto has been partially funded by the Pro[2]Haters - Proactive Profiling of Hate Speech Spreaders (CDTi IDI-20210776), the XAI-DisInfodemics: eXplainable AI for disinformation and conspiracy detection during infodemics (MICIN PLEC2021-007681), the OBULEX - *OBservatorio del Uso de*

Lenguage sEXista en la red (IVACE IMINOD/2022/106), and the ANDHI - ANomalous Diffusion of Harmful Information (CPP2021-008994) R&D grants. The work of Areg Mikael Sarvazyan has been partially developed with the support of valgrAI - Valencian Graduate School and Research Network of Artificial Intelligence and the Generalitat Valenciana, and co-founded by the European Union. The work of Paolo Rosso was done in the framework of the research project on Fairness and Transparency for equitable NLP applications in social media (Grant PID2021-124361OB-C31 funded by MCIN/AEI/10.13039/501100011033 and by ERDF, EU A way of making Europe).

References

1. Bakhtin, A., Gross, S., Ott, M., Deng, Y., Ranzato, M., Szlam, A.: Real or fake? learning to discriminate machine from human generated text. arXiv preprint arXiv:1906.03351 (2019)
2. Brown, T., et al.: Language models are few-shot learners. In: Advances in Neural Information Processing Systems, pp. 1877–1901 (2020)
3. Conneau, A., et al.: Unsupervised cross-lingual representation learning at scale. In: Proceedings of the 58th Annual Meeting of the Association for Computational Linguistics, pp. 8440–8451 (2020)
4. Eloundou, T., Manning, S., Mishkin, P., Rock, D.: GPTs are GPTs: an early look at the labor market impact potential of large language models. arXiv preprint arXiv:2303.10130 (2023)
5. Fandiño, A.G., et al.: MarIA: Spanish language models. Procesamiento del Lenguaje Natural (2022)
6. Flesch, R.: A new readability yardstick. J. Appl. Psychol. **32**, 221 (1948)
7. Gehrmann, S., Strobelt, H., Rush, A.: GLTR: statistical detection and visualization of generated text. In: Proceedings of the 57th Annual Meeting of the Association for Computational Linguistics: System Demonstrations, pp. 111–116 (2019)
8. Gu, C., Huang, C., Zheng, X., Chang, K.W., Hsieh, C.J.: Watermarking pre-trained language models with backdooring. arXiv preprint arXiv:2210.07543 (2022)
9. He, P., Gao, J., Chen, W.: DeBERTav3: improving deBERTa using ELECTRA-style pre-training with gradient-disentangled embedding sharing. In: The Eleventh International Conference on Learning Representations (2023)
10. Ippolito, D., Duckworth, D., Callison-Burch, C., Eck, D.: Automatic detection of generated text is easiest when humans are fooled. In: Proceedings of the 58th Annual Meeting of the Association for Computational Linguistics, pp. 1808–1822 (2020)
11. Jawahar, G., Abdul-Mageed, M., Lakshmanan, V.S., L.: Automatic detection of machine generated text: a critical survey. In: Proceedings of the 28th International Conference on Computational Linguistics, pp. 2296–2309 (2020)
12. Kasneci, E., et al.: ChatGPT for good? on opportunities and challenges of large language models for education. Learn. Individ. Differ. **103**, 102274 (2023)
13. Kirchenbauer, J., Geiping, J., Wen, Y., Katz, J., Miers, I., Goldstein, T.: A watermark for large language models. arXiv preprint arXiv:2301.10226 (2023)
14. Liang, P., et al.: Holistic evaluation of language models. arXiv preprint arXiv:2211.09110 (2022)
15. Maronikolakis, A., Schütze, H., Stevenson, M.: Identifying automatically generated headlines using transformers. In: Proceedings of the Fourth Workshop on NLP for Internet Freedom: Censorship, Disinformation, and Propaganda, pp. 1–6 (2021)

16. Mitchell, E., Lee, Y., Khazatsky, A., Manning, C.D., Finn, C.: DetectGPT: zero-shot machine-generated text detection using probability curvature. arXiv preprint arXiv:2301.11305 (2023)
17. OpenAI: AI text classifier. OpenAI Blog (2023)
18. Ouyang, L., et al.: Training language models to follow instructions with human feedback. In: Advances in Neural Information Processing Systems (2022)
19. Radford, A., Wu, J., Child, R., Luan, D., Amodei, D., Sutskever, I.: Language models are unsupervised multitask learners. In: OpenAI (2019)
20. Rodriguez, J., Hay, T., Gros, D., Shamsi, Z., Srinivasan, R.: Cross-domain detection of GPT-2-generated technical text. In: Proceedings of the 2022 Conference of the North American Chapter of the Association for Computational Linguistics: Human Language Technologies, pp. 1213–1233 (2022)
21. Sadasivan, V.S., Kumar, A., Balasubramanian, S., Wang, W., Feizi, S.: Can AI-generated text be reliably detected? arXiv preprint arXiv:2303.11156 (2023)
22. Sarvazyan, A.M., González, J.Á., Franco-Salvador, M., Rangel, F., Chulvi, B., Rosso, P.: Overview of AuTexTification at IberLEF 2023: detection and attribution of machine-generated text in multiple domains. In: Procesamiento del Lenguaje Natural (2023)
23. Scao, T.L., et al.: BLOOM: A 176B-parameter open-access multilingual language model. arXiv preprint arXiv:2211.05100 (2022)
24. Solaiman, I., et al.: Release strategies and the social impacts of language models. arXiv preprint arXiv:1908.09203 (2019)
25. Tourille, J., Sow, B., Popescu, A.: Automatic detection of bot-generated tweets. In: Proceedings of the 1st International Workshop on Multimedia AI against Disinformation, p. 44–51 (2022)
26. Touvron, H., et al.: LLaMA: open and efficient foundation language models. arXiv preprint arXiv:2302.13971 (2023)
27. Uchendu, A., Le, T., Shu, K., Lee, D.: Authorship attribution for neural text generation. In: Proceedings of the 2020 Conference on Empirical Methods in Natural Language Processing (EMNLP), pp. 8384–8395 (2020)
28. Vaswani, A., et al.: Attention is all you need. In: Advances in Neural Information Processing Systems (2017)
29. Wolf, T., et al.: Transformers: state-of-the-art natural language processing. In: Proceedings of the 2020 Conference on Empirical Methods in Natural Language Processing: System Demonstrations, pp. 38–45 (2020)
30. Zellers, R., et al.: Defending against neural fake news. In: Advances in Neural Information Processing Systems (2019)

Best of CLEF 2022 Labs

Cross-Lingual Candidate Retrieval and Re-ranking for Biomedical Entity Linking

Florian Borchert[(✉)] [iD], Ignacio Llorca [iD], and Matthieu-P. Schapranow [iD]

Hasso Plattner Institute for Digital Engineering, University of Potsdam,
Prof.-Dr.-Helmert-Str. 2-3, 14482 Potsdam, Germany
{florian.borchert,schapranow}@hpi.de, llorcarodriguez@uni-potsdam.de

Abstract. Biomedical entity linking is an essential building block for various clinical applications and downstream NLP tasks. However, only few annotated biomedical datasets with grounded entity mentions for non-English languages are available for training supervised machine learning models. Moreover, the majority of concept aliases in medical vocabularies are also only available in English.

In this work, we consider the problem of linking disease mentions in Spanish clinical case reports to concept identifiers in SNOMED CT, a comprehensive medical terminology system. For these concepts, only a limited number of aliases in the source language are given, but many more can be obtained from other languages and medical vocabularies. We propose a system that utilizes these multilingual aliases to retrieve candidate concepts for a given entity mention and re-ranks retrieved candidates using a trainable cross-encoder. We evaluate our system on the DISTEMIST shared task dataset of the 10th BIOASQ challenge.

Our results show that supervised re-ranking outperforms the previously best-performing rule-based system, while requiring much less task-specific hyperparameter tuning. Detailed ablation experiments demonstrate that multilingual aliases are highly beneficial to improve recall during candidate generation, but hardly affect re-ranking performance.

Keywords: Clinical NLP · Entity Linking · Spanish · BioASQ

1 Introduction

Biomedical entity linking (EL) is an essential task for extracting structured metadata from medical text documents and a building block for various downstream tasks. The target knowledge bases (KB) for biomedical EL are typically derived from medical terminology systems or subsets thereof, e.g., ontology-based systems like SNOMED CT or the Unified Medical Language System (UMLS) Metathesaurus [4,9]. However, the majority of terms in such terminology systems as well as the largest datasets with annotations of grounded

© The Author(s), under exclusive license to Springer Nature Switzerland AG 2023
A. Arampatzis et al. (Eds.): CLEF 2023, LNCS 14163, pp. 135–147, 2023.
https://doi.org/10.1007/978-3-031-42448-9_12

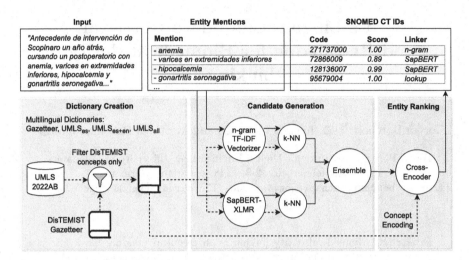

Fig. 1. Overview of our entity linking system. We obtain dictionaries with increasing numbers of aliases in Spanish and other languages for the 111K SNOMED CT concepts relevant for the DISTEMIST shared task. For each dictionary, we evaluate different candidate generation approaches: TF-IDF scores based on character n-grams, dense retrieval with the cross-lingual version of SAPBERT, and an ensemble of both. The generated candidates are re-ranked using a trainable cross-encoder, for which we also evaluate different concept encodings based on mono- and multilingual aliases.

entity mentions are only available in English. Moreover, even the largest annotated English-language datasets only cover a tiny fraction of the concepts found in typical biomedical terminologies [17]. Therefore, linking entity mentions in non-English text to comprehensive biomedical terminologies is still considered as challenging.

The DISTEMIST shared task of the 10th BIOASQ lab addressed the problem of identifying and linking disease mentions in Spanish clinical case reports [16]. In the task, a subset of SNOMED CT with disease-related concepts was considered as the target KB. We have participated in the task with an EL system built upon an ensemble of candidate generators, followed by a rule-based re-ranking step, which obtained the highest F_1 score in the EL sub-track [6].

In this work, we build on our experience in the DISTEMIST task and propose an enhanced EL system, shown in Fig. 1. We extend our previous unsupervised, cross-lingual candidate generation approach through supervised re-ranking with a cross-encoder. In contrast to our previous system, the re-ranker is easier to adapt to other datasets: it has fewer task-specific hyperparameters to consider, while being trainable on given ground truth annotations. Moreover, we present a detailed investigation on the impact of adding multilingual aliases (i.e., synonyms and equivalents in other languages) for both candidate generation and ranking.

The remainder of this work is structured as follows: in Sect. 2, we set our approach in the context of related work. In Sect. 3, we outline details about incorporated datasets, created dictionaries, and components of our EL system. In Sect. 4, we share the results of our approach in the context of the DISTEMIST shared task. Findings, limitations, and potential improvements are discussed in Sect. 5. Our work concludes with an outlook in Sect. 6.

2   Related Work

The prevalent architecture of EL systems consists of a candidate generator followed by an entity ranker, with the goal to link textual entity mentions to a target KB [24]. While the KB considered in general-domain EL is often Wikipedia, the target in biomedical EL are usually concepts in controlled vocabularies, ontologies, or other medical terminology systems. In this context, we consider the task-relevant subsets of concept identifiers, aliases, and metadata (such as semantic type information or definitions) from these sources as our KB.

Historically, representations based on TF-IDF scores, or variants like BM25, have been the dominant approach for candidate generation [18,21]. These are still widely used as components in current, neural systems [17,25]. More recently, dense retrieval based on (contextualized) word embeddings has been employed for candidate generation, with a variety of techniques for fine-tuning Transformer-based encoders for this purpose. Liu et al. propose *self-alignment pretraining* (SAP) based on aliases in the UMLS to improve semantic similarity of concepts in the embedding space [13]. Examples of supervised fine-tuning approaches of candidate generators are multi-class classification models for small, restricted label spaces or the bi-encoder architecture for zero-shot EL [28,29].

A common formulation of the candidate ranking problem is the computation of a similarity score between mentions and candidate concepts, e.g., with Transformer-based cross-encoders [1,15,28]. Xu et al. propose to frame the problem as a multiple-choice task with representations that resemble the aforementioned cross-encoder architectures [29]. Apart from these trainable neural approaches, entity ranking can often be improved by simple heuristics, e.g., by setting thresholds or employing semantic type information [6,18,22,26].

The aforementioned works primarily concern English-language text and rely on target KB aliases or entity descriptions in the source language, which are often not or only partially available for languages other than English. Roller et al. apply neural machine translation models to both KB aliases and entity mentions, obtaining strong performance on the Quaero corpus [19,22]. Wajsbürt et al. treat the problem as a classification task and improve performance on the same corpus through distant supervision from the UMLS and the incorporation of English-language aliases [27]. In contrast, the cross-lingual version of SAPBERT is pre-trained on multilingual UMLS aliases to embed terms from different languages in the same embedding space, allowing for a simple nearest neighbor lookup [13]. SAPBERT has shown competitive performance on several biomedical EL benchmarks, even without fine-tuning on task-specific data [2].

In the context of the DISTEMIST shared task, the three best performing teams in the EL sub-track employed dense representations of mentions and concepts [3, 6, 8]. Our own system was based on a hybrid candidate generation approach with TF-IDF vectors over mono-lingual character n-grams and cross-lingual SAPBERT embeddings, followed by a rule-based re-ranking step [6]. In this work, we improve this pipeline by combining it with a trainable cross-encoder for re-ranking. Additionally, we investigate the impact of using multilingual aliases for both candidate generation and ranking.

3   Materials and Methods

In the following, we describe the dataset, different instances of dictionaries we have assembled, and components of our EL system as depicted in Fig. 1.

3.1   DisTEMIST Dataset

The DISTEMIST dataset consists of Spanish-language clinical case reports with expert annotations of disease mentions. The full training set contains 750 documents, of which 583 were also grounded with SNOMED CT codes (EL sub-track). We have sampled 20% (117) of these documents as an internal validation set, and used the remaining 466 documents for training. These are the same splits as in our DISTEMIST submission [6]. After the shared task, the test set of 250 documents was released, allowing us to consider them in our experiments. While the task included a sub-track for disease NER, we focus our analysis solely on the EL sub-track and assume gold mention spans are given in all our experiments. For data loading, we rely on the BIGBIO framework, to which we have contributed an implementation for loading the EL-relevant part of DISTEMIST [10].

3.2   Dictionaries and Multilingual Aliases

A gazetteer of 111,179 SNOMED CT concepts with 147,280 Spanish aliases was provided as part of the DISTEMIST task. While SNOMED CT is much more extensive, only this limited number of concepts were considered during the annotation phase and used for evaluation. To extend the number of available aliases, we employ the UMLS Metathesaurus (release 2022AB) [4] and obtain increasingly large sets of aliases that can be mapped to the target concepts, i.e., all terms belonging to the same UMLS concept unique identifier as the respective SNOMED CT concept. We consider all vocabularies included in the UMLS and the following language subsets:

- **UMLS$_{es}$:** Spanish synonyms only (493,545 aliases)
- **UMLS$_{es+en}$:** Spanish and English equivalents (1,518,833 aliases)
- **UMLS$_{all}$:** all languages in the UMLS (2,429,879 aliases).

3.3   Candidate Generation

In the following, we describe two different approaches for candidate generation and their combination in an ensemble.

TF-IDF with Character N-Grams. As a first, simple approach to calculate surface form similarity, we encode all candidate aliases as·TF-IDF vectors based on character 3-grams. To this end, we have adapted the implementation from SCISPACY to work with non-English UMLS subsets [18]. At inference time, the same encoding is applied to mention spans, followed by an approximate nearest neighbor search to generate a ranked list of k candidates. Although this approach has a few adaptable parameters (such as the number of characters n for the n-gram index), we keep the default values from SCISPACY in our experiments.

Cross-Lingual SapBERT. While the (sparse) TF-IDF encoding is based on surface form similarity, dense representations are potentially more appropriate to capture the semantic similarity of entity mentions and target concepts. To this end, we employ the cross-lingual version of SAPBERT [13]. We obtain representations for all aliases and mentions by using the embedding of the [CLS] token in the last hidden layer of the BERT model. For efficient retrieval, we use the FAISS library to create an index into the dictionary and perform an approximate nearest neighbor search at inference time [12]. For the ranked candidate list, we use the cosine similarity of concept and mention embeddings as a candidate score.

Ensemble. Additionally, we combine the scored candidate lists by merging and re-sorting them based on their candidate scores. When a candidate is part of both candidate lists, we keep the maximum of the respective scores. Although the candidate scores are not calibrated (i.e., the distribution of scores can be very different between both candidate generators), we do not apply any re-weighting or thresholds at this point and leave it to the following re-ranking step.

3.4   Entity Ranking

We use a cross-encoder with a linear output layer to assign a score to each mention-candidate-pair for re-ranking of generated candidates, which is similar to the approach proposed by Wu et al. [28]. However, we adapted the concept representation to account for the variety of aliases in medical terminologies. In contrast to the candidate generators, the cross-encoder is trainable and makes use of the ground-truth concept labels in the training data.

Mention and Context Encoding. Each mention is encoded together with its context to the left and to the right as follows:

$$[\text{CLS}] \ \text{context}_l \ [\text{START}] \ \text{mention} \ [\text{END}] \ \text{context}_r$$

with [START] and [END] denoting the beginning/end of the mention string that shall be linked. The context length is a hyperparameter, which we fixed to 128 characters for all experiments.

Concept Encoding. We obtain a representation for each concept by concatenating its canonical name with all its aliases ($SYN_{1..n}$) similar to the encoding proposed by Xu et al. [29]. In addition, we encode the concept's semantic type to obtain the following representation:

semantic type [TYPE] canonical name [TITLE] SYN_1 [SEP] ... [SEP] SYN_n.

In our experiments, we evaluate different variations of the concept encoding by considering all available aliases in the dictionaries introduced in Sect. 3.2, i.e., ranging from very few synonyms in the DISTEMIST gazetteer to aliases from all available languages in the UMLS Metathesaurus.

Training and Model Selection. For training, we use batches of $k = 64$ candidate concepts for each mention and encode the concatenation of the mention (and context) with the concept representation using the encoder of a BERT-based Transformer. We have chosen $k = 64$ as suggested by Wu et al. [28]. For the UMLS$_{all}$ setting, larger batches would also not fit into the 48 GB of GPU memory available in our system. We use the cross-encoder implementation from the Sentence Transformers framework and employ a softmax loss to maximize the score of the correct candidate within each batch [20]. The model is trained for 20 epochs on a single NVIDIA A40 GPU. As the final model for each experiment, we keep the checkpoint that maximizes recall@1 on the validation set. We did not perform any hyperparameter optimization - instead we have used the default values from the Sentence Transformers framework (learning rate of 2×10^{-5} with linear decay and warmup, weight decay of 1×10^{-2}).

In our experiments, we compare the initialization of the encoder from two different checkpoints: 1) the same cross-lingual SAPBERT model, which we have used for unsupervised candidate generation and 2) a mono-lingual RoBERTa model pre-trained on a large corpus of Spanish biomedical-clinical documents, which we refer to as PlanTL-GOB-ES [7,14].

3.5   Evaluation

Our primary evaluation metric is recall for different numbers k of candidates (recall@k). When a prediction has to be made for each entity mention, recall@1 is equivalent to accuracy. The DISTEMIST shared task evaluation also considered precision (and the resulting F_1 score), e.g., abstaining from making a prediction was allowed for participating systems. While we focus our analysis on recall, a different trade-off can usually be obtained by varying prediction thresholds, which we have extensively tuned in our original DISTEMIST contribution [6].

4   Results

In the following, we share the results of our system.

Table 1. Candidate generation performance on the <u>validation set</u>. We report the recall for different numbers k of generated candidates for increasingly large dictionaries and different **candidate generators**.

Dictionary	Aliases	Cand. Gen.	Recall @ k						
			1	2	4	8	16	32	64
Gazetteer	147,280	TF-IDF	.397	.489	.541	.595	.641	.680	.717
		SAPBERT	**.444**	.554	.630	.687	.737	.779	.795
		Ensemble	.418	**.565**	.635	.690	.738	.788	.805
UMLS$_{es}$	493,545	TF-IDF	.290	.407	.579	.663	.707	.749	.774
		SAPBERT	.351	.493	.639	.728	.769	.799	.816
		Ensemble	.409	.522	.645	.728	.775	.811	.830
UMLS$_{es+en}$	1,518,833	TF-IDF	.286	.414	.586	.673	.721	.762	.797
		SAPBERT	.363	.524	.659	.744	.786	.808	.826
		Ensemble	.427	.541	**.660**	**.749**	**.795**	**.815**	**.841**
UMLS$_{all}$	2,429,879	TF-IDF	.292	.409	.577	.664	.719	.764	.795
		SAPBERT	.351	.525	.645	.735	.772	.802	.819
		Ensemble	.419	.532	.655	.744	.781	.812	.836

4.1 Candidate Generation

Table 1 shows the results of the candidate generation phase on the validation set. Note that we do not report test set results because we cannot choose the best performing candidate generation model based on test set labels without biasing the final evaluation. The best scores for recall@1 and recall@2 are achieved when using the DISTEMIST gazetteer only, i.e., with the smallest set of aliases. Although SAPBERT consistently outperforms the TF-IDF-based candidate generator, it is beneficial to combine both candidate lists in the ensemble in the vast majority of cases. For larger values of k, including multilingual aliases consistently improves recall. For instance, recall@64 of the ensemble with aliases from UMLS$_{es+en}$ is 3.6pp better than the ensemble with aliases from the DISTEMIST gazetteer only. However, we note that including too many aliases is slightly detrimental, with the UMLS$_{all}$ models performing marginally worse than UMLS$_{es+en}$. For the re-ranking phase, we have used the 64 candidates generated for each mention by the best performing generator, i.e., the ensemble with aliases from UMLS$_{es+en}$.

4.2 Re-ranking

As shown in Table 2, the supervised re-ranking step improves recall for all values of k on the test set. In particular, recall@1 is improved by up to 19.4pp compared to the raw candidate generator output. We find that the general purpose clinical BERT model generally outperforms SAPBERT when used to initialize the cross-encoder - we suppose that the former is better at handling context in the mention

Table 2. Recall at different value of k on the test set after re-ranking. We evaluate two different Transformer models, from which we initialize the cross-encoder and use aliases from increasingly large dictionaries to obtain the entity representations. Comparison to the raw candidate generation performance highlights the benefits of re-ranking, especially for smaller value of k. Recall@64 is identical in all settings, as this is the maximum number of candidates that are subject to re-ranking.

Model	Dictionary	Recall @ k						
		1	2	4	8	16	32	64
SAPBERT	Gazetteer	.584	.648	.696	.729	.763	.783	.798
	UMLS$_{es}$	.576	.649	.700	.733	.760	.783	.798
	UMLS$_{es+en}$	.578	.652	.706	.739	.765	.783	.798
	UMLS$_{all}$	.585	.657	.705	.735	.762	.782	.798
PlanTL-GOB-ES	Gazetteer	.586	.658	.711	.739	.770	**.791**	.798
	UMLS$_{es}$	.590	**.667**	.711	.744	.767	.784	.798
	UMLS$_{es+en}$	.591	.658	**.712**	.749	**.772**	.789	.798
	UMLS$_{all}$	**.592**	.659	.711	**.750**	.770	.785	.798
Cand. Gen. (Ensemble/UMLS$_{es+en}$)		.398	.528	.624	.700	.739	.769	.798

encoding, as SAPBERT was optimized for representing single aliases rather than sentences. However, while multilingual aliases slightly improve recall in some cases, the impact is generally negligible. For representing concepts in the re-reranking phase, few aliases appear to be sufficient for good performance, while resulting in a drastically reduced memory footprint and training time.

5   Evaluation and Discussion

In this section, we discuss our results, including a root-cause analysis of errors and limitations of our work.

5.1   Comparison with Baseline

During the original DISTEMIST shared task, test set labels were not available to participants and the official task results were determined based on predicted, rather than gold-standard mention spans. Thus, we cannot directly compare our results in Table 2 to the other DISTEMIST participants. However, we previously carried out some ablations on the gold-standard labels of our own validation set, where our best performing, highly tuned rule-based system achieved a recall@1 of 62.5% [6]. In comparison, the best performing trained cross-encoder achieves a recall@1 of 63.7% on the same validation set without any post-processing (not shown in Table 2). After applying the same post-processing as before (looking up exact matches of entities in the training set), recall@1 improves by another 3pp. to 66.7%, which is a substantial improvement over our best performing system in the original shared task evaluation. Interestingly, the impact of the training set

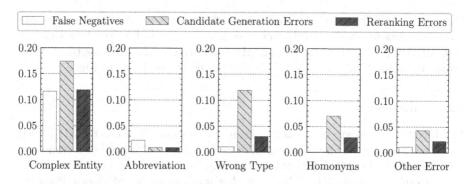

Fig. 2. Errors as a fraction of the total number of linkable entities in the validation set. *False Negatives* refer to the subset of entities that were not among the 64 retrieved candidates. *Candidate Generation Errors* and *Reranking Errors* refer to the remaining entities, where the correct concept was not the top 1 prediction, before and after reranking respectively.

lookup in our system is much less pronounced compared to the previous system, where it accounted for more than 9pp. improvement in recall. This suggests that the cross-encoder learns to rank the majority of the affected cases correctly, even without this heuristic.

5.2   Error Analysis

The recall@64 of the best performing candidate generator achieves a high value, with 84.1% of gold candidates on the validation set (and slightly below on the test set). Nevertheless, it is still far from the near-total recall that is desirable during candidate generation. We investigate the remaining entities, which were missed by our candidate generation step in Fig. 2. A reduction in recall of 11.6% occurs due to *complex entities*, which we define as entity mentions consisting of three or more tokens, occurring commonly among the annotated spans in DISTEMIST. These include noun phrases with multiple modifiers, commonly found in medical diagnoses (e.g., "linfangioma renal bilateral"/*bilateral renal lymphangioma*), but also coordinated clauses, which can be linked to multiple concepts (e.g., "alergia a ácaros, gramíneas y epitelio de animales"/*mite, grass and animal epithelium allergy*). Another 2.2% of errors refer to abbreviations, which are generally challenging to link without prior abbreviation expansion (e.g., "DM" can refer to "diabetes mellitus" or "distrofia miotónica"/*myotonic dystrophy* depending on the context).

Complex entities also account for a large fraction of ranking errors, both before and after re-ranking. Another frequent error scenario is that the top 1 entity has an incorrect semantic type (e.g., *morphological abnormality* instead of *disorder*). Moreover, both candidate generators consider all aliases independently to make linking decisions. This is a common source of ranking errors because many aliases are shared between concepts in the UMLS Metathesaurus (homonyms), although it does not affect the overall recall. For instance, the term "carcinoma renal"/*renal carcinoma* is a potential alias for three different SNOMED CT codes in the DISTEMIST gazetteer. The number of all these errors are dramatically reduced through re-ranking, as shown in Fig. 2.

5.3   Limitations

Our investigations have been carried out in the context of the DISTEMIST task. Thus, our findings may be partly specific to the dataset at hand. For instance, we have found that candidate generators using multilingual aliases from $UMLS_{all}$ are slightly outperformed by $UMLS_{es+en}$ with a smaller number of aliases. However, it is likely that this applies to Spanish language datasets only and the former approach might generalize much better across datasets in different languages.

We have shown that complex (long) entity mentions have a large impact on our EL system, affecting both candidate generation recall and ranking performance. While it is challenging to link such complex mentions to concepts in terminologies (also for human annotators), the annotation of long entity spans is a specific decision made for the DISTEMIST corpus. Out of 5,136 linkable mentions in the DISTEMIST training set, 1,964 (38.2%) consist of three or more tokens. In comparison, the training set of the Quaero corpus with a similar size (5,689 entities) contains only 495 (8.7%) of such complex entities [19].

Our system has a few hyperparameters that we did not systematically optimize, e.g., the context length of the mention encoding and number of retrieved candidates before re-ranking, but also hyperparameters affecting the training, such as learning rate, learning rate schedule, weight decay, or label smoothing. Tuning these parameters could potentially increase performance, although it might bias our system even more towards DISTEMIST. In contrast, we expect the current default values to work reasonably well also for other EL datasets.

6   Conclusion and Outlook

We proposed a system for biomedical EL, consisting of an ensemble of unsupervised candidate generators and a trainable re-ranker. While the goal of the candidate generation step is to obtain a high overall recall, re-ranking improves recall@1 by almost 20pp. on the DISTEMIST shared task dataset. These findings highlight the importance of re-ranking for adapting candidate lists to specific datasets and annotation policies. Interestingly, a simple trainable cross-encoder outperforms our complex rule-based baseline that won the EL sub-track of the DISTEMIST shared task. This is encouraging for other low-resource languages and domains, as the number of annotated entities in DISTEMIST are still relatively small compared to the largest, English-language benchmarks [17].

Our pipeline makes very few assumptions about the source language, as it does not require aliases in this language in the target KB - although they are certainly helpful. Therefore, our approach is applicable to a wide range of natural languages. For instance, results from preliminary experiments suggest that it can also be successfully used together with an existing NER pipeline for German clinical entities [5], even though the number of German aliases in the UMLS is much lower compared to the available Spanish aliases. It can also be easily configured to link entities to controlled vocabularies in specialized medical domains, e.g., for fine-grained semantic classes in clinical notes from kidney patients [23]. To enable other researchers to reproduce our results and adapt our pipeline to their use cases, we have made the source code of our project publicly available [11].

Acknowledgment. Parts of this work were generously supported by grants of the German Federal Ministry of Research and Education (01ZZ1802H, 01ZZ2314N) and the German Federal Ministry of Economic Affairs and Climate Action (01MJ21002A).

References

1. Agarwal, D., Angell, R., Monath, N., McCallum, A.: Entity linking via explicit mention-mention coreference modeling. In: Proceedings of the 2022 Conference of the North American Chapter of the Association for Computational Linguistics: Human Language Technologies. Association for Computational Linguistics, Seattle (2022)
2. Alekseev, A., Miftahutdinov, Z., Tutubalina, E., et al.: Medical crossing: a cross-lingual evaluation of clinical entity linking. In: Proceedings of the Thirteenth Language Resources and Evaluation Conference, pp. 4212–4220. European Language Resources Association, Marseille (2022)
3. Bernik, M., Tovornika, R., Fabjana, B., Marco-Ruizb, L.: Diagñoza: a natural language processing tool for automatic annotation of clinical free text with SNOMED-CT. In: Working Notes of CLEF. CEUR Workshop Proceedings, pp. 235–243 (2022)
4. Bodenreider, O.: The unified medical language system (UMLS): integrating biomedical terminology. Nucleic Acids Res. **32**, D267–D270 (2004)
5. Borchert, F., Lohr, C., Modersohn, L., et al.: GGPONC 2.0-the German clinical guideline corpus for oncology: Curation workflow, annotation policy, baseline NER taggers. In: Proceedings of the Thirteenth Language Resources and Evaluation Conference, pp. 3650–3660 (2022)
6. Borchert, F., Schapranow, M.P.: HPI-DHC @ BioASQ DisTEMIST: Spanish biomedical entity linking with pre-trained Transformers and cross-lingual candidate retrieval. In: Working Notes of CLEF. CEUR Workshop Proceedings, pp. 244–258. Italy, Bologna (2022)
7. Carrino, C.P., Armengol-Estapé, J., Gutiérrez-Fandiño, A., et al.: Biomedical and clinical language models for Spanish: on the benefits of domain-specific pretraining in a mid-resource scenario. arXiv preprint arXiv:2109.03570 (2021)
8. Chizhikova, M., Collado-Montañez, J., López-Úbeda, P., et al.: SINAI at CLEF 2022: leveraging biomedical transformers to detect and normalize disease mentions. In: Working Notes of CLEF. CEUR Workshop Proceedings, pp. 265–273 (2022)
9. Donnelly, K.: SNOMED-CT: the advanced terminology and coding system for eHealth. In: Medical and Care Compunetics 3. No. 121 in Studies in Health Technology and Informatics, pp. 279–290. IOS Press (2006)
10. Fries, J., Weber, L., Seelam, N., et al.: BigBIO: a framework for data-centric biomedical natural language processing. In: Advances in Neural Information Processing Systems, vol. 35, pp. 25792–25806. Curran Associates, Inc. (2022)

11. HPI Digital Health Cluster on GitHub: xMEN (2023). https://github.com/hpi-dhc/xmen. Accessed 23 June 2023
12. Johnson, J., Douze, M., Jégou, H.: Billion-scale similarity search with GPUs. IEEE Trans. Big Data **7**(3), 535–547 (2019)
13. Liu, F., Vulić, I., Korhonen, A., Collier, N.: Learning domain-specialised representations for cross-lingual biomedical entity linking. In: Proceedings of the 59th Annual Meeting of the Association for Computational Linguistics and the 11th International Joint Conference on Natural Language Processing (Volume 2: Short Papers), pp. 565–574. Association for Computational Linguistics, Online (2021)
14. Liu, Y., Ott, M., Goyal, N., et al.: RoBERTa: a robustly optimized BERT pre-training approach. arXiv preprint arXiv:1907.11692 (2019)
15. Logeswaran, L., Chang, M.W., Lee, K., et al.: Zero-shot entity linking by reading entity descriptions. In: Proceedings of the 57th Annual Meeting of the Association for Computational Linguistics, pp. 3449–3460. Association for Computational Linguistics, Florence (2019)
16. Miranda-Escalada, A., Gascó, L., Lima-López, S., et al.: Overview of DisTEMIST at BioASQ: automatic detection and normalization of diseases from clinical texts: results, methods, evaluation and multilingual resources. In: Working Notes of CLEF. CEUR Workshop Proceedings (2022)
17. Mohan, S., Li, D.: MedMentions: a large biomedical corpus annotated with UMLS concepts. In: Proceedings of the 2019 Conference on Automated Knowledge Base Construction, Amherst, Massachusetts, USA (2019)
18. Neumann, M., King, D., Beltagy, I., Ammar, W.: scispaCy: fast and robust models for biomedical natural language processing. In: Proceedings of the 18th BioNLP Workshop and Shared Task, pp. 319–327. Association for Computational Linguistics, Florence (2019)
19. Névéol, A., Cohen, K.B., Grouin, C., et al.: Clinical information extraction at the CLEF eHealth evaluation lab 2016. In: CEUR Workshop Proceedings, vol. 1609, p. 28 (2016)
20. Reimers, N., Gurevych, I.: Sentence-BERT: sentence embeddings using Siamese BERT-networks. In: Proceedings of the 2019 Conference on Empirical Methods in Natural Language Processing and the 9th International Joint Conference on Natural Language Processing, pp. 3982–3992. Association for Computational Linguistics, Hong Kong (2019)
21. Robertson, S., Zaragoza, H.: The Probabilistic Relevance Framework: BM25 and Beyond. Now Publishers Inc. (2009)
22. Roller, R., Kittner, M., Weissenborn, D., Leser, U.: Cross-lingual candidate search for biomedical concept normalization. In: MultilingualBIO: Multilingual Biomedical Text Processing, p. 16 (2018)
23. Roller, R., Uszkoreit, H., Xu, F., et al.: A fine-grained corpus annotation schema of German nephrology records. In: Proceedings of the Clinical Natural Language Processing Workshop, Osaka, Japan, pp. 69–77 (2016)
24. Sevgili, Ö., Shelmanov, A., Arkhipov, M., et al.: Neural entity linking: a survey of models based on deep learning. Semant. Web **13**(3), 527–570 (2022)
25. Sung, M., Jeon, H., Lee, J., Kang, J.: Biomedical entity representations with synonym marginalization. In: Proceedings of the 58th Annual Meeting of the Association for Computational Linguistics, pp. 3641–3650. Association for Computational Linguistics, Online (2020)
26. Vashishth, S., Newman-Griffis, D., Joshi, R., et al.: Improving broad-coverage medical entity linking with semantic type prediction and large-scale datasets. J. Biomed. Inform. **121**, 103880 (2021)

27. Wajsbürt, P., Sarfati, A., Tannier, X.: Medical concept normalization in French using multilingual terminologies and contextual embeddings. J. Biomed. Inform. **114**, 103684 (2021)
28. Wu, L., Petroni, F., Josifoski, M., et al.: Scalable zero-shot entity linking with dense entity retrieval. In: Proceedings of the 2020 Conference on Empirical Methods in Natural Language Processing, pp. 6397–6407. Association for Computational Linguistics, Online (2020)
29. Xu, D., Zhang, Z., Bethard, S.: A generate-and-rank framework with semantic type regularization for biomedical concept normalization. In: Proceedings of the 58th Annual Meeting of the Association for Computational Linguistics, pp. 8452–8464. Association for Computational Linguistics, Online (2020)

Humour Translation with Transformers

Farhan Dhanani[(✉)] [iD], Muhammad Rafi[iD], and Muhammad Atif Tahir[iD]

National University of Computer and Emerging Sciences (NUCES-FAST),
Karachi, Pakistan
k214808@nu.edu.pk

Abstract. This paper presents the solution proposed by team FAST-
MT to the shared tasks of JOKER CLEF 2022 Automatic pun and
humour translation. State-of-the-art Transformer-based models are used
to solve the three tasks introduced in the JOKER CLEF workshop. The
Transformer model is a kind of neural network that tries to learn the contextual information from the sequential data by implicitly comprehending the existing relationships. In task 1, given a piece of text, we need to
classify/explain any instance of wordplay is present in it or not. The proposed solution to task 1 combines the pipeline of token classification, text
classification, and text generation. In task 2, we need to translate single
words (nouns) containing a wordplay. This task is mapped to the problem
of question answering (Q/A) on programmatically extracted texts from
the OPUS parallel corpus. In task 3, contestants are required to translate
the entire phrase containing the wordplay. Sequence to sequence translation models are used to solve this task. The team has adopted different
strategies for each task as they suited to the requirements therein. The
paper reports proposed solutions, implementation details, experimental
studies, and results obtained in JOKER CLEF 2022 automatic pun and
humour translation tasks.

Keywords: Text Classification · Token Classification · Question
Answering · Machine Translation · Transformers

1 Introduction

In our daily communications, humour is one of the most ubiquitous elements that
we, as, a human comprehend comfortably with the help of pre-occupied cultural
experiences and social understandings. But, for computers, this still remains one
of the most daunting jobs as it is extremely difficult even for the current expensive deep-learning-based solutions to correctly apprehend the double-meaning
words, which is one of the most prominent features of humour in almost all
languages. The JOKER CLEF-2022 workshop has come up with a unique set
of challenges under the natural language processing domain. The workshop has
brought professional translators and computer scientists together by presenting
three different tasks to evaluate their perceptions and understandings of humour
and its translations. This paper presents our strategy to solve the three problems
introduced by the workshop with the help of transformer-based pre-trained deep
learning models, along with their implementation and the obtained results.

A. Arampatzis et al. (Eds.): CLEF 2023, LNCS 14163, pp. 148–160, 2023.
https://doi.org/10.1007/978-3-031-42448-9_13

2   Materials and Methods

2.1   TASK-1: Explain and Classify Instances of Wordplay

The JOKER CLEF 2022 [3] team has shared two versions of 10-column-based tabular training and test sets for task 1. First, they have provided a smaller version of both the data sets. Later as the competition timeline grew, they released a new pair of training and test sets with additional records. Both versions of the data sets contain the same ten columns, two source input columns, and eight target columns. The challenge in this task is to construct a model that consumes values from source input columns and predicts the value for the target columns. The following list presents details of each column provided in the data sets, along with our selected approaches to predict the values of target columns based on English text and its associated id from the source input *Wordplay* and *Id* column, respectively.

1. **ID:** An input value that uniquely identifies the associated wordplay text.
   - *Type: Source Input Column*
   - *Example: pun_193*
2. **WORDPLAY:** An input English text that contains a wordplay.
   - *Type: Source Input Column*
   - *Example: Airline pilots make many friends in high places.*
3. **LOCATION:** Words in the given English, which constructs the wordplay.
   - *Prediction Strategy: Token Classification,* **Type:** *Target Column*
   - *Example: Airline pilots make many friends in high places.*
   - *Location: high*
4. **INTERPRETATION:** A possible explanation for the given wordplay in the English text.
   - *Type:* **Prediction Strategy:** *Sequence Generation,* **Type:** *Target Column*
   - *Example: Airline pilots make many friends in high places.*
   - *Interpretation: high (height)/high (addicted)/high (superior)*
5. **HORIZONTAL/VERTICAL:** A binary categorical column to detect whether the target and source of the wordplay co-occur in the given English text.
   - *Prediction Strategy: Sequence Classification,* **Type:** *Target Column*
   - **Example of Horizontal (source and target co-occur):** *They're called lessons (source) because they lessen (target) from day to day.*
   - **Example of Vertical (source and target collapse into a single word):** *Airline pilots make many friends in high (source+target) places.*
6. **MANIPULATION_TYPE:** A categorical variable to detect that the source and target of the wordplay are exact equivalents of each other (**Identity**), or weakly resemble each other (**Similarity**), or both possess different ordering (**Permutation**), or its a group of initials that forms funny meaning (**Abbreviation**).
   - *Prediction Strategy: Sequence Classification,* **Type:** *Target Column*
   - **Example of Identity:** *Airline pilots make many friends in high places.*

- *Example of Similarity*: *They're called* <u>lessons</u> *(source) because they* <u>lessen</u> *(target) from day to day.*
- *Example of Permutation*: *What a dormitory could be- A "Dirty Room."*
  Dormitory = Dirty room. (The word dormitory is an anagram of the word dirty room. In that sense, the re-arrangement of letters is forming a pun.)
- *Example of Abbreviation*: *BRAINS: Bio-Behavioral Research Awards for Innovative New Scientists. (Here* <u>"BRAINS"</u> *has a dual meaning. It can get interpreted as the mind of human and the name of a research award.)*

7. **MANIPULATION_LEVEL:** A categorical variable to detect that the wordplay given in the English text is a kind of phonological manipulation (**Sound**), or it is a kind of textual-based written manipulation (**Written**), or if the detected wordplay is of some other form.
   - *Prediction Strategy*: *Sequence Classification*, *Type*: *Target Column*
   - *Example of Sound*: *Airline pilots make many friends in high places.*
   - *Example of Written*: *We have* <u>tBRAINS.</u> *(Again here* <u>"BRAINS"</u> *has a dual meaning. It can get interpreted as the mind of human and the name of a research award.)*
   - *Example of Other*: *We shape our buildings, and afterwards our buildings shape us.*

8. **CULTURAL_REFERENCE:** A boolean variable (**true/false**) to detect the existence of cultural reference in the given wordplay of the English text.
   - *Prediction Strategy*: *Sequence Classification*, *Type*: *Target Column*
   - *Example of False*: *Airline pilots make many friends in* <u>high</u> *places.*

9. **CONVENTIONAL_FORM:** A boolean variable (**true/false**) to detect whether the given wordplay in the English text belongs to conventional form or not.
   - *Prediction Strategy*: *Sequence Classification*, *Type*: *Target Column*
   - *Example of False*: *Airline pilots make many friends in* <u>high</u> *places.*

10. **OFFENSIVE:** This is a non-evaluated categorical variable. And we have ignored it throughout our experiments. Its purpose is to classify the given wordplay in the English text into offensive categories (**None, Racist, Possibly, Sexist, Other**).
    - *Type*: *Target Column*

We have ignored the offensive column, as it was optional, and trained seven distinct models independently for each of the remaining target columns. All of the prepared seven models process the value of the *"wordplay"* in their input and emit the value for its respective target column. We have used the token-classification-styled training to locate the words forming a pun in the given *"wordplay."* We have treated the English text from the *"wordplay"* columns as a series of space-separated tokens and then prepared the model to classify tokens containing a pun into the following three categories.

- **word_play_token_B:** *To locate the word which begins the pun.*
- **word_play_token_I:** *To locate the additional words included in the pun.*

– **other_token:** *To identify all the words that are not part of the pun.*

Next, we used the auto-regressive technique to construct a sequence generation model for generating the interpretation of the located pun in the input *"wordplay"*. Lastly, we have employed the sequence classification scheme to build five separate models for inferring the values for the remaining five target columns.

2.2 TASK-2: Translate Single Words Containing Wordplay

The second task specifies predicting an equivalent French version of a given English noun. During the translation of popular movies, anime, and video games authors try extremely hard to make the narration relevant to the target audience by inducing the cultural background of the target language while preserving the original emotions attached to characters and other nouns of the story. Automation of this phenomenon is the main purpose of this task. For example, consider a character from the Pokemon [8] series shown in Fig. 1. In the English version the name of the character is "EKANS," but in the French version, it is named "ABO." It's easy to observe that the character visually looks like a small snake. The authors wanted to promote this resemblance linguistically through the name of the character to educate the audience about its nature. Therefore, they named the character "EKANS" in the English version to make it an anadrome of "SNAKE". In the French language, the boa means a masculine snake. Thus, in the French version, the writers have translated the name of this character to "ABO", which is an anadrome of "BOA". Because the french audience doesn't understand the word "SNAKE", the authors have renamed the character to keep them engaged with the vocabulary of their own native language. The challenge in this task is to learn this style of translation between the named entities from the English to French language and predict an appropriate French translations for the given English nouns from the test sets. Figure 2 illustrates another example from the Asterix series [1] to understand this task.

Fig. 1. The image displays a Pokemon [8] character. It is named "EKANS" in the English version of the series, but French version, it's called "ABO".

Fig. 2. The image displays a character from the Astrix comic series [1]. Its name is "Dogmatrix" in English version, but in French, it's called "Idéfix".

The JOKER CLEF 2022 team shared a training data set containing 1164 example translations between English and French nouns for the second task and a test set of only 284 English nouns for which French translation needs to be generated. We have transformed the shared training data set into extractive Question/Answer (Q/A) problem-styled data sets and mapped the task of learning one-to-one relation between English nouns and their corresponding French translations into the extractive Question/Answer problem. To accomplish this transformation, we have utilized OPUS open-source parallel corpus [18] to artificially develop the context for all English/French noun pairs provided in the task 2 training set. We have iteratively selected each English/French noun pair listed in the provided data set. Then extracted those English/French parallel sentence pairs from the OPUS open-source parallel corpus [18] that contains the selected English noun in its English version and the translated French noun in its French version. In this fashion, we have collected contexts for all English/French noun pairs. And transformed the task 2 data set where each record is composed of an English noun and its French translation, along with a list of extracted English/French parallel sentence pairs in the form of contexts. To visualize the transformation, suppose we have only one record in the task 2 training data set, as shown in the Table 1. Given such a scenario, we can pull the following sentence (English/French) pair listed below from the OPUS parallel corpus [18] to generate the extractive Question/Answer problem-styled data set shown in the Table 2.

- **English Version:** *asterix and obelix should stay in the village and not go in the forest!*
- **French Version:** *astérix et obélix ne devraient plus quitter le village.*

Table 1. The table shows the structure of a single record in the training set of task 2.

Id	En	Fr
4	Obelix	Obélix

In such a way, we can transform the entire training data set for task-2 and then utilize popular pre-trained extractive question-answering models from the hugging face [19] repositories to predict the French translation for a given English noun. The models will use the English nouns from the JOKER CLEF [3] task 2 training's data set as the input question, along with the corresponding French sentence pulled from the OPUS parallel corpus [18] as their context. And now, the task for the extractive Q/A models is to learn to locate the exact position of the French translation in the French text for the queried English noun.

After the training completes, we again transformed the test set for task 2 into the extractive Question/Answer styled test data set by applying a similar strategy. The test set of task 2 holds test records for which we don't know the correct French translation of the given English noun. Because of this, we have

Table 2. Conversion of the Table 1 records into the extractive Q/A styled data set.

Id	Context	Question	Answers
4	astérix et obélix ne devraient plus quitter le village	Obelix	{ "text": [Obélix], "answer_start": [11]}

only ensured the existence of the given English noun in the English version of the extracted (English/French) sentence pairs from the OPUS corpus [18] and assumed that the corresponding French translation must also hold its equivalent French version. It's a weak assumption, but we have made this architectural choice to design the solution.

2.3   TASK-3: Translate Entire Phrases Containing Wordplay

The problem description of task 3 is a classical example of sequence to sequence prediction, where the model needs to predict an equivalent French translation for the given English text. It's important to note that multiple valid French translations may exist for a given English sentence containing a wordplay, and the task is to predict any one of it correctly. For example, consider the following scenario where both French translations are correct for the given English sentence.

- **English:** *"Be still my hart" she murmured, thinking how magnificent and stag - like he was.*
- **French-1:** *"Mon cœur se cerf", murmura-t-elle en voyant ce beau et majestueux mâle.*
- **French-2:** *Elle murmura "Calme-toi mon destrier" en pansant combien il était magnifique.*

The JOKER CLEF 2022 team shared a training data set of 5115 sample records along with a test set composed of 2378 English sentences for which French translations need to be generated. We have processed the provided training data set as a JSON dictionary. The key in this dictionary is the English text, and its associated value contains the list of all possible French translations for the keyed English text from the training set. Later, we used the prepared JSON object to train sequence-to-sequence transformer models for learning the mapping between the English text and any of its corresponding valid French translations.

3   Experiments and Results

This section will discuss the transformer architectures we have utilized to implement the approaches discussed in the previous section for solving tasks of the JOKER CLEF 2022 [3] workshop and their results. We have shared our codebase on the public GitHub repository [4]. Thus, all the presented experiments can be easily re-executed to reproduce the mentioned results.

3.1 IMPLEMENTATION OF SOLUTION FOR TASK-1: Explain and Classify Instances of Wordplay

We have used the listed pre-trained transformer models from the hugging face repository and fine-tuned them on the given training data sets for task 1 to make them learn to locate the words forming the wordplay in the given English text through token classification.

- Pre-Trained BERT-BASE [2].
- KEY-BERT [5] with fine-tuned BERT-BASE [2] as its embedder.
- KEY-BERT [5] with pre-trained all-MiniLM-L6-v2 [13] as its as embedder.

The KEY-BERT [5] model can be utilized with different embedders. In this experiment, we have only used it with the fine-tuned "BERT-BASE" [2] model and the pre-trained "all-MiniLM-L6-v2" [13] model. We have processed both the training data sets for task-1 independently and used the hold-out approach to pull aside 9% of the records from both training sets provided for task 1, where the length of the given English text containing the wordplay was more than two. We have kept them hidden from the model throughout its training and used them later to evaluate and rank the predictions of the fine-tuned models in locating the words forming the wordplay. We have fine-tuned independent instances of each of the selected models for less than five epochs on the remaining records of both training sets. And, after fine-tuning, we evaluated the predictions generated from the fine-tuned models on the 9% of records, which we have extracted initially from the training sets to estimate their performance on unknown data points, as shown in the Table 3. Overall our approach has generated comparatively good results for the first data set provided by the JOKER CLEF 2022 [3] team for task 1, and the fine-tuned BERT-BASE [2] model delivers the best performance compared to other variants of KEY-BERT models. We have applied the hold-out approach instead of the K cross-validation to evaluate the performance of token classification transformer models for locating the wordplay in the given English sentences because the provided training data sets contain numerous instances where the length of the English sentences was one, in all such instances, it was apparent that the given English sentence was itself the wordplay. Thus, the token classification transformer models will achieve a perfect score among these records. Hence, using such instances for evaluating BERT-based transformer models will result in an unfair boost in their performance. To mitigate this effect, we have used the hold-out approach and selectively extracted those records from the training set in which the length of the English sentence was more than two.

After locating the target wordplay, we used the GPT-2 [9] model to generate an interpretation for the located wordplay and utilized the pre-trained DistilBERT [14] model from the hugging face [19] repository to predict the categorical label for other target columns of task 1 listed in Table 4. We have made five separate copies of the pre-trained DistilBERT [14] model and fine-tuned them individually to infer the label of each of the five categorical target columns of task 1. In the end, we have selected the fine-tuned BERT-BASE [2] model for locating the words forming the wordplay in the English sentences listed in both

Table 3. The table shows the performance of selected BERT-based transformers models for precisely identifying the words forming the wordplay.

Model Name	Accuracy on the training set 1	Accuracy on the training set 2
BERT-BASE [2] (FINE-TUNED)	71%	31%
KEY-BERT [5] with fine-tuned BERT-BASE [2] as its embedder	33%	16%
KEY-BERT [5] with pre-trained all-MiniLM-L6-v2E [13] as its embedder	15%	3%

Table 4. The table presents the scores allotted by evaluators of the JOKER CLEF 2022 [3] team on our submitted predictions for the test records of test set-2 of task 1.

Column Name	Scores on Test set-2 having 3256 records
LOCATION	1455
MANIPULATION_TYPE	1667
MANIPULATION_LEVEL	2437
HORIZONTAL/VERTICAL	68
CULTURAL_REFERENCE	Not evaluated
CONVENTIONAL_FORM	Not evaluated
OFFENSIVE	Not evaluated

versions of the test sets. Along with five separate copies of the fine-tuned DistilBERT [14] model for predicting labels of the remaining five categorical target columns of task 1 and submitted our predictions to the evaluators of the JOKER CLEF 2022 [3] workshop. We were the only team that successfully submitted the predictions for the test set-1. Thus for the first test set, we have implicitly got the first rank, and the evaluators haven't released any other statistical details or scores for the test set-1 of task 1. Furthermore, for test set-2, we have managed to get ourselves among the top three positions. The Table 4 reveals the scores for the generated predictions from our fine-tuned models to correctly predict the values of all the target columns for each English text listed in the test set-2 of task 1. It's important to note that the evaluators have awarded a score of one point for predicting a correct value for each target column of task 1 against an English text containing a wordplay from test set-2. Plus, the evaluators have not evaluated the predictions for the three columns mentioned in the Table 4.

3.2   IMPLEMENTATION OF SOLUTION FOR TASK-2: Translate Words Containing a Wordplay

We have downloaded pre-trained CamemBERT [6] and DistilBERT [15] models and fine-tuned them to perform extractive question answering for task 2. The CamemBERT [6] model is pre-trained in the French language to retrieve answers for the provided French queries in the French context. Contrastingly, the DistilBERT [15] model is pre-trained in the English language to extract answers in the English context for the given English questions. The reason for choosing these two distinct pre-trained models designed for different languages is because the contents of the transformed data set for task 2 consist of both French and English language. In the previous section, we have observed that our approach has transformed the problem of translating English nouns to their equivalent French versions into the extractive question answering domain. As a result of this transformation, the extractive question-answering models have to process the input question in English and the associated context in the French language to extract the French translation for the given English query from the French context. Because of this heterogeneity of different languages in the transformed data set, we have utilized two different English and French pre-trained extractive question-answering models and compared their performance using 10-fold cross-validation. In our experiments, we have observed that after fine-tuning both models for less than five epochs across the ten-fold of the training data set, the DistilBERT [15] model has provided a far better mean accuracy of 94% compared to the CamemBERT [6] model, which has delivered a mean accuracy of only 59%. Thus, we have used the DistilBERT [15] model to generate the French translations for the English nouns of the test set, and our submission ranked first in the competition. The test set consists of 284 English nouns, out of which 250 were from official sources, and for them, there exists an official French translation. However, the evaluators have also included 34 English nouns in the test set for which official translations were not available. The evaluators have used simple case insensitive string matching to evaluate official French translations for English-named entities with their expected French versions. We can use string matching to evaluate the official translations but not to assess the unofficial translations because they are not part of authentic literature. Thus, the evaluators have manually assessed the unofficial translations based on lexical field preservation, sense preservation, comprehensibility, and the formed wordplay. The Table 5 below demonstrates the obtained score of test submissions against each of the mentioned parameters.

Table 5. The table presents the scores allotted by evaluators of the JOKER CLEF 2022 [3] team on our submitted predictions for the test set of task 2.

Metric	Score	Explanation
Total	284	Total number of records in the test set
Not translated	0	Total number of records that are either missing or not translated in the submission file
Official	250	Number of the official named entities that are correctly translated in the submission file
Not Official	34	Number of translations in the submission file that are unofficial
Lexical Field Preservation	16	Number of translations that preserve the lexical field of the source wordplay in the submission file
Sense Preservation	13	Number of translations that preserve the sense of the source wordplay in the submission file
Comprehensible Terms	26	Number of translations that do not exploit any specialized terms in the submission file
Wordplay form	3	Number of translations that are itself wordplay in the submission file

3.3 IMPLEMENTATION OF SOLUTION FOR TASK-3: Translate Entire Phrases Containing Wordplay

The training data set provided for task 3 only comprised 1,185 unique English sentences for which there exist multiple valid French translations. Because the data set was not huge, we have decided not to fine-tune pre-train models. Our goal was to select the pre-trained model that generates predictions that, on average, provide the highest BLEU [7] scores and least TER [16] scores for the given English phrases in the training set without fine-tuning. We have assumed that the high BLEU [7] scores and low TER [16] scores indicate that the generated French translations by the model for the given English phrases are more similar to expected French translations. It's a weak assumption, but still, we have made this architectural choice to design the solution for task 3. We have downloaded four popular pre-trained sequences to sequence transformer models from the hugging face repositories [19] listed in the Table 6 and evaluated their performance on the provided training set of task 3. The Table 6 also entails that the Helsinki/NLP/opus-mt-en-fr [17] model has given the best performance and produced more desired translation as compared to other models. We have used this model to generate final predictions for English phrases listed in the test set of task 3 and received the second position in the third task.

Table 6. Average BLEU [7] and TER [16] scores achieved by the selected pre-trained sequence-to-sequence transformer models on the training set of task 3.

Model Name	Average BLEU Score	Average TER Score
Helsinki-NLP/opus-mt-en-fr [17]	18.43	0.80
GOOGLE T5 BASE [10]	12.64	0.84
GOOGLE T5 SMALL [12]	11.07	0.85
GOOGLE T5 LARGE [11]	11.90	0.84

The evaluators have manually scored each of the submitted French translations using thirteen different parameters to rate the quality of the generated predictions. Table 7 shows a list of these parameters and explains how the evaluators have used each of them. The table also lists the obtained scores of our submitted predictions against each of the listed thirteen parameters.

Table 7. The table presents the scores allotted by evaluators of the JOKER CLEF 2022 [3] team on our submitted predictions for the test set of task 3.

Metric	Score	Explanation
total	2378	Total records in the test set
valid	2120	Total submitted translations that are valid
not translated	103	Missing or invalid translations in the submission file
nonsense	220	Translations that don't make sense to French speakers
syntax problem	58	Translations that contains syntactical errors
lexical problem	79	Number of translations that contains lexical errors
lexical field preservation	1739	Number of translations in the submission file that have preserved the lexical field of the source wordplay
sense preservation	1453	Number of translations in the submission file that have preserved the sense of the source wordplay
comprehensible terms	867	Number of translations in the submission file that haven't exploited very specialized terms
wordplay form	345	Number of correct translations that are itself wordplay
identifiable wordplay	318	Number of translations provided that are itself wordplay and understandable by the audience
over translation	1	Translations having useless words & are unnecessarily long
style shift	12	Number of translations that have a style shift. For example when vulgarism exists in the source sentence or the produced translation but not in both
hilariousness shift	765	Number of French translations that are much less or much funnier than the source sentence

4   Conclusion

The BERT-BASE model has given the best performance for locating the words forming the wordplay in the English texts of the task 1 data set. We haven't employed the other large BERT variants to locate the wordplay in the given English text of the task 1 data set, but we believe that they will boost the performance of our approach. The DistilBERT model has achieved a good performance in predicting classification labels for the remaining target columns of task 1. We think in the future, the results of the DistilBERT model can be compared with other popular text classification BERT alternatives to rank its performance. The main highlight of our work is our designed technique for solving task 2 in extractive Q/A style, and our submission also ranked top in the competition. An extension of this work can be to test the approach with different language pairs because, in this paper, we have only evaluated it on English/French noun pairs as per task 2 requirements. The DistilBERT model again delivers the best performance for solving task 2. And accurately predicts 96% of the French translations for the given English nouns in the extractive Q/A style. Lastly, we have concluded that Helsinki-NLP/opus-mt-en-fr model has provided the best performance on the task 3 data set by achieving 18.43 and 0.80 averaged BLEU and TER scores. The data set of task 3 doesn't contain a large number of records in it. We think in the future, increasing the size of the English/French parallel corpus containing wordplay and humour will benefit in excelling the research and will immensely help in training better models and establishing new state-of-the-art for the three humour translation tasks of the JOKER CLEF workshop.

References

1. Asterix: Asterix—Wikipedia, the free encyclopedia (2001). https://en.wikipedia. org/wiki/Asterix. Accessed 24 May 2022
2. Devlin, J., Chang, M., Lee, K., Toutanova, K.: BERT: pre-training of deep bidirectional transformers for language understanding. CoRR abs/1810.04805 (2018). http://arxiv.org/abs/1810.04805
3. Ermakova, L., et al.: Overview of JOKER@CLEF 2022: automatic wordplay and humour translation workshop. In: Barrón-Cedeño, A., et al. (eds.) CLEF 2022. LNCS, vol. 13390, pp. 447–469. Springer, Cham (2022). https://doi.org/10.1007/978-3-031-13643-6_27
4. FARHAN: Fast-MT team submission for JOKER CLEF 2022 (2022). https://github.com/FarhanDhanani/joker-clef-22-FAST-MT
5. Grootendorst, M.: KeyBERT: minimal keyword extraction with BERT. (2020). https://doi.org/10.5281/zenodo.4461265
6. Martin, L., et al.: CamemBERT: a tasty French language model. In: Proceedings of the 58th Annual Meeting of the Association for Computational Linguistics (2020)
7. Papineni, K., Roukos, S., Ward, T., Zhu, W.J.: BLEU: a method for automatic evaluation of machine translation. In: Proceedings of the 40th Annual Meeting of the Association for Computational Linguistics, pp. 311–318. Association for Computational Linguistics, Philadelphia, Pennsylvania (2002). https://doi.org/10.3115/1073083.1073135, https://aclanthology.org/P02-1040

8. Pokémon: Pokémon—Wikipedia, the free encyclopedia (2001). https://en.wikipedia.org/wiki/Pok%C3%A9mon. Accessed 24 May 2022
9. Radford, A., Wu, J., Child, R., Luan, D., Amodei, D., Sutskever, I.: Language models are unsupervised multitask learners (2019)
10. Raffel, C., et al.: Exploring the limits of transfer learning with a unified text-to-text transformer (T5 base). J. Mach. Learn. Res. **21**(140), 1–67 (2020), https://huggingface.co/t5-base. Accessed 24 May 2022
11. Raffel, C., et al.: Exploring the limits of transfer learning with a unified text-to-text transformer (T5 large). J. Mach. Learn. Res. **21**(140), 1–67 (2020). https://huggingface.co/t5-large. Accessed 24 May 2022
12. Raffel, C., et al.: Exploring the limits of transfer learning with a unified text-to-text transformer (T5 small). J. Mach. Learn. Res. **21**(140), 1–67 (2020). https://huggingface.co/t5-small. Accessed 24 May 2022
13. Reimers, N., Gurevych, I.: Sentence-BERT: sentence embeddings using Siamese BERT-networks. In: Proceedings of the 2019 Conference on Empirical Methods in Natural Language Processing. Association for Computational Linguistics (2019). http://arxiv.org/abs/1908.10084
14. Sanh, V., Debut, L., Chaumond, J., Wolf, T.: DistilBERT, a distilled version of BERT: smaller, faster, cheaper and lighter. ArXiv abs/1910.01108 (2019). https://huggingface.co/distilbert-base-uncased. Accessed 24 May 2022
15. Sanh, V., Debut, L., Chaumond, J., Wolf, T.: DistilBERT, a distilled version of BERT: smaller, faster, cheaper and lighter (for question answering). In: NeurIPS EMC Workshop (2019). https://huggingface.co/distilbert-base-cased-distilled-squad. Accessed 24 May 2022
16. Snover, M., Dorr, B., Schwartz, R., Micciulla, L., Makhoul, J.: A study of translation edit rate with targeted human annotation. In: Proceedings of the 7th Conference of the Association for Machine Translation in the Americas: Technical Papers, pp. 223–231. Association for Machine Translation in the Americas, Cambridge (2006). https://aclanthology.org/2006.amta-papers.25
17. Tiedemann, J., Thottingal, S.: OPUS-MT—building open translation services for the world. In: Proceedings of the 22nd Annual Conferenec of the European Association for Machine Translation (EAMT). Lisbon, Portugal (2020). https://huggingface.co/Helsinki-NLP/opus-mt-en-fr. Accessed 24 May 2022
18. Tiedemann, J.: Parallel data, tools and interfaces in OPUS. In: Calzolari, N., et al. (eds.) Proceedings of the Eight International Conference on Language Resources and Evaluation (LREC 2012). European Language Resources Association (ELRA), Istanbul (2012)
19. Wolf, T., et al.: Transformers: state-of-the-art natural language processing. In: Proceedings of the 2020 Conference on Empirical Methods in Natural Language Processing: System Demonstrations, pp. 38–45. Association for Computational Linguistics, Online (2020). https://www.aclweb.org/anthology/2020.emnlp-demos.6

Fight Against Misinformation on Social Media: Detecting Attention-Worthy and Harmful Tweets and Verifiable and Check-Worthy Claims

Ahmet Bahadir Eyuboglu, Bahadir Altun, Mustafa Bora Arslan,
Ekrem Sonmezer, and Mucahid Kutlu$^{(\boxtimes)}$

TOBB University of Economics and Technology, Ankara, Turkey
{ialtun,m.kutlu}@etu.edu.tr

Abstract. In this paper, we present our participation in CLEF 2022 CheckThat! Lab's Task 1 on detecting check-worthy and verifiable claims and attention-worthy and harmful tweets. We participated in all subtasks of Task 1 for Arabic, Bulgarian, Dutch, English, and Turkish datasets. We investigate the impact of fine-tuning various transformer models and how to increase training data size using machine translation. We also use feed-forward networks with the Manifold Mixup regularization for the respective tasks. We are ranked first in detecting factual claims in Arabic and harmful tweets in Dutch. In addition, we are ranked second in detecting check-worthy claims in Arabic and Bulgarian.

Keywords: Fact-Checking · Check-worthiness · Attention-worthy tweets · Harmful tweets · Factual Claims

1 Introduction

Social media platforms have emerged as prominent channels for individuals to access and disseminate information by enabling their users to easily share messages and follow others. While these platforms play a crucial role in allowing individuals to express their thoughts, they also possess the potential for detrimental misuse such as spreading misinformation and hateful messages. For instance, the surge in vaccine-related misinformation and conspiracy theories circulating on these platforms has contributed to the growing hesitancy towards vaccination [23]. Moreover, the messages propagated via social media platforms possess the power to shape public opinion on specific matters, compelling governmental entities to respond. For example, numerous countries' governments had to proactively share information about vaccines to mitigate vaccine hesitancy[1].

In this paper, we explain our participation in Task 1 [21] of the CLEF Check That! 2022 Lab [22]. Task 1 covers four subtasks including 1) check-worthy claim detection (Subtask 1A), verifiable factual claim detection (Subtask 1B),

---

[1] https://covid19asi.saglik.gov.tr/?_Dil=2.

harmful tweet detection (Subtask 1C), and attention-worthy tweet detection (Subtask 1D). We participated in all subtasks for Arabic, Bulgarian, Dutch, English, and Turkish languages, resulting in a total of 20 submissions. We explore the impact of i) various transformer models, ii) data augmentation, and iii) the mixup regularization techniques [26]. In particular, we utilize a wide range of different pre-trained transformer models for subtask 1A in Arabic, Bulgarian, Dutch, English, and Turkish, respectively, to explore the impact of transformer model chosen for fine-tuning. Next, we investigate the impact of training data on fine-tuning of transformer models in subtask 1C, and increase the training data by back-translation and machine-translation of datasets in other languages. Subsequently, we explore mixup methods in different levels (i.e., word, sentence, and manifold) using English data in all subtasks. Finally, we pick the most effective method in each approach and compare them in all four subtasks to select models for our official submissions.

In our experiments, we observe that the selection of transformer models significantly influences the overall performance. Furthermore, increasing the training data generally decreases performance for the Bulgarian and Turkish datasets in subtask 1C. Conversely, incorporating additional data into the English and Dutch datasets yields performance improvements.

In the official ranking, our performance varied across different tasks. Considering tasks with at least three participants, we are ranked first in 1B-Arabic and second in 1A-Arabic and 1A-Bulgarian. To ensure the reproducibility of our results, we share our implementation for the mixup method[2].

2   Tasks

CLEF 2022 Check That! Lab offered three tasks, but we participated in only Task 1 which aims to detect relevant tweets that might require fact-checking. Task 1 includes four subtasks:

- Subtask 1A: Predict whether a given tweet is worth fact-checking.
- Subtask 1B: Predict whether a given tweet contains a verifiable claim.
- Subtask 1C: Predict whether a given tweet is harmful to society.
- Subtask 1D: Predict whether a given tweet requires attention of government entities and why.

While subtasks 1A, 1B, and 1C are binary classification tasks, there are nine different labels in subtask 1D. These labels are i) *No*, ii) *Yes, asks question*, iii) *Yes, blame authorities*, iv) *Yes, calls for action*, v) *Yes, Harmful*, vi) *Yes, contains advice*, vii) *Yes, discusses action taken*, viii) *Yes, discusses cure*, and ix) *Yes, other*.

3   Related Work

Many researchers showed great interest in the fight against misinformation and attacked the problem in various ways such as detecting veracity of claims [22],

---

[2] https://github.com/Carnagie/manifold-mixup-text-classification.

tweets that are worth to fact-checked [16], resources needed for fact-checking [13] and others. Our work is related to the detecting tweets to be fact-checked.

As detecting attention worthy claims is recently proposed by Alam et al. [2], the existing studies are the ones that participated in Check That! Lab in 2022 [21]. Regarding detecting "harmful" messages, prior work focused on specific cases of harmful messages such as offensive language detection [24] and rumour detection [6]. In our work, we do not make any distinction among types of harmful messages and attempt to classify messages as harmful or not harmful.

Among these tasks, detecting check-worthy claims might be the most popular one. Check That! Lab has been organizing this task since 2018 [21]. Researchers explored various ways such as developing effective features [18] and deep learning models [16], incorporating various data engineering techniques [12] to address the data requirements of these models. In our work, we investigate various transformer models and data augmentation techniques. In addition, to our knowledge, prior work has not applied mixup regularization methods for these tasks.

4 Approaches

We explore three different approaches for all subtasks including fine-tuning various transformer models, increasing dataset size via machine translation, and employing the Manifold Mixup regularization technique.

4.1 Fine Tuning Various Transformer Models

The most successful systems in previous check-worthy claim detection tasks within the Check That! Lab [25] typically exploited a range of transformer models [29,30]. However, Kartal and Kutlu [16] show that the performance of these models can exhibit significant variations. Hence, we utilize several language-specific transformer models pre-trained with different datasets to explore how to choose the model to be fine-tuned.

4.2 Increasing Training Data via Machine Translation

Previous research on detecting check-worthy claim detection has explored various techniques for augmenting the training data such as back-translation [29], weak supervision [12], and the utilization of datasets from different languages with multi-lingual models [16]. In this approach, we delve into enhancing the training data size through two different methods: 1) leveraging datasets in other languages by machine-translating them into the respective language, and 2) paraphrasing the training data via back-translation and thereby generating additional labeled data for training.

In the first method, we employ a two-step process. We first select a training dataset provided for a different language and utilize Google Translate to machine-translate its corresponding tweets into the target language. Subsequently, we perform fine-tuning on a language-specific transformer model, utilizing both the original data and the machine-translated data collectively. In the case of subtask 1C, we exclusively machine-translate tweets that are labeled as harmful to also address the imbalance in label distribution while simultaneously augmenting the training data size.

In our back-translation method, we first translate the original text to a different language using Google Translate. Next we translate the resulting text back to the original language. This translation approach is anticipated to generate slightly altered texts that convey the same or similar meaning as the originals. We assume that these textual changes will not impact the associated labels. Consequently, we combine the original data with the back-translated data and fine-tune a language-specific transformer model.

4.3 Language Specific BERT with Manifold Mixup

Many of the annotations in the shared task are subjective. For instance, whether a tweet requires attention of government entities might depend on how much the annotators want governments to intervene their life. Similarly, prior work on check-worthiness points out the subjective nature of the task (e.g., [15,16]). In order to focus on this subjectivity problem, we apply the Manifold Mixup regularization proposed by Verma et al. [26]. They demonstrate that their approach yields more robust solutions in image classification. The Manifold Mixup trains neural networks on linear combinations of hidden representations of training examples, yielding flattened class-representations and smoother decision boundaries. In our work, we use BERT embeddings to represent tweets and then train a four-layer feed-forward network with the Manifold Mixup method.

In subtask 1D, we apply a different approach than the other tasks due to its severely imbalanced label distribution. In particular, there are nine labels in subtask 1D, but eight of them are about why a particular tweet is attention-worthy. In addition, the majority of the tweets have "not attention-worthy" label. Therefore, we first binarize labels by merging variants of attention-worthy labels into a single one, yielding only two labels: 1) attention-worthy and 2) not-attention-worthy. Subsequently, we under-sample negative class with the 1/5 ratio and train our Manifold Mixup model. Next, we build another model using eight labels for attention-worthy tweets. If a tweet is classified as attention-worthy, we use the second model to predict why it is attention-worthy.

5   Experiments

5.1 Experimental Setup

We use PyTorch v.1.9.0[3] and Tensorflow[4] to fine-tune and configure transformer models. We import transformer models used in our experiments from Huggingface[5]. In addition, we use Google's SentencePiece library for machine translation[6]. We set the batch size to 32 in all our experiments with fine-tuned

---

[3] https://pytorch.org/.
[4] https://www.tensorflow.org/.
[5] https://huggingface.co/docs/transformers/index.
[6] https://github.com/google/sentencepiece.

transformer models. In experiments on increasing dataset size using machine translation, we train the models for 5 epochs.

We implemented the Manifold Mixup [26] method from scratch using PyTorch v.1.9.0, and set epoch and the batch size to 5 and 2, respectively. We used the following transformer models for each language: AraBERT.v02 [5] for Arabic, RoBERTa-base-bulgarian[7] for Bulgarian, RobBERT [9] for Dutch, the uncased version of BERT-base[8] for English, and DistilBERTurk[9] for Turkish.

We use the official metric for each subtask to evaluate our methods. In particular, we use F_1 score of positive class in subtasks 1A and 1C, accuracy in subtask 1B, and weighted F_1 in subtask 1D.

The shared task organizers provide train, development, test development, and test datasets for each language and subtask. In our experiments during the development phase, we use the train and development datasets for training and validation of the Manifold Mixup model, respectively. In our experiments for fine-tuning various transformer models and increasing dataset size via machine translation, we combine train and development sets for each case and fine-tune models accordingly.

5.2   Experimental Results in the Development Phase

We participate in all subtasks of Task 1 for five languages, yielding 20 different submissions. In addition, we explore three different approaches to determine our final submissions. Therefore, in order to reduce the complexity of experiments, we evaluate the impact of transformer model variants, increasing training data size, and Manifold Mixup technique in subtask 1A, 1C, and 1D, respectively, on the corresponding test development datasets. Next, based on these experiments, we compare three different approaches in all subtasks to determine our submissions for the official evaluation on the test data.

Impact of Transformer Model on Detecting Check-Worthy Claims. In order to observe the impact of transformer models, we identify several transformer models available on the Huggingface platform based on their monthly download scores and evaluate their performance in subtask 1A. The number of transformer models we compare is 9, 3, 5, 13, and 3 for Arabic, Bulgarian, Dutch, English, and Turkish, respectively.

We present the results in Table 1. Firstly, the results for English show the importance of evaluation metric to report the performance of systems. For instance, *distilroberta-base-climate-f* has the worst recall and F_1 scores, but achieves the best accuracy. Secondly, our results suggest that the text used in pre-training has a major impact on the models' performance. For instance, *COVID-Twitter-BERT v1* achieves the best F_1 score among all English models. This should be because it is pretrained with tweets about COVID-19 while the

---

[7] https://huggingface.co/iarfmoose/roberta-base-bulgarian.

[8] https://huggingface.co/bert-base-uncased.

[9] https://huggingface.co/dbmdz/distilbert-base-turkish-cased.

tweets used in the shared task are also about COVID-19. Similarly, *PubMed-BERT*, which is pretrained with research articles on PubMed, yields the second best results for English. However, we also observe some unexpected results in our experiments. For instance, *AraBERT.v1*, which is pre-trained on a smaller dataset compared to other variants of AraBERT (i.e., *AraBERTv0.2-Twitter*, *AraBERTv0.2*, and *AraBERTv2*), outperforms all Arabic specific models. In addition, while *DarijaBERT* is pre-trained with only texts in Moroccan Arabic, it outperforms all other Arabic specific models except *AraBERT.v1*. Furthermore, the best performing model in the Turkish dataset is the one with the smallest vocabulary size. Therefore, our results show that it is not easy to determine a pre-trained model by just comparing models' configurations and texts used in pre-training.

Impact of Training Data in Detecting Harmful Tweets. In this experiment, we use *bert-base-arabertv02* for Arabic, *roberta-small-bulgarian*[10] for Bulgarian, *BERTje* [27] for Dutch, *BERT-base-cased* for English, and *bert-base-turkish-sentiment-cased*[11] for Turkish as language-specific transformer models. Table 2 shows the performance of each model when a different dataset is machine-translated to the corresponding language and respective language-specific model is fine-tuned with the original data and the machine-translated data. We observe that increasing training data does not always improve the performance. In particular, using the original dataset for Arabic, Bulgarian, and Turkish yields the highest results while the performance of models usually increase in English and Dutch datasets by utilizing more labeled samples.

The subjective nature of this task might be one of the reasons for having lower performance by using additional data from other languages. In particular, as each country is dealing with different social issues, it is likely that people living in different countries might disagree on what makes a message harmful for a society. For instance, Turkish annotators might be more sensitive to tweets about refugees compared to annotators for other languages because Turkiye hosts nearly 3.8 million refugees, i.e., the largest refugee population worldwide[12], and thereby, misinformation about refugees might have unpleasant consequences.

In our next experiment, we increase training data using various languages for back-translation which does not deal with social differences across countries. In this experiment, we also use Spanish for back-translation of the Bulgarian dataset[13]. The results are shown in Table 3.

We again observe that we achieve the best result for Arabic and Turkish when we use only the original dataset for training. However, back-translation improves the performance in the Dutch and English datasets. For Bulgarian, back-translation has a minimal impact. We do not observe a particular language

---

[10] https://huggingface.co/iarfmoose/roberta-small-bulgarian.

[11] https://huggingface.co/savasy/bert-base-turkish-sentiment-cased.

[12] https://www.unhcr.org/figures-at-a-glance.html.

[13] We were not able to use Spanish for other languages due to the insufficient time to meet the deadlines of the lab.

Table 1. Results of Various Transformer Models in Detecting Check-Worthy Claims.

	Model	Accuracy	Precision	Recall	F_1
Arabic	AraBERT.v1 [5]	0.413	0.390	**0.932**	**0.550**
	DarijaBERT[a]	0.499	0.420	0.789	0.548
	Ara_DialectBERT[b]	0.431	0.393	0.887	0.545
	arabert_c19 [4]	0.548	0.439	0.627	0.517
	AraBERTv0.2-Twitter [5]	**0.600**	**0.482**	0.526	0.503
	bert-base-arabic [24]	0.481	0.397	0.672	0.5
	CAMeLBERT [14]	0.451	0.372	0.620	0.465
	bert-base-arabertv2[c]	0.534	0.399	0.417	0.408
	bert-base-arabertv02[d]	0.599	0.454	0.206	0.284
Bulg.	RoBERTa-base-bulgarian (see footnote 7)	0.776	**0.451**	0.443	**0.447**
	RoBERTa-small-bulgarian-POS[e]	0.485	0.259	**0.820**	0.394
	bert-base-bg-cased [1]	**0.784**	0.448	0.245	0.317
Dutch	BERTje [27]	0.619	0.516	**0.941**	**0.666**
	RobBERT [9]	**0.650**	0.549	0.764	0.639
	bert-base-nl-cased[f]	0.559	0.469	0.676	0.554
	bert-base-dutch-cased-finetuned-gem[g]	0.638	**0.582**	0.382	0.461
English	COVID-Twitter-BERT v1 [20]	0.721	0.434	0.798	**0.562**
	PubMedBERT [11]	0.745	0.447	0.558	0.496
	BERT base model (uncased) [10]	0.634	0.343	0.689	0.458
	LEGAL-BERT [8]	0.630	0.326	0.604	0.423
	ALBERT Base v2 [17]	0.689	0.353	0.457	0.398
	Bio_ClinicalBERT [3]	0.682	0.337	0.426	0.376
	BERT base model (cased) [10]	0.224	0.224	**1.0**	0.366
	bert-base-uncased-contracts[h]	0.740	0.405	0.333	0.365
	ALBERT Base v1[i]	0.707	0.338	0.317	0.328
	hateBERT [7]	0.770	**0.476**	0.232	0.312
	COVID-Twitter-BERT v2 MNLI[j]	0.667	0.265	0.271	0.268
	RoBERTa base [19]	0.731	0.295	0.139	0.189
	DistilRoBERTa-base-climate-f [28]	**0.783**	0.631	0.093	0.162
Turkish	BERTurk uncased 32K Vocabulary[k]	**0.760**	**0.333**	0.385	**0.357**
	BERTurk uncased 128K Vocabulary[l]	0.337	0.188	**0.859**	0.309
	BERTurk cased 128K Vocabulary[m]	0.562	0.203	0.526	0.293

[a] https://huggingface.co/Kamel/DarijaBERT
[b] https://huggingface.co/MutazYoune/Ara_DialectBERT
[c] https://huggingface.co/aubmindlab/bert-base-arabertv2
[d] https://huggingface.co/aubmindlab/bert-base-arabertv02
[e] https://huggingface.co/iarfmoose/roberta-small-bulgarian-pos
[f] https://huggingface.co/Geotrend/distilbert-base-nl-cased
[g] https://huggingface.co/GeniusVoice/bert-base-dutch-cased-finetuned-gem
[h] https://huggingface.co/nlpaueb/bert-base-uncased-contracts
[i] https://huggingface.co/albert-base-v1
[j] https://huggingface.co/digitalepidemiologylab/covid-twitter-bert-v2-mnli
[k] https://huggingface.co/dbmdz/bert-base-turkish-uncased
[l] https://huggingface.co/dbmdz/bert-base-turkish-128k-uncased
[m] https://huggingface.co/dbmdz/bert-base-turkish-128k-cased

Table 2. Impact of increasing training data by machine-translating another dataset in a different language in detecting harmful tweets. We report F_1 score for each case.

Machine-Translated Data	Bulgarian	Dutch	English	Turkish	Arabic
None	**0.26**	0.26	0.11	**0.55**	**0.68**
Bulgarian	–	**0.39**	0.23	0.13	0.65
Dutch	0.23	–	0.23	0.53	0.60
English	0.21	**0.39**	–	0.48	0.64
Turkish	0.19	0.25	**0.25**	–	0.64
Arabic	0.16	0.27	0.21	0.47	–

Table 3. The impact of increasing train data using various languages for back-translation (BT).

Lang. used for BT	Bulgarian	Dutch	English	Turkish	Arabic
None	0.26	0.26	0.11	**0.55**	**0.68**
Bulgarian	–	0.39	**0.30**	0.51	0.66
Dutch	0.26	–	0.23	0.51	0.67
English	0.26	0.35	–	0.54	0.62
Turkish	0.25	**0.41**	0.25	–	0.61
Arabic	0.13	0.36	0.27	0.49	–
Spanish	**0.27**	–	–	–	–

which yields consistently higher results than others when used as the language for back-translation.

Applying Mixup Technique at Different Levels. In this experiment, we explore variations of the mixup technique. In particular, Manifold Mixup proposed by Verma et al. [26] applies the regularization technique in all hidden layers. When the regularization is applied only in the first layer and only in the last layer, it is called *Word Mixup* and *Sentence Mixup*, respectively. We compare these three methods in all subtasks for English. We use bert-base-cased model in all of our tests. We report accuracy results for all tasks in **Table 4**. We observe that performance of Sentence Mixup is generally lower than others. Word Mixup yields slightly higher accuracy in 1A, 1B, and 1D. In 1C Manifold Mixup outperforms others with a high margin.

Table 4. Results of using different mixup techniques for English in different tasks.

Model	Tasks			
	1A	1B	1C	1D
Word Mixup	**0.768**	**0.642**	0.683	**0.851**
Sentence Mixup	0.767	0.405	0.655	0.765
Manifold Mixup	0.742	0.640	**0.781**	0.850

Selecting Models for Submission. We compare three different approaches for each subtask and language to select the models to submit for official ranking:

1) Fine-tuning the best-performing pre-trained transformer model with the original dataset (FT-BP-TM). We pick the best-performing transformer model for each language in our experiments in which we explore various models for subtask 1A. In particular, we fine-tune *AraBERT.v1, RoBERTa-base-bulgarian, BERTje, COVID-Twitter-BERT v1,* and *BERTurk,* for Arabic, Bulgarian, Dutch, English, and Turkish, respectively, using the corresponding datasets.

2) Fine-tuning a transformer model with back translation (FT-TM-BT). We use the best-performing back-translation setup in our experiments for each language. In particular, we use Spanish, Turkish, Bulgarian, and English for back-translation to increase the size of Bulgarian, Dutch, English, and Turkish datasets, respectively. Note that the back-translation does not improve the performance in the Turkish dataset. However, the FT-BP-TM approach also uses the original dataset for fine-tuning. Therefore, in this approach, we increase the size of Turkish dataset using back-translation. In particular, we use English as the back-translation language because it yields the best results among others (See Table 3).

3) Manifold Mixup. We use the Manifold Mixup method explained in Sect. 4.3. We do not use Word Mixup as their performance is highly similar in two tasks and Manifold Mixup achieves higher accuracy on average across tasks.

Table 5 presents results comparing three approaches for all subtasks. Results for some cases are missing due to technical challenges we encountered. We observe that fine-tuning with the original data usually yields the highest performance.

5.3   Results of Our Submissions

In our submissions, we chose the method for each case according to our results presented in Table 5[14]. Table 6 shows the results. We are ranked first in 1B Arabic and 1C Dutch. Focusing on subtasks with at least four participants, we are ranked second in Arabic 1A and Bulgarian 1A. We also observe that our rankings are generally higher in 1A than other subtasks.

---

[14] We note that some of the results were absent at the time of submission. Therefore, in our submission we chose the results based on the incomplete results.

Table 5. Development Test Results. We report F_1 score for 1A, 1B, and 1C, and average F_1 score for 1D.

Task	Model	Arabic	Bulgarian	Dutch	English	Turkish
1A	FT-BP-TM	**0.47**	**0.47**	0.57	**0.55**	**0.40**
	FT-TM-BT	–	0.42	**0.64**	0.48	**0.40**
	Manifold Mixup	0.14	0.00	0.58	0.48	0.22
1B	FT-BP-TM	**0.89**	**0.87**	0.72	**0.76**	**0.78**
	FT-TM-BT	–	0.86	**0.73**	-	**0.78**
	Manifold Mixup	0.76	0.75	0.49	0.67	0.63
1C	FT-BP-TM	**0.68**	0.24	0.33	**0.35**	0.52
	FT-TM-BT	–	**0.27**	**0.41**	0.30	**0.54**
	Manifold Mixup	0.64	0.00	0.12	0.18	0.30
1D	FT-BP-TM	0.58	**0.80**	**0.72**	**0.86**	**0.83**
	FT-TM-BT	–	0.33	0.31	–	0.28
	Manifold Mixup	**0.65**	**0.80**	0.65	0.78	0.79

Table 6. Results for our official submissions. Results show F_1, accuracy, F_1, and weighted F_1 scores for tasks 1A, 1B, 1C, and 1D, respectively.

Task	Language	Submitted Model	Rank	Score
1A	Arabic	FT-BP-TM	2 (out of 5)	0.495
	Bulgarian	FT-BP-TM	2 (out of 6)	0.542
	Dutch	FT-TM-BT	3 (out of 6)	0.534
	English	FT-BP-TM	4 (out of 14)	0.561
	Turkish	FT-TM-BT	3 (out of 5)	0.118
1B	Arabic	Manifold Mixup	1 (out of 4)	0.570
	Bulgarian	FT-BP-TM	2 (out of 3)	0.742
	Dutch	FT-TM-BT	2 (out of 3)	0.658
	English	FT-BP-TM	9 (out of 10)	0.641
	Turkish	FT-TM-BT	4 (out of 4)	0.729
1C	Arabic	Manifold Mixup	2 (out of 3)	0.268
	Bulgarian	FT-TM-BT	2 (out of 3)	0.054
	Dutch	FT-TM-BT	1 (out of 3)	0.147
	English	FT-BP-TM	5 (out of 12)	0.329
	Turkish	FT-TM-BT	3 (out of 5)	0.262
1D	Arabic	Manifold Mixup	2 (out of 2)	0.184
	Bulgarian	Manifold Mixup	2 (out of 3)	0.887
	Dutch	Manifold Mixup	2 (out of 3)	0.694
	English	Manifold Mixup	4 (out of 7)	0.670
	Turkish	Manifold Mixup	3 (out of 3)	0.806

6   Conclusion

In this paper, we present our participation in CLEF 2022 CheckThat! Lab's Task 1. We participated in all four subtasks of Task1 for Arabic, Bulgarian, Dutch, English, and Turkish, yielding 20 submissions in total. We explore which transformer model yields the highest performance, the impact of increasing training data size by machine translating datasets in other languages and back-translation, and the Manifold Mixup method proposed by Verma et al. [26]. We are ranked first in subtask 1B for Arabic and in subtask 1C for Dutch. In addition, we are ranked second in subtask 1A for Arabic and Bulgarian.

Our observations based on our comprehensive experiments are as follows. Firstly, the performance of transformer models varies dramatically based on the text used for pre-training. Secondly, increasing training data does not always improve the performance. Therefore, it is important to consider biases existing in each dataset. Thirdly, we do not observe that a particular language used for back-translation yields consistently higher performance than others.

In the future, we plan to focus on the subjective nature of the tasks in this lab. In particular, we will first qualitatively analyze the datasets to better understand annotations. Subsequently, we plan to develop a model focusing on dealing with subjective annotations.

References

1. Abdaoui, A., Pradel, C., Sigel, G.: Load what you need: smaller versions of mutlilingual BERT. In: SustaiNLP/EMNLP (2020)
2. Alam, F., et al.: Fighting the COVID-19 infodemic in social media: a holistic perspective and a call to arms. In: Proceedings of the International AAAI Conference on Web and Social Media, vol. 15, pp. 913–922 (2021)
3. Alsentzer, E., et al.: Publicly available clinical BERT embeddings. arXiv preprint arXiv:1904.03323 (2019)
4. Ameur, M.S.H., Aliane, H.: AraCOVID19-MFH: Arabic COVID-19 multi-label fake news and hate speech detection dataset (2021)
5. Antoun, W., Baly, F., Hajj, H.: AraBERT: transformer-based model for arabic language understanding. In: LREC 2020 Workshop Language Resources and Evaluation Conference, p. 9 (2020)
6. Bondielli, A., Marcelloni, F.: A survey on fake news and rumour detection techniques. Inf. Sci. **497**, 38–55 (2019)
7. Caselli, T., Basile, V., Mitrović, J., Granitzer, M.: HateBERT: retraining BERT for abusive language detection in English. In: Proceedings of the 5th Workshop on Online Abuse and Harms (WOAH 2021). Association for Computational Linguistics, Online (2021)
8. Chalkidis, I., Fergadiotis, M., Malakasiotis, P., Aletras, N., Androutsopoulos, I.: LEGAL-BERT: the muppets straight out of law school. In: Findings of the Association for Computational Linguistics: EMNLP 2020. Association for Computational Linguistics, Online (2020)
9. Delobelle, P., Winters, T., Berendt, B.: RobBERT: a Dutch RoBERTa-based language model. In: Findings of the Association for Computational Linguistics: EMNLP 2020. Association for Computational Linguistics, Online (2020)

10. Devlin, J., Chang, M.W., Lee, K., Toutanova, K.: BERT: pre-training of deep bidirectional transformers for language understanding. arXiv preprint arXiv:1810.04805 (2018)
11. Gu, Y., et al.: Domain-specific language model pretraining for biomedical natural language processing (2020)
12. Hansen, C., Hansen, C., Simonsen, J.G., Lioma, C.: Neural weakly supervised fact check-worthiness detection with contrastive sampling-based ranking loss. In: CLEF (Working Notes) (2019)
13. Haouari, F., Elsayed, T., Mansour, W.: Who can verify this? Finding authorities for rumor verification in twitter. Inf. Process. Manage. **60**(4), 103366 (2023)
14. Inoue, G., Alhafni, B., Baimukan, N., Bouamor, H., Habash, N.: The interplay of variant, size, and task type in Arabic pre-trained language models. In: Proceedings of the Sixth Arabic Natural Language Processing Workshop. Association for Computational Linguistics, Kyiv (Online) (2021)
15. Kartal, Y.S., Kutlu, M.: TrClaim-19: the first collection for Turkish check-worthy claim detection with annotator rationales. In: Proceedings of the 24th Conference on Computational Natural Language Learning, pp. 386–395 (2020)
16. Kartal, Y.S., Kutlu, M.: Re-think before you share: a comprehensive study on prioritizing check-worthy claims. IEEE Trans. Comput. Soc. Syst. **10**(1), 362–375 (2023)
17. Lan, Z., Chen, M., Goodman, S., Gimpel, K., Sharma, P., Soricut, R.: ALBERT: a lite BERT for self-supervised learning of language representations. arXiv preprint arXiv:1909.11942 (2019)
18. Lespagnol, C., Mothe, J., Ullah, M.Z.: Information nutritional label and word embedding to estimate information check-worthiness. In: Proceedings of the 42nd International ACM SIGIR Conference on Research and Development in Information Retrieval, pp. 941–944 (2019)
19. Liu, Y., et al.: RoBERTa: a robustly optimized BERT pretraining approach. CoRR abs/1907.11692 (2019). http://arxiv.org/abs/1907.11692
20. Müller, M., Salathé, M., Kummervold, P.E.: COVID-twitter-BERT: a natural language processing model to analyse COVID-19 content on twitter. arXiv preprint arXiv:2005.07503 (2020)
21. Nakov, P., et al.: Overview of the CLEF-2022 CheckThat! Lab task 1 on identifying relevant claims in tweets. In: Working Notes of CLEF 2022–Conference and Labs of the Evaluation Forum, CLEF 2022, Bologna, Italy (2022)
22. Nakov, P., et al.: Overview of the CLEF-2022 CheckThat! Lab on fighting the COVID-19 infodemic and fake news detection. In: Barrón-Cedeño, A., et al. (eds.) CLEF 2022. LNCS, vol. 13390, pp. 495–520. Springer, Cham (2022). https://doi.org/10.1007/978-3-031-13643-6_29
23. Roozenbeek, J., et al.: Susceptibility to misinformation about COVID-19 around the world. Roy. Soc. Open Sci. **7**(10), 201199 (2020)
24. Safaya, A., Abdullatif, M., Yuret, D.: KUISAIL at SemEval-2020 task 12: BERT-CNN for offensive speech identification in social media. In: Proceedings of the Fourteenth Workshop on Semantic Evaluation, pp. 2054–2059. International Committee for Computational Linguistics (2020)
25. Shaar, S., et al.: Overview of the CLEF-2021 CheckThat! Lab task 1 on check-worthiness estimation in tweets and political debates. In: CLEF (Working Notes) (2021)
26. Verma, V., et al.: Manifold mixup: better representations by interpolating hidden states. In: International Conference on Machine Learning, pp. 6438–6447. PMLR (2019)

27. de Vries, W., van Cranenburgh, A., Bisazza, A., Caselli, T., Noord, G.V., Nissim, M.: BERTje: a dutch BERT model. arXiv:1912.09582 (2019)
28. Webersinke, N., Kraus, M., Bingler, J., Leippold, M.: ClimateBERT: a pretrained language model for climate-related text. arXiv preprint arXiv:2110.12010 (2021)
29. Williams, E., Rodrigues, P., Tran, S.: Accenture at CheckThat! 2021: interesting claim identification and ranking with contextually sensitive lexical training data augmentation. arXiv preprint arXiv:2107.05684 (2021)
30. Zengin, M., Kartal, Y., Kutlu, M.: TOBB ETU at CheckThat! 2021: data engineering for detecting check-worthy claims. In: CEUR Workshop Proceedings (2021)

A Re-labeling Approach Based on Approximate Nearest Neighbors for Identifying Gambling Disorders in Social Media

Hermenegildo Fabregat[1,3] , Andres Duque[1,2(✉)] , Lourdes Araujo[1,2] ,
and Juan Martinez-Romo[1,2]

[1] NLP & IR Group, Dpto. Lenguajes y Sistemas Informáticos,
Universidad Nacional de Educación a Distancia (UNED),
Juan del Rosal 16, 28040 Madrid, Spain
{gildo.fabregat,aduque,lurdes,juaner}@lsi.uned.es
[2] IMIENS: Instituto Mixto de Investigación, Escuela Nacional de Sanidad,
Monforte de Lemos 5, 28019 Madrid, Spain
[3] Avature Machine Learning, Madrid, Spain

Abstract. This paper describes a novel approach based on Approximate
Nearest Neighbors (ANN) techniques for modifying the granularity of the
label schema in the training dataset of a classification task, from a user-
based annotation to a message-based one. In particular, we tackle Task
1 of the CLEF 2022 eRisk Workshop which consists in the processing of
messages written by Social Media users, in order to detect early signs of
pathological gambling. Our proposal is based on the calculation of the
nearest neighbors of the vectorial representations of the given messages,
originally annotated at user-level. This way, we obtain a re-labeled training dataset in which messages from the same user can be either positive
or negative. We then use this re-labeled dataset for performing the final
classification on test instances. Compared to other systems participating
in the task, our approach achieves the best average performance in the
proposed evaluation frameworks, and shows to be the fastest one in terms
of time needed to process the whole test dataset. This indicates that the
proposed relabeling scheme allows us to capture more easily the textual
information that leads to a correct detection of pathological gambling.

Keywords: Pathological gambling detection · Approximate Nearest
Neighbors · Vector representations · Re-labeling

1 Introduction

In the Internet era, social media analysis for the early detection of potential
health risks is a particularly interesting research area. In this context, both
methodologies and practical approaches have been developed for the early detection of different types of health risks, such as eating disorders, self-harm or
depression, through the textual analysis of posts and messages of social media
users.

A. Arampatzis et al. (Eds.): CLEF 2023, LNCS 14163, pp. 174–185, 2023.
https://doi.org/10.1007/978-3-031-42448-9_15

In this research, we tackle the detection of early detection of signs of pathological gambling in social media posts. For this aim, we explore the use of vector-based representation of users messages through sentence embeddings, for subsequently detect positive messages using methods based on Approximate Nearest Neighbors (ANN) techniques. One of the most important contributions of this work is the transformation of the training dataset, originally labeled at user level, in order to generate a message-level labeled dataset that should help improving the final results of the proposed system. Although ANNs can be seen as a simple machine learning technique, we show in the paper how this adequate pre-processing of the training dataset based on the reduction of the original label granularity allows us to obtain interesting and promising results. The context in which this system has been developed and tested is Task 1 of the CLEF eRisk 2022 Workshop: Early Detection of Signs of Pathological Gambling [17]. In this task, given a set of users and their messages posted in Reddit forums, a system must decide whether each user should be classified as a pathological or a non-pathological gambler.

The rest of the paper is structured as follows: an overview of previous work related to the task considered and the techniques used in this work is shown in Sect. 2. Section 3 is devoted to describe the addressed task, including the available dataset and evaluation metrics, while the developed system is presented in Sect. 4. The achieved results are shown and discussed in Sect. 5. Finally, Sect. 6 presents the main conclusions and future lines of work.

2   Related Work

Gambling disorder (GD) is characterized by a persistent and recurrent pattern of gambling that is associated with significant distress or substantial upset. The prevalence of GD has been estimated at 0.5% of the adult population in the United States, with comparable or even higher estimates in other countries [18]. However, people with GD are often not treated or even recognized as such. Moreover, GD often co-occurs with other psychiatric disorders: high rates of mood, anxiety, attention deficit disorders and substance use disorders have been reported in people with GD [19]. GD is also often accompanied by a higher rate of unemployment, economic difficulties, divorce, and poorer health, and is also closely related to other addictive disorders, being the first non-substance addictive behavior to be recognized [20].

Social networks are an excellent source of information where studies can be carried out for the early detection of people with gambling problems. In this line, the CLEF eRisk competition considered the problem of pathological gambling for the first time in 2021 [16]. In this edition, no training data was provided to the participating teams, hence different approaches were employed by those teams in order to bypass this limitation. The UPV-Symmanto team [2] crawled gambling-related Reddit forums and annotated users as pathological or non-pathological gamblers. Then, they built a Transformer-based classifier [21] and created an alert-emitting system for determining whether an user belonged to the positive class after a number of messages. The RELAI team [15] also made use of external resources: Reddit forums were crawled for building an Embedding Topic Model (ETM) [8]. Test users were then mapped into a vector of topic probabilities according to the topics extracted by the ETM. Other external resources such as gambling testimonials and questionnaires were then used for comparing similarities with the test vectors and determining which test users are more likely to be positive. The BLUE team [4] and the CeDRI team [11] also built and annotated a custom dataset extracted from Reddit forums in order to deal with the lack of training data. The BLUE team then used Transformer-based models for the final classification, while the CeDRI team employed different approaches, such as TFIDF characteristics or LSTM networks. Finally, the best performing system was UNSL [13]. This team also generated a dataset based on Reddit forums for training its models. Then, different representations such as Bag-of-Words and doc2vec, as well as different classification algorithms (SVMs and LSTM-based neural networks) were proposed. The best performing configuration for this team used a Bag-of-Words representation and a SVM classifier, together with an early alert policy which emitted an alarm after a minimum of 10 posts related to an user and a probability of belonging to the positive class over a certain threshold.

As we previously mentioned, our system is based on a simple approach that has proven to be very effective. Considering that the training dataset is annotated as user level, this is, only information about the user being classified as pathological or non-pathological gambler is given, the idea is to carry out a re-labeling of users' messages using a method based on Approximate Nearest Neighbor (ANN) search. This way, we intend to generate a new version of the training dataset in which all the messages are annotated as being or not at risk of belonging to a pathological gambler. Then, from this re-labeled dataset we should be able to develop an alert-emitting system based on the analysis of individual messages.

Regarding nearest neighbors algorithms, the exact nearest neighbor search (NNS) for the point corresponding to a given query is defined as the point corresponding to the shortest distance to the query. A generalization of the nearest neighbor search is the k-nearest neighbor search (k-NNS), which targets the k nearest vectors for the query. Due to the cost associated with dimensionality, many proposals have been developed focusing on the approximate solution of the NNS and k-NNS problem. A recent work [10] has presented a comparison and evaluation of different approaches to the problem. According to this work, state-of-the-art ANN methods can be classified into three types: Hashing-based, Partition-based and Graph-based. Hashing-based methods transform data points to a low-dimensional representation, where each point is represented by a short code (hash code). Partition-based methods can be seen as the division of high-dimensional space into multiple disjoint regions. The partitioning process is usually done recursively, hence these methods often use a tree- or forest-based representation. We have used one of these methods in this work, Annoy [3], a hyperplane partitioning method that recursively divides the space by the hyperplane with random direction. Finally, graph-based methods construct a proximity graph in which each datum corresponds to a node and the edges connecting some nodes define the neighborhood relationship. The main idea of these methods is that a neighbor's neighbor is likely to also be a neighbor. The search can be performed efficiently by iteratively extending neighbors of neighbors in a best-first search strategy. Depending on the structure of the graph, different graph-based methods can be distinguished. In this work we have used the Hierarchical Navigable Small World (HNSW) graphs method [14], implemented in the Non-Metric Space Library (NMSLIB).

3   Early Detection of Signs of Pathological Gambling

As previously mentioned, the main evaluation framework in which the research presented in this paper has been tested is Task 1 of the eRisk 2022 competition [17], denoted "Early detection of signs of pathological gambling". This is the second edition of the task, which was first introduced in the CLEF 2021 eRisk Workshop [16]. In this task, participating systems are asked to determine whether an individual can be classified as a pathological gambler (positive users) or a non-pathological gambler (negative users) based on the user's Social Media messages. Systems must sequentially analyse chronological posts written by each user in order to detect early traces of pathological gambling. This way, the main objective of the task is to help developing systems for effectively monitor user interaction in different types of online media.

3.1   Dataset

The dataset used in the task is composed of a set of XML documents, each of them containing chronologically ordered Social Media posts belonging to a particular user. The training dataset refers to a total of 2,348 users (164 pathological gamblers and 2,184 control users). The total number of test users is 2,079 (81 pathological gamblers and 1,998 control users). The number of posts per user are shorter for positive users (i.e. pathological gamblers), while the posts are normally longer for this type of users than for non-pathological gamblers. Additional statistics of the dataset can be found in [17].

3.2   Metrics

System evaluation is twofold: decision-based and ranking-based. More information about the complete set of metrics employed in the evaluation can be found in the overviews of the different eRisk competitions [12,16].

4   Proposed Model

The following sections describe the main components of the proposed model and the configurations that have been explored.

4.1 Data Representation

We use Universal Sentence Encoder [6] to encode each user's messages. Such models are trained and optimized for encoding texts longer than words e.g. sentences, phrases or short paragraphs. The model we use is trained with a deep average network [9] (DAN) using data from different sources in English. Although DAN approaches produce unordered representations of the information by averaging the terms in a given text, these models are able to capture subtle differences between similar texts. In short, for each message encoded by this model, a 512-dimensional vector is generated.

4.2 Approximate Nearest Neighbors

Although nearest neighbor retrieval is a conceptually simple procedure, in domains such as social networks, where a large amount of information is available, it is a difficult problem to address. In this domain the use of brute force-based search techniques is replaced by the use of non-exact techniques based on the use of more complex structures e.g. graphs and trees. Currently there are different tools and approaches that have proven to be very successful when analysing recall results and queries per second [1]. Due to their popularity and performance we have explored two different types of ANN approaches, as mentioned in Sect. 2: Annoy[1] and Non-Metric Space Library [14] (NMSLIB).

4.3 Relabeling Process

The corpus provided by the organizers presents a user-based labeling, i.e., each user is labeled as positive if at least a positive message can be found within his/her posts, and negative otherwise. However, positive/negative annotations for each message in the corpus are not provided. We consider that the correct classification of positive and negative messages is crucial for achieving a good performance in this task. Hence, we propose an approach to re-annotate the training corpus in order to generate a message-level labeling. For this purpose, we first consider all messages of a positive user to be positive, and all messages of a negative user to be negative. From the training set, we consider only those messages that contain title information. This indicates that the message represents the opening of a Reddit thread. Through this initial filtering, we intend to give preference to those discussions originally initiated by the subject user (e.g. calls for help or topic-related questions). Once the approximate nearest neighbor query index is generated, we iteratively process each message from each positive user of the training set, and re-annotate its class according to the similarity with the nearest neighbors of the considered message. The specific formula employed for determining whether a particular message is re-annotated or not is presented in Sect. 4.4. We assume that only positive users may contain negative messages, since if negative users contained positive messages, they would have been labeled

---

[1] https://github.com/spotify/annoy.

as positive. Hence, in each iteration of the algorithm, the number of positive messages is reduced if the algorithm re-labels them as negative. After processing the training set, if modifications have been made, the same method is applied again until convergence is reached, this is, until there are no changes in the training set labels. This final re-labeled version of the training dataset is then used for classifying each of the message from a test user.

There are various scenarios related to the original training dataset that could hinder the generation of a useful re-labeled dataset for solving the addressed task. For example, considering all messages from a positive user as positive would generate a large number of false positives that ideally should be transformed into negative messages. However, if these messages are very close to each other, it could happen that the re-labeling algorithm fails to transform these messages, and many false positives persist in the final training dataset. Similarly, an aggressive re-labeling process could result in all messages being labeled as negative, hence the final dataset would be useless. To avoid these situations, a 30% partition of the training dataset has been used as a development dataset to explore and define a set of parameters used in the re-labeling process. Through these parameter study we try to ensure that the final training dataset is balanced enough to generate useful results on the test dataset. This parameter set is closely related to the labeling and scoring functions and is explained in the following section.

4.4 Tag and Scoring Function

Once the training set is transformed using Universal Sentence Encoder, and after generating the nearest neighbor index using either the Annoy or NMSLIB libraries, we propose a labeling or tagging function through which the new label of each message in the training dataset is determined, as mentioned in Sect. 4.3. Given a message M from a user U we consider U_M to be positive if the k nearest neighbors retrieved include at least j positive messages. As previously mentioned, we used the development dataset to explore different values of these parameters in order to guarantee the convergence of the algorithm on a non-zero set of positive training instances after applying the relabeling process. In this step, the best values for these parameters were $k = 10$, $j = 7$.

Regarding the final classification step, a different set of values for parameters k and j was determined after a tuning evaluation stage. In this step, $k = 20$ and $j = 20$, this is, the 20 nearest neighbors of a test message are retrieved from the training dataset, and all of them must be positive in order to classify this test message as positive.

Finally, and following the same idea behind the tagging function, we considered as scoring function the mean distance from U_M to the nearest recovered neighbors: $(1 - \frac{1}{j} \sum_{x=1}^{j} cosine(U_M, M_x))$, where M_x is each of the retrieved nearest messages. As the scoring function is only used in the final classification step for the ranking-based evaluation, the values of the mentioned parameters are $k = 20$, $j = 20$. In this case, the scoring function is calculated only for those messages classified as positive, this is, those messages whose 20 nearest neighbors are all positive. Otherwise, the scoring function will return zero. The risk of pathological gambling of a particular user, each time a new message is analysed, is the maximum scoring obtained by any of the (positive) messages of that user, from those messages processed up to that point.

4.5 Crawling New Positive Instances

In order to reduce the impact on recall that the relabeling algorithm could have, we have collected additional data from gamblers' help associations: 234 testimonial facts from websites[2] with more structured and longer texts, and 232 forum messages[3] with a similar format and structure to Reddit messages. No specific pre-processing techniques such as text size limitation or language control have been employed.

Five configurations of the proposed system were submitted as runs for task 1 of the CLEF eRisk competition. The different characteristics of each configuration are shown in Table 1. Universal Sentence Encoder has been used as encoder while Annoy and Non-Metric Space Library (NMSLIB) have been explored as methods for k-nearest neighbor retrieval. On the other hand, we studied a relabeling process of the training set and the consideration of new data collected automatically.

Table 1. Submitted Runs: Configurations explored in the test phase.

	ANN Library	Relabeling	New data
Run 0	Annoy	No	No
Run 1	Annoy	Yes	No
Run 2	Annoy	No	Yes
Run 3	Annoy	Yes	Yes
Run 4	NMSLIB	Yes	No

5   Results and Discussion

The results obtained by our approach are shown and discussed below.

Execution time: As can be seen in Table 2, the proposed batch of experiments achieved the best execution times among the systems that processed the whole test set. Other systems presenting better execution times only processed a very reduced subset of the test dataset. Regarding the approximate

---

[2] https://gamblershelp.com.au; http://getgamblingfacts.ca; https://gamtalk.org; https://gamcare.org.uk.
[3] https://www.gamtalk.org/groups/community/.

nearest neighbor algorithms employed in this work, while Annoy uses tree-like structures for the representation of nodes and random projections for the division of the subspace between adjacent nodes, NMSLIB uses a graph-based structure and the projection of the different nodes onto a skip-list. Both algorithms include customizable parameters to optimize their performance, e.g. number of trees (Annoy) or number of Zero node links (NMSLIB). Although we do not perform an exhaustive study of these parameters, we try to limit their growth.

Table 2. Test results: Comparison of the execution times required by the participating systems.

Team	#runs	#user writings processed	lapse of time (from 1st to last response)
UNED-NLP (Ours)	**5**	**2001**	**17:58:48**
SINAI	3	46	4 days 12:54:03
BioInfo_UAVR	5	1002	22:35:47
RELAI	5	109	7 days 15:27:25
BLUE	3	2001	3 days 13:15:25
BioNLP-UniBuc	5	3	00:37:33
UNSL	5	2001	1 day 21:53:51
NLPGroup-IISERB	5	1020	15 days 21:30:48
stezmo3	5	30	12:30:26

Decision-based performance: Table 3 shows the results obtained during the decision-based evaluation. This table shows the set of metrics analysed by the task organizers: Precision, Recall, $F1$, $ERDE_5$, $ERDE_{50}$, latency, speed and latency-weigthed $F1$. In addition to the results of our runs, the best run of each team participating in the competition is shown. As it can be seen in the table, considering the latency-weighted $F1$ metric as the summary metric, our R4 configuration obtained the best results, achieving the highest precision/recall ratio. If we analyse the achieved results in terms of latency, i.e., delay shown by the system expressed as the median number of messages that need to be processed before detecting a positive case, as we used the same inference process in all the runs, no great differences can be found between the different submitted runs. However, if we compare runs R0 and R1, which are differentiated by the application of the relabeling process in R1, we find improvements in precision of around 27% with no excessive penalization of other metrics such as recall. The relabeling process presents a high impact on the corpus since the label of more than 90% of the positive instances is modified after applying it. On the other hand, and seeking to reduce the effect on recall produced by the relabeling process, the inclusion of new data automatically collected was considered in the R2 and R3 runs. The obtained results indicate that our approach to collect and process the new data was not the most efficient one. Finally, R1 and R4 differ by the algorithm for nearest neighbor retrieval used (R1: Annoy, R4: NMSLIB). These algorithms include a parameter space that has not been studied in depth. For this reason, and although the NMSLIB algorithm performs significatively better than Annoy, we consider that a more thorough study on the parameters of the latter technique should be performed before discarding its use.

Table 3. Test results: Results of the decision-based evaluation for task T1. For the models included in the comparison, the best results are shown in bold.

	Prec	Rec	F1	ERDE5	ERDE50	latency	speed	latency-weighted F1
R0	0.285	0.975	0.441	0.019	0.010	2.0	0.996	0.4405
R1	0.555	0.938	0.697	0.019	0.009	2.5	0.994	0.693
R2	0.296	0.988	0.456	0.019	0.009	2.0	0.996	0.454
R3	0.536	0.926	0.679	0.019	0.009	3.0	0.992	0.673
R4	0.809	0.938	**0.869**	0.020	**0.008**	3.0	0.992	**0.862**
SINAI R2	**0.908**	0.728	0.808	0.016	0.011	1.0	1.000	0.808
BioInfo_UAVR R1	0.067	**1.000**	0.126	0.047	0.024	5.0	0.984	0.124
RELAI R2	0.052	0.963	0.099	0.036	0.029	1.0	1.000	0.099
BLUE R0	0.260	0.975	0.410	**0.015**	0.009	1.0	1.000	0.410
BioNLP_UniBuc R4	0.046	**1.000**	0.089	0.032	0.031	1.0	1.000	0.089
UNSL R1	0.461	0.938	0.618	0.041	**0.008**	11	0.961	0.594
NLPGroup-IISERB R3	0.140	**1.000**	0.246	0.025	0.014	2.0	0.996	0.245
stezmo3 R4	0.160	0.901	0.271	0.043	0.011	7.0	0.977	0.265

Ranking-based performance: Table 4 shows results obtained in the ranking-based evaluation. During this evaluation, the performance of the system is measured after processing 1, 100, 500 and 1000 messages. As shown in the Table, our R4 run obtains the best results during this evaluation for all metrics in almost all stages. Comparing the differences between R4 and the best runs presented by BLUE and UNSL (the other two systems that processed the whole dataset), our system outperforms in most aspects except for NDCG@100 when analysing 1 and 100 writings. These results indicate that the scoring function described in Sect. 4.4 is an effective heuristic for assessing the risk of pathological gambling after processing each user message.

Qualitative analysis: A small qualitative analysis has been conducted to explore the functioning of the proposed system. To do this, the most representative words and expressions have been extracted from messages in the test set labeled as positive and negative by our system, using the tool YAKE [5]. Regarding positive messages, some of the most informative expressions are:

Table 4. Test results: Results of the ranking-based evaluation for task T1. For the models included in the comparison, the best results are shown in bold.

	1 writing			100 writings			500 writings			1000 writings		
	P@10	NDCG@10	NDCG@100	P@10	NDCG@10	NDCG@100	P@10	NDCG@10	NDCG@100	P@10	NDCG@10	NDCG@100
Run 0	0.9	0.88	0.75	0.4	0.29	0.7	0.3	0.2	0.56	0.3	0.19	0.48
Run 1	0.9	0.81	0.68	0.80	0.73	0.83	0.5	0.43	0.80	0.5	0.37	0.75
Run 2	0.9	0.88	**0.76**	0.60	0.58	0.79	0.4	0.33	0.55	0.3	0.24	0.46
Run 3	0.9	0.81	0.71	0.70	0.66	0.84	0.4	0.35	0.78	0.5	0.42	0.73
Run 4	1	1	0.56	1	1	0.88	1	1	**0.95**	1	1	**0.95**
BLUE Run 1	1	1	0.76	1	1	0.89	1	1	0.91	1	1	0.91
UNSL Run 0	1	1	0.68	1	1	**0.9**	1	1	0.93	1	1	0.95

"gambling", "money", "bet", "stop", "mistake", "lose", "urge" or "control". All of these words might indicate that the users are experiencing problems for controlling their gambling. On the other hand, the analysis of messages labeled as negative also offers some words related to gambling such as "play" or "players", however, most of them are non-informative words and expressions such as "post", "numbers", "game", "area" or "time'.

6 Conclusions and Future Work

This paper describes a novel method based on the use of approximate nearest neighbor algorithms for re-labeling a user-based annotated dataset in order to generate a message-based annotated version of it, and its application to Task 1 of the CLEF eRisk 2022 competition [17], devoted to an early detection of signs of pathological gambling. These algorithms are also employed in the classification phase for retrieving subsets of similar messages to a given one, previously transformed into a vectorial space using sentence embeddings. The final classification of each test message is done by analysing these similar subsets and determining whether the message is more likely to be positive (the user is a pathological gambler) or negative.

The use of algorithms such as Annoy or NMSLIB for large scale nearest neighbor retrieval has been of great help for the fast processing of the data. As shown in Table 2 and having processed all the messages from the test set, our system obtained the best execution times. On the other hand, as shown in Tables 3 and 4, our model has obtained the best results for the $F1$, $ERDE_{50}$ and F-latency metrics in the decision-based evaluation, as well as the best overall results in the ranking-based evaluation. Most of these results are due to the application of the iterative re-labeling process of the corpus described in Sect. 4.3. Through this process we have also validated the use of the vector space generated by Universal Sentence Encoder to analyse the similarity between messages of different classes.

The following lines of future work are being currently considered and explored:

- Study of encoders based on more complex approaches such as BERT [7], or trained with in-domain information.

- Use of alternative methods for performing the final classification after the re-labeling process of the training dataset. For instance, models based on Recurrent Neural Networks (RNN) could be trained with the re-labeled dataset and then used for classifying the test messages. These networks could also benefit with an initialization based on the most similar messages.
- Re-labeling of the training dataset using RNNs: an RNN model trained from the re-labeled dataset could be used for performing a new re-labeling on the original dataset or even to refine an already re-labeled dataset, in order to compare their performance on the final classification.
- Deeper exploration of the parameters used for the construction of the ANN index.
- Analysis of the impact of different thresholds within the scoring function in the ranking-based evaluation (e.g. distance of retrieved neighbors).
- Application of the proposed system to similar tasks.
- Deeper analysis of identified positive and negative messages, which should exhibit easily identifiable features and characteristics that could help in the profiling of this type of pathology.
- Deeper analysis of the re-labeling process in order to analyse its convergence and the possibility of generate the final re-labeled dataset through simpler one-pass algorithms. Comparison with clustering algorithms such as DBSCAN.

Acknowledgments. This work has been partially supported by the Spanish Ministry of Science and Innovation within the DOTT-HEALTH Project (MCI/AEI/FEDER, UE) under Grant PID2019-106942RB-C32 and OBSER-MENH Project (MCIN/AEI/10.13039/501100011033 and NextGenerationEU/PRTR) under Grant TED2021-130398B-C21 as well as project RAICES (IMIENS 2022) and the research network AEI RED2018-102312-T (IA-Biomed).

References

1. Aumüller, M., Bernhardsson, E., Faithfull, A.J.: ANN-benchmarks: a benchmarking tool for approximate nearest neighbor algorithms. CoRR abs/1807.05614 (2018). http://arxiv.org/abs/1807.05614
2. Basile, A., et al.: UPV-symanto at eRisk 2021: mental health author profiling for early risk prediction on the internet. Working Notes of CLEF (2021)
3. Bernhardsson, E.: Annoy: approximate nearest neighbors in C++/Python (2018). https://pypi.org/project/annoy/. Python package version 1.13.0
4. Bucur, A.M., Cosma, A., Dinu, L.P.: Early risk detection of pathological gambling, self-harm and depression using BERT. Working Notes of CLEF (2021)
5. Campos, R., Mangaravite, V., Pasquali, A., Jorge, A., Nunes, C., Jatowt, A.: YAKE! Keyword extraction from single documents using multiple local features. Inf. Sci. **509**, 257–289 (2020)
6. Cer, D., et al.: Universal sentence encoder. CoRR abs/1803.11175 (2018). http://arxiv.org/abs/1803.11175
7. Devlin, J., Chang, M., Lee, K., Toutanova, K.: BERT: pre-training of deep bidirectional transformers for language understanding. In: Burstein, J., Doran, C., Solorio,

T. (eds.) Proceedings of the 2019 Conference of the North American Chapter of the Association for Computational Linguistics: Human Language Technologies, NAACL-HLT 2019, Minneapolis, MN, USA, 2–7 June 2019, Volume 1 (Long and Short Papers), pp. 4171–4186. Association for Computational Linguistics (2019). https://doi.org/10.18653/v1/n19-1423

8. Dieng, A.B., Ruiz, F.J.R., Blei, D.M.: Topic modeling in embedding spaces. Trans. Assoc. Comput. Linguist. **8**, 439–453 (2020). https://doi.org/10.1162/tacl_a_00325

9. Iyyer, M., Manjunatha, V., Boyd-Graber, J., Daumé III, H.: Deep unordered composition rivals syntactic methods for text classification. In: Proceedings of the 53rd Annual Meeting of the Association for Computational Linguistics and the 7th International Joint Conference on Natural Language Processing (Volume 1: Long Papers), pp. 1681–1691. Association for Computational Linguistics, Beijing (2015). https://doi.org/10.3115/v1/P15-1162, https://aclanthology.org/P15-1162

10. Li, W., et al.: Approximate nearest neighbor search on high dimensional data-experiments, analyses, and improvement. IEEE Trans. Knowl. Data Eng. **32**(8), 1475–1488 (2019)

11. Lopes, R.P.: Cedri at erisk 2021: A naive approach to early detection of psychological disorders in social media. In: CEUR Workshop Proceedings, pp. 981–991. CEUR Workshop Proceedings (2021)

12. Losada, D.E., Crestani, F., Parapar, J.: Overview of eRisk at CLEF 2020: early risk prediction on the internet (extended overview). In: Working Notes of CLEF 2020 - Conference and Labs of the Evaluation Forum, Thessaloniki, Greece, vol. 2696 (2020). http://ceur-ws.org/Vol-2696/paper_253.pdf

13. Loyola, J.M., Burdisso, S., Thompson, H., Cagnina, L., Errecalde, M.: UNSL at eRisk 2021: a comparison of three early alert policies for early risk detection. In: Working Notes of CLEF 2021-Conference and Labs of the Evaluation Forum, Bucarest, Romania (2021)

14. Malkov, Y.A., Yashunin, D.A.: Efficient and robust approximate nearest neighbor search using hierarchical navigable small world graphs. CoRR abs/1603.09320 (2016). http://arxiv.org/abs/1603.09320

15. Maupomé, D., Armstrong, M.D., Rancourt, F., Soulas, T., Meurs, M.J.: Early detection of signs of pathological gambling, self-harm and depression through topic extraction and neural networks. In: Proceedings of the Working Notes of CLEF (2021)

16. Parapar, J., Martín-Rodilla, P., Losada, D.E., Crestani, F.: Overview of erisk at CLEF 2021: early risk prediction on the internet (extended overview). In: Proceedings of the Working Notes of CLEF 2021 - Conference and Labs of the Evaluation Forum, Bucharest, Romania, vol. 2021, no. 2936, pp. 864–887 (2021). http://ceur-ws.org/Vol-2936/paper-72.pdf

17. Parapar, J., Martín Rodilla, P., Losada, D.E., Crestani, F.: Overview of eRisk 2022: early risk prediction on the internet. In: Crestani, F., et al. (eds.) CLEF 2019. LNCS, vol. 11696, pp. 347–357. Springer, Cham (2022). https://doi.org/10.1007/978-3-030-28577-7_27

18. Potenza, M.N., et al.: Gambling disorder. Nat. Rev. Dis. Primers **5**(1), 1–21 (2019)

19. Potenza, M.N., Kosten, T.R., Rounsaville, B.J.: Pathological gambling. Jama **286**(2), 141–144 (2001)

20. Rash, C.J., Weinstock, J., Van Patten, R.: A review of gambling disorder and substance use disorders. Sugaku Exposit. **7**, 3 (2016)

21. Vaswani, A., et al.: Attention is all you need. CoRR abs/1706.03762 (2017). http://arxiv.org/abs/1706.03762

Touché 2022 Best of Labs: Neural Image Retrieval for Argumentation

Tobias Schreieder[(✉)] and Jan Braker

Leipzig University, 04109 Leipzig, Germany
{fp83rusi,jb64vyso}@studserv.uni-leipzig.de

Abstract. Given a text query on a controversial topic, the task of Image Retrieval for Argumentation is to rank images according to how well they can be used to support a discussion on the topic. This paper provides a detailed investigation of the challenges of this task by means of a novel and modular retrieval pipeline. All findings relate to our work from last year's CLEF Touché'22 lab and a reproducibility study based on it. There, we demonstrate the unified retrieval pipeline NeurArgs and provide improved stance models. This work presents the approaches of the two papers, regarding the problems identified and the solutions provided. Herewith, we achieve an effectiveness improvement in argumentative image detection of up to 0.832 precision@10. However, despite this success, our study also revealed a previously unknown negative result: when it comes to stance detection, none of the tested stance models can convincingly beat a random baseline. Therefore, we conduct a thorough error analysis to understand the inherent challenges of image stance detection and provide insight into potential new approaches to this task.

Keywords: Argumentation · Image retrieval · Image stance detection

1 Introduction

Several years ago, social media discussions changed from being mainly text-focused towards including more images or videos. Specific platforms that focus on images, like Instagram, then became increasingly popular and are still today. In discussions on social media, people thus also often include images to illustrate their stance and arguments on the topic in question, or to support written arguments. Whether images can be "argumentative," i.e., whether they can represent arguments in their own right, is controversial [5]. However, their usefulness for argumentation is obvious: Kjeldsen [7] notes that images can underpin and support arguments, clarify facts, and convey facts more effectively than words.

Although retrieval systems for textual arguments exist [19], none exist yet specifically for image retrieval for argumentation. A search engine dedicated to the retrieval of images that are relevant to controversial topics can be useful for finding images to support one's stance on social media or elsewhere, and to get a "visual" overview of the landscape of opinions at-a-glance for personal deliberation. The first shared task to present pioneering approaches to this research

A. Arampatzis et al. (Eds.): CLEF 2023, LNCS 14163, pp. 186–197, 2023.
https://doi.org/10.1007/978-3-031-42448-9_16

question was the CLEF Touché lab "Image Retrieval for Arguments" in 2022. There, three different teams presented different approaches to solving the task, which were evaluated independently and uniformly with Tira. This software tries to solve the problem of scientific reproducibility, especially for shared tasks [12].

To pave the way for more effective image retrieval systems for arguments, in this paper we first briefly present our approaches from the CLEF Touché'22 lab, which are referred to subsequently as Aramis [2]. Based on this, weaknesses of the approaches are made visible and the improved retrieval pipeline NeurArgs, as well as other compatible stance models, which we introduced with Carnot et al. [4], are discussed. Inspired by the three-stage evaluation of image retrieval for arguments proposed by Kiesel et al. [6], we propose NeurArgs, a modular retrieval system with three AI models to unify approaches: a topic model to identify images relevant to a query, an argument model to identify images suitable for argumentation, and a stance model to sort images into pros and cons. By employing the system to combine the approaches submitted to the CLEF Touché'22 lab, we improve over the lab's best score by 0.064 in the lab's precision metric, reaching a score of 0.832. However, none of the 8 stance models we evaluated convincingly improves stance detection over a random baseline.

The paper is structured as follows: Sect. 2 reviews related work. Section 3 provides a brief overview of the Touché22-Image-Retrieval-for-Arguments dataset, and Sect. 4 shows the development of the Aramis models published in 2022 to NeurArgs and details the different models that we employ in our analyses. Section 5 presents the results of the reproduced and newly developed models (all code linked there), which successfully reproduces the state-of-the-art but also unveils our main negative result in the comparison with naive baselines. Section 6 then provides a qualitative analysis of the challenges for stance detection to aid researchers in overcoming our current result.

2   Related Work

Several former works exist on argument retrieval from text collections. The first systems were *args.me* [19], *ArgumenText* [18], and *IBM debater* [9]. For their evaluation, Potthast et al. [11] suggest employing the retrieved arguments' query relevance as well as rhetorical, logical, and dialectical quality in Cranfield style experiments. However, more detailed aspects of argument quality are discussed in the literature [19] and could be used to evaluate argument search engines.

Approaches for image retrieval have been explored for many years, mostly in content-based image retrieval. There, the query is itself an image and relevant results are similar to other images. Therefore, the content of the images needs to be analyzed. Smeulders et al. [15] provide an overview of the conducted research in the field in the early years. One of the important early projects regarding content-based image retrieval was presented by Rui et al. [13]. They used image feature vectors to establish a connection between images and terms. The works of Latif et al. [8] and Meharban and Priya [10] give a more recent overview of approaches and features for web image search. For example, Shao et al. [14]

propose to reduce the number of colors of images to a few representative ones in order to search more effectively for images containing a certain color-base. Color features seem to be highly promising when retrieving images for arguments due to colors evoking specific emotions [17]. A relatively new approach in image retrieval is to employ optical character recognition software like Tesseract [16] to extract the text from the images and then to extract standard features from the text for indexing. This approach seems especially promising for meme images and other images containing written arguments.

The retrieval of images for arguments has been sparsely explored so far. The pioneering work by Kiesel et al. [6] attempted this task by simply extending the search query with different terms to get different results for each stance. In their most effective approach, the query was either extended with the word "good" (for the pro stance) or the word "anti" (for the con stance). This method achieved good results overall, but was not able to improve upon a random classifier with regard to stance detection. The same authors then organized a shared task at the CLEF Touché'22 lab [1]. We employ the lab's data and the most effective participating approaches in our system comparison (Sects. 3 and 5).

3   The Touché'22 Dataset

For our research into image retrieval for argumentation, we employ the dataset of the corresponding Touché'22 shared task [1], which was located at the CLEF 2022 conference. The data is freely accessible online.[1] The dataset contains 23,841 images for 50 controversial topics. The topics include, for example, "is golf a sport?" or "should education be free?" The images were crawled using regular image search engine queries related to the 50 topics. In addition to the image itself, the dataset contains, for example, a screenshot of the web page it appeared on, the text from that web page, or the image's rank in the regular search engine's result list. For our analysis (Sect. 5) we employ the queries, the image pixel values and recognized text, and the corresponding web page's HTML source code. The dataset also contains three relevance ratings (on-topic, pro, con) for each of the 6607 images that the participants retrieved for the 2022 lab. The images shown in this paper, except the schematic of our modular system in Fig. 1, are taken from this dataset. The Aramis group evaluated 20 topics, resulting in 9559 evaluated images. A comparison between the two evaluations doesn't show much of a difference, so we decided to use the ratings from the Touché'22 lab.

4   Development of the NeurArgs Approach

Inspired by the three-stage evaluation of image retrieval for arguments by Kiesel et al. [6], we propose the retrieval system NeurArgs with three AI models. Each

---

[1] https://touche.webis.de/data.html#touche22-image-retrieval-for-arguments.

stage has its own model, which is illustrated in Fig. 1. An image is considered relevant if it fits the topic (topic-relevance), provides a statement on some topic (argumentativeness) and fits a previously specified stance (pro/support or con/attack) on the topic (stance-relevance). In this paper, we want to showcase the development from the Aramis approaches to the new NeurArgs framework.

Using a modular architecture allows us to investigate each stage separately. The previous work shows, that stance detection is the most challenging subtask for now. Therefore, we compare different stance models based on the unified topic and argument model of NeurArgs. The following sections first introduce both the models in general and the specific models used in our analysis (Sect. 5). Table 1 provides an overview of the features each model employs.

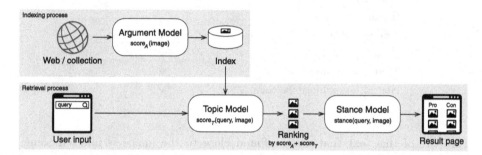

Fig. 1. Schematic of NeurArgs: Images from the web or a collection are, together with the web pages they appeared on, scored by the argument model for argumentativeness (score$_A$) and indexed. In the retrieval process, the user issues a query, which is used to score the images for topicality (score$_T$), rank images by the sum of the two scores, and classify their stance to sort them into two result lists (Pro vs. Con) for display.

4.1 The Topic Model

The topic model of NeurArgs ranks images by their relevance to the user's query by assigning a score to each image in the index (cf. Fig. 1). As the score depends on the query, the topic model must be part of the retrieval process. As a first naive approach, Aramis proposed a DirichletLM model based on the HTML page where the image is located and the preprocessed query as input [2]. The Touché'22 evaluation showed that this approach achieved poorer topic-relevance precision@10 compared to the baseline [1]. Therefore, we adapt the Elasticsearch BM25 retrieval from the best-performing system of Boromir [3].

The NeurArgs topic model combines our introduced pipeline and extends them in several places. Specifically, we employ textual matching of the query, text from the image's context (web page) and text from the image itself. The query and the recognized text on the image are preprocessed using standard stopword and punctuation removal, and lowercasing. The text from the HTML source code of the image's web page gets extracted and is also being preprocessed.

Table 1. Input features employed by the respective models detailed in Sect. 4: search query (topic), image pixels, recognized text (via OCR), and HTML source code of the web page on which the image was originally found.

Model	Query	Image features		
	Text	Image file		HTML
		Pixels	Text	
Topic model	✓		✓	✓
Argument model		✓	✓	
Stance models				
Oracle				
NeurArgs baseline		✓	✓	
Random baseline				
Aramis Formula	✓	✓	✓	✓
Aramis Neural	✓	✓	✓	✓
Neural text+image 3class	✓	✓	✓	
Neural text+image 2x2class	✓	✓	✓	

The part of this text that can be found close to the image is indexed using Elasticsearch's BM25. Additionally, the recognized text on the image is used for retrieval boosting. As this topic model already considerably improved over the best approach in Touché'22 (cf. Sect. 5), we did not investigate further models but focused our attention on different stance models instead.

4.2 The Argument Model

Our NeurArgs argument model ranks images by their suitability for argumentation by assigning a score to each image in the index (cf. Fig. 1). Conceptually, an image that shows either critical or supportive attitudes should receive a high argument score. Unlike for the topic model, this score does not depend on the query. Therefore, the model's score for each image is indexed alongside the image, and directly used in the retrieval function. The argument model employs the query-independent features that are previously employed by the Aramis approach for the Touché'22 lab [2]. Furthermore, we employ the same neural network classifier for NeurArgs as Aramis for calculating the argumentativeness score from the features. We detail those features below for completeness.

The first set of features are color properties, with the intent to capture the overall mood of the image. We calculate the average and dominant color of the image as RGB values, as well as the area share of red, green, blue, and yellow. Other features used for the neural network are the image type (graphic or photography) and diagram-likeness. We adopt the simple common color heuristic of Aramis for image type classification, and the approach to diagram-likeness based on horizontal kernels [2]. Additionally, general text features are used: text length, sentiment, the area percentage of the image occupied by text, and the position of the text in a 8×8 grid [2]. Here, the text position is used as a hint to identify memes and image quotes. The text is extracted using Tesseract OCR[2] after converting the image to gray scale and adjusting Tesseract's configuration for maximum text recognition. Afterward, only words that occur in a standard English dictionary are kept to improve detection precision.

4.3   The Stance Model

The stance model sorts the ranked images into pros and cons (cf. Fig. 1). To this end, stance models label each image for a topic as pro, con, both, or neither (cf. Kiesel et al. [6]). Only images labeled as pro or con are placed on the result page in the respective column in decreasing score order. Note that, according to the Touché task, an image can be both pro and con, in which case it is considered a relevant image if placed in either one or both result lists. As the score depends on the query, the stance model must be part of the retrieval process.

The Touché'22 results show that none of the participating models achieved a high precision for stance detection [1]. In our reproducibility study, we focus our investigations on the stance detection subtask and compare 14 approaches, including two baseline approaches and the oracle. In this paper, we compare the stance models of Aramis and selected further developments with the results of the best approach of Touché'22 (Boromir). Boromir used a sentiment detection BERT-model to classify the image based on the sentiment of the title of the image's original web page [3]. All other models are explained and evaluated in the corresponding paper of Carnot et al. [4]. In the following, we briefly discuss the modular stance models that classify the previously generated ranking results:

Oracle is a theoretic approach that uses the ground-truth stance labels and thus provides the upper limit. As the ground truth contains only stance labels for topic-relevant and argumentative images, the oracle's scores are the overall achievable maximum for our setting. However, as the dataset contains less than 10 images for some topic and stance combinations, this score is less than 100%.

NeurArgs baseline classifies each image as both pro and con, which results in an identical result list for each stance.

Random baseline classifies images as either pro or con with equal probability.

Aramis Formula uses the heuristic formula developed by team Aramis that is based on the same features used in the argument model [2]. Additionally, the query, the interrelation, and sentiments of the mentioned texts are used as features. The weights for each feature were set manually by Aramis.

Aramis Neural is a neural network, also developed by team Aramis [2], that uses the same features as Aramis Formula to classify images as either pro, neutral, or con. The neutral images are not further used in the results.

---

[2] https://github.com/tesseract-ocr/tesseract.

Neural text+image 3class employs a feedforward neural network classifier using the image resized to 256×256 pixels, the query text, and the recognized text of the images as input. The network combines a BERT model with a ResNet50V2 extended by some dropout layers to prevent overfitting. It has three output neurons that represent pro, neutral, and con.

Neural text+image 2x2class employs the same architecture as the neural text+image 3class approach, but with a single output neuron. The architecture is trained twice, once for pro and once for con images. Both are entirely independent of each other. The network calculates a score for the entry which shows if the image fits the stance. It needs to be above half of the highest score of the current query to be accepted in the respective category.

5   Evaluation of the NeurArgs Approach

Table 2 shows the results of our extended analysis. For consistency with the existing evaluation, we only use images where rating already exists and refrain from annotating images ourselves. Hence, the retrieved lists are condensed, a 5-fold cross-validation is used for evaluating the machine-learning-based approaches. The code for this study is available online.[3] Besides comparing more approaches, our evaluation also goes deeper than the original one of Bondarenko et al. [1] in that it shows results also for pro and con separately. The Touché'22 lab only used precision@10, arguing that this was closest to the setting of a user looking at a single page of result images. Additionally, we calculated NDCG@10 scores, which performed very similar to precision@10 and are therefore not separately shown. The exact values can be found in the work of Carnot et al. [4].

5.1   Topic and Argument Retrieval

We first detail the results for the retrieval of topic-relevant and argumentative images. This setup corresponds to omitting the stance model in Fig. 1 from the NeurArgs retrieval. The NeurArgs topic and argument models are used for all shown stance models. Since the assignment to the classes pro or con is based on the images with the highest score, the stance model can influence the topic relevance and argumentativeness scores. At this point, we also tested different weightings for the topic model's $score_T$ and the argument model's $score_A$ than the simple sum, but none lead to improvements for the different approaches.

As seen in Table 2, with a precision@10 of 92.6% for topic-relevance and 83.2% for argumentativeness, the NeurArgs baseline outperforms all methods from the CLEF Touché'22 lab. For reference, the most effective method from the lab, developed by Boromir, only achieved a topic-relevance precision score of 87.8% (-4.8%) and an argumentativeness score of 76.8% (-6.4%). Note that the baseline uses the same images for both stances and thus always retrieves only 10 images total, whereas other approaches might retrieve up to 20 images. However, most

---

[3] https://github.com/webis-de/SIGIR-23.

Table 2. The table shows the precision@10 scores on condensed lists for all 50 topics, sorted by stance-relevance (both) for all stance detection models. For this purpose, topic-relevance, argumentativeness and stance-relevance are always evaluated in relation to the overall system for the 20 images retrieved (10 pro and 10 con). The "both" scores are the averages for the 10 pro and 10 con images. In each case, the best results were highlighted in bold. All stance models follow the NeurArgs topic model and the argument model as described in Sect. 4, except for Best of Touché'22 and the Oracle.

Stance Model	Precision@10								
	Topic-relevance			Argumentativeness			Stance-relevance		
	Pro	Con	Both	Pro	Con	Both	Pro	Con	Both
Oracle	1.000	1.000	1.000	1.000	1.000	1.000	1.000	0.802	0.901
Neural text+image 2x2class	0.924	0.822	0.873	0.830	0.766	0.798	0.660	**0.310**	**0.485**
Aramis Formula	0.920	0.814	0.867	**0.838**	0.742	0.790	**0.690**	0.216	0.453
NeurArgs baseline	**0.926**	**0.926**	**0.926**	0.832	**0.832**	**0.832**	0.662	0.232	0.447
Neural text+image 3class	0.924	0.866	0.895	0.830	0.800	0.815	0.660	0.226	0.443
Random baseline	0.894	0.888	0.891	0.816	0.812	0.814	0.664	0.222	0.443
Aramis Neural	0.694	0.676	0.685	0.668	0.640	0.654	0.588	0.278	0.433
Best of Touché'22 (Boromir)	0.884	0.872	0.878	0.782	0.754	0.768	0.594	0.256	0.425

other models achieve almost the same performance as the NeurArgs baseline, with only slight losses in terms of topic relevance and argumentativeness. Moreover, Table 2 shows that the scores for topic-relevance and argumentativeness are very similar between images retrieved for pros and cons, with only a few exceptions like for Neural text+image 2x2class. Thus, the images retrieved for both pro and con are equally argumentative for most approaches.

5.2 Comparison of Stance Detection Models

Table 2 shows that stance detection is a challenge in image retrieval for argumentation. The best result that possibly could have been achieved for stance-relevance precision@10 lies at 90.1% as shown by the oracle. This is because not every topic has ten images on each side in the evaluation data, and missing images will be treated in the same way as an incorrect image. We find that the neural text+image 2x2class model that uses the image and associated text as input is the most effective, with a precision@10 of 48.5%. In this comparison, however, the NeurArgs baseline, which outputs an identical list of images for pro and con, comes in third with 44.7% (-3.8%). On the pro side, other models achieved the highest precision results: the Aramis Formula model exceed 69.0%. In general, results on the pro side range from 58.8% to 69.0%. Unfortunately, none of the models were able to classify the majority of con images correctly. The precision range for the con side lies between 21.6% and 31.0%. The reason for this drop in precision on the con side can be found in the dataset. For a number of topics, there are not enough con images annotated to retrieve ten images, which makes it impossible to achieve high precision scores. Therefore, the best theoretically possible result is 80.2% (Oracle).

However, Table 2 also shows the main negative result of our study: none of the approaches can convincingly beat our baselines. With a stance-relevance (both) precision@10 of 44.3%, the random baseline is about half a percentage point below the NeurArgs baseline. When we conducted significance tests (Student's t-test with Bonferroni correction at p=0.05) to detect if our approaches improve significantly upon the baseline in terms of precision@10, we found that only the oracle improves over it significantly. Worse, a number of models, such as Aramis Neural and Best of Touché'22 (Boromir), were not able to outperform the random model or the NeurArgs baseline. Especially when considering that one of the baselines is purely random, we thus have to conclude that, so far, stance detection in image retrieval for argumentation is an unsolved problem.

6   Insights into Image Stance Detection

Although our analysis in Sect. 5 confirms the seemingly good results of the approaches from the Touché lab, our analysis also revealed that no approach can convincingly beat naive baselines such as random or both-sides classification in detecting the image stance. This negative result suggests that the analyzed approaches fail to account for key challenges of the stance detection task. To uncover these challenges, we performed a qualitative analysis of the images the approaches retrieved and misclassified. Specifically, we identified nine challenges:

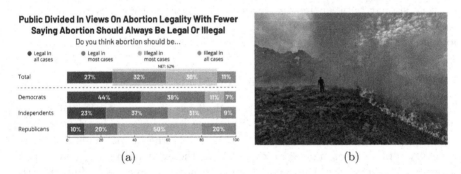

(a)   (b)

Fig. 2. (a) Different valuations cause stance ambiguity. The image could be pro "should abortion be legal?" if one supports the Democrats, but con if not. (b) Image understanding depends on background knowledge. The image could be pro "is human activity primarily responsible for global climate change?" depending on the viewer's expertise.

Different Valuations Cause Stance Ambiguity. Images or diagrams may contain several pieces of information that lead to different or opposite conclusions for different audiences. Specifically, a person's background, socialization, and opinions influence how they interpret the image stance. Figure 2a illustrates this by the political party affiliation. Someone who supports the Democrats sees in the chart that their favorite party is in favor of legal abortion (pro stance).

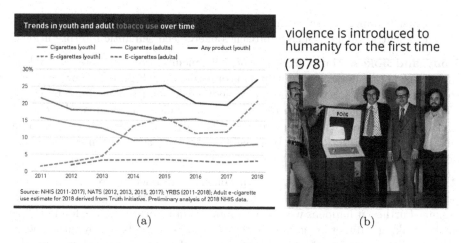

(a)   (b)

Fig. 3. (a) Neutral images. The image is neither clearly pro nor con "is vaping with e-cigarettes safe?". (b) Irony and Jokes. The image is con "do violent video games contribute to youth violence?" if one gets the joke about "pong" being a violent game.

Republicans, instead, might see the image as con. This problem is challenging for both algorithms and annotation campaigns. To solve it for algorithms, one could identify images with this problem and either not show them in the results or classify them based on a user-provided audience profile. For annotation campaigns, one could provide special training for annotators for such cases.

Image Understanding Depends on Background Knowledge. Some images require the viewer to have certain background knowledge to understand their stance. The image in Fig. 2b is pro "is human activity primarily responsible for global climate change?" for viewers who connect the burning of forests and the climate impact. Without that, the image is not even topic-relevant. This problem provides a challenge for algorithms and annotation campaigns. Analyzing the context of the image web page could provide hints on the relevant knowledge.

Unbalanced Image Stance Distribution. For some topics, there are much more pro images available than con images, or vice versa, which can result in biased stance detectors if one does not pay attention to such skewed data in the training process. For example, the dataset contains only very few con images for the topic "should bottled water be banned?". One solution is to balance the training dataset and remove topics with overly skewed distributions.

Neutral Images. Some images, like diagrams, contain thought-provoking impulses on a topic, but are not evidently pro or con. However, they can be visually very similar to arguments with a unique stance, which can be a problem. For example, the image in Fig. 3a is very informative without clearly being pro or con "is vaping with e-cigarettes safe?". Nevertheless, one can imagine visually very similar images that are clearly pro or con, which provides a challenge in classifier training. To solve this, it might be necessary to develop a classifier to detect neutral images. Such approaches likely need semantic interpretations.

Irony and Jokes. Many images, especially memes, contain irony and jokes, which may not be understood by humans or algorithms. Figure 3b shows a meme that was retrieved for the topic "do violent video games contribute to youth violence?". The image is a joke on the idea that video games created violence, as if violence had not existed before. We expect irony detection for images to be very challenging. Still, it might be possible to transfer advances in textual irony detection (e.g., [20]) to visual irony detection.

Additional Problems. Besides the problems mentioned, there are additional problems such as regional images that are only relevant for people in certain regions. Further, it happens with several topics that both stances are found in one image, making a direct assignment difficult. This is exacerbated by images with more than two stances, which makes the choice of a binary classifier unsuitable. Another problem is understanding the semantics of diagrams by algorithms.

7   Conclusion

For the task of image retrieval for argumentation, we compared 8 approaches (including the previous state-of-the-art, two baselines, and the oracle) while emphasizing stance detection. To compare different approaches, we proposed the modular image retrieval system NeurArgs: a topic model to identify images relevant to a query, an argument model to identify images suitable for argumentation, and a stance model to sort images into pros and cons. The approaches shown in our paper employ features of the query, the image file or the web page an image was indexed on. The NeurArgs approach for the topic and argument model, which we have derived from the experience of the Touché'22 lab, provide a new state-of-the-art for the respective parts of the task, reaching 0.926 precision@10 for topic-relevance and 0.832 precision@10 for argumentativeness.

However, the extended analysis also uncovers a strong negative result: none of the analyzed approaches can convincingly beat a random baseline (or a both-sides baseline) when it comes to stance detection. We thus conclude that stance detection in image retrieval for argumentation is so far an unsolved problem. To pave the way for future approaches, we identified nine different challenges for stance detection and provided some examples and possible solutions.

References

1. Bondarenko, A., et al.: Overview of touché 2022: argument retrieval. In: Barrón-Cedeño, A., et al. (eds.) CLEF 2022. LNCS, vol. 13390, pp. 311–336. Springer, Heidelberg (2022). https://doi.org/10.1007/978-3-031-13643-6_21
2. Braker, J., Heinemann, L., Schreieder, T.: Aramis at touché 2022: argument detection in pictures using machine learning. Working Notes Papers of the CLEF (2022)

3. Brummerloh, T., Carnot, M.L., Lange, S., Pfänder, G.: Boromir at touché 2022: combining natural language processing and machine learning techniques for image retrieval for arguments. Working Notes Papers of the CLEF (2022)
4. Carnot, M.L., et al.: On stance detection in image retrieval for argumentation. In: Proceedings of the SIGIR. ACM (2023)
5. Champagne, M., Pietarinen, A.V.: Why images cannot be arguments, but moving ones might. Argumentation **34**(2), 207–236 (2019)
6. Kiesel, J., Reichenbach, N., Stein, B., Potthast, M.: Image retrieval for arguments using stance-aware query expansion. In: Al-Khatib, K., Hou, Y., Stede, M. (eds.) ArgMining 2021 at EMNLP. ACL (2021)
7. Kjeldsen, J.E.: The rhetoric of thick representation: how pictures render the importance and strength of an argument salient. Argumentation **29**(2), 197–215 (2014). https://doi.org/10.1007/s10503-014-9342-2
8. Latif, A., et al.: Content-based image retrieval and feature extraction: a comprehensive review. Math. Probl. Eng. **2019** (2019)
9. Levy, R., Bogin, B., Gretz, S., Aharonov, R., Slonim, N.: Towards an argumentative content search engine using weak supervision. In: Bender, E.M., Derczynski, L., Isabelle, P. (eds.) Proceedings of the COLING. ACL (2018)
10. Meharban, M., Priya, D.: A review on image retrieval techniques. Bonfring Int. J. Adv. Image Process. **6** (2016)
11. Potthast, M., et al.: Argument search: assessing argument relevance. In: Piwowarski, B., Chevalier, M., Gaussier, É., Maarek, Y., Nie, J., Scholer, F. (eds.) Proceedings of the SIGIR. ACM (2019)
12. Potthast, M., Gollub, T., Wiegmann, M., Stein, B.: TIRA integrated research architecture. In: Ferro, N., Peters, C. (eds.) Information Retrieval Evaluation in a Changing World. TIRS, vol. 41, pp. 123–160. Springer, Cham (2019). https://doi.org/10.1007/978-3-030-22948-1_5
13. Rui, Y., Huang, T.S., Mehrotra, S.: Content-based image retrieval with relevance feedback in mars. In: Proceedings of the ICIP, vol. 2. IEEE (1997)
14. Shao, H., Wu, Y., Cui, W., Zhang, J.: Image retrieval based on MPEG-7 dominant color descriptor. In: Proceedings of the ICYCS. IEEE (2008)
15. Smeulders, A.W., Worring, M., Santini, S., Gupta, A., Jain, R.: Content-based image retrieval at the end of the early years. IEEE Trans. Pattern Anal. Mach. Intell. **22**(12), 1349–1380 (2000)
16. Smith, R.: An overview of the tesseract OCR engine. In: Proceedings of the ICDAR. IEEE (2007)
17. Solli, M., Lenz, R.: Color emotions for multi-colored images. Color Res. Appl. **36** (2011)
18. Stab, C., et al.: ArgumenText: searching for arguments in heterogeneous sources. In: Liu, Y., Paek, T., Patwardhan, M.S. (eds.) Proceedings of the NAACL-HLT. ACL (2018)
19. Wachsmuth, H., et al.: Building an argument search engine for the web. In: Habernal, I., et al. (eds.) Proceedings of the ArgMining@EMNLP. ACL (2017)
20. Zhang, S., Zhang, X., Chan, J., Rosso, P.: Irony detection via sentiment-based transfer learning. Inf. Process. Manag. **56**(5) (2019)

SimpleText Best of Labs in CLEF-2022: Simplify Text Generation with Prompt Engineering

Shih-Hung Wu[✉][iD] and Hong-Yi Huang

Chaoyang University of Technology, Taichung, Taiwan
shwu@cyut.edu.tw, s11027604@gm.cyut.edu.tw

Abstract. This paper reports our approach to the SimpleText@CLEF-2022. For the task 1: what is in (or out)?, we designed a two-stage filtering scheme that utilizes the traditional keyword finding approach TF-IDF score to find the important documents in the first stage and the important sentences in the second stage. The result is comparable to manual run and ranked first in task 1. For the Task 3: Rewrite this!, our system adopts the T5 generation model to rewrite the original sentences. We fine-tuned the model to generate simplified sentence. The result ranked second in task 3. The simplified sentence generated by T5 model cannot fully express the meaning of the original sentence, in a following further experiments, we adopted the GPT3.5 and GPT4 models to generate simple text, and they give better results according to in our evaluation metrics based on readability and vocabulary simplicity.

Keywords: Simple Text Generation · TF-IDF · T5 model · GPT3.5 model · GPT4 model

1 Introduction

Interpreting scientific texts requires solid background knowledge and uses tricky terminology so that the scientific texts are hard to understand. How to simplify complex text in an automatic way is the key point of research. In CLEF-2022 SimpleText Lab [1] provides tasks to promote the research of text simplification. The goal of research is to make scientific texts more comprehensible to the general public in an automatic manner. SimpleText provides challenges of automatic text simplification in the following tasks:

- TASK 1: What is in (or out)? The goal of task 1 is given a query, a system has to find passages to include in a simplified summary.
- TASK 2: What is unclear? Given a passage and a query, a system has to rank terms that are required to be explained for understanding this passage.
- TASK 3: Rewrite this! Given a passage from scientific abstracts, a system has to rewrite it into a simplify passage.

SimpleText aims find the textual expression carrying information that should be simplified, the background information should be provided and the most relevant

A. Arampatzis et al. (Eds.): CLEF 2023, LNCS 14163, pp. 198–208, 2023.
https://doi.org/10.1007/978-3-031-42448-9_17

or helpful. Also system should try to improve the readability of a given short text.

In year 2022, we focus on Task1 and Task3 with the techniques from other related works. Here we report our approach, dataset, system, and results to the tasks in Sect. 2 to 5. In Sect. 6, we make more discussion on further experiments of task3 and show promising direction of simple text generation.

2   Techniques in Our Approach

Our system uses the TF-IDF score to find the important sentences in a two-stage filtering scheme for task 1, and adopts the T5 generation model to rewrite the original sentences for task 3, the detail is given in Sect. 4. Here, we will give a brief introduction to TF-IDF and T5 model.

2.1   Term Frequency Inverse Document Frequency (TF-IDF)

Term Frequency Inverse Document Frequency (TF-IDF) is a statistical measure that evaluates how relevant a word is to a document in a collection of documents. TF-IDF is calculated by multiplying two different metrics. The term frequency (TF) means the number of times the word appears in a document. The inverse document frequency (IDF) means, how common or rare a word is in the entire document set. The TF-IDF score for the word t in the document d from the document set D is calculated as follows:

$$tfidf(t, d, D) = tf(t, d) \cdot idf(t, D) \tag{1}$$

$$tf(t, d) = \log\left(+freq(t, d)\right) \tag{2}$$

$$idf(t, D) = \log\left(\frac{N}{count(d \in D : t \in d)}\right) \tag{3}$$

2.2   Exploring the Limits of Transfer Learning with a Unified Text-to-Text Transformer(T5)

Transformer-based models have achieved state-of-the-art performance for abstractive summarization [2–4]. T5, or Text-to-Text Transfer Transformer [2], is a Transformer based architecture that uses a text-to-text approach. T5 can convert all NLP tasks into Text-to-Text. The framework is shown in Fig. 1. Our Task3 system is built on T5 model.

3   Data Set

SimpleText's data use the Citation Network Dataset: DBLP+Citation, ACM Citation network (12th version) as source of scientific documents to be simplified. The data is two-fold: Medicine and Computer Science. Scientific textual content and authorship on any topic related to computer science can be extracted from this corpus. Detail description please read the overview paper [1].

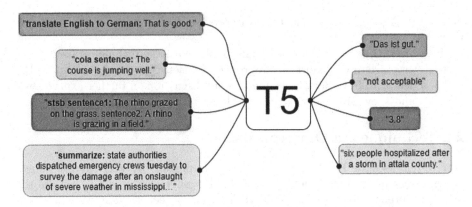

Fig. 1. Text-to-text framework of T5 model.

4   System

Since we focus only Task1 and Task3, here we give the detail of our system in task 1 in Subsect. 4.1 and task 3 in Subsect. 4.2.

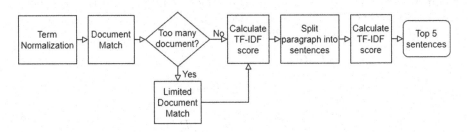

Fig. 2. Flowchart for Task 1.

4.1   Search Passage Using a Two-Stage TF-IDF Filter

In Task1, the system uses TF-IDF score to filter the article and find the top 5 sentences matched by the query term. The flowchart is shown in Fig. 2. The query term is normalized, and then the abstract matches it is extracted. If there are too many matched files, our system will ranking them by TF-IDF score to find the Top 5 files. Since only single sentence in the article is required, the TF-IDF score is calculated again after separating each sentence in the article, and the sentence with the highest TF-IDF score in each file is found, and the Top 5 sentence is obtained. Note that we limit article matching because we want to reduce the number of files, and when the conditions are true, the matching criteria are changed to abstracts and titles instead of just abstracts. In addition, when there is no matching document, we will split the query terms into single words to match the file, and when calculating the TFIDF scores, our system will calculate them separately and then take the sum.

Table 1. Training parameters.

Parameter	value
Model	t5-base
TRAIN_BATCH_SIZE	4
VALID_BATCH_SIZE	1
TRAIN_EPOCHS	3
LEARNING_RATE	1e−4

4.2 T5 Model for Summarization

In Task 3, our system adopt the T5 model to generate simplified sentences. We use the 648 data in the training set to fine-tune the T5 model with a ratio of 8:2 between the training set and the validation set, and the hyper-parameters are shown in Table 1. In addition, when generating sentences, we set the generated token to 0.78 times the source sentence token. This ratio is based on the average sample sentence token and source sentence token ratio of the data set, as shown in Table 2.

Figure 3 shows the training flowchart of our Task3 system. The first step, we prepend the input sequence with 'summarize:' (task_prefix) before encoding it. This will help in improving the performance, as this task prefix was used during T5's pre-training. Then uses the T5Tokenizer encoding sequence and train the model with parameters in Table 3. Finally generate the summary.

Table 2. The generated examples.

source sentence	generated sentence
We describe a PDA (Personal Digital Assistant) based CSCW system called NewsMate, which provides mobile and distributed news journalists with timely information	NewsMate provides mobile and distributed news journalists with timely information
A CDA is a mobile user device, similar to a Personal Digital Assistant (PDA)	A CDA is a mobile user device, similar to a Personal Digital Assistant

Fig. 3. T5 model Training flowchart.

5   Results and Discussion

We participated in the SimpleText challenge under the name "CYUT Team2". Our reported results in this section are obtained from the SimpleText official report [1].

For the Task1 evaluation, our team CYUT comes out on top of the ranking list by achieving a score of nDCG@5= 0.3322. Table 3 presents in more details the achievements of each run in more details. These values show that the automatic run made by CYUT and the manual run significantly outperform other automatic runs in terms of selecting the abstracts with a high relevance. Besides, it is important to note that the pooling method only kept articles chosen by at least two participants and gave a relevance score on a scale of 0 to 5. This method of evaluation will be detrimental to teams that find unique documents.

For the Task3 evaluation, we are ranked second with an score of 0.122 in Table 4. Scores are evaluated by the average harmonic mean of normalized opposite values of Lexical Complexity, Syntactic Complexity and Distortion Level. In Table 5 shown information distortion in evaluated runs. It should be noted that most of the results generated by our method are truncated. The reason is we controlled the length ratio of our generation model to 0.8. This is not a necessary setting. We cancelled the setting in the further experiment and eliminated the truncated sentence problem.

Table 3. SimpleText Task 1: Ranking of official submissions on combined score [1].

Team	Score	#Docs	Doc Avg	#Queries	Query Avg	nDCG@5
CYUT	125	44	0.53	77	1.62	0.3322
UAMS-MF*	163	54	0.87	99	1.65	0.2761
UAMS	52	17	0.22	40	1.30	0.1048
NLP@IISERB	26	7	0.35	13	2.00	0.0290

Manual run.

As we can observe from Table 6. The main weakness of our approach lies in the indicator of [Omission of Essential Details] in the table. This is a distortion hard to deal, since the objective of the simplification is to facilitate perception of the main idea of the source. To distinguish which details and concepts are essential is a crucial task in the text simplification. On the other hand, our approach shows less information distortion on other criteria.

Table 4. SimpleText Task 1: Ranking of official submissions on combined score [1].

Run	Score
PortLinguE full	0.149
CYUT Team2	**0.122**
CLARA-HD	0.119

Table 5. SimpleText Task 3: General results of official runs [1].

Run	Total	Unchanged	Truncated	Valid	Longer	Length Ratio	Evaluated	Uncorrect Syntax	Unresolved Anaphora	Minors	Syntax Complexity	Lexical Complexity	Information Loss
CLARA-HD	116,763	128	2,292	111,627	201	0.61	851	28	3	68	2.10	2.42	3.84
CYUT Team2	116,763	549	101,104	111,818	49	0.81	126	1		32	2.25	2.30	2.26
PortLinguE_full	116,763	42,189	852	111,589	3,217	0.92	564			5	2.94	3.06	1.50

Table 6. SimpleText Task 3: Information distortion in evaluated runs [1].

Run	Evaluated	Non-Sense	Contra sense	Topic Shift	Wrong Synonym	Ambiguity	Omission Of Essential Details	Overgeneralization	Oversimplification	Unsupported Information	Unnecessary Details	Redundancy	Style
CLARA-HD	851	162	68	37	20	80	314	59	203	26	10	29	13
CYUT Team2	126	2	1			4	42	4	5				4
PortLinguE_full	564	9	3	4	3	19	94	9	13	2	2	5	1

6   Further Evaluation on the Large Language Model

Our approach to Task 3 is based on the T5 pre-trained large language model (LLM). To our knowledge, the best pre-trained model is GPT4 [5]. To find the potential of LLM on the text simplification, we conducted a minor experiment on GPT3.5 [6] and GPT4 [5] with a subset of SimpleText 2022 task dataset. We sampled 100 sentences from the SimpleText 2022 dataset and used three models to generate simplified text. Since the prompt will affect the result of text generation of LLM, we find that prompting engineering is an important issue [7]. The prompts used in our experiment are listed in Table 7.

It is important to note that {text} represents the input text, {simple text} represents the simplified text, {text1}, {text2}, {simple text1}, {simple text2} all refer to the input texts that will not change. Please refer to the footnote below the table for more details. Additionally, the GPT3.5 prompt and GPT4 prompt are the same.

In order to find how well the models simplified the sentences, we evaluate in two aspects: The readability and vocabulary. Since we cannot reproduce the official evaluation, we used other available metrics to evaluate the generated text.

Table 7. Prompts used in our evaluation.

Model	Prompts
T5	"summarize: {text}, {simple text}"
GPT4 prompt1	"Original: {text} Simplified:"
GPT4 prompt2	"Simplify the following sentences to make them easier to understand Here are some examples: Example 1: Original: {text1} Simplified: {simple text1} Example 2: Original: {text2} Simplified: {simple text2} Please simplify this sentence: {text}"
GPT4 prompt3	"Your task is to simplify the following sentences to make them easier to understand. Please note that your response should be flexible enough to allow for various relevant and creative simplifications, as long as they accurately convey the intended meaning Please simplify this sentence: {text}"
GPT3.5 prompt3	#gpt4 prompt3

*text1: Our model is made of residual convolutional blocks with hierarchical dilated skip connections joined in steps.
*simple text1: This model is based on enhanced convolutional neural networks (a deep-learning approach often used for image recognition).
*text2: However, instead of collecting large datasets once again, we collect a number of smaller datasets containing a few hundred frames each and use transfer learning techniques on the CNN trained on UR robots to adapt it to a new robot having different shapes and visual features.
*simple text2: We collect a number of smaller datasets and use transfer learning techniques on the CNN trained on UR robots to adapt it to a robot that looks different

6.1   Flesch-Kincaid Readability Tests

Flesch reading ease is an index that measure the level of sentences, the higher the easier for the reader [8]. In the following table we can see that the readability increased from 26.67 to 35.11 for T5 model, and around 50 for the GPT models. The Flesch-Kincaid grade level [9] is an index that measure the corresponding reader level, the lower the younger. This is a grade index in that a score of 10.x means that a tenth grader would be able to read the text. GPT model gives lower level than the source and the T5 model. The evaluation results in Table 8 show that the model can increase the readability of the sentences.

Table 8. Flesch-Kincaid readability for different models.

Model & prompt	Flesch reading ease	Flesch-Kincaid grade level
source	26.67	14.96
T5	35.11	12.8
GPT4 prompt 3	**50.57**	10.91
GPT3.5 prompt 3	49.97	**10.88**

We test the GPT4 with three different prompts. The results in the following Table 9 shows that the prompts really affect the result of readability. A detailed prompt can give better result on text simplify task. Here the prompt 1 is the shortest prompt that only as the model to simplify the text. The prompt 2 give examples on what is a simplify text. The prompt 3 give more detail on how to simplify a sentence.

Table 9. Flesch-Kincaid readability of GPT4 in three different prompts.

Model & prompt	Flesch reading ease	Flesch-Kincaid grade level
GPT4 prompt 1	35.31	12.84
GPT4 prompt 2	47.23	11.25
GPT4 prompt 3	**50.57**	**10.91**

To analyze sentence simplification, we have selected a sample sentence, which is presented in the Table 10. From the information presented in the table, it is clear that the original sentence uses the term "serodiagnosis" as a noun in the medical domain to describe a diagnosis made through the examination of blood. However, the T5 model produces a shortened sentence, which is a known limitation in our SimpleText 2022 experiments. On the contrary, the sentences generated by GPT models offer improved simplification outcomes by preserving the crucial keyword "biosensors" and substituting the complex term "serodiagnosis" with "detect", "diagnosis", or "testing".

6.2   The Vocabulary Profile

We utilize Capel's vocabulary, as presented in his works [10,11], to determine the vocabulary profile within the datasets. This vocabulary is aligned with the principles and descriptions of the Common European Framework of Reference for Languages (CEFR), offering more precise vocabulary grading and detailed descriptions. CEFR categorizes language proficiency into six levels: A1 (Beginner), A2 (Elementary), B1 (Intermediate), B2 (Upper Intermediate), C1 (Advanced), and C2 (Proficient). Each level signifies increasing proficiency and competence in language comprehension, speaking, reading, and writing. A1 represents the foundational level, while C2 denotes the highest level of proficiency.

Table 10. Example in our evaluation.

Model & prompt	Sentence	FRE	FKGL
source	The excellent performances of these biosensors provide a prospective space for future antibody-detection-based disease **serodiagnosis**	−2.98	17.4
T5	excellent performances of these biosensors provide a prospective space for future antibody-detection-based disease ser	14.97	14.7
GPT4 prompt 1	These great biosensors offer potential for future disease diagnosis using antibody detection methods	7.52	15.4
GPT4 prompt 2	The great results from these biosensors offer potential for future disease testing using antibody detection	30.87	12.7
GPT4 prompt 3	These great biosensors offer a promising area for future disease testing using antibody detection	40.35	11.1
GPT3.5 prompt 3	The biosensors work really well and could be used in the future to detect diseases with antibodies	**62.68**	**8.7**

*FRE: Flesch reading ease
*FKGL: Flesch-Kincaid grade level

The distribution of usage of vocabulary usage is very different among different models. As we can see from the following Fig. 4. The source sentences used more unlisted vocabulary, which means difficult words for student from A1-C2. The percentage of vocabulary in the A1-A2 level has increased to 62% for the GPT models, and the usage of unlisted words drop to 10%. We can conclude that the GPT model indeed uses simpler words and reduces the use of unlisted vocabulary.

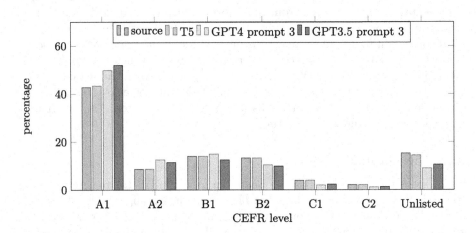

Fig. 4. The vocabulary profile for different models.

6.3   Academic Word List

Another way to measure the usage of vocabulary is to count the percentage of academic words in the text. In the following Table 11, we can find that the percentage of academic words according to a list provide by Coxhead [12]. The percentage of academic words drops from the 15.64% of the source dataset to around 11% for the GPT models. The percentage is coherent to the unlisted vocabulary in the previous table. Therefore, we can find that decrease the percentage of academic words is taking placed for the GPT models.

Table 11. The Academic Word Profile of different models.

Model & prompt	Percentage of academic words
source	15.64%
T5	15.50%
GPT4 prompt 3	11.67%
GPT3.5 prompt 3	**11.36%**

Table 12. The Academic Word Profile to the three prompts for GPT4.

Model & prompt	Percentage of academic words
GPT4 prompt 1	15.75%
GPT4 prompt 2	13.02%
GPT4 prompt 3	**11.67%**

The different prompts also matter on the usage of academic words, as shown in the Table 12. Prompt 1, which is the shortest, does not result in any improvement. However, prompt 3, which provides the most detailed information, leads to the greatest improvement.

7   Conclusion and Future Works

This paper reports our approach to the SimpleText lab. In terms of information retrieval for Task 1, we achieve top of results using the TF-IDF filter. However, the polysemy problem of TF-IDF will cause difficult to find extended topic document. From our perspective, it would be more beneficial for Task 1 to have a better information retrieval model. In terms of generating sentences for Task 3, the generation result of T5 is not satisfactory. The simplified sentence cannot fully express the meaning of the original sentence. It is not suitable using the T5 model, or the number of training data is insufficient. In the future, we consider using other models and increase the size of the dataset to improve the performance. The results in our further evaluation on the GPT3.5 and GPT4 models show improvement on using LLM.

References

1. Ermakova, L., et al.: Automatic simplification of scientific texts: SimpleText lab at CLEF-2022. In: Hagen, M., et al. (eds.) ECIR 2022. LNCS, vol. 13186, pp. 364–373. Springer, Cham (2022). https://doi.org/10.1007/978-3-030-99739-7_46
2. Raffel, C., et al.: Exploring the limits of transfer learning with a unified text-to-text transformer (2020)
3. Lewis, M., et al.: BART: denoising sequence-to-sequence pre-training for natural language generation, translation, and comprehension. In: Proceedings of the 58th Annual Meeting of the Association for Computational Linguistics, pp. 7871–7880. Association for Computational Linguistics (2020)
4. Devlin, J., Chang, M.-W., Lee, K., Toutanova, K.: BERT: pre-training of deep bidirectional transformers for language understanding. In: Proceedings of the 2019 Conference of the North American Chapter of the Association for Computational Linguistics: Human Language Technologies, Minneapolis, Minnesota (Volume 1: Long and Short Papers), pp. 4171–4186. Association for Computational Linguistics (2019)
5. OpenAI. GPT-4 technical report (2023)
6. Brown, T.B., et al.: Language models are few-shot learners (2020)
7. Huang, J., et al.: Large language models can self-improve. arXiv preprint arXiv:2210.11610 (2022)
8. Flesch, R.: How to Write Plain English: A Book for Lawyers and Consumers. Harper & Row (1979)
9. Kincaid, J.P., Fishburne, R.P., Jr., Rogers, R.L., Chissom, B.S.: Derivation of new readability formulas (automated readability index, fog count and flesch reading ease formula) for navy enlisted personnel. Inst. Simul. Training **56** (1975)
10. Capel, A.: A1–B2 vocabulary: insights and issues arising from the English profile wordlists project. Engl. Profile J. **1**, e3 (2010)
11. Capel, A.: Completing the English vocabulary profile: C1 and C2 vocabulary. Engl. Profile J. **3**, e1 (2012)
12. Coxhead, A.: A new academic word list. TESOL Q. **34**(2), 213–238 (2000)

Answer Retrieval for Math Questions Using Structural and Dense Retrieval

Wei Zhong$^{(\boxtimes)}$ ⓘ, Yuqing Xie$^{(\boxtimes)}$ ⓘ, and Jimmy Lin$^{(\boxtimes)}$ ⓘ

David R. Cheriton School of Computer Science, University of Waterloo, Waterloo, Canada
{w32zhong,yuqing.xie,jimmylin}@uwaterloo.ca

Abstract. Answer retrieval for math questions is a challenging task due to the complex and structured nature of mathematical expressions. In this paper, we combine a structure retriever and a domain-adapted ColBERT retriever to improve the effectiveness of math answer and formula retrieval. We find these two approaches generate highly effective outcomes because structure search can use unsupervised structure similarity as a strong prior signal to math document relevance, and the ColBERT retriever is able to capture contextual similarity and semantic matching effectively by finding additional relevant math contents even if they are using different formulas.

Keywords: Mathematics information retrieval · structure search · dense retrieval · math-aware search

1 Introduction

Recently, the Mathematics Information Retrieval (MIR) field has gained attention due to the increasing number of scientific publications and the need to retrieve math-related content by formulas. The primary task in MIR is to retrieve relevant information from documents that contain math formulas. However, the heterogeneous nature of math content, which includes rich-structured formulas and their textual context, requires special treatment to create an effective search engine. This is because math languages have special semantic properties such as expression commutativity and symbol substitution equivalence that require consideration different from traditional retrieval models. Moreover, retrieving math content poses a general challenge as classic retrieval models do not demonstrate enough power to capture the similarities hidden behind the contextual connections and abstract meaning in math language.

With the surge of deep neural retrievers, we witness the capacity of deep models being able to boost in-domain effectiveness to a new level. Although the deep model may still fall behind the *scaling law* [18] when it comes to understanding and reasoning of math language [22,43]. Using neural retrievers powered by deep learning can be a good alternative middle ground between a complete comprehension of mathematics, and ignoring the presence of math language in

A. Arampatzis et al. (Eds.): CLEF 2023, LNCS 14163, pp. 209–223, 2023.
https://doi.org/10.1007/978-3-031-42448-9_18

documents at all – because information retrieval techniques can retrieve existing user-generated math knowledge directly by *similarity search* without the need to understand them in depth. At the same time, the good modeling power offered by deep language models (e.g., BERT [7]) can presumably discover more semantic matches than previous bag-of-word models based on exact lexical matches which are commonly seen in traditional IR.

In this work, we combine a structure-aware search engine with a bi-encoder dense retriever (See [21] for this classification) to capture math formula similarity from the signal of structure matching and capture other semantic similarities through supervised semantic matching. According to our recent findings [62], we adopt the ColBERT learned dense retriever [19] due to its high effectiveness demonstrated in our evaluation of previous MIR tasks. Compared to the previous work [61], we further introduce a domain-adapted retriever backbone named Coco-MAE [59]. We demonstrate an improved effectiveness by incorporating structure search and a ColBERT model built on top of this backbone.

2   The ARQMath Lab

The ARQMath Tasks [24,25,30] have been one of the few tasks trying to address the problem of retrieving math questions containing structured formulas. The ARQMath Lab includes multiple tasks, with Task 1 being a Community Question and Answer (CQA) task that requires the retrieval of relevant answer posts from a limited set of Math StackExchange (MSE)[1] corpus spanning from 2010 to 2018. The queries consist of real-world questions sampled from later-year MSE threads. Meanwhile, Task 2 focuses on retrieving relevant formulas, including their context, in documents given a specified query formula from Task 1. To ensure formula diversity, this task requires the return of no more than five visually distinct formulas. Failing to do so will result in an unjudged result. In ARQMath-3 [25], an Open Domain QA task (Task 3) was introduced, which challenges participants to return a single answer for each Task 1 topic. The answer is not limited to extractive approaches but can also include automatically generated answers from models potentially trained using data outside the ARQMath collection.

3   Related Work

Early work in the field of MIR simply applies specialized tokenizers to handle math formulas [31]. Later, different intermediate tree representations are utilized to extract unsupervised features for capturing structure similarities.

In particular, the OPT (Operator Tree) representation is first utilized by extracting features from leaf-root paths [13,14]. Following work in this direction [11,52,57] expand the leaf-root path set by extracting prefixes and suffixes and use them for additionally retrieving sub-expressions in formulas. However,

---

[1] https://math.stackexchange.com/.

without whole structure comparison, a naive leaf-root path matching offers high recall but lacks good precision in retrieval. More recently, Approach0 [60,63] approximately matches structure holistically without overlapping features by grouping the root-end nodes of leaf-root paths on the fly. Using a structure-based dynamic pruning [60], this first-stage structure search in OPT representations can efficiently identify a maximum common structure in the first stage of retrieval. However, matching the OPT structure more strictly in the re-ranking stage is more common [6,26,27] and practical as fewer candidates need to be considered.

On the other hand, the Symbol Layout Tree (SLT) [48,54] represents a lower-level structure semantics for math formulas. Similar to the LaTeX representation, it only captures the layout or the topology of a formula. This creates an advantage of little ambiguity in parsing. SLT is adopted as the main representation by a line of MIR works, e.g., the Tangent and Tangent-S systems [39,55,56], and the Tangent-L or the MathDowsers system [5,8,32–34]. Local features such as symbols on adjacent nodes or nodes within a distance window and their spatial relations are together tokenized into *math tuples* and used for retrieval.

Different from structure matching, data-driven methods for MIR discover semantic matches without resorting to formula structure match constraints. Methods in this direction have been explored initially using linear neighbor tokens with limited success [9]. However, models such as Tangent-CFT [29], NTFEM [4], and FORTE [50] first construct structure representation(s) and learn word embeddings or tree embeddings from structure features have later shown effective.

Recent advances in natural language processing have led to the use of the Transformer model [49] for the MIR domain. Although it has been shown that the Transformer-based language model may still be relatively weak at math tasks [12,43], these Transformer models have nevertheless demonstrated their effectiveness at MIR. For instance, The MathBERT model [40] introduces structure mask in addition to BERT objectives and has been evaluated on the NTCIR-12 collection [53]. In addition, SentenceBERT [44] has been domain-adapted for MIR by regressing QA pair scores based on user-generated data in the original Math StackExchange thread [36,38]. Notably, The DPRL QASim method [27] uses two Transformers as similarity assessors, one question-question SentenceBERT [44] assessor pretrained on the Quora website and fine-tuned using related/duplicate links on the MSE website, as well as a question-answer TinyBERT [15] assessor pretrained on the MS-MARCO dataset [1] and fine-tuned on the ARQMath training data. The similarity produced by QASim is a product of these two assessor scores where the question-question model evaluates the topic question and the question to which the document answer is given. Lastly, the TU_DBS system utilized an ALBERT-based cross encoder as a primary model [45,46] which was further pre-trained on the ARQMath corpus directly with a maximum token input of 512.

4   Structural and Dense Retrieval for MIR

4.1   Structure Search Using Approach0

Built on the Approach0 system [60,63], we construct an enhanced OPT representation where a variable with or without superscript/subscript can still match each other using the leaf-root paths extracted from the OPT, and this customized OPT is designed to maximize retrieval recall. Figure 1 shows an example of OPT we use for representing an ARQMath topic formula.

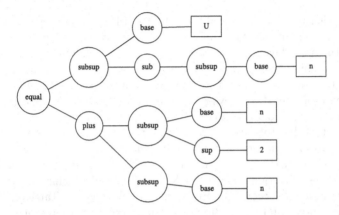

Fig. 1. Operator Tree representation for formula "$U_n = n^2 + n$" (Topic B.285). Operator and leaf (i.e., operand) are denoted by circle and box respectively. In order to improve recall, operands with or without subscript (**sub**) or superscript (**sup**) are represented canonically under a **subsup** token.

To further improve search recall, Approach0 tokenizes all nodes along the extracted leaf-root paths (and their prefixes). These paths are used similarly as text keywords in regular IR settings, however, the query processing is a specialized version where query and document formula substructures are recovered at run time by grouping root-end node IDs stored in the inverted index.

More specifically, assume any subtree rooted at m in the query formula Q is denoted as $Q^{(m)}$ (similar notation used for a document formula D). If the complete set of leaf-root path prefixes that the subtree contains is $\mathfrak{T}(Q^{(m)})$, then each path $t \in \mathfrak{T}(Q^{(m)})$ will be uniquely mapped to a posting list key in the inverted index. Assuming that a match among two sets of leaf-root paths implies that their corresponding OPT structures are also aligned, then any query-document maximum matching with respect to a given path will be easily identified by comparing the cardinality of leaf-root paths associated to t, i.e., $|Q_t^{(m)}|$ and $|D_t^{(n)}|$, and taking their minimum. As a result, we can find the maximum matching between $Q^{(m)}$ and $D^{(n)}$ by summing the maximum matching numbers among all keys in $\mathfrak{T}(Q^{(m)})$, and the maximum matching between Q and D is then

$$w^*(Q, D) = \max_{m,n} \sum_{t \in \mathfrak{T}(Q^{(m)})} \min\left(\left|Q_t^{(m)}\right|, \left|D_t^{(n)}\right|\right). \tag{1}$$

In practice, we also weigh each matched by a path idf [64], i.e.,

$$w^*(Q, D) = \max_{m,n} \sum_{t \in \mathfrak{T}(Q^{(m)})} \min\left(\left|Q_t^{(m)}\right|, \left|D_t^{(n)}\right|\right) \cdot \log \frac{N}{df_t} \tag{2}$$

where N is the total number of paths in the index and df_t is the document frequency of path t. This resembles the tf–idf scoring except for the "term frequency" here counts for common structure "width", i.e., the number of matched leaves in the common subtrees.

To accelerate query processing, a dynamic pruning strategy [60] has been applied to help the retriever skip evaluating some documents. This involves calculating a set of posting lists where their upperbound is insufficient to generate a top candidate in search results such that the associated postings can be skipped if a document formula's paths only occur in these posting lists.

In addition to the aforementioned structure weight, we take into consideration a few other factors for the final similarity score. First, to discern the difference between $E = mc^2$ and $y = ax^2$, paths in the maximum common structure are paired with their operand symbols and *fingerprints*. They together will determine the *symbol score* by greedily matching as many as points using the following rules:

- 1 point if both the operand symbol and the fingerprint match
- a lower point b_1 if only operand symbol match
- a nonzero base point b_2 otherwise $(b_2 < b_1)$
- variables of the same symbol cannot be matched to variables of different symbols

where the fingerprint is a hash value of the symbols of up to 4 operator nodes on top of the path leaf, it also takes into account the sign value (i.e., $1, -1$) induced for each operator (this includes the sign of the operand itself since the sign of an operand is induced into the subsup node which is always placed on top of an operand).

We normalize the symbol score with the number of paths and produce the (normalized) symbol score S'_{sym}. For query formula Q and document formula D, the final symbol score factor we multiply is a rescaled version:

$$S_{sym}(Q, D \mid w^*) = \frac{1}{1 + (1 - S'_{sym})^2} \tag{3}$$

Second, we add penalties to long formulas in a document. Assume the original formula length is L_D, the penalty $P(D; \eta)$ is parameterized by $\eta \in [0, 1]$:

$$P(D; \eta) = 1 - \eta + \eta \cdot \frac{1}{\log(1 + L_D)} \tag{4}$$

Finally, the overall score for math formula similarity $S(Q, D)$ is given by

$$S(Q, D) = w^*(Q, D) \cdot S_{sym}(Q, D \mid w^*) \cdot P(D; \eta) \tag{5}$$

To handle text keywords, we adopt the BM25+ scoring schema [23]. The overall score in Approach Zero is a weighted sum of all partial scores obtained by BM25+ in normal text keywords and those by formula similarity scoring in math formula keywords. Text words are processed with the Porter stemmer [41].

4.2 Dense Semantic Search Using ColBERT

The ColBERT model [19] is a dense retrieval model that employs a bi-encoder architecture based on the BERT backbone. A bi-encoder model contains two independent encoders: a query encoder and a document encoder, which encodes passages independently until the similarity scoring stage, i.e., the interaction between the two encoders is deferred until the similarity scoring stage. The ColBERT model stands out from other dense retrievers in that it preserves all output embeddings associated with each token, as opposed to using passage-level [CLS] embedding. However, to reduce the space footprint and speed up indexing, ColBERT pools each BERT output into a smaller dimension embedding with a default size of 128. These output embeddings are pretrained for the MLM objective [7], and therefore capture fine-grained contextualized semantics for individual tokens.

Given a query represented as a token sequence $q = q_0, q_1, ...q_l$, and a document passage represented as a token sequence $p = d_1, d_2, ...d_n$, the ColBERT model computes the token-level score $s(q_i, d_j)$ for the complete token pairs $i \in [E(q)]$ and $j \in [E(p)]$, where $E(q)$ and $E(p)$ denote the sets of output embeddings for the query and passage, respectively. The vanilla ColBERT model calculates this token-level score using either the dot product or L2 distance between the normalized output embeddings associated with each token. To compute the overall similarity score between a query q and a passage p, the ColBERT model employs the MaxSim operation where the overall similarity score is the sum of the maximum similarity scores between each query token q_i and its best matching token d_j in the passage, specifically,

$$S(q,p) = \sum_{i \in [E(q)]} \max_{j \in [E(p)]} s(q_i, d_j). \tag{6}$$

During training, the ColBERT model takes a triple of a query and a contrastive passage pair, i.e., (q, p^+, p^-), as input to optimize a pairwise cross-entropy loss. To differentiate between the encoding of a query and a passage, the model always prepends an unused token, either [Q] for query or [D] for document, at the beginning of each input sequence. To further enhance the model's performance, the authors employ a technique called *query augmentation* by replacing padding query tokens [PAD] with [MASK] tokens before query encoding.

The index used by the ColBERT model contains all the encoded passage tokens, along with document IDs and their lengths, which are used to locate the offset of each passage token. To facilitate efficient query processing, a two-stage retrieval process is employed. First, an approximate nearest neighbor (ANN) search is performed to filter a pool of top candidate tokens for each query token, individually. In the second stage, unique documents associated with the top candidate tokens are located, and their entire passage embeddings are loaded into the GPU for fast MaxSim operation. It is worth noting that the candidate selection stage comes at a cost and has made the end-to-end retrieval an approximation of what is originally defined in Eq. 6.

We constructed our version of ColBERT by utilizing the HuggingFace Transformers [51] library. To perform the first-stage approximate nearest neighbors (ANN) search, we relied on the Faiss package [16]. In order to overcome memory constraints, we divided the index into several shards, each shard's size was adjusted to fit into memory and GPU without exceeding capacity. We sequentially loaded each shard as required.

4.3   Combing Structural and Dense Retrieval

For our retrieval purposes, we opt to merge the top results from each component. The final score s is interpolated by tuning the hyperparameter λ as follows:

$$s = \lambda s_a + (1 - \lambda)s_d \tag{7}$$

where s_a and s_d are scores generated from the structure search and the ColBERT model respectively. Studies have demonstrated that this fusion scheme is more empirically robust than other alternatives for retrieval purposes [3,62].

5   Experimental Setup

5.1   Dataset and Evaluation Protocol

We report our results on the most recent ARQMath lab collection, i.e., the ARQMath-3 [25]. The evaluation of our runs in ARQMath adheres to the official protocols, with the top 1000 hits being considered for each run. The metrics employed to measure effectiveness include NDCG', MAP', and P'@10. Notably, the prime versions of these metrics only consider the judged documents, making for a fairer comparison between participant systems and post-hoc systems that were not part of the judgment pool. Regarding relevance levels, the official protocol includes four categories: High, Medium, Low, and Irrelevant, with scores ranging from 3 to 0 in descending order. To facilitate binary metrics such as MAP' and P'@10, we used a H+M binarization approach by combining High and Medium relevance into 1, and Low and Irrelevant into 0.

5.2   Training Data

For learning the ColBERT model, the data for further pertaining is made by ourselves.[2] This dataset comprises 1.69 million documents that were crawled

---

[2] To download our raw corpus: https://vault.cs.uwaterloo.ca/s/G36Mjt55HWRS NRR.

from the MSE and the Art of Problem-Solving community (AoPS) website. We generate both sentence pairs and in-context spans used for training the NSP [7] and ICT objectives [20]. These data are created following the same process by Zhong et al. [59,61].

To fine-tune our model, we utilize the ARQMath training data, which comprises Q&A posts of MSE from before 2018. We generate training triplets consisting of a question, a positive answer sourced from an accepted answer, a duplicate question, or any answer post that received more than 7 upvotes for the query. For each triplet, we sample a random answer passage as a hard negative from a question that shares a tag. In total, we generated 594,000 triplets for fine-tuning.

5.3 Preprocessing for Math

In the structure search, LaTeX are parsed into OPT using the PyA0 toolkit [58] before leaf-root paths are extracted and indexed for structure retrieval. To enhance the ColBERT model's performance with mathematical expressions, we add LaTeX math tokens generated from the PyA0 lexer as additional vocabulary after further pretraining. Additionally, we tokenize numbers using the CHARACTER scheme proposed by [35]. This orthography is robust enough to handle various user-created content.

5.4 Training Configurations

We further pretrain a new math-domain BERT model [61] and additionally a Coco-MAE model [59] both from the bert-base checkpoints from HuggingFace [51]. The domain-adapted BERT is trained using the original BERT objectives (MLM + NSP) [7] while the Coco-MAE model uses ICT objectives for further pretraining and adapts an autoencoding architecture to improve the learned retrieval representation. We follow the original training scheme and configurations for both models, using half-precision, and are fine-tuned against the generated ARQMath training triples for 7 epochs.

Compared to our original participant system in ARQMath-3 [25], the new Coco-MAE backbone is able to simulate retrieval tasks as early as in the further pretraining stage. And the autoencoding architecture used in the pretraining injects more token-level information for the [CLS] embedding by creating a challenging task to perform the MLM task at a weak decoder. This backbone has been shown to improve ColBERT effectiveness as the MaxSim scoring operation also relies on [CLS] embedding by default.

5.5 Task-Specific Settings

For task-specific settings, we follow our ARQMath-3 task participant methods [61]. Baseline systems in our experiments are chosen from the most effective submissions from the ARQMath-3 [25].

In Task 1 and Task 2, we prepend the original question text to each answer post before indexing in order to improve recall. Furthermore, we utilize manually extracted and modified text and formula keywords in our unsupervised retrieval. This is done to prevent the model from filtering too aggressively using strict structure matches. In addition, we limit the number of formulas that can be queried by the structure search in order to achieve reasonable efficiency. There are also formula queries that can be improved substantially by rewriting manually to construct a clear OPT representation, those improvements is hard to be exploited by automatic algorithm. For example, we replace the text mode "m++" to "m+1" so that our parser can handle the rare increment expression correctly. However, for anyone who wants to compare our results directly, our manual topics are available.[3]. When implementing supervised retrieval with ColBERT, we leverage the entire topic content as input. When combining both results, we fix $\lambda = 0.5$ for all tasks.

However, Task 1 and Task 2 are handled differently in a number of cases. We use two different backbones (i.e., BERT and Coco-MAE) in the ColBERT pass to handle Task 1. For Task 2, we only use ColBERT (BERT) due to the large number of formulas need to be encoded, however, we have tried two encoding methods. In the first case, we consider isolated formula without any context words. In the second case, formula context is injected in the following way: We use the query formula ID (i.e., `qid`) to identify the specified formula in the full-text question, and mask every other formula except the query formula by rewriting each one to the special `[MASK]` token. And we index the full-text embeddings of a document, each formula and its context in the document will get indexed, except we mask out all other formulas. To make sure we do not return more than 5 visually distinct formulas for a topic (as required by Task 2), we simply index at most 5 document formulas.

For Task 3, we use our Task 1 ColBERT (BERT) run for producing a set of candidate answers. Because a post-hoc run for Task 3 is hard to be fairly compared with submitted runs, we do not consider the more updated ColBERT (Coco-MAE) model in Task 3.

We have designed three strategies for selecting an answer:

- **Original:** A straightforward way to narrow down answer posts is to take the top-1 result from Task 1, we will use this as our first strategy
- **Re-rank:** The next strategy is simply considering a larger set of top results – here we use the top-20 results from Task 1 – and select one of them from the windowed snippets reranked by ColBERT.
- **Re-map:** We first try to retrieve the most similar corpus question post given a topic question (through the methods we use in Task 1), then we pick the accepted answer post, if non-exists, the top voted answer post as our candidate answer post.

A window of sentences will go through the beginning of a post to its end to select a combination of sentence spans to form candidate snippets. We also add a

---

[3] https://github.com/approach0/pya0/blob/arqmath3/topics-and-qrels.

selection strategy that includes every answer without a window limited. Finally, we use the same ColBERT model to score the snippets and pick the top 1 snippet as the final answer. Moreover, we also double-check the produced answers manually, if there are cases the final selected snippet is still too short, or there is no candidate available, or the math delimiters are unpaired, we will randomly copy an answer from a parallel run. As a result, our runs for all 3 tasks are considered as manual runs.

6 Evaluation

Table 1. Results for the ARQMath-3 Task 1 evaluation. In addition to the official measurements, we have also reported the BPref metric as well as the average number of judged hits per topic.

Runs	ARQMath-3 Task 1				
	NDCG'	MAP'	P'@10	BPref	Judged
Others (team/run)					
MSM/Ensemble_RRF	0.504	0.157	0.241	0.138	154.9
MIRMU/MiniLM+RoBERTa	0.498	0.184	0.267	0.169	120.8
Ours					
ColBERT (BERT)	0.418	0.162	0.251	0.165	89.0
ColBERT (Coco-MAE)	0.490	0.202	0.310	0.197	99.9
Struct. Search (Porter)	0.397	0.159	0.271	0.164	76.9
Struct. + ColBERT (BERT)	0.508	0.216	0.345	0.207	110.0
Struct. + ColBERT (Coco-MAE)	**0.546**	**0.237**	**0.360**	**0.221**	115.5

We present the results for ARQMath-3 in Table 1, 2, and 3. In this ARQMath Lab, a total of 9 teams have participated, and we achieve the best results in all three tasks. Here we show only the top submitted runs. For a complete overview of other participant systems, please refer to the ARQMath-3 overview paper [25].

Task 1 Baselines: The MSM run [10] is produced from an ensemble model in which each method is mainly developed as part of an Information Retrieval course taught at the Faculty of Informatics, Masaryk University, Brno, Czech Republic. And the MIRMU run [10,37] is produced by a dense retriever pipeline that uses a miniLM as a bi-encoder (first-stage) retriever and a RoBERTa model as a cross-encoder reranker.

Task 2 Baselines: The DPRL Tangent-CFTED [28] uses tree edit distance to rerank a set of candidates retrieved by formula FastText embeddings trained on structure features. And the latex_L8_a040 from MathDowsers team [17] is the default configuration for a newly rewritten and improved system on their

Table 2. Results for the ARQMath-3 Task 2 evaluation. In addition to the official measurements, we have also reported the BPref metric as well as the average number of judged hits per topic.

Runs	ARQMath-3 Task 2				
	NDCG'	MAP'	P'@10	BPref	Judged
Others (team/run)					
DPRL/Tangent-CFTED	0.694	0.480	0.611	0.471	61.7
MathDowsers/latex_L8_a040 †	0.640	0.451	0.549	0.443	60.3
Ours					
ColBERT (formula only)	0.604	0.436	0.622	0.446	42.8
ColBERT (in-context)	0.152	0.080	0.218	0.093	6.4
Struct. Search	0.639	0.501	0.615	0.505	45.9
Struct. + ColBERT	**0.720**	**0.568**	**0.688**	**0.560**	56.2

Table 3. Effectiveness evaluation for ARQMath-3 Labs (**Task 3**). The *Type* column lists the type of the QA model: 'E' for extractive and 'G' for generative.

Runs/Description	Type	AP'	P'@1
Others (team/run)			
GPT-3 (baseline)	G	**1.346**	**0.500**
DPRL/SBERT-SVMRank	E	0.462	0.154
TU_DBS/amps3_sel_hints	G	0.325	0.078
Ours			
Re-map ColBERT	E	0.949	0.282
Re-rank Struct. + ColBERT	E	1.179	0.372
Original Struct. + ColBERT	E	1.231	0.397
Re-rank Struct. (Porter) + ColBERT	E	**1.282**	**0.436**

previous Tangent-L system, with a relative weight of 0.40 on math tuples (over text terms).

Task 3 Baselines: In Task3, *text-davinci-002*, GPT-3 [2] is used as the baseline system. Another generative run *amps3_sel_hints* by the TU_DBS team [47] uses the GPT-2 model [42] but is further fine-tuned on the AMPS dataset [12]. On the other hand, the DPRL run, SBERT-SVMRank [28], uses an extractive approach based on SVM and Sentence-BERT models.

The ColBERT model based on BERT which we have submitted for the ARQMath-3 Lab is comparable to the ensemble MSM and has a higher precision than the Tangent-CFTED which further reranks top candidates using tree edit distance. Using an enhanced backbone, we can further outperform other submitted runs in Task 1 using ColBERT retriever alone. Consistently, the combination of ColBERT and structure search has been shown very complementary, boosting the top precision as much as 37% in Task 1. However, ColBERT has not achieved proportionally boosts in Task 2, this is reasonable as Task 1 input information is sufficient for ColBERT to utilize its context window and modeling power. However, the in-context matching of formulas does not produce effective scores in Task 2, we assume this could be a result from the different distributions for passages in training and inference.

The Task 3 effectiveness of our model is largely attributed to the ability for our hybrid search using structure search and data-driven model to generate high top precision in Task 1. However, it is intriguing to note that the text-davinci-002 model, which is a variant of the powerful GPT-3 model, has exhibited a superior ability to answer math questions compared to our extractive approach that relies on a highly effective retriever (while it remains uncertain whether the GPT-3 model merely recalls certain answers from its training data).

7   Conclusion and Future Work

Combining a dense retriever with a manual structure search method has shown highly effective for math answer retrieval tasks. However, the need to maintain multiple embedding vectors can be resource-intensive. As such, we have recently addressed the efficiency issue using better single-vector representations [59]. In this paper, we adopted the updated Coco-MAE backbone for ColBERT and further advanced effectiveness of the results in our ARQMath-3 submission. A future direction for us is to automatically select structure keywords without restricting too much the search results. Excitingly, large language models such as GPT-3 and others, hold the potential to enable the active selection and complete automation of query processes in our model. It remains to be seen whether a generative approach will eventually become the dominant method for directly and comprehensively handling queries in this domain.

Acknowledgments. This research was supported in part by the Natural Sciences and Engineering Research Council (NSERC) of Canada. Computational resources were provided by Compute Ontario and Compute Canada.

References

1. Bajaj, P., et al.: MS Marco: a human generated machine reading comprehension dataset (2016)
2. Brown, T., et al.: Language models are few-shot learners. In: NeurIPS (2020)
3. Bruch, S., Gai, S., Ingber, A.: An analysis of fusion functions for hybrid retrieval. arXiv:2210.11934 (2022)
4. Dai, Y., Chen, L., Zhang, Z.: An N-ary tree-based model for similarity evaluation on mathematical formulae. In: 2020 IEEE International Conference on Systems, Man, and Cybernetics (SMC). IEEE (2020)
5. Dallas, F.: Math information retrieval using a text search engine. Master's thesis, University of Waterloo (2018)

6. Davila, K., Zanibbi, R.: Layout and semantics: combining representations for mathematical formula search. In: SIGIR (2017)
7. Devlin, J., Chang, M.W., Lee, K., Toutanova, K.: BERT: pre-training of deep bidirectional transformers for language understanding. arXiv:1810.04805 (2019)
8. Fraser, D., Kane, A., Tompa, F.: Choosing math features for BM25 ranking with Tangent-L. In: DocEng (2018)
9. Gao, L., Jiang, Z., Yin, Y., Yuan, K., Yan, Z., Tang, Z.: Preliminary exploration of formula embedding for mathematical information retrieval: can mathematical formulae be embedded like a natural language? arXiv:1707.05154 (2017)
10. Geletka, M., Kalivoda, V., Štefánik, M., Toma, M., Sojka, P.: Diverse semantics representation is King: MIRMU and MSM at ARQMath 2022. In: CLEF (2022)
11. Hagino, H., Saito, H.: Partial-match retrieval with structure-reflected indices at the NTCIR-10 math task. In: NTCIR. Citeseer (2013)
12. Hendrycks, D., et al.: Measuring mathematical problem solving with the math dataset. arXiv preprint arXiv:2103.03874 (2021)
13. Hijikata, Y., Hashimoto, H., Nishida, S.: An investigation of index formats for the search of MathML objects. In: 2007 IEEE/WIC/ACM International Conferences on Web Intelligence and Intelligent Agent Technology-Workshops (2007)
14. Hijikata, Y., Hashimoto, H., Nishida, S.: Search mathematical formulas by mathematical formulas. In: SHI (Symposium on Human Interface) (2009)
15. Jiao, X., et al.: TinyBERT: distilling BERT for natural language understanding (2019)
16. Johnson, J., Douze, M., Jégou, H.: Billion-scale similarity search with GPUs. IEEE Trans. Big Data **7**, 535–547 (2019)
17. Kane, A., Ng, Y.K., Tompa, F.: Dowsing for answers to math questions. Doing better with less. In: CLEF (2022)
18. Kaplan, J., et al.: Scaling laws for neural language models. arXiv preprint arXiv:2001.08361 (2020)
19. Khattab, O., Zaharia, M.: ColBERT: efficient and effective passage search via contextualized late interaction over BERT. In: SIGIR (2020)
20. Lee, K., Chang, M.W., Toutanova, K.: Latent retrieval for weakly supervised open domain question answering. arXiv:1906.00300 (2019)
21. Lin, J.: A proposed conceptual framework for a representational approach to information retrieval (2021)
22. Lu, P., Qiu, L., Yu, W., Welleck, S., Chang, K.W.: A survey of deep learning for mathematical reasoning. arXiv preprint arXiv:2212.10535 (2022)
23. Lv, Y., Zhai, C.: Lower-bounding term frequency normalization. In: CIKM (2011)
24. Mansouri, B., Agarwal, A., Oard, D., Zanibbi, R.: Finding old answers to new math questions: the ARQMath Lab at CLEF 2020. In: Jose, J.M., et al. (eds.) Advances in Information Retrieval (2020)
25. Mansouri, B., Novotný, V., Agarwal, A., Oard, D.W., Zanibbi, R.: Overview of ARQMath-3 (2022): third CLEF lab on answer retrieval for questions on math (working notes version). In: Working Notes of CLEF 2022 - Conference and Labs of the Evaluation Forum (2022)
26. Mansouri, B., Oard, D.W., Zanibbi, R.: DPRL systems in the CLEF 2020 ARQMath lab. In: CLEF (2020)
27. Mansouri, B., Oard, D.W., Zanibbi, R.: DPRL systems in the CLEF 2021 ARQMath lab: sentence-BERT for answer retrieval, learning-to-rank for formula retrieval. In: CLEF (2021)
28. Mansouri, B., Oard, D.W., Zanibbi, R.: DPRL systems in the CLEF 2022 ARQMath lab: introducing MathAMR for math-aware search. In: CLEF (2021)

29. Mansouri, B., Rohatgi, S., Oard, D.W., Wu, J., Giles, C.L., Zanibbi, R.: Tangent-CFT: an embedding model for mathematical formulas. In: SIGIR (2019)
30. Mansouri, B., Zanibbi, R., Oard, D.W., Agarwal, A.: Overview of ARQMath-2 (2021): second CLEF lab on answer retrieval for questions on math (working notes version). In: Working Notes of CLEF 2021 - Conference and Labs of the Evaluation Forum (2021)
31. Miller, B.R., Youssef, A.: Technical aspects of the digital library of mathematical functions. In: AMAI (2003)
32. Ng, Y.K.: Dowsing for math answers: exploring MathCQA with a math-aware search engine. Master's thesis, University of Waterloo (2021)
33. Ng, Y.K., Fraser, D., Kassaie, B., Tompa, F.: Dowsing for answers to math questions: ongoing viability of traditional MathIR. In: CLEF (2021)
34. Ng, Y.K., et al.: Dowsing for math answers with Tangent-L. In: CLEF (2020)
35. Nogueira, R., Jiang, Z., Lin, J.: Investigating the limitations of transformers with simple arithmetic tasks. arXiv:2102.13019 (2021)
36. Novotný, V., Sojka, P., Štefánik, M., Lupták, D.: Three is better than one: ensembling math information retrieval systems. In: CLEF (2020)
37. Novotný, V., Štefánik, M.: Combining sparse and dense information retrieval. In: CLEF (2022)
38. Novotný, V., Štefánik, M., Lupták, D., Geletka, M., Zelina, P., Sojka, P.: Ensembling ten math information retrieval systems. In: CLEF (2021)
39. Pattaniyil, N., Zanibbi, R.: Combining TF-IDF text retrieval with an inverted index over symbol pairs in math expressions: the Tangent math search engine at NTCIR 2014. In: NTCIR (2014)
40. Peng, S., Yuan, K., Gao, L., Tang, Z.: MathBERT: a pre-trained model for mathematical formula understanding. arXiv:2105.00377 (2021)
41. Porter, M.F.: An algorithm for suffix stripping. Program (1980)
42. Radford, A., et al.: Language models are unsupervised multitask learners. OpenAI blog (2019). https://openai.com/blog/better-language-models/
43. Rae, J.W., et al.: Scaling language models: methods, analysis & insights from training gopher. arXiv:2112.11446 (2021)
44. Reimers, N., Gurevych, I.: Sentence-BERT: sentence embeddings using siamese BERT-networks. arXiv:1908.10084 (2022)
45. Reusch, A., Thiele, M., Lehner, W.: An ALBERT-based similarity measure for mathematical answer retrieval. In: SIGIR (2021)
46. Reusch, A., Thiele, M., Lehner, W.: TU_DBS in the ARQMath lab 2021, CLEF. In: CLEF (2021)
47. Reusch, A., Thiele, M., Lehner, W.: Transformer-encoder and decoder models for questions on math. In: CLEF (2022)
48. Schellenberg, T., Yuan, B., Zanibbi, R.: Layout-based substitution tree indexing and retrieval for mathematical expressions. In: DRR (2012)
49. Vaswani, A., et al.: Attention is all you need. In: NIPS (2017)
50. Wang, Z., Lan, A.S., Baraniuk, R.G.: Mathematical formula representation via tree embeddings. In: iTextbooks@ AIED (2021)
51. Wolf, T., et al.: Transformers: state-of-the-art natural language processing. In: EMNLP (2020)
52. Yokoi, K., Aizawa, A.: An approach to similarity search for mathematical expressions using MathML. In: DML (Digital Mathematics Library) (2009)
53. Zanibbi, R., Aizawa, A., Kohlhase, M., Ounis, I., Topic, G., Davila, K.: NTCIR-12 MathIR task overview. In: NTCIR (2016)

54. Zanibbi, R., Blostein, D.: Recognition and retrieval of mathematical expressions. In: IJDAR (2012)
55. Zanibbi, R., Davila, K., Kane, A., Tompa, F.: The Tangent search engine: improved similarity metrics and scalability for math formula search. arXiv:1507.06235 (2015)
56. Zanibbi, R., Davila, K., Kane, A., Tompa, F.: Multi-stage math formula search: using appearance-based similarity metrics at scale. In: SIGIR (2016)
57. Zhong, W.: A novel similarity-search method for mathematical content in LaTeX markup and its implementation. Ph.D. thesis, University of Delaware (2015)
58. Zhong, W., Lin, J.: PyA0: a Python toolkit for accessible math-aware search. In: SIGIR (2021)
59. Zhong, W., Lin, S.C., Yang, J.H., Lin, J.: One blade for one purpose: advancing math information retrieval using hybrid search. In: Proceedings of the 46th International ACM SIGIR Conference on Research and Development in Information Retrieval (2023)
60. Zhong, W., Rohatgi, S., Wu, J., Giles, C.L., Zanibbi, R.: Accelerating substructure similarity search for formula retrieval. In: Jose, J.M., et al. (eds.) ECIR 2020. LNCS, vol. 12035, pp. 714–727. Springer, Cham (2020). https://doi.org/10.1007/978-3-030-45439-5_47
61. Zhong, W., Xie, Y., Lin, J.: Applying structural and dense semantic matching for the ARQMath lab 2022, CLEF. In: CLEF (2022)
62. Zhong, W., Yang, J.H., Lin, J.: Evaluating token-level and passage-level dense retrieval models for math information retrieval (2022)
63. Zhong, W., Zanibbi, R.: Structural similarity search for formulas using leaf-root paths in operator subtrees. In: Azzopardi, L., Stein, B., Fuhr, N., Mayr, P., Hauff, C., Hiemstra, D. (eds.) ECIR 2019. LNCS, vol. 11437, pp. 116–129. Springer, Cham (2019). https://doi.org/10.1007/978-3-030-15712-8_8
64. Zhong, W., Zhang, X., Xin, J., Lin, J., Zanibbi, R.: Approach Zero and Anserini at the CLEF-2021 ARQMath track: applying substructure search and BM25 on operator tree path tokens. In: CLEF (2021)

CLEF 2023 Lab Overviews

Overview of BioASQ 2023: The Eleventh BioASQ Challenge on Large-Scale Biomedical Semantic Indexing and Question Answering

Anastasios Nentidis[1,2(✉)], Georgios Katsimpras[1], Anastasia Krithara[1], Salvador Lima López[3], Eulália Farré-Maduell[3], Luis Gasco[3], Martin Krallinger[3], and Georgios Paliouras[1]

[1] National Center for Scientific Research "Demokritos", Athens, Greece
{tasosnent,gkatsibras,akrithara,paliourg}@iit.demokritos.gr
[2] Aristotle University of Thessaloniki, Thessaloniki, Greece
[3] Barcelona Supercomputing Center, Barcelona, Spain
{salvador.limalopez,eulalia.farre,lgasco,martin.krallinger}@bsc.es

Abstract. This is an overview of the eleventh edition of the BioASQ challenge in the context of the Conference and Labs of the Evaluation Forum (CLEF) 2023. BioASQ is a series of international challenges promoting advances in large-scale biomedical semantic indexing and question answering. This year, BioASQ consisted of new editions of the two established tasks b and Synergy, and a new task (MedProcNER) on semantic annotation of clinical content in Spanish with medical procedures, which have a critical role in medical practice. In this edition of BioASQ, 28 competing teams submitted the results of more than 150 distinct systems in total for the three different shared tasks of the challenge. Similarly to previous editions, most of the participating systems achieved competitive performance, suggesting the continuous advancement of the state-of-the-art in the field.

Keywords: Biomedical knowledge · Semantic Indexing · Question Answering

1 Introduction

The BioASQ challenge has been focusing on the advancement of the state-of-the-art in large-scale biomedical semantic indexing and question answering (QA) for more than 10 years [34]. In this direction, it organizes different shared tasks annually, developing respective benchmark datasets that represent the real information needs of experts in the biomedical domain. This allows the participating teams from around the world, who work on the development of systems for biomedical semantic indexing and question answering, to benefit from the publicly available datasets, evaluation infrastructure, and exchange of ideas in the context of the BioASQ challenge and workshop.

© The Author(s), under exclusive license to Springer Nature Switzerland AG 2023
A. Arampatzis et al. (Eds.): CLEF 2023, LNCS 14163, pp. 227–250, 2023.
https://doi.org/10.1007/978-3-031-42448-9_19

Here, we present the shared tasks and the datasets of the eleventh BioASQ challenge in 2023, as well as an overview of the participating systems and their performance. The remainder of this paper is organized as follows. First, Sect. 2 presents a general description of the shared tasks, which took place from January to May 2023, and the corresponding datasets developed for the challenge. Then, Sect. 3 provides a brief overview of the participating systems for the different tasks. Detailed descriptions for some of the systems are available in the proceedings of the lab. Subsequently, in Sect. 4, we present the performance of the systems for each task, based on state-of-the-art evaluation measures or manual assessment. Finally, in Sect. 5 we draw some conclusions regarding the 2023 edition of the BioASQ challenge.

2   Overview of the Tasks

The eleventh edition of the BioASQ challenge (BioASQ 11) consisted of three tasks: (1) a biomedical question answering task (task b), (2) a task on biomedical question answering on developing medical problems (task Synergy), both considering documents in English, and (3) a new task on semantic annotation of medical documents in Spanish with clinical procedures (MedProcNER) [26]. In this section, we first describe this year's editions of the two established tasks b (task 11b) and Synergy (Synergy 11) with a focus on differences from previous editions of the challenge [23, 24]. Additionally, we also present the new MedProc-NER task on clinical procedure semantic recognition, linking, and indexing in Spanish medical documents.

2.1   Biomedical Semantic QA - Task 11b

The eleventh edition of task b (task 11b) focuses on a large-scale question-answering scenario in which the participants are required to develop systems for all the stages of biomedical question answering. As in previous editions, the task examines four types of questions: "yes/no", "factoid", "list" and "summary" questions [7]. In this edition, the training dataset provided to the participating teams for the development of their systems consisted of 4,719 biomedical questions from previous versions of the challenge annotated with ground-truth relevant material, that is, articles, snippets, and answers [14]. Table 1 shows the details of both training and test datasets for task 11b. The test data for task 11b were split into four independent bi-weekly batches. These include two batches of 75 questions and two batches of 90 questions each, as presented in Table 1.

As in previous editions of task b, task 11b was also divided into two phases which run for two consecutive days for each batch: (phase A) the retrieval of the relevant material and (phase B) providing the answers to the questions. In each phase, the participants have 24 h to submit the responses generated by their systems. In particular, a test set consisting of the bodies of biomedical questions, written in English, was released for phase A and the participants were expected to identify and submit relevant elements from designated resources, namely PubMed/MEDLINE-article abstracts, and snippets extracted from these

Table 1. Statistics on the training and test datasets of task 11b. The numbers for the documents and snippets refer to averages per question.

Batch	Size	Yes/No	List	Factoid	Summary	Documents	Snippets
Train	4,719	1,271	901	1,417	1,130	9.03	12.04
Test 1	75	24	12	19	20	2.48	3.28
Test 2	75	24	12	22	17	2.96	4.33
Test 3	90	24	18	26	22	2.66	3.77
Test 4	90	14	24	31	21	2.80	3.91
Total	5,049	1,357	967	1,515	1,210	8.62	11.5

resources. Then, some relevant articles and snippets for these questions, which have been manually selected by the experts, were also released in phase B and the participating systems were challenged to respond with *exact answers*, that is entity names or short phrases, and *ideal answers*, that is, natural language summaries of the requested information.

2.2 Task Synergy 11

The task Synergy was introduced two years ago [23] envisioning a continuous dialog between the experts and the systems. In task Synergy, the motivation is to make the advancements of biomedical information retrieval and question answering available to biomedical experts studying open questions for developing problems, aiming at a synergy between automated question-answering systems and biomedical experts. In this model, the systems provide relevant material and answers to the experts that posed some open questions. The experts assess these responses and feed their assessment back to the systems. This feedback is then exploited by the systems in order to provide more relevant material, considering more recent material that becomes available in the meantime, and improved responses to the experts as shown in Fig. 1. This process proceeds with new feedback and new responses from the systems for the same open questions that persist, in an iterative way, organized in rounds.

After eight rounds of the task Synergy in the context of BioASQ9 [15] and four more in the context of BioASQ10 [27], all focusing on open questions for the developing problem of the COVID-19 pandemic, in BioASQ11 we extended the Synergy task (Synergy 11) to open questions for any developing problem of interest for the participating biomedical experts [26]. In this direction, the four bi-weekly rounds of Synergy 11 were open to any developing problem, and a designated version of the PubMed/MEDLINE repository was considered for the retrieval of relevant material in each round. As in previous versions of the task, and contrary to task b, the open questions were not required to have definite answers and the answers to the questions could be more volatile. In addition, a set of 311 questions on COVID-19, from the previous versions of the Synergy task, were available, together with respective incremental expert feedback and answers, as a development set for systems participating in this edition of the task. Table 2 shows the details of the datasets used in task Synergy 11.

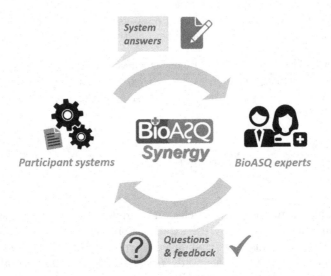

Fig. 1. The iterative dialogue between the experts and the systems in the BioASQ Synergy task on question answering for developing biomedical problems.

Table 2. Statistics on the datasets of Task Synergy. "Answer ready" stands for questions marked as having enough relevant material to be answered after the assessment of material submitted by the systems in the respective round. In round 2, ten questions were omitted from the test, as no feedback was available for them from the respective expert for the material retrieved by the systems in round 1. This feedback become available in round three, hence these questions were again included in the test set.

Round	Size	Yes/No	List	Factoid	Summary	Answer ready
1	53	12	17	11	13	14
2	43	11	14	7	11	32
3	53	12	17	11	13	37
4	53	12	17	11	13	42

Similar to task 11b, four types of questions are examined in Synergy 11 task: yes/no, factoid, list, and summary, and two types of answers, *exact* and *ideal*. Moreover, the assessment of the systems' performance is based on the evaluation measures used in task 11b. However, contrary to task 11b, Synergy 11 was not structured into phases, with both relevant material and answers received together. For new questions, only relevant material, that is relevant articles and snippets, was required until the expert considered that enough material has been gathered and marked the questions as "ready to answer". Once a question is marked as "ready to answer", the systems are expected to respond to the experts with both new relevant material and answers in subsequent rounds.

2.3   Medical Semantic Annotation in Spanish - MedProcNER

Clinical procedures play a critical role in medical practice, being an essential tool for the diagnosis and treatment of patients. They are also a difficult information type to extract, often being made up of abbreviations, multiple parts, and even descriptive sections. Despite their importance, there are not many resources that focus in-depth on the automatic detection of clinical procedures, and even fewer, if any, consider concept normalization.

With this in mind, this year we introduced the MedProcNER (Medical Procedure Named Entity Recognition) shared task as part of BioASQ11 as summarized in Fig. 2. The task challenges participants to create automatic systems that can extract different aspects of information about clinical procedures. These aspects are divided into three different sub-tasks:

– **Clinical Procedure Recognition:** This is a named entity recognition (NER) task where participants are challenged to automatically detect mentions of clinical procedures in a corpus of clinical case reports in Spanish.
– **Clinical Procedure Normalization:** In this entity linking (EL) task, participants must create systems that are able to assign SNOMED CT codes to the mentions retrieved in the previous sub-task.
– **Clinical Procedure-based Document Indexing:** This is a semantic indexing challenge in which participants automatically assign clinical procedure SNOMED CT codes to the full clinical case report texts so that they can be indexed. In contrast to the previous sub-task, participants do not need to rely on any previous systems, making this an independent sub-task.

To enable the development of clinical procedure recognition, linking and indexing systems, we have released the MedProcNER/ProcTEMIST corpus, a Gold Standard dataset of 1,000 clinical case reports manually annotated by multiple clinical experts with clinical procedures. The case reports were carefully selected by clinical experts and belong to various medical specialties including, amongst others, oncology, odontology, urology, and psychiatry. They are the same text documents that were used for the corpus and shared task on diseases DisTEMIST [21], building towards a collection of fully-annotated texts for clinical concept recognition and normalization. The MedProcNER corpus is publicly available on Zenodo[1].

In addition to the text annotations, the mentions in the corpus have been normalized to SNOMED CT. SNOMED CT (Systematized Nomenclature of Medicine Clinical Terms) is a comprehensive clinical terminology and coding system designed to facilitate the exchange and communication of health-related information across different healthcare settings and systems, which makes it fit for the normalization of varied clinical concepts. For the task, only a subset of 250 normalized documents was released as training data. The complete normalized dataset will be released as post-workshop material.

---

[1] https://doi.org/10.5281/zenodo.7817745.

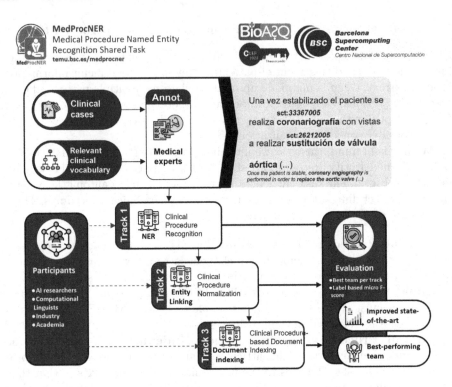

Fig. 2. Overview of the MedProcNER Shared Task.

Annotation and normalization guidelines were specifically created for this task. The current version of the guidelines includes 31 pages and a total of 60 rules that describe how to annotate different procedure types ranging from simple explorations to complex surgical descriptions. They also include a discussion of the task's importance and use cases, basic information about annotation process, a description of different procedure types and comparisons with similar clinical entity types, and indications and resources for the annotators. As with the DisTEMIST corpus, the guidelines were refined via multiple rounds of inter-annotator agreement (IAA) through parallel annotation of a section of the corpus. The final IAA score (computed as the pairwise agreement between two independent annotators) is of 81.2. The MedProcNER guidelines are available in Zenodo[2].

In addition to the corpus and guidelines, some additional resources have been released as part of the task. First, a SNOMED CT gazetteer was released containing official terms and synonyms from the relevant branches of SNOMED CT for the grounding of procedure mentions. The MedProcNER gazetteer has been built using the 31/10/2022 version of the Spanish edition of SNOMED CT, which is composed than 300,000 concepts organized in 19 different hierar-

---

[2] https://doi.org/10.5281/zenodo.7817666.

chies including "procedure", "substance" and "regime/therapy". To simplify the entity linking and indexing task, we compiled a reduced subset of the terminology with a smaller set of concepts to which the mentions can be mapped. The gazetteer consists of 234,674 lexical entries, out of which 130,219 are considered main terms. Within these entries, there are 130,219 unique codes originating from 19 hierarchies.

Next, to foster the advancement of document indexing with other terminologies and boost the reusability of MedProcNER data, we have created cross-mappings that connect the SNOMED CT mentions found in the corpus to MeSH and ICD-10. These mappings were achieved using the UMLS Meta-thesaurus. Finally, a Multilingual Silver Standard similar to last year's DisTEMIST [21] and LivingNER [20] was created in six different languages: English, French, Italian, Portuguese, Romanian and Catalan. This Silver Standard was automatically generated using a lexical annotation transfer approach in which the corpus' texts and Gold Standard annotations are translated separately and then mapped onto each other using a look-up system. This look-up takes into account individual annotations in each file, their translations and also a lemmatized version of the entities (obtained using spaCy[3]). Transferred annotations carry over the SNOMED CT code originally assigned to the Spanish annotation. All additional resources are available in Zenodo together with the Gold Standard corpus (See Footnote 2).

As for the task evaluation, all three MedProcNER sub-tasks are evaluated using micro-averaged precision, recall and F1-score. It is important to highlight that the evaluation of entity linking systems is not conducted in isolation but rather in an end-to-end manner. Instead of being provided an exhaustive list of mentions to be normalized, participants had to rely on their predictions from the named entity recognition stage. Consequently, the obtained scores might not accurately represent the overall performance of the systems. However, this type of evaluation does offer a more comprehensive assessment of complete systems, closely resembling their performance in real-world applications.

MedProcNER is promoted by the Spanish Plan for the Advancement of Language Technology (Plan TL)[4] and organized by the Barcelona Supercomputing Center (BSC) in collaboration with BioASQ. A more in-depth analysis of the MedProcNER Gold Standard, guidelines and additional resources is presented in the MedProcNER overview paper [17].

3 Overview of Participation

3.1 Task 11b

19 teams competed this year in task 11b submitting the responses of 76 different systems for both phases A and B, in total. In particular, 9 teams with 37 systems participated in Phase A, while in Phase B, the number of participants

---

[3] https://spacy.io/.
[4] https://plantl.mineco.gob.es.

and systems were 16 and 59 respectively. Six teams engaged in both phases. An overview of the technologies employed by the teams is provided in Table 3 for the systems for which a description was available. Detailed descriptions for some of the systems are available at the proceedings of the workshop.

Table 3. Systems and approaches for task 11b. Systems for which no information was available at the time of writing are omitted.

Systems	Phase	Ref	Approach
bioinfo	A, B	[4]	BM25, PubMedBERT, monoT5, reciprocal rank fusion (RRF), ALPACA-LoRA, OA-Pythia, OA-LLaMA
UR-gpt	A, B	[5]	GPT-3.5-turbo, GPT-4
ELECTROBERT	A, B	[29]	ELECTRA, ALBERT, BioELECTRA, BERT, GANBERT, BM25, RM3
MindLab	A, B	[32]	BM25, CNN, SBERT
dmiip	A, B	–	BM25, GPT-3.5, PubMedBERT, BioBERT, BioLinkBERT, ELECTRA
A&Q	A	[33]	BM25, PubMedBERT
IRCCS	A	–	transformers, cosine similarity, BM25
MarkedCEDR	A	[16]	BM25, BERT, CEDR
ELErank	B	[1]	BioM-ELECTRA, S-PubMedBERT
NCU-IISR	B	[12]	GPT-3, GPT-4
AsqAway	B	[30]	BioBERT, BioM-Electra
MQ	B	–	GPT-3.5, sBERT, DistilBERT
DMIS-KU	B	[13]	BioLinkBERT, GPT-4, data augmentation
MQU	B	[10]	BART, BioBART

The ("*bioinfo*") team from the University of Aveiro participated in both phases of the task with five systems. In phase A, they developed a two-stage retrieval pipeline. The first stage adopted the traditional BM25 model. In contrast, the second stage implemented transformer-based neural re-ranking models from PubMedBERT and monoT5 checkpoints. Additionally, synthetic data were used to augment the training regimen. The reciprocal rank fusion (RRF) was utilized to ensemble the outputs from various models. For Phase B, their systems utilized instruction-based transformer models, such as ALPACA-LoRA, OA-Pythia, and OA-LLaMA, for conditioned zero-shot answer generation. More specifically, given the most relevant article from Phase A, the model was designed to generate an *ideal answer* based on the information contained in the relevant article.

Another team participating in both phases is the team from the University of Regensburg. Their systems ("*UR-gpt*") relied on two commercial versions of the GPT Large Language model (LLM). Specifically, their systems experimented with both GPT-3.5-turbo and GPT-4 models. In phase A, their systems used zero-shot learning for query expansion, query reformulation and re-ranking. For Phase B, they used zero-shot learning, grounded with relevant snippets.

The BSRC Alexander Fleming team also participated in both phases with the systems "*ELECTROBERT*". Their systems are built upon their previously developed systems [31] and also adapted the semi-supervised method GAN-BERT [9] for document relevance classification. Furthermore, for the initial document selection phase their systems utilize BM25 combined with RM3 query expansion with optimized parameters.

The '*MindLab*" team competed in both phases of the task with five systems. For document retrieval their systems used the BM25 scoring function and semantic-similarity as a re-ranking strategy. For passage retrieval their systems used a metric learning method which fuses different similarity measures through a siamese convolutional network.

The "*dmiip*" team from the Fudan University participated in both phases of the task with five systems. In phase A, their systems used BM25 and GPT for the retrieval stage, and a cross-encoder ranker based on different biomedical PLMs, such as PubMedBERT, BioBERT, BioLinkBERT and ELECTRA for the ranking stage. Biomedical PLMs and GPT-3.5 are also utilized in Phase B. The systems are initially finetuned on SQuAD and then trained with the BioASQ training dataset.

In phase A, the "*A&Q*" team participated with five systems. Their systems are based on a multi-stage approach which incorporates a bi-encoder model in the retrieval stage, and a cross-encoder model at the re-ranking stage. At the retrieval stage, a hybrid retriever that combines dense and sparse retrieval, where the dense retrieval is implemented with the bi-encoder and the sparse retrieval is implemented with BM25. Both encoders are initialized with PubMedBERT and further trained on PubMed query-article search logs.

The IRCCS team participated with five systems ("*IRCCS*") in phase A. Their systems follow a two-step methodology. First, they score the documents using the BM25 ranking function. Then, the second step is to re-rank them based on cosine similarity between the query and each document, which are encoded using various transformers models.

The IRIT lab team competed also in phase A with two systems ("*Marked-CEDR*"). Their systems adopt a two-stage retrieval approach composed of a retriever and a re-ranker. The former is based on BM25. The later is an implementation of a BERT cross-encoder named CEDR.

In phase B, the Ontotext team participated with two systems ("*ELErank*"). Their systems used BioM-ELECTRA as a backbone model for both yes/no and factoid questions. For yes/no questions, it was fine-tuned in a sequence classification setting, and for factoid questions, it was fine-tuned in a token classification setting (for extractive QA). Before applying classification, the sentences were ranked based on their cosine similarity to the question. Top-5 most relevant sentences were used for classification. Sentence embeddings for ranking were calculated with S-PubMedBERT.

The National Central Uni team competed with five systems "*NCU-IISR*" in phase B. Their systems utilized OpenAI's ChatCompletions API, incorporating

Prompt Engineering techniques to explore various prompts. Specifically, their systems used GPT-3 and GPT-4 for answer generation.

The CMU team participated with four different systems (*"AsqAway"*) in phase B. Their system adopt an ensembling approach using transformer models. For factoid and list questions they use a BioBERT and BioM-Electra ensemble. For yes/no questions, they employ BioM-Electra.

The Korea University team participated with five systems (*"DMIS-KU"*). They employed different pre-processing, training, and data augmentation methods and different QA models. For the yes/no type, the systems utilized the "full-snippet" pre-processing method, where all snippets were concatenated into a single context. The BioLinkBERT-large model was used as the embedding model. For the factoid type, the "single-snippet" method was used, which involved processing one snippet at a time. The BioLinkBERT-large was trained using the SQuAD dataset and fine-tuned using the BioASQ training data. For the list type, the full-snippet method was used again. Additionally, their systems employed a dataset generation framework, called LIQUID, to augment the training data. Also, the GPT-4 was utilized to answer list questions in a one-shot manner. In all question types, the final predictions are produced by combining the results from multiple single models using an ensemble method.

There were two teams from the Macquarie University. The first team participated with five systems (*"MQ"*) in phase B and focused on finding the *ideal answers*. Three of their systems employed GPT-3.5 and various types of prompts. The rest of the systems were based on their previously developed systems [22]. The second tean (*"MQU"*) competed with five systems in phase B which utilised BART and BioBART that were fine-tuned for abstractive summarisation.

As in previous editions of the challenge, a baseline was provided for phase B *exact answers*, based on the open source OAQA system [37]. This system relies on more traditional NLP and Machine Learning approaches, used to achieve top performance in older editions of the challenge, and now serves as a baseline. The system is developed based on the UIMA framework. In particular, question and snippet parsing is done with ClearNLP. Then, MetaMap, TmTool [36], C-Value, and LingPipe [6] are employed for identifying concepts that are retrieved from the UMLS Terminology Services (UTS). Finally, the relevance of concepts, documents, and snippets is identified based on some classifier components and some scoring and ranking techniques are also employed.

Furthermore, this year we introduced two more baselines for phase B *ideal answers*, BioASQ Baseline ZS and BioASQ Baseline FS, which are based on zero-shot prompting of Biomedical LMs. Both systems utilized the BioGPT, a language model trained exclusively on biomedical abstracts and papers, with the former using as input only the question body, and the latter using the concatenation of the question body and the relevant snippets until the input length is exceeded.

3.2 Task Synergy 11

In this edition of the task Synergy (Synergy 11) 5 teams participated submitting the results from 12 distinct systems. An overview of systems and approaches employed in this task is provided in Table 4, for the systems for which a description was available. More detailed descriptions for some of the systems are available at the proceedings of the workshop.

Table 4. Systems and their approaches for task Synergy. Systems for which no description was available at the time of writing are omitted.

System	Ref	Approach
dmiip	–	BM25, GPT-3.5, PubMedBERT, BioBERT, BioLinkBERT, ELECTRA
bio-answerfinder	[28]	Bio-ELECTRA++, BERT, weighted relaxed word mover's distance (wRWMD), pyserini with MonoT5, SQuAD, GloVe
ELECTROBERT	[29]	ELECTRA, ALBERT, BioELECTRA, BERT, GANBERT, BM25, RM3

The Fudan University (*"dmipp"*) competed in task Synergy with the same models they used for task 11b. Additionally, they expanded the query with the shortest relevant snippet in the provided feedback.

The *"UCSD"* team competed in task Synergy with two systems. Their systems (*"bio-answerfinder"*) used the Bio-AnswerFinder end-to-end QA system they had previously developed [28] with few improvements, including the use of the expert feedback data in retraining of their model's re-ranker.

The BSRC Alexander Fleming team participated with two systems. Similar to task b, their systems (*"ELECTROBERT"*) built upon their previously developed systems [31] and also adapted the semi-supervised method GANBERT [9].

3.3 Task MedProcNER

Among the 47 teams registered for the MedProcNER task, 9 teams submitted at least one run of their predictions. Specifically, all 9 teams engaged in the entity recognition sub-task, while 7 teams participated in the entity linking sub-task. Additionally, 4 teams took part in the document indexing sub-task. Overall, a total of 68 runs were submitted, reflecting the collective efforts and contributions of the participating teams.

Table 5 gives an overview of the methodologies used by the participants in each of the sub-tasks. As is the case in many modern NLP approaches, the majority of the participants used transformers-based models. RoBERTa [19] and SapBERT [18] models were the most popular for named entity recognition and entity linking respectively. In addition to this, in order to boost the systems' performance some teams also relied on recurrent classifiers such as CRFs (e.g.

Table 5. Systems and approaches for task MedProcNER. Systems for which no description was available at the time of writing are omitted. In the Task column, NER stands for named entity recognition (i.e. sub-task 1), EL for entity linking (i.e. sub-task 2) and DI for document indexing (i.e. sub-task 3)

Team	Ref	Task	Approach
BIT.UA	[3]	NER	Transformer-based solution with masked CRF and data augmentation
		EL	Semantic search with pretrained transformer-based models using an unsupervised approach
		DI	Indexing of codes found in EL step
Fusion	[35]	NER	Different BERT-family models fine-tuned for token classification
		EL	Cross-lingual SapBERT (SapBERT-UMLS-2020AB-all-lang-from-XLMR-large)
KFU NLP Team	–	NER	Ensemble of different SapBERT and mGEBERT models with and without adapters
		EL	Synonym Marginalization loss function and UniPELT adapters
		DI	Indexing of codes found in EL step
NLP-CIC-WFU	–	NER	Fine-tuned RoBERTa models combined with different pre-processing and post-processing techniques
Onto-NLP	[2]	NER	Fine-tuned RoBERTa models + lexical matching
		EL	SapBERT models + lexical matching + majority voting
University of Regensburg	[5]	All	In-context few-shot (3) learning with GPT-3.5-turbo and GPT-4
SINAI	[8]	NER	Clinical transformer models + recurrent classifiers (GRU, CRF)
		EL	Combination of matching techniques with token similarity based normalization
Vicomtech	[38]	NER	Seq2seq systems with BERT-like models
		EL	Semantic search techniques based on transformer models (SapBERT) and cross-encoders
		DI	Combination of first two systems

BIT.UA [3], SINAI team [8]), adapters (e.g. KFU NLP team), model ensembling/voting (e.g. KFU NLP team, Onto-text [2]) and data augmentation (e.g. BIT.UA [3]). Interestingly, one of the participants (Samy Ateia from the University of Regensburg [5]) proposes an approach based on Generative Pre-trained Transformers (GPT) models for all three sub-tasks.

4   Results

4.1   Task 11b

Phase A: The Mean Average Precision (MAP) was the official measure for evaluating system performance on document retrieval in phase A of task 11b, which is based on the number of ground-truth relevant elements. For snippet retrieval, however, the situation is more complicated as a ground-truth snippet may overlap with several distinct submitted snippets, which makes the interpretation of MAP less straightforward. For this reason, since BioASQ9 the F-measure is used for the official ranking of the systems in snippet retrieval, which is calculated based on character overlaps[5] [23].

Since BioASQ8, a modified version of Average Precision (AP) is adopted for MAP calculation. In brief, since BioASQ3, the participant systems are allowed to return up to 10 relevant items (e.g. documents or snippets), and the calculation of AP was modified to reflect this change. However, some questions with fewer than 10 golden relevant items have been observed in the last years, resulting in relatively small AP values even for submissions with all the golden elements. Therefore, the AP calculation was modified to consider both the limit of 10 elements and the actual number of golden elements [25].

Table 6. Preliminary results for document retrieval in batch 1 of phase A of task 11b. Only the top-10 systems are presented, based on MAP.

System	Mean Precision	Mean Recall	Mean F-measure	MAP	GMAP
bioinfo-0	0.2118	0.6047	**0.2774**	**0.4590**	**0.0267**
bioinfo-1	**0.2152**	0.5964	0.2769	0.4531	**0.0267**
bioinfo-2	0.1498	0.5978	0.2192	0.4522	0.0233
bioinfo-3	0.1712	**0.6149**	0.2418	0.4499	0.0239
dmiip3	0.1133	0.6127	0.1823	0.4462	0.0240
A&Q4	0.1027	0.5816	0.1667	0.4404	0.0215
A&Q3	0.1027	0.5816	0.1667	0.4404	0.0215
A&Q5	0.1000	0.5738	0.1627	0.4397	0.0191
dmiip5	0.1120	0.5993	0.1799	0.4391	0.0209
bioinfo-4	0.1529	0.5944	0.2183	0.4275	0.0164

Tables 6 and 7 present some indicative preliminary results for the retrieval of documents and snippets in batch 1. The full results are available online on the result page of task 11b, phase A[6]. The final results for task 11b will be available after the completion of the manual assessment of the system responses by the BioASQ team of biomedical experts, which is still in progress, therefore the results reported here are currently preliminary.

Phase B: In phase B of task 11b, the competing systems submit exact and *ideal answers*. As regards the *ideal answers*, the official ranking of participating systems is based on manual scores assigned by the BioASQ team of experts that assesses each *ideal answer* in the responses [7]. The final position of systems providing *exact answers* is based on their average ranking in the three question types where *exact answers* are required, that is "yes/no", "list", and "factoid".

---

[5] http://participants-area.bioasq.org/Tasks/b/eval_meas_2022/.
[6] http://participants-area.bioasq.org/results/11b/phaseA/.

Table 7. Preliminary results for snippet retrieval in batch 1 of phase A of task 11b. Only the top-10 systems are presented, based on F-measure.

System	Mean Precision	Mean Recall	Mean F-measure	MAP	GMAP
dmiip3	**0.1109**	**0.4309**	**0.1647**	**0.4535**	**0.0104**
dmiip4	0.1099	0.4144	0.1628	0.4142	0.0065
dmiip2	0.1075	0.4023	0.1589	0.4234	0.0077
dmiip5	0.1075	0.4053	0.1589	0.4327	0.0089
dmiip1	0.1027	0.3863	0.1518	0.4038	0.0061
MindLab QA Reloaded	0.0833	0.1918	0.0991	0.1389	0.0023
MindLab Red Lions++	0.0816	0.1807	0.0944	0.1228	0.0013
Deep ML methods for	0.0808	0.1485	0.0904	0.1208	0.0007
MindLab QA System	0.0838	0.1245	0.0887	0.0995	0.0003
MindLab QA System ++	0.0838	0.1245	0.0887	0.0995	0.0003

Table 8. Results for batch 2 for *exact answers* in phase B of task 11b. Only the top-15 systems based on Yes/No F1 and the BioASQ Baseline are presented.

System	Yes/No		Factoid			List		
	F1	Acc.	Str. Acc.	Len. Acc.	MRR	Prec.	Rec.	F1
IISR-2	**1.0000**	**1.0000**	**0.5455**	0.6364	**0.5909**	**0.5099**	0.3577	0.3980
IISR-1	**1.0000**	**1.0000**	0.5000	0.5455	0.5227	0.4861	0.3310	0.3678
DMIS-KU-4	**1.0000**	**1.0000**	0.3636	0.5909	0.4697	0.2983	0.3683	0.2871
DMIS-KU-1	0.9524	0.9577	0.3182	**0.6818**	0.4773	0.3349	0.3623	0.3080
DMIS-KU-2	0.9524	0.9577	0.3182	**0.6818**	0.4561	0.3486	0.3456	0.3087
DMIS-KU-3	0.9524	0.9577	0.3636	0.5909	0.4621	0.2818	0.4058	0.3178
UR-gpt4-zero-ret	0.9474	0.9564	**0.5455**	0.5909	0.5682	0.3742	0.4369	0.3828
dmiip5	0.9474	0.9564	0.4091	0.4545	0.4242	0.1413	0.2200	0.1676
capstone-1	0.9091	0.9161	0.4091	0.5455	0.4561	0.2085	0.3810	0.2617
DMIS-KU-5	0.9000	0.9143	0.3636	0.5909	0.4621	0.2534	0.4593	0.3022
dmiip1	0.9000	0.9143	0.3182	0.5909	0.4318	0.2271	0.3760	0.2501
UR-gpt3.5-turb	0.8889	0.9111	**0.5455**	0.5909	0.5682	0.4598	**0.4671**	**0.4316**
dmiip3	0.8571	0.8730	0.3182	0.5455	0.3992	0.2851	0.2464	0.2232
AsqAway_1	0.8421	0.8693	0.4545	0.4545	0.4545	0.1780	0.1968	0.1756
AsqAway_2	0.8421	0.8693	0.4545	0.4545	0.4545	0.2023	0.3226	0.2327
BioASQ_Baseline	0.6000	0.4667	0.0909	0.1364	0.1136	0.1185	0.2784	0.1613

Summary questions for which no *exact answers* are submitted are not considered in this ranking. In particular, the mean F1 measure is used for the ranking in list questions, the mean reciprocal rank (MRR) is used for the ranking in factoid questions, and the F1 measure, macro-averaged over the classes of yes and no, is used for yes/no questions. Table 8 presents some indicative preliminary results on *exact answer* extraction from batch 2. The full results of phase B of task 11b are available online[7]. These results are preliminary, as the final results for task 11b will be available after the manual assessment of the system responses by the BioASQ team of biomedical experts.

---

[7] http://participants-area.bioasq.org/results/11b/phaseB/.

TaskB Phase B results : Exact answers

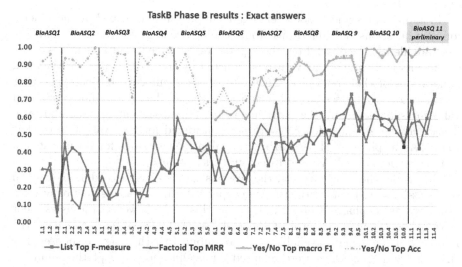

Fig. 3. The evaluation scores of the best-performing systems in task B, Phase B, for *exact answers*, across the eleven years of the BioASQ challenge. Since BioASQ6 the official measure for Yes/No questions is the macro-averaged F1 score (macro F1), but accuracy (Acc) is also presented as the former official measure. The black dots in 10.6 highlight that these scores are for an additional batch with questions from new experts [23].

The top performance of the participating systems in *exact answer* generation for each type of question during the eleven years of BioASQ is presented in Fig. 3. The preliminary results for task 11b, reveal that the participating systems keep improving in answering all types of questions. In batch 2, for instance, presented in Table 8, several systems manage to correctly answer literally all yes/no questions. This is also the case for batch 3 and batch 4. Some improvements are also observed in the preliminary results for factoid questions compared to the previous years, but there is still more room for improvement, as done for list questions where the preliminary performance is comparable to the one of the previous year.

4.2   Task Synergy 11

In task Synergy 11 the participating systems were expected to retrieve documents and snippets, as in phase A of task 11b, and, at the same time, provide answers for some of these questions, as in phase B of task 11b. In contrast to task 11b, however, due to the developing nature of the relevant knowledge, no answer is currently available for some of the open questions. Therefore only the questions indicated to have enough relevant material gathered from previous rounds ("Answer ready") require the submission of *exact* and *ideal answers* by the participating systems.

In addition, no golden documents and snippets were provided by the experts for new questions. For questions from previous rounds, on the other hand, a separate file with feedback from the experts was provided, that is elements of the documents and snippets previously submitted by the participants with manual annotations of their relevance. Therefore, these documents and snippets, that have already been assessed and included in the feedback, were not considered valid for submission by the participants in the subsequent rounds, and even if accidentally submitted, they were not considered for the evaluation of that round. As in phase A of task 11b, the evaluation measures for document and snippet retrieval are MAP and F-measure respectively.

Regarding the *ideal answers*, the systems were ranked according to manual scores assigned to them by the BioASQ experts during the assessment of systems responses as in phase B of task B [7]. In this task, however, the assessment took place during the course of the task, so that the systems can have the feedback of the experts available, prior to submitting their new responses. For the *exact answers*, which were required for all questions except the summary ones, the measure considered for ranking the participating systems depends on the question type. For the yes/no questions, the systems were ranked according to the macro-averaged F1-measure on the prediction of no and yes answers. For factoid questions, the ranking was based on mean reciprocal rank (MRR), and for list questions on mean F1-measure.

Table 9. Results for document retrieval of the first round of the Synergy 11 task.

System	Mean precision	Mean Recall	Mean F-Measure	MAP	GMAP
dmiip2	**0.3026**	**0.3772**	**0.2803**	**0.2791**	**0.0572**
dmiip4	0.3000	0.3714	0.2760	0.2788	0.0512
dmiip5	0.2667	0.3230	0.2466	0.2525	0.0578
dmiip1	0.2256	0.2668	0.2034	0.2001	0.0151
dmiip3	0.1974	0.2116	0.1741	0.1708	0.0080
bio-answerfinder	0.1575	0.1390	0.1244	0.1468	0.0021
bio-answerfinder-2	0.1575	0.1247	0.1232	0.1236	0.0011
ELECTROLBERT-3	0.0893	0.0180	0.0285	0.0212	0.0000

Some indicative results for the Synergy task are presented in Table 9. The full results of Synergy 11 task are available online[8]. Overall, the collaboration between participating biomedical experts and question-answering systems allowed the progressive identification of relevant material and extraction of *exact* and *ideal answers* for several open questions for developing problems, such as COVID-19, Colorectal Cancer, Duchenne Muscular Dystrophy, Alzheimer's Disease, and Parkinson's Disease. In particular, after the completion of the four rounds of the Synergy 11 task, enough relevant material was identified for providing an answer to about 79% of the questions. In addition, about 42% of the

---

[8] http://participants-area.bioasq.org/results/synergy_v2023/.

questions had at least one *ideal answer*, submitted by the systems, which was considered satisfactory (ground truth) by the expert that posed the question.

4.3   Task MedProcNER

All in all, the top scores for each sub-task were:

- **Clinical Procedure Recognition.** The BIT.UA team attained all top 5 positions with their transformer-based solution that also uses masked CRF and data augmentation. They achieved the highest F1-score, 0.7985, highest precision (0.8095) and highest recall (0.7984). Teams Vicomtech and SINAI also obtained F1-scores over 0.75.
- **Clinical Procedure Normalization.** The highest F1-score (0.5707), precision (0.5902) and recall (0.5580) were obtained by Vicomtech. Teams SINAI and Fusion were also above 0.5 F1-score using token similarity techniques and a cross-lingual SapBERT, respectively.
- **Clinical Procedure-based Document Indexing.** The Vicomtech team also obtained the highest F1-score (0.6242), precision (0.6371) and recall (0.6295), with the KFU NLP Team coming in second place (0.4927 F1-score). In this sub-task, all participating teams reused their systems and/or output from previous sub-tasks.

The complete results for the entity recognition, linking and document indexing are shown in Tables 10, 11 and 12, respectively.

Overall, the performance of the systems presented for the MedProcNER shared task is very diverse, with scores ranging from 0.759 F-score (by the BIT.UA team on the entity recognition task) to 0.126 (University of Regenburg on the entity linking task). This gap evidences mainly two things: the multitude of approaches and the difficulty of the corpus. On the one hand, the systems presented for the task were very varied. Even amongst BERT-based models, participants tried different strategies such as using models pre-trained on different domains (biomedical, clinical) and languages (Spanish, multilingual), implementing different pre/post-processing techniques, data augmentation and using multiple output layers (CRF, GRU, LSTM). Again, it is remarkable that one of the participants (Samy Ateia from the University of Regensburg) used GPT3.5 (ChatGPT) and GPT4 for their submissions. Even though the overall performance is not too good (especially in terms of recall), this is partly to be expected since the system was fine-tuned for the task using a few-shot approach. On the other hand, the Gold Standard corpus is very varied in terms of mentions, with many mentions being quite long and descriptive (especially surgical mentions). Additionally, the text documents span multiple medical specialties, which introduces not only more variety in clinical procedures but also possible ambiguities due to the use of specialized abbreviations. In the future, we will expand the corpus with more annotated documents to address this issue.

Table 10. Results of MedProcNER Entity Recognition sub-task. The best result is bolded, and the second-best is underlined.

Team Name	Run name	P	R	F1
BIT.UA	run4-everything	**0.8095**	<u>0.7878</u>	**0.7985**
BIT.UA	run0-lc-dense-5-wVal	<u>0.8015</u>	<u>0.7878</u>	<u>0.7946</u>
BIT.UA	run1-lc-dense-5-full	0.7954	**0.7894**	0.7924
BIT.UA	run3-PlanTL-dense	0.7978	0.787	0.7923
BIT.UA	run2-lc-bilstm-all-wVal	0.7941	0.7823	0.7881
Vicomtech	run1-xlm_roberta	0.8054	0.7535	0.7786
Vicomtech	run2-roberta_bio	0.7679	0.7629	0.7653
SINAI	run1-fine-tuned-roberta	0.7631	0.7505	0.7568
Vicomtech	run3-longformer_base	0.7478	0.7588	0.7533
SINAI	run4-fulltext-LSTM	0.7538	0.7353	0.7444
SINAI	run2-lstmcrf-512	0.7786	0.7043	0.7396
SINAI	run5-lstm-BIO	0.7705	0.7049	0.7362
KFU NLP Team	predicted_task1	0.7192	0.7403	0.7296
SINAI	run3-fulltext-GRU	0.7396	0.711	0.725
Fusion	run4-Spanish-RoBERTa	0.7165	0.7143	0.7154
Fusion	run3-XLM-RoBERTA-Clinical	0.7047	0.6916	0.6981
NLP-CIC-WFU	Hard4BIO...postprocessing	0.7188	0.654	0.6849
NLP-CIC-WFU	Hard4BIO_RoBERTa	0.7132	0.6507	0.6805
Fusion	run1-BioMBERT	0.6948	0.6599	0.6769
Fusion	run2-BioMBERT	0.6894	0.6599	0.6743
Fusion	run5-Adapted-ALBERT	0.6928	0.6264	0.658
NLP-CIC-WFU	Lazy4BIO...postprocessing	0.6301	0.6002	0.6148
Onto-NLP	run1-...voting-filtered	0.7425	0.4374	0.5505
Onto-NLP	run1-...voting	0.7397	0.4374	0.5497
University Regensburg	run2-gpt-4	0.6355	0.3874	0.4814
saheelmayekar	predicted_data	0.3975	0.535	0.4561
Onto-NLP	run1-...exact_match	0.3296	0.6104	0.428
University Regensburg	run1-gpt3.5-turbo	0.523	0.2106	0.3002

Compared to last year's DisTEMIST task, which had a very similar setting, results are overall a bit higher but still quite similar. In terms of named entity recognition methodologies, transformers and BERT-like models were the most popular in both tasks, with RoBERTa not only being the most widely used but also achieving some of the best results. In the entity linking sub-task systems that use SapBERT seem to have gained popularity, being used by at least 3 teams, including the top-scoring system, with very good results. In contrast, in

Table 11. Results of MedProcNER Entity Linking sub-task. The best result is bolded, and the second-best is underlined.

Team Name	Run name	P	R	F1
Vicomtech	run1-xlm_roberta	**0.5902**	0.5525	**0.5707**
Vicomtech	run2-roberta_bio	<u>0.5665</u>	<u>0.5627</u>	<u>0.5646</u>
Vicomtech	run3-roberta_bio	0.5662	0.5625	0.5643
Vicomtech	run5-longformer_base	0.5498	**0.558**	0.5539
Fusion	run4-Spanish-RoBERTa	0.5377	0.5362	0.5369
Fusion	run1-BioMBERT	0.5432	0.516	0.5293
Fusion	run3-XLM-RoBERTA	0.5332	0.5235	0.5283
SINAI	run1-fine-tuned-roberta	0.531	0.5224	0.5267
Vicomtech	run4-roberta_bio	0.5248	0.5213	0.523
Fusion	run2-BioMBERT	0.5332	0.5105	0.5216
Fusion	run5-Adapted-ALBERT	0.5461	0.4939	0.5187
SINAI	run2-lstmcrf-512	0.5455	0.4936	0.5183
SINAI	run5-lstm-BIO	0.5352	0.4898	0.5115
SINAI	run4-fulltext-LSTM	0.5173	0.5047	0.5109
SINAI	run3-fulltext-GRU	0.5079	0.4884	0.498
KFU NLP Team	predicted_task2	0.3917	0.4033	0.3974
Onto-NLP	run1-pharmaconer-top1	0.2742	0.508	0.3562
Onto-NLP	run1-pharmaconer-voter	0.2723	0.5044	0.3536
Onto-NLP	run1-cantemist-top1	0.2642	0.4895	0.3432
Onto-NLP	run1-ehr-top1	0.263	0.4873	0.3416
BIT.UA	run4-everything	0.3211	0.3126	0.3168
BIT.UA	run3-PlanTL-dense	0.3188	0.3145	0.3166
BIT.UA	run0-lc-dense-5-wVal	0.318	0.3126	0.3153
BIT.UA	run1-lc-dense-5-full	0.3143	0.3121	0.3132
BIT.UA	run2-lc-bilstm-all-wVal	0.3133	0.3087	0.311
University Regensburg	run2-gpt-4	0.4304	0.1282	0.1976
University Regensburg	run1-gpt-3.5-turbo	0.4051	0.0749	0.1264

last year's DisTEMIST only one team (HPI-DHC) used it, and actually achieved the best entity linking score using an ensemble of SapBERT and TF-IDF with re-ranking and a training data lookup.

Table 12. Results of MedProcNER Indexing sub-task. The best result is bolded, and the second-best is underlined.

Team Name	Run name	P	R	F1
Vicomtech	run5_roberta_bio	0.619	**0.6295**	**0.6242**
Vicomtech	run4_xlm_roberta	**0.6371**	0.6109	0.6239
Vicomtech	run1_roberta_bio	0.6182	**0.6295**	0.6238
Vicomtech	run3_longformer	0.6039	0.6288	0.6161
Vicomtech	run2_roberta_bio	0.5885	0.5917	0.5901
KFU NLP Team	predicted_task3	0.4805	0.5054	0.4927
BIT.UA	run3-PlanTL-dense	0.3544	0.3654	0.3598
BIT.UA	run4-everything	0.3551	0.3619	0.3585
BIT.UA	run0-lc-dense-5-wVal	0.3517	0.3619	0.3567
BIT.UA	run1-lc-dense-5-full	0.3475	0.3612	0.3542
BIT.UA	run2-lc-bilstm-all-wVal	0.3484	0.3593	0.3537
University Regensburg	run2-gpt-4	0.5266	0.1811	0.2695
University Regensburg	run1-gpt3.5-turbo	0.506	0.1083	0.1785

5   Conclusions

This paper provides an overview of the eleventh BioASQ challenge. This year, the challenge consisted of three tasks: (1) Task 11b on biomedical semantic question answering in English and (2) task Synergy 11 on question answering for developing problems, both already established from previous years of the challenge, and (3) the new task MedProcNER on retrieving medical procedure information from medical content in Spanish.

The preliminary results for task 11b reveal some improvements in the performance of the top participating systems, mainly for yes/no and factoid answer generation. However, room for improvement is still available, particularly for factoid and list questions, where the performance is less consistent.

The new edition of the Synergy task in an effort to enable a dialogue between the participating systems with biomedical experts revealed that state-of-the-art systems, despite still having room for improvement, can be a useful tool for biomedical experts that need specialized information for addressing open questions in the context of several developing problems.

The new task MedProcNER introduced three new challenging subtasks on annotating clinical case reports in Spanish. Namely, Named Entity Recognition, Entity Linking, and Semantic Indexing for medical procedures. Due to the importance of semantic interoperability across data sources, SNOMED CT was the target terminology employed in this task, and multilingual annotated resources have been released. This novel task on medical procedure information indexing in Spanish highlighted the importance of generating resources to develop and evaluate systems that (1) effectively work in multilingual and non-English scenarios and (2) combine heterogeneous data sources.

The ever-increasing focus of participating systems on deep neural approaches, already apparent in previous editions of the challenge, is also observed this year. Most of the proposed approaches built on state-of-the-art neural architectures (BERT, PubMedBERT, BioBERT, BART etc.) adapted to the biomedical domain and specifically to the tasks of BioASQ. This year, in particular, several teams investigated approaches based on Generative Pre-trained Transformer (GPT) models for the BioASQ tasks.

Overall, several systems managed competitive performance on the challenging tasks offered in BioASQ, as in previous versions of the challenge, and the top performing of them were able to improve over the state-of-the-art performance from previous years. BioASQ keeps pushing the research frontier in biomedical semantic indexing and question answering for eleven years now, offering both well-established and new tasks. Lately, it has been extended beyond the English language and biomedical literature, with the tasks MESINESP [11], DisTEMIST [21], and this year with MedProcNER. In addition, BioASQ reaches a more and more broad community of biomedical experts that may benefit from the advancements in the field. This has been done initially for COVID-19, through the introductory versions of Synergy, and was later extended into more topics with the collaborative batch of task 10b and the extended version of Synergy 11, introduced this year. The future plans for the challenge include a further extension of the benchmark data for question answering through a community-driven process, extending the community of biomedical experts involved in the Synergy task, as well as extending the resources considered in the BioASQ tasks, both in terms of documents types and language.

Acknowledgments. Google was a proud sponsor of the BioASQ Challenge in 2022. The eleventh edition of BioASQ is also sponsored by Ovid. Atypon Systems Inc. is also sponsoring this edition of BioASQ. The MEDLINE/PubMed data resources considered in this work were accessed courtesy of the U.S. National Library of Medicine. BioASQ is grateful to the CMU team for providing the *exact answer* baselines for task 11b, as well as to Georgios Moschovis and Ion Androutsopoulos, from the Athens University of Economics and Business, for providing the *ideal answer* baselines. The MedProcNER track was partially funded by the Encargo of Plan TL (SEDIA) to the Barcelona Supercomputing Center. Due to the relevance of medical procedures for implants/devices specially in the case cardiac diseases this project is also supported by the European Union's Horizon Europe Coordination & Support Action under Grant Agreement No 101058779 (BIOMATDB) and DataTools4Heart Grant Agreement No. 101057849. We also acknowledge the support from the AI4PROFHEALTH project (PID2020-119266RA-I00).

References

1. Aksenova, A., Asamov, T., Boytcheva, S., Ivanov, P.: Improving Biomedical Question Answering with Sentence-Based Ranking at BioASQ-11b (2023)

2. Aksenova, A., Ivanov, P., Asamov, T., Boytcheva, S.: Leveraging biomedical ontologies for clinical procedures recognition in Spanish at BioASQ MedProcNER. In: Working Notes of CLEF 2023 - Conference and Labs of the Evaluation Forum (2023)

3. Almeida, T., Jonker, R.A.A., Poudel, R., Silva, J.M., Matos, S.: BIT.UA at MedProcNER: discovering medical procedures in Spanish using transformer models with MCRF and augmentation. In: Working Notes of CLEF 2023 - Conference and Labs of the Evaluation Forum (2023)

4. Almeida, T., Jonker, R.A.A., Poudel, R., Silva, J.M., Matos, S.: Two-stage IR with synthetic training and zero-shot answer generation at BioASQ 11 (2023)

5. Ateia, S.: Is ChatGPT a Biomedical Expert? - Exploring the Zero-Shot Performance of Current GPT Models in Biomedical Tasks (2023)

6. Baldwin, B., Carpenter, B.: Lingpipe (2003). Available from World Wide Web: http://alias-i.com/lingpipe

7. Balikas, G., et al.: Evaluation framework specifications. Project deliverable D4.1, UPMC (2013)

8. Chizhikova, M., Collado-Montañez, J., Díaz-Galiano, M.C., Ureña-López, L.A., Martín-Valdivia, M.T.: Coming a long way with pre-trained transformers and string matching techniques: clinical procedure mention recognition and normalization. In: Working Notes of CLEF 2023 - Conference and Labs of the Evaluation Forum (2023)

9. Croce, D., Castellucci, G., Basili, R.: GAN-BERT: generative adversarial learning for robust text classification with a bunch of labeled examples. In: Proceedings of the 58th Annual Meeting of the Association for Computational Linguistics, pp. 2114–2119 (2020)

10. Galat, D., Rizoiu, M.A.: Enhancing Biomedical Text Summarization and Question-Answering: On the Utility of Domain-Specific Pre-training (2023)

11. Gasco, L., et al.: Overview of BioASQ 2021-MESINESP track. Evaluation of advance hierarchical classification techniques for scientific literature, patents and clinical trials (2021)

12. Hsueh, C.Y., Zhang, Y., Lu, Y.W., Han, J.C., Meesawad, W., Tsai, R.T.H.: NCU-IISR: Prompt Engineering on GPT-4 to Stove Biological Problems in BioASQ 11b Phase B (2023)

13. Kim, H., Hwang, H., Lee, C., Seo, M., Yoon, W., Kang, J.: Exploration of Various Techniques in Biomedical Question Answering: From Pre-processing to GPT-4 (2023)

14. Krithara, A., Nentidis, A., Bougiatiotis, K., Paliouras, G.: BioASQ-QA: a manually curated corpus for biomedical question answering. Sci. Data 10(1), 170 (2023)

15. Krithara, A., Nentidis, A., Paliouras, G., Krallinger, M., Miranda, A.: BioASQ at CLEF2021: large-scale biomedical semantic indexing and question answering. In: Hiemstra, D., Moens, M.-F., Mothe, J., Perego, R., Potthast, M., Sebastiani, F. (eds.) ECIR 2021. LNCS, vol. 12657, pp. 624–630. Springer, Cham (2021). https://doi.org/10.1007/978-3-030-72240-1_73

16. Lesavourey, M., Hubert, G.: BioASQ 11B: integrating domain specific vocabulary to BERT-based model for biomedical document ranking (2023)

17. Lima-López, S., et al.: Overview of MedProcNER task on medical procedure detection and entity linking at BioASQ 2023. In: Working Notes of CLEF 2023 - Conference and Labs of the Evaluation Forum (2023)

18. Liu, F., Shareghi, E., Meng, Z., Basaldella, M., Collier, N.: Self-alignment pre-training for biomedical entity representations. In: Proceedings of the 2021 Conference of the North American Chapter of the Association for Computational

Linguistics: Human Language Technologies, pp. 4228–4238. Association for Computational Linguistics (2021). https://doi.org/10.18653/v1/2021.naacl-main.334. https://aclanthology.org/2021.naacl-main.334

19. Liu, Y., et al.: RoBERTa: a robustly optimized BERT pretraining approach. arXiv preprint arXiv:1907.11692 (2019)

20. Miranda-Escalada, A., Farré-Maduell, E., Lima-López, S., Estrada, D., Gascó, L., Krallinger, M.: Mention detection, normalization & classification of species, pathogens, humans and food in clinical documents: overview of LivingNER shared task and resources. Procesamiento del Lenguaje Natural (2022)

21. Miranda-Escalada, A., et al.: Overview of DISTEMIST at BioASQ: automatic detection and normalization of diseases from clinical texts: results, methods, evaluation and multilingual resources (2022)

22. Mollá, D.: Query-focused extractive summarisation for biomedical and Covid-19 complex question answering. In: 2022 Conference and Labs of the Evaluation Forum, CLEF 2022, pp. 305–314 (2022)

23. Nentidis, A., et al.: Overview of BioASQ 2021: the ninth BioASQ challenge on large-scale biomedical semantic indexing and question answering. In: Candan, K.S., et al. (eds.) CLEF 2021. LNCS, vol. 12880, pp. 239–263. Springer, Cham (2021). https://doi.org/10.1007/978-3-030-85251-1_18

24. Nentidis, A., et al.: Overview of BioASQ 2022: the tenth BioASQ challenge on large-scale biomedical semantic indexing and question answering. In: Barrón-Cedeño, A., et al. (eds.) CLEF 2022. LNCS, vol. 13390, pp. 337–361. Springer, Cham (2022). https://doi.org/10.1007/978-3-031-13643-6_22

25. Nentidis, A., et al.: Overview of BioASQ 2020: the eighth BioASQ challenge on large-scale biomedical semantic indexing and question answering. In: Arampatzis, A., et al. (eds.) CLEF 2020. LNCS, vol. 12260, pp. 194–214. Springer, Cham (2020). https://doi.org/10.1007/978-3-030-58219-7_16

26. Nentidis, A., Krithara, A., Paliouras, G., Farre-Maduell, E., Lima-Lopez, S., Krallinger, M.: BioASQ at CLEF2023: the eleventh edition of the large-scale biomedical semantic indexing and question answering challenge. In: Kamps, J., et al. (eds.) ECIR 2023. LNCS, vol. 13982, pp. 577–584. Springer, Cham (2023). https://doi.org/10.1007/978-3-031-28241-6_66

27. Nentidis, A., Krithara, A., Paliouras, G., Gasco, L., Krallinger, M.: BioASQ at CLEF2022: the tenth edition of the large-scale biomedical semantic indexing and question answering challenge. In: Hagen, M., et al. (eds.) ECIR 2022. LNCS, vol. 13186, pp. 429–435. Springer, Cham (2022). https://doi.org/10.1007/978-3-030-99739-7_53

28. Ozyurt, I.B.: End-to-end biomedical question answering via bio-answerfinder and discriminative language representation models (2021)

29. Panou, D., Reczko, M.: Semi-supervised training for biomedical question answering (2023)

30. R., R., Rauchwerk, J., Rajwade, P., Gummadi, T.: Biomedical Question Answering using Transformer Ensembling (2023)

31. Reczko, M.: ELECTROLBERT: combining replaced token detection and sentence order prediction. In: CLEF (Working Notes) (2022)

32. Rosso-Mateus, A., Montes-Y-Gómez, M., Munoz Serna, L.A., Gonzalez, F.: Deep Metric Learning for Effective Passage Retrieval in the BioASQ Challenge (2023)

33. Shin, A.D., Jin, Q., Lu, Z.: Multi-stage Literature Retrieval System Trained by PubMed Search Logs for Biomedical Question Answering (2023)

34. Tsatsaronis, G., et al.: An overview of the BioASQ large-scale biomedical semantic indexing and question answering competition. BMC Bioinform. **16**, 138 (2015). https://doi.org/10.1186/s12859-015-0564-6
35. Vassileva, S., Grazhdanski, G., Boytcheva, S., Koychev, I.: Fusion @ BioASQ MedProcNER: transformer-based approach for procedure recognition and linking in Spanish clinical text. In: Working Notes of CLEF 2023 - Conference and Labs of the Evaluation Forum (2023)
36. Wei, C.H., Leaman, R., Lu, Z.: Beyond accuracy: creating interoperable and scalable text-mining web services. Bioinformatics (Oxford, England) **32**(12), 1907–10 (2016). https://doi.org/10.1093/bioinformatics/btv760
37. Yang, Z., Zhou, Y., Eric, N.: Learning to answer biomedical questions: OAQA at BioASQ 4b. In: ACL 2016, p. 23 (2016)
38. Zotova, E., García-Pablos, A., Cuadros, M., Rigau, G.: VICOMTECH at MedProcNER 2023: transformers-based sequence-labelling and cross-encoding for entity detection and normalisation in Spanish clinical texts. In: Working Notes of CLEF 2023 - Conference and Labs of the Evaluation Forum (2023)

Overview of the CLEF–2023 CheckThat! Lab on Checkworthiness, Subjectivity, Political Bias, Factuality, and Authority of News Articles and Their Source

Alberto Barrón-Cedeño[1]([✉])[ID], Firoj Alam[2][ID], Andrea Galassi[1][ID],
Giovanni Da San Martino[3][ID], Preslav Nakov[4][ID], Tamer Elsayed[5][ID],
Dilshod Azizov[5][ID], Tommaso Caselli[6][ID], Gullal S. Cheema[7][ID],
Fatima Haouari[5][ID], Maram Hasanain[2][ID], Mucahid Kutlu[8][ID], Chengkai Li[9][ID],
Federico Ruggeri[1][ID], Julia Maria Struß[10][ID], and Wajdi Zaghouani[11][ID]

[1] Università di Bologna, Bologna, Italy
a.barron@unibo.it
[2] Qatar Computing Research Institute, Ar-Rayyan, Qatar
[3] University of Padua, Padua, Italy
[4] Mohamed bin Zayed University of Artificial Intelligence, Masdar, UAE
[5] Qatar University, Doha, Qatar
[6] University of Groningen, Groningen, The Netherlands
[7] L3S Research Center, Leibniz University of Hannover, Hannover, Germany
[8] TOBB University of Economics and Technology, Ankara, Turkey
[9] University of Texas at Arlington, Arlington, USA
[10] University of Applied Sciences Potsdam, Potsdam, Germany
[11] Hamad bin Khalifa University, Ar-Rayyan, Qatar
https://checkthat.gitlab.io

Abstract. We describe the sixth edition of the CheckThat! lab, part of the 2023 Conference and Labs of the Evaluation Forum (CLEF). The five previous editions of CheckThat! focused on the main tasks of the information verification pipeline: check-worthiness, verifying whether a claim was fact-checked before, supporting evidence retrieval, and claim verification. In this sixth edition, we zoom into some new problems and for the first time we offer five tasks in seven languages: Arabic, Dutch, English, German, Italian, Spanish, and Turkish. Task 1 asks to determine whether an item —text or text plus image— is check-worthy. Task 2 aims to predict whether a sentence from a news article is subjective or not. Task 3 asks to assess the political bias of the news at the article and at the media outlet level. Task 4 focuses on the factuality of reporting of news media. Finally, Task 5 looks at identifying authorities in Twitter that could help verify a given target claim. For a second year, CheckThat! was the most popular lab at CLEF-2023 in terms of team registrations: 127 teams. About one-third of them (a total of 37) actually participated.

Keywords: Fact Checking · Check-Worthiness · Subjectivity · Political Bias · Factuality of Reporting · Authority Finding

© The Author(s), under exclusive license to Springer Nature Switzerland AG 2023
A. Arampatzis et al. (Eds.): CLEF 2023, LNCS 14163, pp. 251–275, 2023.
https://doi.org/10.1007/978-3-031-42448-9_20

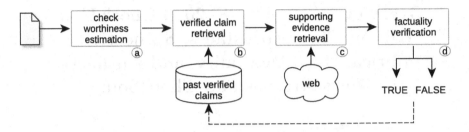

Fig. 1. The `CheckThat`! verification pipeline, featuring the four core tasks. Task 1 on check-worthiness this year is the only one that belongs to these core tasks.

1   Introduction

From its conception, the `CheckThat`! lab has been dedicated to promoting and fostering the development of technology to assist investigative journalists to perform fact-checking, focusing on political debates, social media posts, and news articles. The five previous editions of the lab have been held annually from 2018 to 2022, targeting diverse Natural Language Processing (NLP) and Information Retrieval (IR) tasks, part of the `CheckThat`! pipeline [19,20,36,37,67–69,72,73] is shown in Fig. 1.

For the first time, `CheckThat`! 2023 [18] zooms out of the core pipeline and focuses on *auxiliary* tasks that help in addressing the different steps of the pipeline. For that, it challenged the research community with five tasks in seven languages: Arabic, Dutch, English, German, Italian, Spanish, and Turkish. Task 1 [2], the only one that follows up from previous editions and the only one that is part of the core pipeline, asks systems to find out whether a given claim in a tweet is worth fact-checking. This year, for the first time, Task 1 offers a multimodal track. Task 2 [41] requires to determine whether a sentence from a news article is objective or conveys a subjective point of view, influenced by personal feelings, tastes, or opinions. Task 3 [29] asks systems to measure the level of political bias of news reporting at the article and at the media level. Task 4 [65] asks to assess the factuality of reporting at the news media level. Task 5 [46] challenges models to retrieve a set of authority Twitter accounts for a given rumor propagating in Twitter.

Task 1 is what professionals take the most advantage of, since the amount of information online is impossible for one to keep up with. Task 2 helps check-worthiness, by spotting opinionated snippets that might not be relevant for fact-checking. Task 3 could help factuality verification by contributing information about both the stance of a claim and a piece of evidence. Task 4 could help to determine whether the information from a news outlet can be trusted a priori. Finally, Task 5 could help in identifying people/institutions that can challenge a claim. Table 1 showcases the language coverage and the type of documents included in this year's tasks.

Table 1. Overview of the 2023 tasks: language coverage and type of document.

task	ar	de	du	en	es	it	tr	documents
Task 1	■			■	■			tweets (incl. multimodal), debates and speeches
Task 2	■	■	■	■		■	■	news articles and tweets
Task 3				■				news articles
Task 4				■				news articles
Task 5	■							tweets

Table 2. Overview of the tasks offered in the previous editions of the lab.

tasks	years					domains					languages						
	2018	2019	2020	2021	2022	debates	speeches	tweets	web pages	news articles	English	Arabic	Bulgarian	Spanish	Turkish	German	Dutch
check-worthiness estimation	■	■	■	■	■	■	■	■			■	■	■	■	■		■
verified claim retrieval			■	■	■	■	■	■			■	■					
supporting evidence retrieval		■	■						■		■						
claim verification	■	■	■								■	■					
fake news detection				■	■					■	■						■
topic identification					■						■						

2 Previously on the CheckThat! Lab

During the previous five iterations of the CheckThat! lab, it has focused on various tasks from the claim verification pipeline, in a multitude of languages and in different domains (cf. Table 2).

The first iteration of CheckThat! in 2018 [10, 21] focused on check-worthiness and claim verification of political debates and speeches in Arabic and English. Both tasks were then continued in the following iteration, with an additional focus on fact-checking by a task on classifying and ranking supporting evidence from the web [11, 37, 48]. The 2020 edition [20] of the CheckThat! lab covered the full claim verification pipeline, with check-worthiness estimation, verified claim and supporting evidence retrieval, and claim verification; social media data was first included in that iteration of the lab [47, 91]. The fourth edition of the lab in 2021 put focus on multilinguality by offering tasks in five languages [73, 89, 90]. The edition also featured a new fake news detection task [92], where the focus was not on a claim, but on an article; this task was quite popular and it was continued in the 2022 edition of the CheckThat! lab.

The 2022 year year's edition of the CheckThat! lab [67] paid special attention to the various sub-aspects of check-worthiness estimation, namely factuality, harmfulness, and attention-worthiness estimation, again in a multitude of languages. Transformer-based models were extensively used.

The highest-ranking systems additionally implemented data augmentation and supplementary preprocessing measures [66]. The second task in the 2022 edition of the lab asked to detect previously fact-checked claims from tweets, political debates, and speeches [71]. The best system used the Sentence T5 transformer and GPT-Neo models. The third task in the 2022 edition of the CheckThat! lab asked to predict the veracity of the main claim in an English news article, with English or German training data. The most successful approaches fine-tuned a BERT-based model. The cross-language nature of the task has mainly been addressed using machine translation [54].

3   Description of the 2023 Tasks

The 2023 edition of the CheckThat! lab is organized around five tasks, four of which are run for the first time (cf. Sects. 3.2 to 3.5). Moreover, two tasks have two subtasks (cf. Sects. 3.1 and 3.3).

3.1   Task 1: Check-Worthiness in Multimodal and Multigenre Content

The goal of this task is to assess whether a given statement, in a tweet or from a political debate, is worth fact-checking [2]. In order to make that decision, one would need to ponder about questions, such as "does it contain a verifiable factual claim?" or "is it harmful?", before deciding on the final check-worthiness label [4]. Task 1 is divided into two subtasks. Subtask 1A is offered in Arabic and English, whereas subtask 1B is offered in Arabic, English and Spanish.

Subtask 1A: Multimodality. Given a tweet with the text and its corresponding image, predict whether it is worth fact-checking. Here, answers to the questions relevant for deriving a label are based on both the image and the text. The image plays two roles for check-worthiness estimation: (i) there is a piece of evidence (e.g., an event, an action, a situation, a person's identity, etc.) or illustration of certain aspects from the textual claim, and/or (ii) the image contains overlaid text that contains a claim (e.g. misrepresented facts and figures) in a textual form.

Subtask 1B: Multigenre. A text snippet alone, either from a tweet or from a political debate or speech, has to be assessed for check-worthiness.

3.2   Task 2: Subjectivity in News Articles

Given a sentence from a news article or a tweet (in the case of Turkish), Task 2 asks systems to determine whether the sentence is subjective or objective [41]. A sentence is **subjective** if its contents are based on or influenced by personal feelings, tastes, or opinions; otherwise, it is considered **objective**. This task pays the most attention to multilinguality this year. It is offered in Arabic, Dutch, English, Italian, German, and Turkish, with an additional multilingual setting.

3.3   Task 3: Political Bias of News Articles and News Media

The goal of the task is to detect political bias of news reporting at the article and at the media level. This is an ordinal classification task and it is offered in English [29]. It includes two subtasks:

Subtask 3A: Political Bias of News Articles. Given an article, classify its leaning as left, center, or right.

Subtask 3B: Political Bias of News Media. Given a news outlet, predict its overall political bias as left, center, or right.

3.4   Task 4: Factuality of Reporting of News Media

In this task, we specifically target media credibility. The goal is to predict the factuality of reporting at the media level, given a set of articles from the target news outlet: low, mixed, and high. This is another ordinal classification task, and it is offered in English only [65].

3.5   Task 5: Authority Finding in Twitter

The task asks systems to retrieve authority Twitter accounts for a given rumor that propagates in Twitter [46]. Given a tweet spreading a rumor, the participating systems need to retrieve a ranked list of authority Twitter accounts that can help verify that rumor, as such accounts may tweet evidence that supports or denies the rumor [43]. This task is offered in Arabic.

4   Datasets

4.1   Task 1: Check-Worthiness of Multimodal and Multigenre Content

Subtask 1A: Multimodality. The dataset used for subtask 1A was derived from [25], with the existing data repurposed for training and development purposes. We followed the schema from [25] to produce a new testing set.

The dataset focused on three topics: COVID-19, climate change and technology. Each tweet was labeled using both the image and the text, with Optical Character Recognition (OCR) performed using the Google Vision API to extract the text from the images. We provided 3,175 annotated examples and around 110k unlabeled tweets of text–image pairs and OCR output. Two annotators, one of them an expert, annotated the new test data. The non-expert went through a dry run of 50 examples, where disagreements were discussed and resolved. For the final test set of 736 examples, the Cohen's Kappa inter-annotator agreement [27] was 0.49 for the check-worthy label, indicating moderate agreement. The expert annotator resolved any remaining disagreements.

Table 3. Task 1: Check-worthiness in multimodal and multigenre content.
Statistics about the CT–CWT–23 corpus for all three languages.

Subtask	Class	Train	Dev	Dev-Test	Test	Total
1A Arabic	No	1,421	207	402	792	2,822
	Yes	776	113	220	203	1,312
	Total	2,197	320	622	995	**4,134**
1A English	No	1,536	184	374	459	2,553
	Yes	820	87	174	277	1,358
	Total	2,356	271	548	736	**3,911**
1B Arabic	No	4,301	789	682	123	5,895
	Yes	1,758	485	411	377	3,031
	Total	6,059	1,274	1,093	500	**8,926**
1B English	No	12,818	4,270	794	210	18,092
	Yes	4,058	1,355	238	108	5,759
	Total	16,876	5,625	1,032	318	**23,851**
1B Spanish	No	14,805	2,157	4,190	4,491	25,643
	Yes	2,682	391	759	509	4,341
	Total	17,487	2,548	4,949	5,000	**29,984**

For subtask 1A Arabic, we followed several steps for training, development, dev-test, and test datasets. For the former three partitions, we used the CT–CWT–21 [90] and the CT–CWT–22 [66] datasets annotated for check-worthiness with topics focusing on COVID-19 and politics. The labelling of the datasets follows the annotation schema discussed in [3,4]. To develop multimodal datasets based on these datasets, we crawled images associated with tweets. For tweets with multiple images, we retrieved only the first one. For the former three partitions, we derived the label for multimodality from the textual modality, and thus these can be seen as weakly labeled annotations. For the test set, we crawled tweets using similar keywords to those reported in [3,4]. For the annotation, three annotators followed the same annotation schema, but for multimodality We used majority voting to select the final labels.

Subtask 1B: Multigenre. The dataset for Subtask 1B consists of tweets in Arabic and Spanish as well as statements from political debates in English. The Arabic tweets are collected using keywords related to COVID-19 and vaccines, using the annotation schema in [4]. The training, the development, and the dev-test partitions of the dataset come from CT–CWT–21 [90] and CT–CWT–22 [66]. For the test set, we used the same approach as for subtask 1A. The dataset for English consists of transcribed sentences from candidates during the US presidential election debates and annotated by human annotators [9]. For the first three partitions, we used essentially the same dataset reported in [9], with some updates that reflect improved annotation accuracy. The test set contains sentences that were not included in [9].

Table 4. Task 2: Subjectivity in news articles. Statistics about the datasets for all six languages and the multilingual setting, and the distribution of objective (Obj) and subjective (Subj) examples.

Language	Training		Dev		Test		Total
	Obj	Subj	Obj	Subj	Obj	Subj	
Arabic	905	280	227	70	363	82	**1,927**
Dutch	489	311	107	93	263	237	**1,500**
English	532	298	106	113	116	127	**1,292**
German	492	308	123	77	194	97	**1,291**
Italian	1,231	382	167	60	323	117	**2,280**
Turkish	422	378	100	100	129	111	**1,140**
Multilingual	4,371	2,257	300	300	300	300	**7,828**

The Spanish dataset is also a combination of CT-CWT-21 [90], CT-CWT-22 [66] and newly collected content. It is composed of tweets collected from Twitter accounts and transcriptions from Spanish politicians, which are manually annotated by professional journalists who are experts in fact-checking.

Table 3 shows statistics about the datasets for Task 1. Across the different subtasks, dataset sizes range from 3,911 to 29,984, which are the largest so far across different languages over the years for the check-worthiness task.

4.2 Task 2: Subjectivity in News Articles

The datasets for all languages in Task 2 were produced on the basis of the subjectivity identification guidelines outlined in [7]. The sentences were extracted from news articles, with the exception of Turkish, in which each sentence is manually extracted from tweets about politics. Table 4 shows the label distribution. The training set of the multilingual dataset is the union of the training material from all languages. The development and the testing sets, on the other hand, are formed by randomly selecting 50 subjective and 50 objective sentences from the respective development and testing sets in all languages. In total, we annotated 9,351 sentences covering six languages.

4.3 Task 3: Political Bias of News Articles and News Media

Table 5 reports the label distribution of the datasets for Task 3.

For Subtask 3A. we release a collection of 55k articles from 1,023 media sources annotated for bias at the article level. The articles were crawled from AllSides.[1] To make sure the data is up to date and pertinent to the present political environment, the dataset includes news articles published from the end of 2022 to the beginning of 2023.

---

[1] https://www.allsides.com.

Table 5. Task 3: Political bias of news articles and news media. Statistics about the CT–Bias–23 datasets [29].

Class	Train	Dev	Test	Total		Class	Train	Dev	Test	Total
left	12,073	1,342	2,589	**16,004**		left	216	31	25	**272**
center	15,449	1,717	1,959	**19,125**		center	296	34	29	**359**
right	17,544	1,949	650	**20,143**		right	305	39	48	**392**
	19,125	**16,044**	**20,143**	**55,272**			**817**	**104**	**102**	**1,023**

<table>
<tr><td align="center">Task 3A: Bias of Articles</td><td align="center">Task 3B: Bias of News Media</td></tr>
</table>

Table 6. Task 4: Factuality of reporting of news media. Statistics about the CT–Factuality–23 dataset [65].

Class	Train	Dev	Test	Total
High	233	32	31	**296**
Mixed	593	72	72	**737**
Low	121	16	19	**156**
	947	**120**	**122**	**1,189**

AllSides curates articles from a variety of reputable national and international news sources to ensure a balanced representation across different political spectra. The articles are annotated following a rigorous scheme that involves expert reviewers. For the data split, we divide them into 80%, 10% and 10% for the training, development, and test, respectively.

For Subtask 3B. Our dataset is sourced from Media Bias/Fact Check,[2] which follows a meticulous approach to characterize media sources, conducted by experts. This dataset contains a subset of data used in previous research [15]. Note that we remap the bias from a 7-point scale (extreme-left, left, center-left, center, center-right, right, and extreme-right) to a 3-point scale: left, center, and right (for this, we exclude center-left and center-right). For the data split, we divide them by news media as the same splits as subtask 3A, but we randomly select up to 11 news articles from each news medium. Finally, we release annotated articles for each medium, which are to be used for the classification.

4.4   Task 4: Factuality of Reporting of News Media

We use the same kind of data as for Task 3, but with labels for factuality (again on an ordinal scale). We obtain the annotations and the analysis of the factuality of reporting from Media Bias/Fact Check, where they are manually labeled by professional fact-checkers. We use a 3-point scale: low, mixed, and high factuality.

---

[2] https://www.mediabiasfactcheck.com.

Table 7. Task 5: Authority finding. Rumor collection and relevance judgments statistics.

Data split	Rumors	Authorities
Training	120	849
Development	30	195
Testing	30	172
Total	**180**	**1,216**

The dataset consists of 1,189 news media: see Table 6 for detailed statistics. For each new medium, we include approximately ten articles.

4.5   Task 5: Authority Finding in Twitter

For training, we adopted the AuFIN [44] collection, which comprises 150 rumors expressed in tweets, associated with 1,044 authority Twitter accounts, and a user collection of 395,231 accounts along with their Twitter lists (1,192,284 unique lists). Each authority is graded as *highly relevant* or *relevant* to the rumor, i.e., having a higher priority to be contacted for verification or not. The rumors cover three categories: politics, sports, and health; 50 from each category. We split the rumors into 120 for training and 30 for development.

For testing, we collected 30 new rumors from AraFacts [5], where we focused on the ones collected from Misbar[3] and Fatabyyano,[4] which were used recently to construct Arabic rumor verification [45] and fake news detection [53] datasets. We selected 10 rumors from each of the three categories. For each rumor, two annotators separately identified all authority Twitter accounts that can help support or debunk the rumor following the same annotation guidelines as for AuFIN. The Cohen's Kappa inter-annotator agreement [27] was 0.91 and 0.42 for the authority label and the graded relevance, which correspond to almost perfect and moderate agreements, respectively [56]. Table 7 presents the overall statistics about the rumor collection and the relevance judgments for each data split. For an overall summary of the user collection, we refer the reader to [44].

5   Results

In this section, we present the top-performing submissions for each of the five tasks. For details about all participating approaches, refer to the corresponding task paper: Task 1 [2], Task 2 [41], Task 3 [29], Task 4 [65], and Task 5 [46].

5.1   Task 1: Check-Worthiness in Multimodal and Unimodal Contents

A total of 14 teams participated in Task 1 and submitted 35 runs.

---

[3] https://misbar.com/.
[4] https://fatabyyano.net/.

Table 8. Overview of the approaches for **subtask 1A**. The numbers in the language box show the position of the team in the official ranking; ☑ = part of the official submission; ✔ = considered in internal experiments.

Team	Arabic	English	BERT	RoBERTa	mBERT	XLM RoBERTa	AraBERT	BERTweet	GPT-3	Electra	Vision Transformer	ConvNext	ResNet	VGG	EfficientNet	Text	Image	OCR Output	CLIP	LSTM	CatBoost	Preprocessing
CSECU-DSG [12]	1	4				☑	☑	☑			☑					☑	☑				☑	
ES-VRAI [81]	-		☑								✔	☑				☑	☑				✔	
Fraunhofer SIT [40]		1	☑								✔				✔	☑	✔	☑				☑
Mtop*	2	6														☑	☑					☑
Z-Index [97]	3	5	☑			☑	✔				☑					☑	☑					☑
ZHAW-CAI [105]		2	☑	☑				☑		☑	☑					☑	☑					☑

- Run submitted after the deadline.   *No working note submitted.

Table 9. Subtask 1A: Multimodal check-worthiness estimation. Shown are the top-3 submissions for Arabic and English. The F1 score is computed with respect to the positive class.

Team	F1	Team	F1
Arabic		**English**	
1 CSECU-DSG [12]	0.399	1 Fraunhofer SIT [40]	0.712
2 Mtop*	0.312	2 ZHAW-CAI [105]	0.708
3 Z-Index [97]	0.301	- ES-VRAI [81]	0.704

- Run submitted after the deadline.   *No working note submitted.

Subtask-1A. A total of 7 and 4 teams submitted their runs for English and for Arabic, respectively, out of which four made submissions for both languages. Table 8 gives an overview of the submitted models per language. This was a binary classification task, and we used the F1 score for the positive class as the official evaluation measure. Table 9 shows the performance of the top official submissions on the test set.

Starting with the best-performing system for English: team *Fraunhofer SIT* [40] tackled the problem by fine-tuning individual text classifiers on the tweet text and on the OCR text, respectively. They further used pre-processing for the tweet text and extracted the text from images using *easyOCR*.[5] Two BERT [33] models were fine-tuned on each input, and the final label for each example in the test set was a re-weighted combination of the two predictions based on the validation loss.

---

[5] https://github.com/JaidedAI/EasyOCR.

Team *ZHAW-CAI* [105] submitted official runs for the English track only. They trained different unimodal and multimodal systems and then combined them using a kernel-based ensemble. This ensemble was trained using an SVM for classification. For the text-based model, n-gram features were extracted separately from the tweet text, and prompt response from GPT-3 (Open AI's text-davinci-003), and SVMs were trained on these features. In addition, an Electra [26] model was fine-tuned over the tweet text for classification. For the multimodal model, features from Twitter-based RoBERTa [59] and ViT were extracted, fused via pooling, and passed through a dense layer for classification. The submission model is an ensemble of the four features described earlier with their individual kernels and combined with an average kernel to be used in an SVM for classification.

Team *ES-VRAI* [81] comprehensively evaluated several pre-trained vision and text models, different classifiers, and several early and late fusion strategies to select the best model for the English data. Their submitted model combined BERT and ResNet50 [49] features in an early fusion mode.

Team *CSECU-DSG* [12] participated in both the Arabic and the English tracks. They used a model that jointly fine-tunes two transformers. A language-specific BERT is used to represent the tweet text, and ConvNext [61] is used for image feature extraction. They uses BERTweet [74] for English data, and AraBERT [8] for Arabic. In addition, a BiLSTM was used on top of the text features to handle long-term contextual dependency. Finally, the features from the BiLSTM and the ConvNext were concatenated and followed up by a multisample dropout [51] to predict the final label.

Team *Z-Index* [97] also participated in both languages. They used BERT for English tweet text and ResNet50 for images, and a feed-forward neural network for fusion and classification. In addition, mBERT [33] was used for Arabic text. The backbone networks were fine-tuned along with the feed-forward network to train the model for the task. In their internal evaluation, they also experimented with XLM-Roberta [28], which performed better by 4% than the BERT variant for both languages.

To summarize all the contributions of the participating teams: one common theme across the methods was the use of large pre-trained models and their features for semantic information extraction. Only team *Fraunhofer SIT* used two separate classifiers, while the rest used fusion and ensemble strategies. Two of the teams further used OCR.

Subtask-1B. A total of 11, 6, and 7 teams submitted their runs for English, Arabic, and Spanish, respectively, out of which 6 teams submitted runs for all languages. Table 10 gives an overview of the submitted systems per language. This was a binary classification task, and we used F1 score with respect to the positive class as the official evaluation measure. Table 11 shows the performance of the top official submissions on the test set.

The best-performing team on English is *OpenFact* [85], who fine-tuned GPT-3[6] using 7.7K examples of sentences from debates and speeches annotated for

---

[6] https://platform.openai.com/docs/models/gpt-3.

Table 10. Overview of the systems for **subtask 1B**. The numbers in the language box refer to the position of the team in the official ranking; ☑ = part of the official submission.

Team	English	Arabic	Spanish	BERT	RoBERTa	XLM RoBERTa	MarBERT	AraBERT	BERTweet	BETO	BERTIN	Spanish RoBERTa	GPT	MultinomialNB	LSTM	Data augmentation	Preprocessing
Accenture [99]	3	2	5	☑	☑												☑
CSECU-DSG [12]	7	4	3			☑		☑	☑	☑				☑			
DSHacker [64]	9	5	1			☑						☑				☑	
ES-VRAI [81]	5	1	2	☑	☑		☑										
Fraunhofer SIT [40]	2			☑													☑
OpenFact [85]	1											☑					
Z-Index [97]	6	3	6			☑											☑

Table 11. Subtask 1B: Multigenre (unimodal) check-worthiness estimation. Shown are the top-3 submissions for English debates, and for Arabic and Spanish tweets. The F1 score is calculated with respect to the positive class.

Team	F1	Team	F1	Team	F1
English		Arabic		Spanish	
1 OpenFact [85]	0.898	1 ES-VRAI [81]	0.809	1 DSHacker [64]	0.641
2 Fraunhofer SIT [40]	0.878	2 Accenture [99]	0.733	2 ES-VRAI [81]	0.627
3 Accenture [99]	0.860	3 Z-Index [97]	0.710	3 CSECU-DSG [12]	0.599

check-worthiness, extracted from an already existing dataset [9]. In addition to that, during internal experiments, the team also experimented with fine-tuning a variety of BERT models and found that fine-tuning DeBERTaV3 [50] leads to near-identical performance to that of the model based on GPT-3.

Team *Fraunhofer SIT* [40] fine-tuned a BERT model [33] three times starting with a different seed for model initialization, resulting in three models. The team used ensemble learning using a model souping technique that adaptively adjusts the influence of each individual model based on its performance on the dev subset.

Team *Accenture* [99] also fine-tuned large pre-trained models: RoBERTa [60] for English and GigaBERT for Arabic [55]. They further proposed to extend the training subset with examples resulting from back-translating the same subset using AWS translation.[7]

Team *ES-VRAI* [81] achieved the best and the second best performance for Arabic and for Spanish, respectively. After comprehensive evaluation of several language-specific pre-trained models, their official submission for Arabic was based on fine-tuning MARBERT [1] using the training subset, after downsampling examples from the majority class. Fine-tuned XLM-RoBERTa model was used to produce the official submitted run for the Spanish test set.

Team *Z-Index* [97] participated in all three languages using the same system architecture. Their system includes a feed forward network, where input is represented using embeddings.[8] The network was trained using the training set released per language.

Team *DSHacker* [64] achieved the best performance for Spanish. Their system is based on fine-tuning XLM-RoBERTa [28] using the available train data, and additional datasets obtained by data augmentation. For data augmentation, they used GPT-3.5[9] to translate the input train set to English and to Arabic resulting in two additional training subsets. GPT-3.5 was also used to paraphrase the original Spanish training data, resulting in a third augmented training subset.

Team *CSECU-DSG* [12] also participated in all three languages. Their model includes jointly fine-tuning two transformers: a language-specific BERT and Twitter XLM-RoBERTa [17] to represent the input text. In addition, a BiLSTM module was used on top of the text features to handle long-term contextual dependency. Finally, the features from the BiLSTM were followed by a multisample dropout strategy [51] to produce the final prediction.

5.2 Task 2: Subjectivity in News Articles

Task 2 has seen the participation of 12 teams, with a total of 45 runs. The majority of the participants (seven out of 12) submitted runs for more than one language, with four teams participating in all languages.

Table 12 offers a snapshot of the approaches, whereas Table 13 reports the performance results for the top-three submissions per task, ranked on the basis of macro-averaged F_1 (cf. [41] for the whole ranking, including submissions after the deadline).

All systems used neural networks. Two teams, Fraunhofer SIT [39] and TOBB ETU [31], based their submissions on GPT-3* models, structuring the task via prompts in zero-shot or few-shot settings. All other participants fine-tuned encoder-based Transformers mostly using multi-lingual models (e.g., mBERTaV3, XML-R, and mBERT). Generative models based on the GPT-3* family were mostly used to augment the training data, rather than adopting standard upsampling and downsampling methods.

---

[7] https://aws.amazon.com/translate/.

[8] No enough details were available about the source of these embeddings.

[9] https://platform.openai.com/docs/models/gpt-3-5.

Table 12. Task 2 Overview of the approaches. The numbers in the language box refer to the position of the team in the official ranking.

Team	Multilingual	Arabic	Dutch	German	English	Italian	Turkish	BERT	RoBERTa	XLM RoBERTa	GigaBERT	M-BERT	M-DeBERTa	S-BERT	SetFit	ChatGPT	GPT-3	BART	LSTM	Gradient Boosting	Multi-lingual training	Data augmentation	Feature Selection	Ensemble
Accenture [100]		3	5	7	8	3	4	✓	✓	✓												✓		
Awakened					10											✓	✓							✓
DWReCo [86]				4	1		2	✓	✓								✓					✓		
ES-VRAI [82]	-									✓														
Fraunhofer SIT [39]					5	6											✓							
Gpachov [76]					2					✓			✓	✓										✓
KUCST					4			✓															✓	✓
NN [34]	1	1	2	2	5	2	3	✓														✓		
tarrekko	-	-	-	-	-	-	-																	
Thesis Titan [57]	2	2	1	1	3	1	1								✓							✓		
TOBB ETU [31]	4	5	3	3	9	5	6					✓												
TUDublin [95]					11	6					✓						✓							✓

- Run submitted after the deadline.

Only two teams, Thesis Titan [57] and NN [34] achieved consistently good results across all languages, with a ranking in the top-3 positions. As in previous editions of the CheckThat! lab, using all languages helped to boost the performance.

All the top systems are in the range [0.75, 0.80] in terms of macro-F1, well above the baselines. The only language outlier is Turkish, where the top approach by Thesis Titan [57] achieved an outstanding F1 score of 0.89. In the majority of the languages, the distance between the first and the second system tends to be higher than two points. In the Arabic, English, and multilingual settings, the distances range between 0.05 and 0.01 points. Rather than pointing only to differences in the annotation of the data, this may suggest that some approaches have found optimal language-specific hyper-parameters.

5.3 Task 3: Political Bias of News Articles and News Media

Table 14 shows the results for Task 3, in which four teams participated. Two teams participated in both subtasks, and all teams outperformed the baseline.

Team Accenture [102] used back-translation to augment the minority classes in the datasets that label the article and the news source bias into three cate-

Table 13. Task 2 Top-3 performing models per language.

Team		F1	Team		F1	Team		F1
Multilingual			**English**			**Italian**		
1	NN [34]	81.97	1	DWReCo [86]	78.18	1	Thesis Titan [57]	75.75
2	Thesis Titan [57]	81.00	2	Gpachov [76]	77.34	2	NN [34]	71.01
3	*baseline*	73.56	3	Thesis Titan [57]	76.78	3	Accenture [100]	65.52
Arabic			**German**			**Turkish**		
1	NN [34]	78.75	1	Thesis Titan [57]	81.52	1	Thesis Titan [57]	89.94
2	Thesis Titan [57]	77.53	2	NN [34]	74.13	2	DWReCo [86]	84.11
3	Accenture [100]	72.53	3	TOBB ETU [31]	71.19	3	NN [34]	81.21
Dutch								
† 1	Thesis Titan [57]	81.43						
2	NN [34]	75.57						
3	TOBB ETU [31]	73.01						

† Team involved in the preparation of the data.

Table 14. Task 3: Top-3 performing models when identifying political bias of news articles and news media (MAE score).

Subtask 3A			Subtask 3B		
Team		**MAE**	**Team**		**MAE**
1	Accenture [102]	0.473	1	Accenture [102]	0.549
2	TOBB ETU [31]	0.646	2	Awakened [103]	0.765
3	KUCST	0.736	3	*baseline*	0.902

gories: left, center, and right. Then, they used this augmented data to fine-tune RoBERTa transformer models. Team TOBB ETU [31] explored zero-shot and few-shot classification by using ChatGPT exclusively for subtask 3A.

5.4   Task 4: Factuality of Reporting of News Media

Five teams participated in this task, with participants proposing three distinct approaches to predict the veracity of the news outlets.

In an effort to reduce the influence of redundant data and to enhance the model resilience, team CUCPLUS [58] used RoBERTa coupled with regularized adversarial training. Team Accenture [101] aimed at maximizing the amount of training data and developed a RoBERTa model that learns the factual reporting patterns of news articles and news sources.

5.5   Task 5: Authority Finding in Twitter

Two teams participated in this task, submitting four runs. Both teams adopted the Twitter profile name and descriptions, and the Twitter lists as a user pre-

Table 15. Task 4: Top-3 models on the factuality of reporting of news media outlets task (MAE score).

	Team	MAE
1	CUCPLUS [58]	0.295
2	NLPIR-UNED	0.344
3	Accenture [101]	0.467

Table 16. Task 5 Evaluation results, in terms of P@1, P@5, and nDCG@5, ranked by P@5. Teams with a + sign include task organisers.

	Team (run ID)	P@5	P@1	nDCG@5
1	+bigIR (Hybrid3)	0.260	0.367	0.297
2	+bigIR (Hybrid1)	0.247	0.367	0.282
3	+bigIR (Hybrid2)	0.227	0.333	0.247
	BM25 baseline	0.087	0.133	0.104
4	ES-VRAI [83] (Model1)	0.067	0.067	0.071

sentation. Moreover, both teams incorporated the lexical matching between the rumor and the users in addition to the users network features to retrieve the corresponding authorities.

Team bigIR further used semantic matching by adopting Arabic BERT [84] fine-tuned on the full training data and deployed in the Tahaqqaq real-time system [94]. As shown in Table 16, all runs by the bigIR team managed to outperform the baseline by a sizable margin.

6 Related Work

Related work has focused on detecting misinformation and fact-checking across a variety of sources: news articles, forums, and social media [13,15,75,108]. This has given rise to variety of tasks, such as claim extraction [80], checkworthiness estimation [52,68], relevant document retrieval [70,107], detecting previously fact-checked claims [62,87,88], profiling articles and news outlets for their bias [14,96] and factuality [15,16,77], and recommendation systems to encourage people to engage in fact-checking [104]. There have been also a number of related shared tasks, which focused on rumour veracity [32], fact-checking in community question answering forums [63], propaganda techniques and framing in text and images [30,35,78,93], and fact verification and evidence finding for tabular data in scientific documents [106]. Other initiatives include FEVER [98], the Fake News Challenge [42], and the multimodal task at MediaEval [79].

7   Conclusions and Future Work

We presented the 2023 edition of the `CheckThat!` lab, which was once again one of the most popular CLEF labs, attracting a total of 37 active participating teams. This year, `CheckThat!` offered five tasks in seven languages: Arabic, Dutch, English, German, Italian, Spanish, and Turkish.

Task 1 focused on determining the check-worthiness of an item, whether it is a text or a combination of a text and image. Task 2 asked to predict the subjectivity or the objectivity of sentences. Task 3 aimed at detecting the political bias both at the level of a news article and of a news medium. Task 4 asked to measure the level of factuality of reporting of a news medium. Finally, Task 5 tasked the participants to identify authoritative sources on Twitter that could assist in verifying a given input claim. Tasks 2, 3, 4, and 5 were organized this year for the first time. For Task 1, most teams used large pre-trained models, OCR and data augmentation. In Task 2, most teams relied on transformers, and some used generative models (GPT*) to augment the training data or to flag subjective sentences. In Task 3, the most successful participants used RoBERTa and ChatGPT. In Task 4, most participants used RoBERTa, and some used stylistic features. In Task 5, the best team used lexical and semantic matching.

In the future, we plan to continue this year's trend of considering tasks that could play a relevant role in the analysis of journalistic and social media posts, and that go beyond factuality.

Acknowledgments. The work of F. Alam, M. Hasanain and W. Zaghouani is partially supported by NPRP 13S-0206-200281 and NPRP 14C-0916-210015 from the Qatar National Research Fund (a member of Qatar Foundation). The work of A. Galassi is supported by the European Commission NextGeneration EU programme, PNRR-M4C2-Investimento 1.3, PE00000013-"FAIR" Spoke 8. The work of Fatima Haouari was supported by GSRA grant #GSRA6-1-0611-19074 from the Qatar National Research Fund. The work of Tamer Elsayed was made possible by NPRP grant #NPRP-11S-1204-170060 from the Qatar National Research Fund.

The findings achieved herein are solely the responsibility of the authors.

References

1. Abdul-Mageed, M., Elmadany, A., et al.: Arbert & marbert: deep bidirectional transformers for Arabic. In: Proceedings of the 59th Annual Meeting of the Association for Computational Linguistics and the 11th International Joint Conference on Natural Language Processing, vol. 1: Long Papers, pp. 7088–7105 (2021)
2. Alam, F., et al.: Overview of the CLEF-2023 CheckThat! lab task 1 on check-worthiness in multimodal and multigenre content. In: Aliannejadi, M., Faggioli, G., Ferro, N., Vlachos, Michalis (eds.) Working Notes of CLEF 2023. Conference and Labs of the Evaluation Forum CLEF 2023, Thessaloniki, Greece (2023)
3. Alam, F., et al.: Fighting the COVID-19 infodemic in social media: a holistic perspective and a call to arms. In: Proceedings of the Conference on Web and Social Media, pp. 913–922. ICWSM 2021 (2021)

4. Alam, F., et al.: Fighting the COVID-19 infodemic: modeling the perspective of journalists, fact-checkers, social media platforms, policy makers, and the society. In: Findings of EMNLP 2021, pp. 611–649 (2021)

5. Ali, Z.S., Mansour, W., Elsayed, T., Al-Ali, A.: AraFacts: the first large Arabic dataset of naturally occurring claims. In: Proceedings of the Sixth Arabic Natural Language Processing Workshop, pp. 231–236 (2021)

6. Aliannejadi, M., Faggioli, G., Ferro, N., Vlachos, Michalis (eds.): Working Notes of CLEF 2023. Conference and Labs of the Evaluation Forum CLEF 2023, Thessaloniki, Greece (2023)

7. Antici, F., et al.: A corpus for sentence-level subjectivity detection on English news articles (2023)

8. Antoun, W., Baly, F., Hajj, H.: AraBERT: transformer-based model for Arabic language understanding. In: Proceedings of the 4th Workshop on Open-Source Arabic Corpora and Processing Tools, pp. 9–15. OSAC 2020, Marseille, France (2020)

9. Arslan, F., Hassan, N., Li, C., Tremayne, M.: A benchmark dataset of check-worthy factual claims. In: Proceedings of the International AAAI Conference on Web and Social Media, vol. 14, pp. 821–829 (2020)

10. Atanasova, P., et al.: Overview of the CLEF-2018 CheckThat! lab on automatic identification and verification of political claims. Task 1: check-worthiness. In: Cappellato, L., Ferro, N., Nie, J.Y., Soulier, L. (eds.) Working Notes of CLEF 2018-Conference and Labs of the Evaluation Forum. CEUR Workshop Proceedings (2018)

11. Atanasova, P., Nakov, P., Karadzhov, G., Mohtarami, M., Da San Martino, G.: Overview of the CLEF-2019 CheckThat! lab on automatic identification and verification of claims. Task 1: check-worthiness. In: Cappellato, L., Ferro, N., Losada, D., Müller, H. (eds.) Working Notes of CLEF 2019 Conference and Labs of the Evaluation Forum. CEUR Workshop Proceedings (2019)

12. Aziz, A., Hossain, M., Chy, A.: CSECU-DSG at CheckThat! 2023: transformer-based fusion approach for multimodal and multigenre check-worthiness. In: Aliannejadi, M., Faggioli, G., Ferro, N., Vlachos, Michalis (eds.) Working Notes of CLEF 2023. Conference and Labs of the Evaluation Forum CLEF 2023, Thessaloniki, Greece (2023)

13. Ba, M.L., Berti-Equille, L., Shah, K., Hammady, H.M.: VERA: a platform for veracity estimation over web data. In: Proceedings of the 25th International Conference Companion on World Wide Web, pp. 159–162 (2016)

14. Baly, R., Da San Martino, G., Glass, J., Nakov, P.: We can detect your bias: Predicting the political ideology of news articles. In: Proceedings of the 2020 Conference on Empirical Methods in Natural Language Processing, EMNLP 2020, pp. 4982–4991 (2020)

15. Baly, R., et al.: What was written vs. who read it: news media profiling using text analysis and social media context. In: Proceedings of the 58th Annual Meeting of the Association for Computational Linguistics, pp. 3364–3374 (2020)

16. Baly, R., Karadzhov, G., Saleh, A., Glass, J., Nakov, P.: Multi-task ordinal regression for jointly predicting the trustworthiness and the leading political ideology of news media. In: Proceedings of the 17th Annual Conference of the North American Chapter of the Association for Computational Linguistics: Human Language Technologies, pp. 2109–2116. NAACL-HLT 2019, Minneapolis, MN, USA (2019)

17. Barbieri, F., Espinosa Anke, L., Camacho-Collados, J.: XLM-T: a multilingual language model toolkit for twitter. arXiv e-prints, pp. arXiv-2104 (2021)

18. Barrón-Cedeño, A., et al.: The CLEF-2023 CheckThat! Lab: checkworthiness, subjectivity, political bias, factuality, and authority. In: Kamps, J., et al. (eds.) ECIR 2023. LNCS, vol. 13982, pp. 506–517. Springer, Cham (2023). https://doi.org/10.1007/978-3-031-28241-6_59

19. Barrón-Cedeño, A.: CheckThat! at CLEF 2020: enabling the automatic identification and verification of claims in social media. In: Jose, J.M., et al. (eds.) ECIR 2020. LNCS, vol. 12036, pp. 499–507. Springer, Cham (2020). https://doi.org/10.1007/978-3-030-45442-5_65

20. Barrón-Cedeño, A.: Overview of CheckThat! 2020: automatic identification and verification of claims in social media. In: Arampatzis, A., et al. (eds.) CLEF 2020. LNCS, vol. 12260, pp. 215–236. Springer, Cham (2020). https://doi.org/10.1007/978-3-030-58219-7_17

21. Barrón-Cedeño, A., et al.: Overview of the CLEF-2018 CheckThat! lab on automatic identification and verification of political claims. Task 2: factuality. In: Cappellato, L., Ferro, N., Nie, J.Y., Soulier, L. (eds.) Working Notes of CLEF 2018-Conference and Labs of the Evaluation Forum. CEUR Workshop Proceedings (2018)

22. Cappellato, L., Eickhoff, C., Ferro, N., Névéol, A. (eds.): CLEF 2020 Working Notes. CEUR Workshop Proceedings (2020)

23. Cappellato, L., Ferro, N., Losada, D., Müller, H. (eds.): Working notes of CLEF 2019 conference and labs of the evaluation forum. CEUR Workshop Proceedings (2019)

24. Cappellato, L., Ferro, N., Nie, J.Y., Soulier, L. (eds.): Working Notes of CLEF 2018-Conference and Labs of the Evaluation Forum. CEUR Workshop Proceedings (2018)

25. Cheema, G.S., Hakimov, S., Sittar, A., Müller-Budack, E., Otto, C., Ewerth, R.: MM-claims: a dataset for multimodal claim detection in social media. In: Findings of the Association for Computational Linguistics: NAACL 2022, pp. 962–979. Seattle, Washington (2022)

26. Clark, K., Luong, M., Le, Q.V., Manning, C.D.: ELECTRA: pre-training text encoders as discriminators rather than generators. In: Proceedings of the 8th International Conference on Learning Representations. ICLR 2020 (2020)

27. Cohen, J.: A coefficient of agreement for nominal scales. Educ. Psychol. Measur. 20(1), 37–46 (1960)

28. Conneau, A., et al.: Unsupervised cross-lingual representation learning at scale. CoRR abs/1911.02116 (2019)

29. Da San Martino, G., Alam, F., Hasanain, M., Nandi, R.N., Azizov, D., Nakov, P.: Overview of the CLEF-2023 CheckThat! lab task 3 on political bias of news articles and news media. In: Aliannejadi, M., Faggioli, G., Ferro, N., Vlachos, Michalis (eds.) Working Notes of CLEF 2023. Conference and Labs of the Evaluation Forum CLEF 2023, Thessaloniki, Greece (2023)

30. Da San Martino, G., Barrón-Cedeño, A., Wachsmuth, H., Petrov, R., Nakov, P.: SemEval-2020 task 11: detection of propaganda techniques in news articles. In: Proceedings of the Workshop on Semantic Evaluation, pp. 1377–1414 (2020)

31. Deniz Türkmen, M., Coşgun, G., Kutlu, M.: TOBB ETU at CheckThat! 2023: utilizing chatGPT to detect subjective statements and political bias. In: Aliannejadi, M., Faggioli, G., Ferro, N., Vlachos, Michalis (eds.) Working Notes of CLEF 2023. Conference and Labs of the Evaluation Forum CLEF 2023, Thessaloniki, Greece (2023)

32. Derczynski, L., Bontcheva, K., Liakata, M., Procter, R., Hoi, G.W.S., Zubiaga, A.: SemEval-2017 task 8: RumourEval: determining rumour veracity and support for rumours. In: Proceedings of the 11th International Workshop on Semantic Evaluation (SemEval-2017), pp. 69–76 (2017)
33. Devlin, J., Chang, M.W., Lee, K., Toutanova, K.: BERT: pre-training of deep bidirectional transformers for language understanding. In: Proceedings of the Conference of the North American Chapter of the Association for Computational Linguistics: Human Language Technologies, pp. 4171–4186. NAACL-HLT 2019, Minneapolis, MN, USA (2019)
34. Dey, K., Tarannum, P., Hasan, M.A., Noori, S.R.H.: Nn at CheckThat! 2023: Subjectivity in news articles classification with transformer based models. In: Aliannejadi, M., Faggioli, G., Ferro, N., Vlachos, Michalis (eds.) Working Notes of CLEF 2023. Conference and Labs of the Evaluation Forum CLEF 2023, Thessaloniki, Greece (2023)
35. Dimitrov, D., et al.: SemEval-2021 Task 6: detection of persuasion techniques in texts and images. In: Proceedings of the International Workshop on Semantic Evaluation, pp. 70–98. SemEval 2021 (2021)
36. Elsayed, T., et al.: CheckThat! at CLEF 2019: automatic identification and verification of claims. In: Azzopardi, L., Stein, B., Fuhr, N., Mayr, P., Hauff, C., Hiemstra, D. (eds.) ECIR 2019. LNCS, vol. 11438, pp. 309–315. Springer, Cham (2019). https://doi.org/10.1007/978-3-030-15719-7_41
37. Elsayed, T., et al.: Overview of the CLEF-2019 CheckThat! lab: automatic identification and verification of claims. In: Crestani, F., et al. (eds.) CLEF 2019. LNCS, vol. 11696, pp. 301–321. Springer, Cham (2019). https://doi.org/10.1007/978-3-030-28577-7_25
38. Faggioli, G., Ferro, N., Joly, A., Maistro, M., Piroi, F. (eds.): CLEF 2021 working notes. Working Notes of CLEF 2021-Conference and Labs of the Evaluation Forum (2021)
39. Frick, R.A.: Fraunhofer sit at CheckThat! 2023: can LLMs be used for data augmentation & few-shot classification? detecting subjectivity in text using chatGPT. In: Aliannejadi, M., Faggioli, G., Ferro, N., Vlachos, Michalis (eds.) Working Notes of CLEF 2023. Conference and Labs of the Evaluation Forum CLEF 2023, Thessaloniki, Greece (2023)
40. Frick, R.A., Vogel, I., Choi, J.E.: Fraunhofer SIT at CheckThat! 2023: enhancing the detection of multimodal and multigenre check-worthiness using optical character recognition and model souping. In: Aliannejadi, M., Faggioli, G., Ferro, N., Vlachos, Michalis (eds.) Working Notes of CLEF 2023. Conference and Labs of the Evaluation Forum CLEF 2023, Thessaloniki, Greece (2023)
41. Galassi, A., et al.: Overview of the CLEF-2023 CheckThat! lab task 2 on subjectivity in news articles. In: Aliannejadi, M., Faggioli, G., Ferro, N., Vlachos, Michalis (eds.) Working Notes of CLEF 2023. Conference and Labs of the Evaluation Forum CLEF 2023, Thessaloniki, Greece (2023)
42. Hanselowski, A., et al.: A retrospective analysis of the fake news challenge stance-detection task. In: Proceedings of the 27th International Conference on Computational Linguistics, pp. 1859–1874 (2018)
43. Haouari, F., Elsayed, T.: Detecting stance of authorities towards rumors in Arabic tweets: a preliminary study. In: Kamps, J., et al. (eds.) ECIR 2023. LNCS, vol. 13981, pp. 430–438. Springer, Cham (2023). https://doi.org/10.1007/978-3-031-28238-6_33
44. Haouari, F., Elsayed, T., Mansour, W.: Who can verify this? finding authorities for rumor verification in Twitter. Inf. Process. Manage. **60**(4), 103366 (2023)

45. Haouari, F., Hasanain, M., Suwaileh, R., Elsayed, T.: ArCOV19-rumors: Arabic COVID-19 twitter dataset for misinformation detection. In: Proceedings of the Sixth Arabic Natural Language Processing Workshop, pp. 72–81 (2021)

46. Haouari, F., Sheikh Ali, Z., Elsayed, T.: Overview of the CLEF-2023 CheckThat! lab task 5 on authority finding in twitter. In: Aliannejadi, M., Faggioli, G., Ferro, N., Vlachos, Michalis (eds.) Working Notes of CLEF 2023. Conference and Labs of the Evaluation Forum CLEF 2023, Thessaloniki, Greece (2023)

47. Hasanain, M., et al.: Overview of CheckThat! 2020 Arabic: automatic identification and verification of claims in social media. In: Cappellato, L., Eickhoff, C., Ferro, N., Névéol, A. (eds.) CLEF 2020 Working Notes. CEUR Workshop Proceedings (2020)

48. Hasanain, M., Suwaileh, R., Elsayed, T., Barrón-Cedeño, A., Nakov, P.: Overview of the CLEF-2019 CheckThat! lab on automatic identification and verification of claims. Task 2: evidence and factuality. In: Cappellato, L., Ferro, N., Losada, D., Müller, H. (eds.) Working notes of CLEF 2019 conference and labs of the evaluation forum. CEUR Workshop Proceedings (2019)

49. He, K., Zhang, X., Ren, S., Sun, J.: Deep residual learning for image recognition. In: Proceedings of the 2016 IEEE Conference on Computer Vision and Pattern Recognition, pp. 770–778. CVPR 2016, Las Vegas, Nevada, USA (2016)

50. He, P., Gao, J., Chen, W.: DeBERTaV3: improving DeBERTa using ELECTRA-style pre-training with gradient-disentangled embedding sharing. In: The Eleventh International Conference on Learning Representations (2023)

51. Inoue, H.: Multi-sample dropout for accelerated training and better generalization. CoRR abs/1905.09788 (2019)

52. Kartal, Y.S., Kutlu, M.: Re-think before you share: a comprehensive study on prioritizing check-worthy claims. In: IEEE Transactions on Computational Social Systems (2022)

53. Khalil, A., Jarrah, M., Aldwairi, M., Jararweh, Y.: Detecting Arabic fake news using machine learning. In: Proceedings of the International Conference on Intelligent Data Science Technologies and Applications, pp. 171–177 (2021)

54. Köhler, J., et al.: Overview of the CLEF-2022 CheckThat! lab task 3 on fake news detection. In: Working Notes of CLEF 2022–Conference and Labs of the Evaluation Forum. CLEF 2022, Bologna, Italy (2022)

55. Lan, W., Chen, Y., Xu, W., Ritter, A.: An empirical study of pre-trained transformers for Arabic information extraction. In: Proceedings of the 2020 Conference on Empirical Methods in Natural Language Processing (EMNLP), pp. 4727–4734 (2020)

56. Landis, J.R., Koch, G.G.: The measurement of observer agreement for categorical data. Biometrics **33**, 159–174 (1977)

57. Leistra, F., Caselli, T.: Thesis titan at CheckThat! 2023: language-specific fine-tuning of mdebertav3 for subjectivity detection. In: Aliannejadi, M., Faggioli, G., Ferro, N., Vlachos, Michalis (eds.) Working Notes of CLEF 2023. Conference and Labs of the Evaluation Forum CLEF 2023, Thessaloniki, Greece (2023)

58. Li, C., Xue, R., Lin, C., Fan, W.: CUCPLUS at CheckThat! 2023: text combination and regularized adversarial training for news media factuality evaluation. In: Aliannejadi, M., Faggioli, G., Ferro, N., Vlachos, Michalis (eds.) Working Notes of CLEF 2023. Conference and Labs of the Evaluation Forum CLEF 2023, Thessaloniki, Greece (2023)

59. Liu, Y., et al.: RoBERTa: a robustly optimized BERT pretraining approach. CoRR abs/1907.11692 (2019)

60. Liu, Y., et al.: RoBERTa: a robustly optimized BERT pretraining approach. arXiv preprint arXiv:1907.11692 (2019)
61. Liu, Z., Mao, H., Wu, C., Feichtenhofer, C., Darrell, T., Xie, S.: A convnet for the 2020s. In: Proceedings of the IEEE/CVF Conference on Computer Vision and Pattern Recognition,pp. 11966–11976. CVPR '22, New Orleans, LA, USA (2022)
62. Mansour, W., Elsayed, T., Al-Ali, A.: This is not new! spotting previously-verified claims over Twitter. Inf. Process. Manage. **60**(4), 103414 (2023)
63. Mihaylova, T., Karadzhov, G., Atanasova, P., Baly, R., Mohtarami, M., Nakov, P.: SemEval-2019 task 8: fact checking in community question answering forums. In: Proceedings of the Workshop on Semantic Evaluation, pp. 860–869 (2019)
64. Modzelewski, A., Sosnowski, W., Wierzbicki, A.: DSHacker at CheckThat! 2023: Check-Worthiness in Multigenre and Multilingual Content With GPT-3.5 Data Augmentation. In: Aliannejadi, M., Faggioli, G., Ferro, N., Vlachos, Michalis (eds.) Working Notes of CLEF 2023. Conference and Labs of the Evaluation Forum CLEF 2023, Thessaloniki, Greece (2023)
65. Nakov, P., et al.: Overview of the CLEF-2023 CheckThat! lab task 4 on factuality of reporting of news media. In: Aliannejadi, M., Faggioli, G., Ferro, N., Vlachos, Michalis (eds.) Working Notes of CLEF 2023. Conference and Labs of the Evaluation Forum CLEF 2023, Thessaloniki, Greece (2023)
66. Nakov, P., et al.: Overview of the CLEF-2022 CheckThat! lab task 1 on identifying relevant claims in tweets. In: Working Notes of CLEF 2022–Conference and Labs of the Evaluation Forum. CLEF 2022, Bologna, Italy (2022)
67. Nakov, P., et al.: Overview of the CLEF-2022 CheckThat! lab on fighting the COVID-19 infodemic and fake news detection. In: Proceedings of the Conference of the CLEF Association: Information Access Evaluation meets Multilinguality, Multimodality, and Visualization. CLEF 2022, Bologna, Italy (2022)
68. Nakov, P., et al.: The CLEF-2022 CheckThat! Lab on Fighting the COVID-19 Infodemic and Fake News Detection. In: Hagen, M., et al. (eds.) ECIR 2022. LNCS, vol. 13186, pp. 416–428. Springer, Cham (2022). https://doi.org/10.1007/978-3-030-99739-7_52
69. Nakov, P., et al.: Overview of the CLEF-2018 lab on automatic identification and verification of claims in political debates. In: Working Notes of CLEF 2018 - Conference and Labs of the Evaluation Forum. CLEF 2018 (2018)
70. Nakov, P., et al.: Automated fact-checking for assisting human fact-checkers. In: Proceedings of the 30th International Joint Conference on Artificial Intelligence, pp. 4551–4558 (2021)
71. Nakov, P., Da San Martino, G., Alam, F., Shaar, S., Mubarak, H., Babulkov, N.: Overview of the CLEF-2022 CheckThat! lab task 2 on detecting previously fact-checked claims. In: Working Notes of CLEF 2022–Conference and Labs of the Evaluation Forum. CLEF 2022, Bologna, Italy (2022)
72. Nakov, P., et al.: The CLEF-2021 CheckThat! Lab on Detecting Check-Worthy Claims, Previously Fact-Checked Claims, and Fake News. In: Hiemstra, D., Moens, M.-F., Mothe, J., Perego, R., Potthast, M., Sebastiani, F. (eds.) ECIR 2021. LNCS, vol. 12657, pp. 639–649. Springer, Cham (2021). https://doi.org/10.1007/978-3-030-72240-1_75
73. Nakov, P., et al.: Overview of the CLEF–2021 CheckThat! Lab on Detecting Check-Worthy Claims, Previously Fact-Checked Claims, and Fake News. In: Candan, K.S., et al. (eds.) CLEF 2021. LNCS, vol. 12880, pp. 264–291. Springer, Cham (2021). https://doi.org/10.1007/978-3-030-85251-1_19

74. Nguyen, D.Q., Vu, T., Nguyen, A.T.: BERTweet: a pre-trained language model for English tweets. In: Proceedings of the 2020 Conference on Empirical Methods in Natural Language Processing: System Demonstrations, pp. 9–14 (2020)
75. Nguyen, V.H., Sugiyama, K., Nakov, P., Kan, M.Y.: FANG: leveraging social context for fake news detection using graph representation. In: Proceedings of the 29th ACM International Conference on Information and Knowledge Management, pp. 1165–1174. CIKM 2020 (2020)
76. Pachov, G., Dimitrov, D., Koychev, I., Nakov, P.: Gpachov at CheckThat! 2023: a diverse multi-approach ensemble for subjectivity detection in news articles. In: Aliannejadi, M., Faggioli, G., Ferro, N., Vlachos, Michalis (eds.) Working Notes of CLEF 2023. Conference and Labs of the Evaluation Forum CLEF 2023, Thessaloniki, Greece (2023)
77. Panayotov, P., Shukla, U., Sencar, H.T., Nabeel, M., Nakov, P.: GREENER: graph neural networks for news media profiling. In: Proceedings of the 2022 Conference on Empirical Methods in Natural Language Processing, pp. 7470–7480. EMNLP 2022, Abu Dhabi, UAE (2022)
78. Piskorski, J., Stefanovitch, N., Da San Martino, G., Nakov, P.: SemEval-2023 task 3: detecting the category, the framing, and the persuasion techniques in online news in a multi-lingual setup. In: Proceedings of the 17th International Workshop on Semantic Evaluation. SemEval 2023, Toronto, Canada (2023)
79. Pogorelov, K., et al.: FakeNews: Corona virus and 5G conspiracy task at MediaEval 2020. In: MediaEval (2020)
80. Reddy, R.G., et al.: NewsClaims: a new benchmark for claim detection from news with attribute knowledge. In: Proceedings of the 2022 Conference on Empirical Methods in Natural Language Processing, pp. 6002–6018. EMNLP 2022, Abu Dhabi, UAE (2022)
81. Sadouk, H.T., Sebbak, F., Zekiri, H.E.: Es-vrai at CheckThat! 2023: lcheck worthiness in multimodal and multigenre contents through fusion and sampling approaches. In: Aliannejadi, M., Faggioli, G., Ferro, N., Vlachos, Michalis (eds.) Working Notes of CLEF 2023. Conference and Labs of the Evaluation Forum CLEF 2023, Thessaloniki, Greece (2023)
82. Sadouk, H.T., Sebbak, F., Zekiri, H.E.: Es-vrai at CheckThat! 2023: Enhancing model performance for subjectivity detection through multilingual data augmentation. In: Aliannejadi, M., Faggioli, G., Ferro, N., Vlachos, Michalis (eds.) Working Notes of CLEF 2023. Conference and Labs of the Evaluation Forum CLEF 2023, Thessaloniki, Greece (2023)
83. Sadouk, H.T., Sebbak, F., Zekiri, H.E.: Es-vrai at CheckThat! 2023: Leveraging bio and lists information for enhanced rumor verification in twitter. In: Aliannejadi, M., Faggioli, G., Ferro, N., Vlachos, Michalis (eds.) Working Notes of CLEF 2023. Conference and Labs of the Evaluation Forum CLEF 2023, Thessaloniki, Greece (2023)
84. Safaya, A., Abdullatif, M., Yuret, D.: KUISAIL at SemEval-2020 task 12: BERT-CNN for offensive speech identification in social media. In: Proceedings of the Fourteenth Workshop on Semantic Evaluation, pp. 2054–2059. International Committee for Computational Linguistics, Barcelona (online) (2020)
85. Sawiński, M., et al.: Openfact at CheckThat! 2023: Head-to-head GPT vs. BERT - a comparative study of transformers language models for the detection of checkworthy claims. In: Aliannejadi, M., Faggioli, G., Ferro, N., Vlachos, Michalis (eds.) Working Notes of CLEF 2023. Conference and Labs of the Evaluation Forum CLEF 2023, Thessaloniki, Greece (2023)

86. Schlicht, I.B., Khellaf, L., Altiok, D.: Dwreco at CheckThat! 2023: enhancing subjectivity detection through style-based data sampling. In: Aliannejadi, M., Faggioli, G., Ferro, N., Vlachos, Michalis (eds.) Working Notes of CLEF 2023. Conference and Labs of the Evaluation Forum CLEF 2023, Thessaloniki, Greece (2023)

87. Shaar, S., Alam, F., Da San Martino, G., Nakov, P.: The role of context in detecting previously fact-checked claims. In: Findings of the Association for Computational Linguistics: NAACL 2022, pp. 1619–1631 (2022)

88. Shaar, S., Babulkov, N., Da San Martino, G., Nakov, P.: That is a known lie: detecting previously fact-checked claims. In: Proceedings of the 58th Annual Meeting of the Association for Computational Linguistics, pp. 3607–3618 (2020)

89. Shaar, S., et al.: Overview of the CLEF-2021 CheckThat! lab task 2 on detecting previously fact-checked claims in tweets and political debates. In: Faggioli, G., Ferro, N., Joly, A., Maistro, M., Piroi, F. (eds.) CLEF 2021 working notes. Working Notes of CLEF 2021-Conference and Labs of the Evaluation Forum (2021)

90. Shaar, S., et al.: Overview of the CLEF-2021 CheckThat! lab task 1 on checkworthiness estimation in tweets and political debates. In: Faggioli, G., Ferro, N., Joly, A., Maistro, M., Piroi, F. (eds.) CLEF 2021 working notes. Working Notes of CLEF 2021-Conference and Labs of the Evaluation Forum (2021)

91. Shaar, S.,et al.: Overview of CheckThat! 2020 English: Automatic identification and verification of claims in social media. In: Cappellato, L., Eickhoff, C., Ferro, N., Névéol, A. (eds.) CLEF 2020 Working Notes. CEUR Workshop Proceedings (2020)

92. Shahi, G.K., Struß, J.M., Mandl, T.: Overview of the CLEF-2021 CheckThat! lab: Task 3 on fake news detection. In: Faggioli, G., Ferro, N., Joly, A., Maistro, M., Piroi, F. (eds.) CLEF 2021 working notes. Working Notes of CLEF 2021-Conference and Labs of the Evaluation Forum (2021)

93. Sharma, S., et al.: Findings of the CONSTRAINT 2022 shared task on detecting the hero, the villain, and the victim in memes. In: Proceedings of the Workshop on Combating Online Hostile Posts in Regional Languages during Emergency Situations, pp. 1–11. Dublin, Ireland (2022)

94. Sheikh Ali, Z., Mansour, W., Haouari, F., Hasanain, M., Elsayed, T., Al-Ali, A.: Tahaqqaq: a real-time system for assisting twitter users in Arabic claim verification. In: Proceedings of the 46th International ACM SIGIR Conference on Research and Development in Information Retrieval (2023)

95. Shushkevich, E., Cardiff, J.: Tudublin at CheckThat! 2023: Chatgpt for data augmentation. In: Aliannejadi, M., Faggioli, G., Ferro, N., Vlachos, Michalis (eds.) Working Notes of CLEF 2023. Conference and Labs of the Evaluation Forum CLEF 2023, Thessaloniki, Greece (2023)

96. Stefanov, P., Darwish, K., Atanasov, A., Nakov, P.: Predicting the topical stance and political leaning of media using tweets. In: Proceedings of the Annual Meeting of the Association for Computational Linguistics, pp. 527–537. ACL 2020 (2020)

97. Tarannum, P., Hasan, M.A., Alam, F., Noori, S.R.H.: Z-Index at CheckThat! 2023: Unimodal and multimodal checkworthiness classification. In: Aliannejadi, M., Faggioli, G., Ferro, N., Vlachos, Michalis (eds.) Working Notes of CLEF 2023. Conference and Labs of the Evaluation Forum CLEF 2023, Thessaloniki, Greece (2023)

98. Thorne, J., Vlachos, A., Christodoulopoulos, C., Mittal, A.: FEVER: a large-scale dataset for fact extraction and VERification. In: Proceedings of the 2018 Conference of the North American Chapter of the Association for Computational

Linguistics: Human Language Technologies, pp. 809–819. NAACL-HLT 2018, New Orleans, Louisiana, USA (2018)

99. Tran, S., Rodrigues, P., Strauss, B., Williams, E.: Accenture at CheckThat! 2023: Identifying claims with societal impact using NLP data augmentation. In: Aliannejadi, M., Faggioli, G., Ferro, N., Vlachos, Michalis (eds.) Working Notes of CLEF 2023. Conference and Labs of the Evaluation Forum CLEF 2023, Thessaloniki, Greece (2023)

100. Tran, S., Rodrigues, P., Strauss, B., Williams, E.: Accenture at CheckThat! 2023: Impacts of back-translation on subjectivity detection. In: Aliannejadi, M., Faggioli, G., Ferro, N., Vlachos, Michalis (eds.) Working Notes of CLEF 2023. Conference and Labs of the Evaluation Forum CLEF 2023, Thessaloniki, Greece (2023)

101. Tran, S., Rodrigues, P., Strauss, B., Williams, E.: Accenture at CheckThat! 2023: Learning to detect factuality levels of news sources. In: Aliannejadi, M., Faggioli, G., Ferro, N., Vlachos, Michalis (eds.) Working Notes of CLEF 2023. Conference and Labs of the Evaluation Forum CLEF 2023, Thessaloniki, Greece (2023)

102. Tran, S., Rodrigues, P., Strauss, B., Williams, E.: Accenture at CheckThat! 2023: Learning to detect political bias of news articles and sources. In: Aliannejadi, M., Faggioli, G., Ferro, N., Vlachos, Michalis (eds.) Working Notes of CLEF 2023. Conference and Labs of the Evaluation Forum CLEF 2023, Thessaloniki, Greece (2023)

103. Truică C.O., Apostol, E.S., Paschke, A.: Awakened at CheckThat! 2022: fake news detection using BiLSTM and sentence transformer. In: Working Notes of CLEF 2022 - Conference and Labs of the Evaluation Forum. CLEF 2022, Bologna, Italy (2022)

104. Vo, N., Lee, K.: The rise of guardians: fact-checking url recommendation to combat fake news. In: Proceedings of the International ACM SIGIR Conference on Research & Development in Information Retrieval, pp. 275–284. SIGIR 2018 (2018)

105. von Däniken, P., Deriu, J., Cieliebak, M.: Zhaw-cai at CheckThat! 2023: Ensembling using kernel averaging. In: Aliannejadi, M., Faggioli, G., Ferro, N., Vlachos, Michalis (eds.) Working Notes of CLEF 2023. Conference and Labs of the Evaluation Forum CLEF 2023, Thessaloniki, Greece (2023)

106. Wang, N.X., Mahajan, D., Danilevsky, M., Rosenthal, S.: SemEval-2021 task 9: fact verification and evidence finding for tabular data in scientific documents (SEM-TAB-FACTS). In: Proceedings of the 15th International Workshop on Semantic Evaluation, pp. 317–326. SemEval 2021 (2021)

107. Yasser, K., Kutlu, M., Elsayed, T.: Re-ranking web search results for better fact-checking: a preliminary study. In: Proceedings of the 27th ACM International Conference on Information and Knowledge Management, pp. 1783–1786 (2018)

108. Zubiaga, A., Liakata, M., Procter, R., Wong Sak Hoi, G., Tolmie, P.: Analysing how people orient to and spread rumours in social media by looking at conversational threads. PloS one **11**(3), e0150989 (2016)

Overview of DocILE 2023: Document Information Localization and Extraction

Štěpán Šimsa[1], Michal Uřičář[1(✉)], Milan Šulc[5], Yash Patel[2],
Ahmed Hamdi[3], Matěj Kocián[1], Matyáš Skalický[1], Jiří Matas[2],
Antoine Doucet[3], Mickaël Coustaty[3], and Dimosthenis Karatzas[4]

[1] Rossum, Prague, Czech Republic
{stepan.simsa,michal.uricar,matej.kocian,stepan.simsa}@rossum.ai
[2] Visual Recognition Group, Czech Technical University in Prague,
Prague, Czech Republic
[3] University of La Rochelle, La Rochelle, France
[4] Computer Vision Center, Universitat Autónoma de Barcelona, Barcelona, Spain
[5] Second Foundation, Prague, Czech Republic
https://rossum.ai

Abstract. This paper provides an overview of the DocILE 2023 Competition, its tasks, participant submissions, the competition results and possible future research directions. This first edition of the competition focused on two Information Extraction tasks, Key Information Localization and Extraction (KILE) and Line Item Recognition (LIR). Both of these tasks require detection of pre-defined categories of information in business documents. The second task additionally requires correctly grouping the information into tuples, capturing the structure laid out in the document. The competition used the recently published DocILE dataset and benchmark that stays open to new submissions. The diversity of the participant solutions indicates the potential of the dataset as the submissions included pure Computer Vision, pure Natural Language Processing, as well as multi-modal solutions and utilized all of the parts of the dataset, including the annotated, synthetic and unlabeled subsets.

Keywords: Information Extraction · Computer Vision · Natural Language Processing · Optical Character Recognition · Document Understanding

1 Introduction

Documents, such as invoices, purchase orders, contracts, and financial statements, are a major form of communication between businesses. Extraction of the key information from such documents is an essential task, as they contain a wealth of valuable information critical for day-to-day decision-making, compliance, and operational efficiency.

Machine learning techniques, particularly those based on deep learning, natural language processing, and computer vision, have shown great promise in

A. Arampatzis et al. (Eds.): CLEF 2023, LNCS 14163, pp. 276–293, 2023.
https://doi.org/10.1007/978-3-031-42448-9_21

a number of document understanding tasks [4,5,8,10,17,25,26,32–34], such as understanding of forms [1,2,36], receipts [4,6], tables [3,20,35], or invoices [11,12, 18]. Another approach to document understanding is question answering [13,14].

The DocILE competition and lab at CLEF 2023 called for contributions to the DocILE benchmark [22], which focuses on the practically oriented tasks of Key Information Localization and Extraction (KILE) and Line Item Recognition (LIR), as defined in [23].

This paper provides an overview of the first run of the DocILE competition, summarizing the participants solutions and their final results, as well as a breakdown of the results with respect to certain information, e.g., with respect to zero-shot/few-shot/many-shot layouts in the training or with respect to text extractions, which are not otherwise checked in the main evaluation metric.

The paper is structured as follows: Sect. 2 describes the DocILE dataset, its acquisition and distribution to individual subsets; Sect. 3 summarizes the DocILE competition tasks and their respective evaluation process; all competing methods submitted to the competition are briefly described in Sect. 4; results from the competition and their breakdown and discussion are provided in Sect. 5; and finally, Sect. 6 concludes the paper.

2 Data

The competition was based on the DocILE dataset [22] of business documents, which consists of three distinct subsets: *annotated*, *unlabeled*, and *synthetic*. The *annotated* set comprises 6,680 real business documents sourced from publicly available platforms, which have been carefully annotated. The *unlabeled* set consists of a massive collection of 932,467 real business documents also obtained from publicly available sources, intended for unsupervised pre-training purposes. The dataset draws its documents from two public data sources: UCSF Industry Documents Library [30] and Public Inspection Files (PIF) [31]. UCSF Industry Documents Library is a digitalized archive of documents created by industries that impact public health, while PIF consists of public files of American broadcast stations, specifically focusing on political campaign ads. The documents were retrieved in a PDF format, and various selection criteria were applied to ensure the quality and relevance of the dataset. The *synthetic* set comprises 100,000 documents generated using a proprietary document generator. These synthetic documents are designed to mimic the layout and structure of 100 fully annotated real business documents from the annotated set.

Participants were allowed to use the 5,180 training samples, 500 validation samples and the full synthetic and unlabeled dataset. The remaining 1,000 documents form the test set. Usage of external document datasets or models pre-trained on such datasets was forbidden in the competition, while datasets and pre-trained models from other domains—such as images from ImageNet [19] or texts from BooksCorpus [37]—were allowed.

For each document, the dataset contains the original PDF file and OCR pre-computed using the DocTR [15] library achieving excellent recognition scores in [16]. Annotations are provided for documents in the annotated and synthetic sets and include field annotations for the two competition tasks, KILE and LIR, as well as additional metadata: original source of the document, layout cluster ID[1], table grid annotation, document type, currency, page count and page image sizes. Annotations for the test set are not publicly available.

3   Tasks and Evaluation

The competition had two tracks, one for each of the two tasks, KILE and LIR, respectively. The goal of both of these tasks is to detect semantic fields in the document, i.e., for each category (field type) localize all the text boxes that have this semantic meaning and extract the corresponding text. For LIR, fields have to be additionally grouped into Line Items, i.e., tuples representing a single item. For a more formal definition, refer to [23], where the tasks were first defined. An example document with annotations for KILE and LIR is illustrated in Fig. 1.

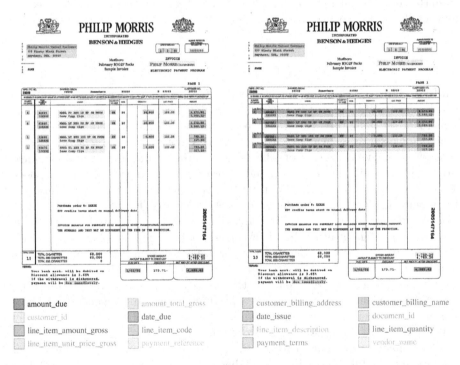

Fig. 1. An example DocILE document with KILE and LIR annotations emphasized. Left: field annotations. Right: Line Item areas displayed by alternating blue ■ and green ▨. Bottom: color legend for KILE and LIR field types. The image is taken from [22]. (Color figure online)

---

[1] Clusters are formed by documents that have similar visual layout and placement of semantic information in this layout.

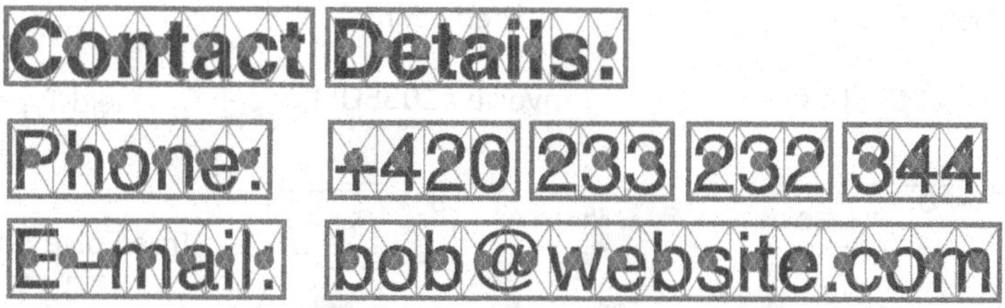

(a) Each pre-computed OCR word is split uniformly into pseudo-character boxes based on the number of characters. Pseudo-Character Centers are the centers of these boxes.

(b) Correct extraction examples.

(c) Incorrect extraction examples.

Fig. 2. Correct and incorrect bounding box predictions of the phone number are shown in 2b and in 2c, respectively. A predicted field matches the location of a ground truth field if their bounding boxes cover the same text. More precisely, the fields must contain exactly the same Pseudo-Character Centers defined in 2a. Note: in 2b, only one of the predictions would be considered correct if all three boxes were predicted. Images are taken from [21].

The DocILE benchmark is hosted on the Robust Reading Challenge portal[2]. As the test set annotations remain private, the only way to compare the solutions on the test set is to make a submission to the benchmark. During the competition, participants did not see the results or even their own score, so they had to select the final solution without gathering any info about the test set.

To focus the competition on the most important part of the two tasks, which is the semantic understanding of the values in the documents, only the localization part was evaluated. This means the tasks can be framed as object detection tasks, with LIR additionally requiring the grouping of the detected objects into Line Items. Therefore, standard object detection metrics are employed, with Average Precision (AP) as the main metric for KILE and F1 as the main metric for LIR. A predicted and a ground truth field are matching if they have the same field type and if they cover the same text in the document, as explained in detail in Fig. 2. For LIR the fields also need to belong to corresponding Line Items, where this correspondence is found with a matching that maximizes the total number of matched fields, as shown in Fig. 3.

Extracting the text of the localized fields is an obvious extension of the two tasks whose precision is also important. Therefore, both tracks in the benchmark have a separate leaderboard, where the extracted text is compared with the

---

[2] https://rrc.cvc.uab.es/?ch=26.

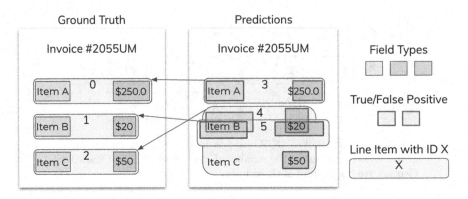

Fig. 3. Visualization of Line Item (LI) matching. Both the annotations and predictions consist of three line items where LI 3 and LI 0 are clearly matched together. The two fields of "Item B" are detected both as part of LI 4 and LI 5, so greedy assignment might assign LI 4 to LI 1, leading to only three matched fields in total. Instead, maximum matching assigns LI 4 to LI 2 and LI 5 to LI 1, leading to four matched fields overall.

annotated text for each matched field pair and an exact match is required to count the pair as a true positive pair.

The benchmark contains additional leaderboards for zero-shot, few-shot and many-shot evaluation. This is the same evaluation as in the main leaderboard but evaluated only on a subset of the test documents. Specifically, it is evaluated only on documents from layout clusters that have zero (zero-shot), one to three (few-shot) or four and more (many-shot) samples available for training (i.e., in the training or validation set). These test subsets contain roughly 250, 250 and 500 documents, respectively. This enables a more detailed analysis of the methods and helps to understand which methods generalize better to new document layouts and which can better overfit to clusters with many examples available for training.

4   Submissions

The competitions received contributions from 5 teams for the KILE task and 4 teams in the LIR task. See Fig. 4 to compare this with the number of dataset downloads and competition registrations. In this section, we briefly present all the submitted methods in an alphabetical order.

GraphDoc—USTC-IFLYTEK, China. Team from the University of Science and Technology of China and iFLYTEK AI Research, China submitted a method [29], which jointly solves both KILE and LIR tasks. Their approach is based on a modified GraphDoc [34] tailored for the purpose of the DocILE competition, pre-trained on the DocILE unlabeled set and consequently fine-tuned on the training set. Both competition tasks are handled like Named Entity

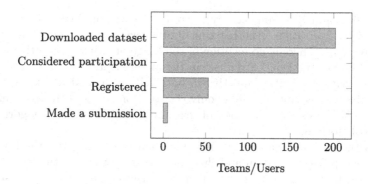

Fig. 4. In the four and half months since its release on February 1, the DocILE dataset was downloaded by over 200 unique users. While 159 of them indicated they are considering participating in the competition, only 53 teams actually registered and 5 of them made a submission to the benchmark. Based on the feedback from a few of these teams, we attribute this to the tight schedule and to the competitiveness of the baselines, as they were not so easy to beat.

Recognition (NER), followed by a special *Merger* module, which operates on the attention layer from the GraphDoc model and the merging strategy is therefore learned, unlike in the baseline method. The authors noticed the inherent nature of the KILE and LIR task and exploited it naturally—the word tokens are merged to instances by the first level Merger module and then the second Merger module operates on these instances for the line item classes and merges them into final line items. The proposed method still uses some level of a rule-based post-processing, which is based on the observation of data: 1) some field annotations contain only part of the detected text boxes from DocTR and need to be manually split (such as `currency_code_amount_due` fields that usually contains only the symbol '$'); 2) some symbols are frequently detected as part of the OCR word box, but excluded from the annotations (such as the symbol '#'); 3) Text boxes that are far apart rarely belong to the same instance, or to the same line item.

Besides the contribution on the model side, the authors also devoted some effort to improve the OCR detections provided, by removing the detections with low confidence and by running DocTR [15] on scaled-up images (1.25×, 1.5×, and 1.75×) and aggregating the found text boxes to improve the recall of the OCR detections. The OCR detections are also re-ordered, similarly as in the baseline methods, in the top-down left-right reading order.

Since the proposed method uses multi-modal input (text, layout, vision), we can put it into a category of NLP + CV.

LiLT—University of Information Technology, Vietnam. Team from University of Information Technology, Vietnam submitted a method based on the baselines with a layout-aware backbone LiLT [28]. The authors decided to re-split the provided dataset to 80% for training and 20% for validation (original ratio was 90% and 10%, respectively), arguing that the original split was leading to a poor generalization. Another contribution was filtering out low-confident OCR detections. There is no mention of the usage of either the synthetic or the unlabeled sets of the DocILE dataset in the manuscript.

Unfortunately, despite competing in both KILE and LIR tasks, the authors submitted a manuscript describing only the solution for the LIR task. Since the backbone LiLT uses a combination of text and layout input, we categorize it as a pure NLP solution.

From the authors' manuscript, it is not clear what data was the LiLT pre-trained checkpoint trained on and therefore we are not sure if the method does not violate the benchmark rules that prohibit the usage of models pre-trained on other document datasets[3].

Union-RoBERTa—University of Information Technology, Vietnam. Team from University of Information Technology, Vietnam submitted a method [27] which is heavily based on the provided baselines. Their method, coined as Union-RoBERTa, is an ensemble of two provided baselines [22] with a plain RoBERTa trained from scratch on the synthetic and training data using Fast Gradient Method. They use the affirmative strategy for the ensemble (hence the Union in the name) and follow it by an additional merging of fields based on distance with a threshold tuned on the validation set. This ensemble is then used to generate pseudo-labels for 10,000 samples from the unlabeled set which are then used for additional pre-training of the three models followed by an additional training on the training set. Although there is not much novelty in the proposed method, it is a nice example how well-established practices can yield significant improvements.

The proposed method participated in the KILE task only. Since the method is based on RoBERTa models, we put it into a pure NLP category.

ViBERTGrid—Ricoh Software Research Center, China. Team from Ricoh Software Research Center, China submitted a method based on token classification with ViBERTGrid [10], followed by a distance-based merging procedure. The team participated in both KILE and LIR tasks. However, the results were below baselines for both tasks and the authors decided not to submit manuscript with further details. We can just guess, based on the ViBERTGrid, that the method was probably a combination of NLP with CV.

We noticed that the method probably suffers from not using the adequate score (all detections were using the same score 1.0) which could explain why AP is significantly lower compared to the other methods, while F1 measure on the KILE task is in the middle of the ranking, as seen in Fig. 5a and discussed more in Sect. 5.4.

---

[3] In the LiLT paper [28], they pre-train the model on the IIT-CDIP [9] dataset which is a document dataset.

YOLOv8—University of West Bohemia, Czech Republic. Team from University of West Bohemia, Czech Republic submitted a method [24] based on the combination of YOLOv8 [7] and CharGrid [8] with modifications, such as splitting the word boxes to pseudo-characters, not using the one number encoding of a character directly but a three numbers encoding instead, and concatenating the image with the CharGrid representation. The authors did not leverage synthetic nor unlabeled parts of the dataset, but they used augmentations during training. Due to the faster training procedure, they decided to use just random translation for augmentation, even though the best results in ablation study were observed when mosaicking was applied. The method works quite well on the KILE task (where it even achieves the highest F1) but falls behind on the LIR task. The latter is attributed to the increased number of false positive detections.

Interestingly, this is the only pure CV based model.

5 Results and Discussion

The results for the KILE and LIR tasks, including the baselines from the DocILE dataset paper [22], are displayed in Figs. 5a and 5b, respectively. We can see that while on KILE task the participants approaches clearly outperform the provided baseline by a big margin on the main evaluation metric (AP), on the LIR task, there is not such a big improvement, except for the GraphDoc based approach. The baseline methods are marked with ⊟ symbol.

Interestingly, for the KILE task, the secondary metric (F1) does not seem to be correlated with the primary metric (AP) and several of the methods, including the baselines, are comparatively much better on F1 than on AP. In fact, the YOLOv8 based approach outperforms the otherwise winning GraphDoc in F1 metric. This might be related to the fact that AP takes into account the score assigned to individual predictions, while F1 does not, and that some teams focused on assigning good scores to predictions more than others, as discussed in Sect. 5.4.

In the LIR task, there is some correlation between the primary metric, which in this task is F1, and the secondary metric (AP), with a slight violation for the GraphDoc based method.

Considering the achieved metric values, we can say that the DocILE benchmark poses very challenging tasks, because the best results on both KILE and LIR tasks are below 80% of the respective quality metric.

5.1 Text Extraction Evaluation

Figure 6 summarizes the results when text extractions are checked in the evaluation. Note, that this was intentionally not done in the main evaluation, which focuses more on the localization part, so that participants do not have to focus on optimizing the OCR solution for text read out. However, in a real-world system, this would likely be the main metric for evaluation and therefore we present results of all of the competing methods when this strict text comparison is employed. By definition, all methods are performing worse on both KILE and LIR task, compared to the main localization-only evaluation. Also both AP and F1 metrics show less variance for all competing methods. Unfortunately, the YOLOv8 based method did not provide the text outputs (which was not required for the competition), so we cannot evaluate this method properly.

The KILE task, summarized in Fig. 6a, shows that the GraphDoc still outperforms the other competitors. However, the margin is not as big as in the final evaluation.

The LIR task is summarized in Fig. 6b. Surprisingly, the GraphDoc based method, which was winning in the main evaluation, and which kept its position for the KILE task, is now lagging behind quite significantly. We believe this might be attributed to the lack of effort invested to the text read-out after merging.

5.2   Evaluation on Zero/Few/Many-Shot Layouts

In this section, we present a break-down of the evaluation with respect to the document layouts seen/unseen during training, hence providing hints about how the particular method generalize. We have three distinct categories for this evaluation: 1) zero-shot, formed by document layouts that were not in the training nor validation sets; 2) few-shot, which is formed by document layouts that have 1–3 samples in the training and validation subset of the DocILE dataset; 3) many-shot, with 4 or more samples in the training and validation subset.

In Fig. 7, we show the results of the first category—zero-shot. For the KILE task (Fig. 7a), we can see that GraphDoc is still a clear winner with a relatively high margin. However, interestingly, YOLOv8 performs much worse, compared to the overall results. This might be attributed to the fact that this method did not leverage the unlabeled part of the DocILE dataset and therefore is more prone to overfitting. The RoBERTa baseline performs better than RoBERTa with supervised pre-training on synthetic data, which might be caused by the fact that synthetic documents are based on selected layouts from the training set and these layouts are not present in the zero-shot test subset, although we do not see the same effect in the case of LayoutLMv3 or the LIR task. Union-RoBERTa gets to the second place; considering it is basically an ensemble of the baselines, this might be an indicator that ensembling can also improve generalization properties. It is also worth mentioning that ViBERTGrid is very good in generalization when the F1 measure is concerned.

The LIR task (Fig. 7b) shows similar results—GraphDoc remains on the first place, LiLT lost its second position to RoBERTa with supervised pre-training on synthetic data and LayoutLMv3 baseline pre-trained on synthetic data swapped its position with RoBERTa baseline. Note, that for both tasks, the results are significantly worse for the zero-shot setup compared to the overall results, showing a room for improvement with respect to generalization of all competing methods.

The results of the few-shot evaluation are in Fig. 8. The KILE task (Fig. 8a) shows that only a few similar layouts during training can help significantly. We see, that YOLOv8 gets back to the second place, RoBERTa+synth baseline improves significantly. It is also worth mentioning that all methods improve both the AP and F1 metrics by roughly 10%, compared to the zero-shot setup, with some exceptions with even a better improvement, and ViBERTGrid, which has a lower improvement.

In the LIR task (Fig. 8b), we can see that all methods get closer to each other, similarly as it was in the overall evaluation. However, what is really surprising is that the results for zero-shot variant were actually slightly better than the results for few-shot. Also, the LiLT benefits from seeing at least a few similar layouts during training much more than GraphDoc and overtakes its first position. Also RoBERTa baseline is slightly better than RoBERTa+synth.

Finally, in Fig. 9, we show the results for the many-shot scenario. For the KILE task (Fig. 9a), it can be seen that the order of competing methods converges to the same one as for the overall results, with the only exception of LayoutLMv3 and LayoutLMv3+synth baselines, which are swapped. We can also see, that the results are roughly 10% better than for the overall case, which is not surprising, since the overall case contains also unseen layout examples. For the LIR task (Fig. 9b), we see a similar trend, but the improvement is not that significant. Interestingly, the LayoutLMv3+synth baseline gets to the second place outperforming both LiLT and RoBERTa+synth baselines. However, we should point out that the results of these methods are very close.

5.3 Using Synthetic and Unlabeled Data

According to the submitted participant papers, only GraphDoc and partly also Union-RoBERTa (they used only 10, 000 samples) leveraged the unlabeled part of the DocILE dataset. We believe that the reason for not using the unlabeled data was mainly relatively tight time constraints. It is visible that GraphDoc-based method wins in almost all comparisons with the exception of few-shot (Fig. 8b) and text extraction (Fig. 6b) LIR tasks. However, it is hard to judge if this could be attributed to the usage of the unlabeled data.

Only the authors of Union-RoBERTa report the usage of the synthetic part of the DocILE dataset. Again, the reason for not using the provided synthetic data might be time constraints. From the baselines point of view, we see that using the synthetic data helps in most situations, with a few exceptions like zero-shot KILE task (Fig. 7a) and few-shot LIR task (Fig. 8b), where RoBERTa performs better than RoBERTa+synth, however, simultaneously the LayoutLMv3+synth outperforms LayoutLMv3. But we should point out that in these cases the differences are not very big.

5.4 Importance of Score for Average Precision

While for the F1 metric the score assigned to individual predictions is ignored, it plays an important role for the AP metric. In AP, predicted fields are first sorted by the score, then the precision-recall pairs are computed iteratively and finally the metric itself is the average precision achieved for different recall thresholds.

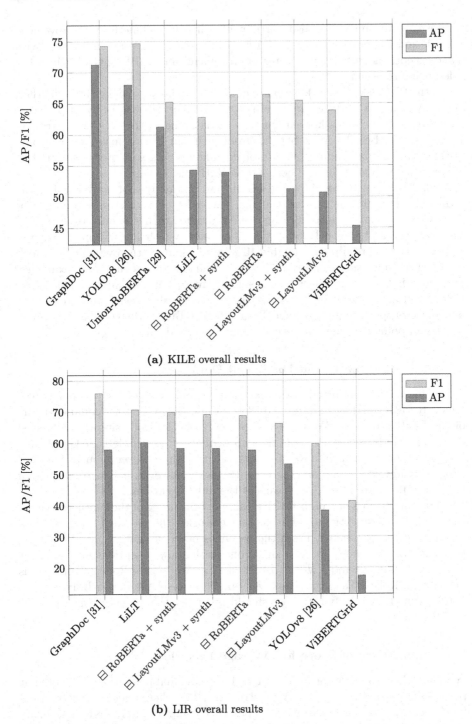

(a) KILE overall results

(b) LIR overall results

Fig. 5. Final results of the DocILE'23 competition for Task 1: KILE (5a) and Task 2: LIR (5b).

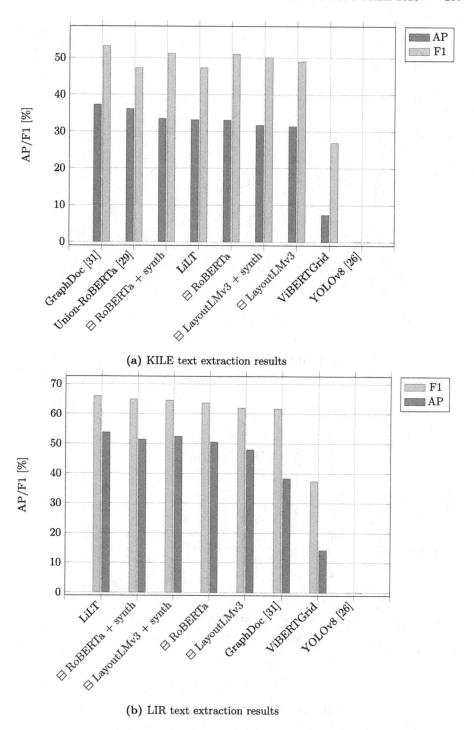

(a) KILE text extraction results

(b) LIR text extraction results

Fig. 6. Text extraction results for Task 1: KILE (6a) and Task 2: LIR (6b).

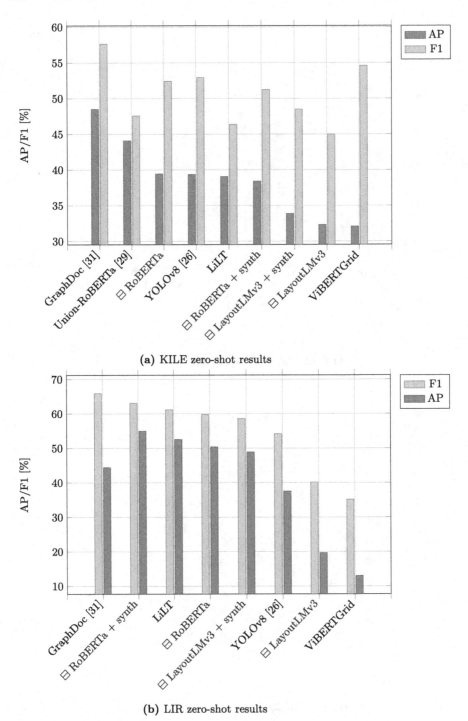

(a) KILE zero-shot results

(b) LIR zero-shot results

Fig. 7. Results on the zero-shot subset for Task 1: KILE (7a) and Task 2: LIR (7b).

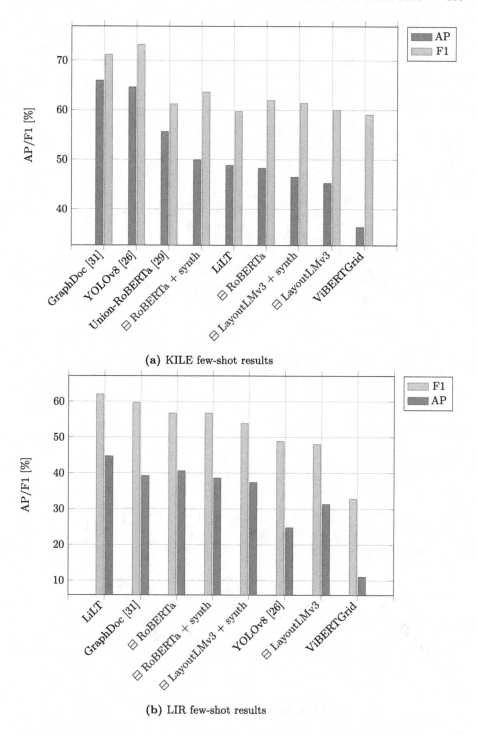

(a) KILE few-shot results

(b) LIR few-shot results

Fig. 8. Results on the few-shot subset for Task 1: KILE (8a) and Task 2: LIR (8b).

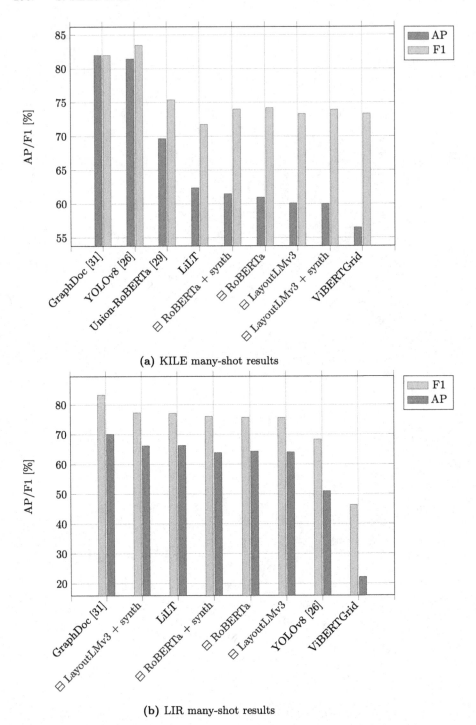

(a) KILE many-shot results

(b) LIR many-shot results

Fig. 9. Results on the many-shot subset for Task 1: KILE (9a) and Task 2: LIR (9b).

Therefore, if we can ensure that there are more true positives among the predictions with higher score than among the predictions with lower score, the precision will increase for lower recall thresholds and remain similar for higher recall thresholds, when compared to the case when scores are random.

To prove this point, we can look at two examples. The ViBERTGrid method used the same score for all predictions and it achieves very poor results on AP compared to its results on F1, as can be seen in Fig. 5. On the other hand, in the participant paper of the GraphDoc method, they argue that the prediction score is important for the AP metric and they show that by using a carefully selected score they achieve a 13.6% higher result on AP on the validation set compared to using the same score for all predictions. We can see in Fig. 5 that for the KILE task GraphDoc has the smallest difference between the AP and F1 metrics of all the methods. This is not the case for LIR, maybe because here AP was not the main evaluation metric and so less focus might have been given to assigning a correct score to the predictions in this case.

6   Conclusion

We presented the first edition of the DocILE 2023 competition, which consisted of two tracks: KILE and LIR. Both tasks consist of detection of pre-defined categories of information in business documents. The latter task additionally requires grouping the information into tuples. In the end, we obtained 5 submissions for KILE and 4 submissions for LIR. The diversity of the chosen approaches shows the potential of the DocILE dataset and benchmark, which spans the computer vision domain, the layout analysis, and natural language processing. Unsurprisingly, some of the submissions even used a multi-modal approach. The values of the respective error metrics indicate that the benchmark is non-trivial and the problems are far from being solved.

The benchmark remains open to new submissions, leaving it as a springboard for future research and for the document understanding community. To point out just a few possible research questions for this benchmark: 1) How to best use the unlabeled and synthetic datasets (as most of the solutions did not focus on these parts of the dataset as much)? 2) Is it possible to better utilize the fact that many documents share the same layout and push the performance on the few-shot subset closer to the performance on the many-shot subset? 3) Which parts of the tasks are better solved by pure NLP solutions (such as the baselines), which are better solved by pure CV solutions (such as YOLOv8) and do the multi-modal solutions (such as GraphDoc) already utilize both of the modalities as much as possible or is one of the modalities still under-utilized?

References

1. Davis, B., Morse, B., Cohen, S., Price, B., Tensmeyer, C.: Deep visual template-free form parsing. In: ICDAR (2019)

2. Hammami, M., Héroux, P., Adam, S., d'Andecy, V.P.: One-shot field spotting on colored forms using subgraph isomorphism. In: ICDAR (2015)
3. Herzig, J., Nowak, P.K., Müller, T., Piccinno, F., Eisenschlos, J.M.: Tapas: weakly supervised table parsing via pre-training. arXiv (2020)
4. Hong, T., Kim, D., Ji, M., Hwang, W., Nam, D., Park, S.: Bros: a pre-trained language model focusing on text and layout for better key information extraction from documents. In: AAAI (2022)
5. Huang, Y., Lv, T., Cui, L., Lu, Y., Wei, F.: LayoutLMv3: pre-training for document AI with unified text and image masking. In: ACM-MM (2022)
6. Huang, Z., et al.: ICDAR2019 competition on scanned receipt OCR and information extraction. In: ICDAR (2019)
7. Jocher, G., Chaurasia, A., Qiu, J.: YOLO by Ultralytics (2023). https://github.com/ultralytics/ultralytics
8. Katti, A.R., et al.: CharGrid: towards understanding 2D documents. In: Riloff, E., Chiang, D., Hockenmaier, J., Tsujii, J. (eds.) Proceedings of the 2018 Conference on Empirical Methods in Natural Language Processing, Brussels, Belgium, 31 October–4 November 2018, pp. 4459–4469. Association for Computational Linguistics (2018). https://aclanthology.org/D18-1476/
9. Lewis, D., Agam, G., Argamon, S., Frieder, O., Grossman, D., Heard, J.: Building a test collection for complex document information processing. In: SIGIR (2006)
10. Lin, W., et al.: ViBERTgrid: a jointly trained multi-modal 2D document representation for key information extraction from documents. In: Lladós, J., Lopresti, D., Uchida, S. (eds.) ICDAR 2021. LNCS, vol. 12821, pp. 548–563. Springer, Cham (2021). https://doi.org/10.1007/978-3-030-86549-8_35
11. Lohani, D., Belaïd, A., Belaïd, Y.: An invoice reading system using a graph convolutional network. In: Carneiro, G., You, S. (eds.) ACCV 2018. LNCS, vol. 11367, pp. 144–158. Springer, Cham (2019). https://doi.org/10.1007/978-3-030-21074-8_12
12. Majumder, B.P., Potti, N., Tata, S., Wendt, J.B., Zhao, Q., Najork, M.: Representation learning for information extraction from form-like documents. In: ACL (2020)
13. Mathew, M., Bagal, V., Tito, R., Karatzas, D., Valveny, E., Jawahar, C.: InfographicVQA. In: WACV (2022)
14. Mathew, M., Karatzas, D., Jawahar, C.: DocVQA: a dataset for VQA on document images. In: WACV (2021)
15. Mindee: docTR: Document Text Recognition. https://github.com/mindee/doctr (2021)
16. Olejniczak, K., Šulc, M.: Text detection forgot about document OCR. In: CVWW (2023)
17. Powalski, R., Borchmann, Ł, Jurkiewicz, D., Dwojak, T., Pietruszka, M., Pałka, G.: Going full-TILT boogie on document understanding with text-image-layout transformer. In: Lladós, J., Lopresti, D., Uchida, S. (eds.) ICDAR 2021. LNCS, vol. 12822, pp. 732–747. Springer, Cham (2021). https://doi.org/10.1007/978-3-030-86331-9_47
18. Riba, P., Dutta, A., Goldmann, L., Fornés, A., Ramos, O., Lladós, J.: Table detection in invoice documents by graph neural networks. In: ICDAR (2019)
19. Russakovsky, O., et al.: ImageNet large scale visual recognition challenge. Int. J. Comput. Vis. 115(3), 211–252 (2015). https://doi.org/10.1007/s11263-015-0816-y
20. Schreiber, S., Agne, S., Wolf, I., Dengel, A., Ahmed, S.: DeepDeSRT: deep learning for detection and structure recognition of tables in document images. In: ICDAR (2017)

21. Šimsa, Š, Šulc, M., Skalický, M., Patel, Y., Hamdi, A.: DocILE 2023 teaser: document information localization and extraction. In: Kamps, J., et al. (eds.) ECIR 2023. LNCS, vol. 13982, pp. 600–608. Springer, Cham (2023). https://doi.org/10.1007/978-3-031-28241-6_69

22. Šimsa, Š., et al.: DocILE benchmark for document information localization and extraction. arXiv preprint arXiv:2302.05658 (2023). Accepted to ICDAR 2023

23. Skalický, M., Šimsa, Š, Uřičář, M., Šulc, M.: Business document information extraction: Towards practical benchmarks. In: Barrón-Cedeño, A., et al. (eds.) CLEF 2022. LNCS, vol. 13390, pp. 105–117. Springer, Cham (2022). https://doi.org/10.1007/978-3-031-13643-6_8

24. Straka, J., Gruber, I.: Object detection pipeline using YOLOv8 for document information extraction. In: Aliannejadi, M., Faggioli, G., Ferro, N., Vlachos, M. (eds.) Working Notes of CLEF 2023 - Conference and Labs of the Evaluation Forum, Thessaloniki, Greece, 18–21 September. CEUR Workshop Proceedings, CEUR-WS.org (2023)

25. Tanaka, R., Nishida, K., Yoshida, S.: VisualMRC: machine reading comprehension on document images. In: AAAI (2021)

26. Tang, Z., et al.: Unifying vision, text, and layout for universal document processing. arXiv (2022)

27. Tran, B.G., Bao, D.N.M., Bui, K.G., Duong, H.V., Nguyen, D.H., Nguyen, H.M.: Union-RoBERTa: RoBERTas ensemble technique for competition on document information localization and extraction. In: Aliannejadi, M., Faggioli, G., Ferro, N., Vlachos, M. (eds.) Working Notes of CLEF 2023 - Conference and Labs of the Evaluation Forum, Thessaloniki, Greece, 18–21 September. CEUR Workshop Proceedings, CEUR-WS.org (2023)

28. Wang, J., Jin, L., Ding, K.: LiLT: a simple yet effective language-independent layout transformer for structured document understanding. In: ACL (2022)

29. Wang, Y., Du, J., Ma, J., Hu, P., Zhang, Z., Zhang, J.: USTC-iFLYTEK at DocILE: a multi-modal approach using domain-specific GraphDoc. In: Aliannejadi, M., Faggioli, G., Ferro, N., Vlachos, M. (eds.) Working Notes of CLEF 2023 - Conference and Labs of the Evaluation Forum, Thessaloniki, Greece, 18–21 September. CEUR Workshop Proceedings, CEUR-WS.org (2023)

30. Web: Industry Documents Library. https://www.industrydocuments.ucsf.edu/. Accessed 20 Oct 2022

31. Web: Public Inspection Files. https://publicfiles.fcc.gov/. Accessed 20 Oct 2022

32. Xu, Y., et al.: LayoutLMv2: multi-modal pre-training for visually-rich document understanding. In: ACL (2021)

33. Xu, Y., Li, M., Cui, L., Huang, S., Wei, F., Zhou, M.: LayoutLM: pre-training of text and layout for document image understanding. In: KDD (2020)

34. Zhang, Z., Ma, J., Du, J., Wang, L., Zhang, J.: Multimodal pre-training based on graph attention network for document understanding. IEEE Trans. Multimed. (2022)

35. Zhong, X., Tang, J., Jimeno-Yepes, A.: PubLayNet: largest dataset ever for document layout analysis. In: ICDAR (2019)

36. Zhou, J., Yu, H., Xie, C., Cai, H., Jiang, L.: iRMP: from printed forms to relational data model. In: HPCC (2016)

37. Zhu, Y., et al.: Aligning books and movies: towards story-like visual explanations by watching movies and reading books. In: ICCV (2015)

Overview of eRisk 2023: Early Risk Prediction on the Internet

Javier Parapar[1](✉)[ID], Patricia Martín-Rodilla[1][ID], David E. Losada[2][ID],
and Fabio Crestani[3][ID]

[1] Information Retrieval Lab, Centro de Investigación en Tecnoloxías da Información
e as Comunicacións (CITIC), Universidade da Coruña, A Coruña, Spain
{javierparapar,patricia.martin.rodilla}@udc.es
[2] Centro Singular de Investigación en Tecnoloxías Intelixentes (CiTIUS),
Universidade de Santiago de Compostela, Santiago de Compostela, Spain
david.losada@usc.es
[3] Faculty of Informatics, Università della Svizzera italiana (USI), Lugano,
Switzerland
fabio.crestani@usi.ch

Abstract. This paper provides an overview of eRisk 2023, the seventh
edition of the CLEF conference's lab dedicated to early risk detection.
Since its inception, our lab has aimed to explore evaluation methodologies, effectiveness metrics, and other processes associated with early
risk detection. The applications of early alerting models are diverse and
encompass various domains, including health and safety. eRisk 2023 consisted of three tasks. The first task involved ranking sentences based on
their relevance to standardised depression symptoms. The second task
focused on early detection of signs related to pathological gambling. The
third task required participants to automatically estimate an eating disorders questionnaire by analysing user writings on social media.

Keywords: Early risk · Depression · Pathological gambling · Eating
disorders

1 Introduction

The primary objective of eRisk is to investigate evaluation methodologies, metrics, and other relevant factors related to the development of research collections
and the identification of signs associated with early risk detection. Early detection technologies have significant potential across various fields, especially those
targeting safety and health applications. In situations where individuals display
symptoms of mental illnesses, infants face interactions with sexual abusers, or
potential criminals publish antisocial threats online, automated systems can play
a vital role by issuing early warnings.

A. Arampatzis et al. (Eds.): CLEF 2023, LNCS 14163, pp. 294–315, 2023.
https://doi.org/10.1007/978-3-031-42448-9_22

Our lab primarily focuses on psychological issues, specifically depression, self-harm, pathological gambling, and eating disorders. Over the years, we have discovered that the interaction between psychological diseases and language use is complex and that there is room for improvement in the effectiveness of automatic language-based screening models. In 2017, we conducted an exploratory task on the early detection of depression, utilising innovative evaluation methods and a test dataset described in [11]. The following year, in 2018, we continued fostering the detection of early signs of depression and introduced a new task for detecting early signs of anorexia [12,13]. The subsequent year, 2019, saw the continuation of the challenge on early identification of anorexia symptoms, the introduction of a new challenge on early detection of self-harm, and the proposal of a third task focused on estimating a user's responses to a depression questionnaire based on their social media interactions [14–16]. In 2020, we further pursued the early detection of self-harm and introduced a task for estimating the severity of depression symptoms [17–19]. In 2021, our focus shifted to two tasks: early detection of pathological gambling and self-harm and a task for severity estimation of depression [28–30]. Finally, last year, we presented three tasks: early pathological gambling detection, early detection of depression, and severity estimation of eating disorders [31–33].

In 2023, eRisk presented three campaign-style tasks [34]. The first task consisted of ranking sentences based on their relevance to each of the 21 symptoms of depression derived from the BDI-II questionnaire. To that end, this novel task provided participants with a collection of sentences extracted from publications of social media users. The second task represented the third edition of early risk detection of pathological gambling. Lastly, the third task was the second edition of the eating disorder severity estimation challenge. Detailed descriptions of these tasks can be found in the subsequent sections of this overview article.

We had 98 teams registered for the lab. We finally received results from 20 of them: 37 runs for Task 1, 48 runs for Task 2 and 20 for Task 3.

2   Task 1: Search for Symptoms of Depression

This was a new task in 2023. Task 1 consists of producing ranking of sentences (derived from user's writings) based on their relevance to specific symptoms of depression. Participants were asked to rank sentences according to their relevance to the 21 standardised symptoms outlined in the BDI-II Questionnaire [3]. In this context, a sentence was considered relevant to a particular symptom if it conveyed information about the user's state in connection with that symptom. It is important to note that a sentence could be deemed as relevant even if it indicates a positive information about the symptom (e.g., "*I feel quite happy lately*" should be considered relevant for symptom 1 "Sadness" of the BDI-II).

```
1 QO sentence-id-121 0001 10 myGroupNameMyMethodName
1 QO sentence-id-234 0002 9.5 myGroupNameMyMethodName
1 QO sentence-id-345 0003 9 myGroupNameMyMethodName
...
21 QO sentence-id-456 0998 1.25 myGroupNameMyMethodName
21 QO sentence-id-242 0999 1 myGroupNameMyMethodName
21 QO sentence-id-347 1000 0.9 myGroupNameMyMethodName
```

Fig. 1. Example of a participant's run.

2.1   Dataset

The corpus provided to the participants was a TREC formatted sentence-tagged dataset (based on eRisk's past data). Table 1 reports some statistics of the corpus.

Table 1. Corpus statistics for Task 1: Search for Symptoms of Depression.

Number of users	3,107
Number of sentences	3,807,115
Average number of words per sentence	13.63

2.2   Assessment Process

Given the corpus of sentences and the description of the symptoms from the BDI-II questionnaire, the participants were free to decide on the best strategy to derive queries for representing the BDI-II symptoms. Each participating team submitted up to 5 variants (runs). Each run included 21 TREC-style formatted rankings of sentences, as shown in Fig. 1. For each symptom, the participants could should submit up to 1000 results sorted by estimated relevance. We received 37 runs from 10 participating teams (see Table 2).

To create the relevance judgments, three expert assessors annotated a pool of sentences associated with each symptom. These candidate sentences were obtained by performing top-k pooling from the relevance rankings submitted by the participants in the task (37 different ranking methods).

The assessors were provided with specific instructions to determine the relevance of candidate sentences. They were instructed to consider a sentence relevant if it addressed the topic and provided explicit information about the individual's state in relation to the symptom. This dual concept of relevance (on-topic and reflective of the user's state with respect to the symptom) introduced a higher level of complexity compared to more standard relevance assessments. Consequently, we developed a robust annotation methodology and formal assessment guidelines to ensure consistency and accuracy.

Table 2. Task 1 (Search for Symptoms of Depression): Number of runs from participants.

Team	# of submissions
BLUE	5
Formula-ML	4
GMU-FAST	2
Manson-NLP	1
NailP	5
OBSER-MENH	5
RELAI	5
UMU	2
UNSL	3
uOttawa	5
Total	37

To create the pool of sentences for assessment, we implemented top-k pooling with $k = 50$. The resulting pool sizes per sentence are reported in Table 3. We observed that certain sentences had identical text but different IDs, possibly due to multiple users writing the same text. To alleviate the assessors' workload, we automated the removal of duplicates. The annotation software, which was specifically developed to support eRisk 2023's annotation, automatically assigned the assessors' relevance labels to all sentences that are identical.

The annotation process involved a team of three assessors with different backgrounds and expertise. One of the assessors has professional training in psychology, while the other two are computer science researchers –a postdoctoral fellow and a Ph.D. student– with a specialisation in early risk technologies.

To ensure consistency and clarity throughout the process, the lab organisers conducted a preparatory session with the assessors. During this session, an initial version of the guidelines was discussed, and any doubts or questions raised by the assessors were addressed. This collaborative effort resulted in the final version of the guidelines[1].

In accordance with these guidelines, a sentence is considered relevant only if it provides "some information about the state of the individual related to the topic of the BDI item". This criterion serves as the basis for determining the relevance of sentences during the annotation process.

Following the initial meeting, the assessors labelled the pools of the first three BDI topics. Subsequently, we organised another meeting to further address any additional concern or question that arose during the process. This collaborative session proved valuable in refining the annotation criteria and ensuring consistency. The final outcomes of the annotation process are presented in Table 3, where the number of relevant sentences per BDI item is reported (last two columns). We marked a sentence as relevant following two aggregation criteria: unanimity and majority.

---

[1] https://erisk.irlab.org/guidelines_task1_erisk23.html.

Table 3. Task 1 (Search for Symptoms of Depression): Size of the pool for every BDI Item

BDI Item (#)	original	unique	# rels (3/3)	# rels (2/3)
Sadness (1)	1110	1069	179	318
Pessimism (2)	1150	1096	104	325
Past Failure (3)	973	918	160	300
Loss of Pleasure (4)	1013	948	97	204
Guilty Feelings (5)	829	794	83	143
Punishment Feelings (6)	1079	1036	21	50
Self-Dislike (7)	1005	957	158	288
Self-Criticalness (8)	1072	1023	76	174
Suicidal Thoughts or Wishes (9)	953	923	260	349
Crying (10)	983	917	230	320
Agitation (11)	1080	1057	69	154
Loss of Interest (12)	1077	1021	70	168
Indecisiveness (13)	1110	1044	61	141
Worthlessness (14)	1067	986	71	144
Loss of Energy (15)	1082	1027	129	204
Changes in Sleeping Pattern (16)	938	904	203	350
Irritability (17)	1047	1008	94	155
Changes in Appetite (18)	984	947	103	224
Concentration Difficulty (19)	1024	981	83	141
Tiredness or Fatigue (20)	1033	994	123	221
Loss of Interest in Sex (21)	971	922	97	158

2.3   Results

The performance results for the participating systems are shown in Tables 4 (majority-based qrels) and 5 (unanimity-based qrels). The tables report several standard performance metrics, such as Mean Average Precision (MAP), mean R-Precision, mean Precision at 10 and mean NDCG at 1000. Remarkably, Formula-ML, from the team NITK Surathkal, achieved the top-ranking performance across all metrics and relevance judgement types. Their effective results demonstrate their exceptional competence in the field.

3   Task 2: Early Detection of Pathological Gambling

This task represents the third edition of the challenge, which aims to develop innovative models for the early identification of pathological gambling risk. Pathological gambling, also referred to as ludomania and commonly known as

Table 4. Ranking-based evaluation for Task 1 (majority voting)

Team	Run	AP	R-PREC	P@10	NDCG
Formula-ML [35]	SentenceTrainsformers_0.25	**0.319**	**0.375**	**0.861**	**0.596**
Formula-ML	SentenceTrainsformers_0.1	0.308	0.359	**0.861**	0.584
Formula-ML	result2	0.086	0.170	0.457	0.277
Formula-ML	word2vec_0.1	0.092	0.176	0.500	0.285
OBSER-MENH [21]	salida-distilroberta-90-cos	0.294	0.359	0.814	0.578
OBSER-MENH	salida-mpnet-90-cos	0.265	0.333	0.805	0.550
OBSER-MENH	salida-mpnet-21-cos	0.120	0.207	0.471	0.365
OBSER-MENH	salida-distilroberta-21-cos	0.158	0.249	0.543	0.418
OBSER-MENH	salida-mini12-21-cos	0.114	0.184	0.305	0.329
uOttawa [41]	USESim	0.160	0.248	0.600	0.382
uOttawa	Glove100Sim	0.017	0.052	0.195	0.105
uOttawa	RobertaSim	0.033	0.080	0.329	0.150
uOttawa	GloveSim	0.011	0.038	0.162	0.075
uOttawa	BertSim	0.084	0.150	0.505	0.271
BLUE [5]	SemSearchOnBDI2Queries	0.104	0.126	0.781	0.211
BLUE	SemSearchOnGeneratedQueriesMentalRoberta	0.029	0.063	0.367	0.105
BLUE	SemSearchOnBDI2QueriesMentalRoberta	0.027	0.044	0.386	0.089
BLUE	SemSearchOnGeneratedQueries	0.052	0.074	0.586	0.139
BLUE	SemSearchOnAllQueries	0.065	0.086	0.629	0.160
NailP [4]	T1_M2	0.095	0.146	0.519	0.226
NailP	T1_M4	0.095	0.146	0.519	0.221
NailP	T1_M3	0.073	0.114	0.471	0.180
NailP	T1_M5	0.089	0.140	0.486	0.223
NailP	T1_M1	0.074	0.114	0.471	0.189
RELAI [22]	bm25|mpnetbase	0.048	0.081	0.538	0.140
RELAI	BM25	0.016	0.061	0.043	0.145
RELAI	bm25|mpnetbase_simcse	0.030	0.066	0.390	0.114
RELAI	bm25|mpnetqa_simcse	0.027	0.063	0.376	0.109
RELAI	bm25|mpnetqa	0.038	0.075	0.438	0.126
UNSL [39]	Prompting-Classifier	0.036	0.090	0.229	0.180
UNSL	Similarity-AVG	0.001	0.008	0.010	0.016
UNSL	Similarity-MAX	0.001	0.011	0.019	0.019
UMU [27]	LexiconMultilingualSentenceTransformer	0.073	0.140	0.495	0.222
UMU	LexiconSentenceTransformer	0.054	0.122	0.362	0.191
GMU-FAST	FAST-DCMN-COS-INJECT	0.001	0.002	0.014	0.004
GMU-FAST	FAST-DCMN-COS-INJECT_FULL	0.001	0.003	0.014	0.005
Mason-NLP [37]	MentalBert	0.035	0.072	0.286	0.117

"gambling addiction", involves an uncontrollable urge to gamble despite the negative consequences. According to the World Health Organization (WHO), the prevalence of adult gambling addiction in 2017 ranged from 0.1% to 6.0% [1]. The objective of this task was to process evidence in a sequential manner and detect early indications of compulsive or disordered gambling as soon as possible. Participating systems were required to analyse user posts on social media in the

Table 5. Ranking-based evaluation for Task 1 (unanimity)

Team	Run	MAP	R-PREC	P@10	NDCG
Formula-ML [35]	SentenceTransformers_0.25	0.268	**0.360**	**0.709**	**0.615**
Formula-ML	SentenceTransformers_0.1	**0.293**	0.350	0.685	0.611
Formula-ML	result2	0.079	0.155	0.357	0.290
Formula-ML	word2vec_0.1	0.085	0.163	0.357	0.299
OBSER-MENH [21]	salida-distilroberta-90-cos	0.281	0.344	0.652	0.604
OBSER-MENH	salida-mpnet-90-cos	0.252	0.337	0.643	0.575
OBSER-MENH	salida-distilroberta-21-cos	0.135	0.216	0.390	0.413
OBSER-MENH	salida-mini12-21-cos	0.099	0.165	0.214	0.329
OBSER-MENH	salida-mpnet-21-cos	0.101	0.189	0.319	0.366
uOttawa [41]	USESim	0.139	0.232	0.438	0.380
uOttawa	GloveSim	0.008	0.028	0.110	0.063
uOttawa	Glove100Sim	0.011	0.042	0.110	0.092
uOttawa	RobertaSim	0.025	0.068	0.190	0.140
uOttawa	BertSim	0.070	0.130	0.357	0.260
BLUE [5]	SemSearchOnBDI2Queries	0.129	0.167	0.643	0.260
BLUE	SemSearchOnAllQueries	0.067	0.105	0.452	0.177
BLUE	SemSearchOnGeneratedQueriesMentalRoberta	0.018	0.059	0.186	0.085
BLUE	SemSearchOnGeneratedQueries	0.052	0.088	0.381	0.147
BLUE	SemSearchOnBDI2QueriesMentalRoberta	0.032	0.058	0.300	0.104
NailP [4]	T1_M2	0.090	0.143	0.410	0.229
NailP	T1_M4	0.090	0.143	0.410	0.224
NailP	T1_M5	0.083	0.139	0.338	0.222
NailP	T1_M1	0.073	0.114	0.343	0.192
NailP	T1_M3	0.073	0.114	0.343	0.181
UMU [27]	LexiconSentenceTransformer	0.044	0.110	0.210	0.175
UMU	LexiconMultilingualSentenceTransformer	0.059	0.125	0.333	0.209
RELAI [22]	BM25	0.012	0.036	0.019	0.135
RELAI	bm25|mpnetbase_simcse	0.026	0.059	0.243	0.103
RELAI	bm25|mpnetqa_simcse	0.023	0.052	0.262	0.097
RELAI	bm25|mpnetqa	0.030	0.065	0.290	0.109
RELAI	bm25|mpnetbase	0.039	0.069	0.343	0.124
UNSL [39]	Similarity-MAX	0.001	0.006	0.010	0.012
UNSL	Prompting-Classifier	0.020	0.063	0.090	0.157
UNSL	Similarity-AVG	0.000	0.005	0.005	0.011
GMU-FAST	FAST-DCMN-COS-INJECT_FULL	0.001	0.003	0.014	0.006
GMU-FAST	FAST-DCMN-COS-INJECT	0.001	0.002	0.010	0.003
Mason-NLP [37]	MentalBert	0.024	0.054	0.190	0.099

order they were written. Successful outcomes from this task could potentially be utilised for sequential monitoring of user interactions across various online platforms such as blogs, social networks, and other forms of digital media.

The test collection utilised for this task followed the same format as the collection described in the work by Losada and Crestani [10]. The collection contains writings, including posts and comments, obtained from a selected group of social media users. Within this dataset, users are categorised into two groups: pathological gamblers and non-pathological gamblers. For each user, the collection contains a sequence of writings arranged in chronological order. To facilitate the task and ensure uniform distribution, we established a dedicated server that systematically provided user writings to the participating teams. Further details regarding the server's setup and functioning are available at the lab's official website[2].

This was a train-test task. For the training stage, the teams had access to training data where we released the whole history of writings for training users. We indicated which users had explicitly mentioned that they are pathological gamblers. The participants could therefore tune their systems with the training data. In 2023, the training data for Task 1 was composed of user from previous editions of the self-harm task.

During the test stage, participants connected to our server and engaged in an iterative process of receiving user writings and sending their responses. At any point within the chronology of user writings, participants had the freedom to halt the process and issue an alert. After reading each user writing, teams were required to decide between two options: i) alerting about the user, indicating a predicted sign of gambling risk, or ii) not alerting about the user. Participants independently made this choice for each user in the test split. It is important to note that once an alert was issued, it was considered final, and no further decisions regarding that particular individual were taken into account. Conversely, the absence of alerts was considered non-final, allowing participants to subsequently submit an alert if they detected signs of risk emerging.

To evaluate the systems' performance, we employed two indicators: the accuracy of the decisions made and the number of user writings required to reach those decisions. These criteria provide valuable insights into the effectiveness and efficiency of the systems under evaluation. To support the test stage, we deployed a REST service. The server iteratively distributes user writings and waits for responses from participants. Importantly, new user data was not provided to a specific participant until the service received a decision from that particular team. The submission period for the task was open from January 16th, 2023, until April 14th, 2023.

To construct the ground truth assessments, we adopted established approaches that aim to optimise the utilisation of assessors' time, as documented in previous studies [25, 26]. These methods employ simulated pooling strategies, enabling the effective creation of test collections. The main statistics of the test collection used for T2 are presented in Table 6.

Table 6. Task 2 (pathological gambling). Main statistics of test collection

	Pathological Gamblers	Control
Num. subjects	103	2071
Num. submissions (posts & comments)	33,719	1,069,152
Avg num. of submissions per subject	327.33	516.25
Avg num. of days from first to last submission	≈675	≈878
Avg num. words per submission	28.9	20.47

3.1 Decision-Based Evaluation

This evaluation approach uses the binary decisions made by the participating systems for each user. In addition to standard classification measures such as Precision, Recall, and F1 score (computed with respect to the positive class), we also calculate ERDE (Early Risk Detection Error), which has been utilised in previous editions of the lab. A detailed description of ERDE was presented by Losada and Crestani in [10]. Essentially, ERDE is an error measure that incorporates a penalty for delayed correct alerts (true positives). The penalty increases with the delay in issuing the alert, measured by the number of user posts processed before making the alert.

Since 2019, we have incorporated additional decision-based metrics in our evaluation toolkit. These metrics aim to address certain limitations of ERDE. Some research teams have analyzed ERDE and proposed alternative evaluation approaches. Trotzek and colleagues [40] introduced $ERDE_o^\%$, a variant of ERDE that normalises the evaluation based on the percentage of user writings seen before the alert. While this approach addresses the issue of user contribution normalisation, it relies on knowledge of the total number of user writings, which may not be available in real-life applications. Another proposed alternative evaluation metric for early risk prediction is $F_{latency}$, proposed by Sadeque and colleagues [36]. This measure aligns well with our objectives. Additionally we also used $latency_{TP}$, the delay in emitting a decision computed for the true positives. More details about the metrics can be found in [32].

3.2 Ranking-Based Evaluation

In addition to the evaluation discussed above, we employed an alternative form of evaluation to further assess the systems. After each data release (new user writing), participants were required to provide the following information for each user in the collection:

- A decision for the user (alert or no alert), which was used to calculate the decision-based metrics discussed previously.
- A score representing the user's level of risk, estimated based on the evidence observed thus far.

The scores were used to create a ranking of users in descending order of estimated risk. For each participating system, a ranking was generated at each data release point, simulating a continuous re-ranking approach based on the observed evidence. In a real-life scenario, this ranking would be presented to an expert user who could make decisions based on the rankings (e.g., by inspecting the top of the rankings).

Each ranking can be evaluated using standard ranking metrics such as P@10 or NDCG. Therefore, we report the performance of the systems based on the rankings after observing different numbers of writings.

3.3 Results

Table 7 shows the participating teams, the number of runs submitted and the approximate lapse of time from the first response to the last response. This time-lapse is indicative of the degree of automation of each team's algorithms. A few of the submitted runs processed the entire thread of messages (2004), but many variants stopped earlier. Some of the teams were still submitting results at the deadline time. Two teams processed the thread of messages reasonably fast (around a day for processing the entire history of user messages). The rest of the teams took several days to run the whole process.

Table 7. Task 2 (pathological gambling): participating teams, number of runs, number of user writings processed by the team, and lapse of time taken for the entire process.

team	#runs	#user writings processed	lapse of time (from 1st to last response)
UNSL	3	2004	1 day 02:17:36.417
ELiRF-UPV	1	2004	1 day 13:03:54.419
Xabi_EHU	5	2004	4 days 23:52:41.454
OBSER-MENH	5	2004	6 days 03:56:44.247
RELAI	5	764	6 days 09:12:00.148
NLP-UNED-2	5	2004	7 days 04:24:33.158
NUS-eRisk	5	2004	9 days 14:39:26.347
BioNLP-IISERB	5	61	10 days 00:49:40.529
SINAI	5	809	10 days 13:00:02.164
UMU	5	2004	14 days 00:29:30.434
NLP-UNED	5	1151	54 days 19:27:42.538

Table 8 reports the decision-based performance achieved by the participating teams. In terms of Precision, $F1$, $ERDE_5$, $ERDE_{50}$, and latency-weighted $F1$ the best performing team was the ELiRF-UPV (run 0). Regarding $latency_{TP}$ and *speed* SINAI (runs 0 and 2) are the ones that having perfect values obtained

Table 8. Decision-based evaluation for Task 2

Team	Run	P	R	$F1$	$ERDE_5$	$ERDE_{50}$	$latency_{TP}$	$speed$	$latency\text{-}weighted\ F1$
UNSL [39]	2	0.752	0.854	0.800	0.048	0.013	14.0	0.949	0.759
UNSL	0	0.752	0.767	0.760	0.048	0.017	15.0	0.945	0.718
UNSL	1	0.79	0.806	0.798	0.048	0.014	13.0	0.953	0.761
ELiRF-UPV [20]	0	**1.000**	0.883	**0.938**	**0.026**	**0.010**	4.0	0.988	**0.927**
Xabi_EHU [9]	0	0.846	0.961	0.900	0.030	0.012	8.0	0.973	0.875
Xabi_EHU	1	0.89	0.864	0.877	0.035	0.017	12.0	0.957	0.839
Xabi_EHU	2	0.79	0.913	0.847	0.036	0.015	13.0	0.953	0.807
Xabi_EHU	3	0.829	0.942	0.882	0.033	0.013	12.0	0.957	0.844
Xabi_EHU	4	0.756	0.961	0.846	0.031	0.013	8.0	0.973	0.823
OBSER-MENH [21]	0	0.048	**1.000**	0.092	0.064	0.049	3.0	0.992	0.092
OBSER-MENH	1	0.048	**1.000**	0.092	0.063	0.050	3.0	0.992	0.091
OBSER-MENH	2	0.048	**1.000**	0.092	0.063	0.050	3.0	0.992	0.091
OBSER-MENH	3	0.048	**1.000**	0.092	0.063	0.049	3.0	0.992	0.091
OBSER-MENH	4	0.048	**1.000**	0.092	0.063	0.050	3.0	0.992	0.091
RELAI [22]	0	0.000	0.000	0.000	0.047	0.047			
RELAI	1	0.058	0.971	0.109	0.048	0.039	**1.0**	**1.000**	0.109
RELAI	2	0.058	0.971	0.109	0.048	0.039	**1.0**	**1.000**	0.109
RELAI	3	0.000	0.000	0.000	0.047	0.047			
RELAI	4	0.047	**1.000**	0.09	0.08	0.046	11.0	0.961	0.087
NLP-UNED-2 [6]	1	0.957	0.883	0.919	0.034	0.016	13.0	0.953	0.876
NLP-UNED-2	2	0.947	0.883	0.914	0.034	0.016	12.0	0.957	0.875
NLP-UNED-2	3	0.896	0.922	0.909	0.030	0.014	10.0	0.964	0.877
NLP-UNED-2	0	0.945	0.844	0.892	0.038	0.019	18.0	0.933	0.833
NLP-UNED-2	4	0.764	0.883	0.819	0.033	0.010	13.0	0.953	0.781
NUS-eRisk	4	0.062	0.951	0.117	0.059	0.040	6.0	0.981	0.114
NUS-eRisk	0	0.063	0.767	0.116	0.068	0.050	27.0	0.899	0.104
NUS-eRisk	1	0.06	0.903	0.113	0.068	0.043	13.0	0.953	0.107
NUS-eRisk	2	0.057	0.971	0.107	0.06	0.042	4.0	0.988	0.106
NUS-eRisk	3	0.067	0.874	0.125	0.065	0.042	17.0	0.938	0.117
BioNLP-IISERB [38]	0	0.933	0.68	0.787	0.038	0.037	62.0	0.766	0.603
BioNLP-IISERB	1	0.938	0.592	0.726	0.042	0.042	62.0	0.766	0.557
BioNLP-IISERB	3	**1.000**	0.049	0.093	0.045	0.045	**1.0**	**1.000**	0.093
BioNLP-IISERB	4	**1.000**	0.039	0.075	0.047	0.046	19.0	0.930	0.070
BioNLP-IISERB	2	0.000	0.000	0.000	0.047	0.047			
SINAI [24]	3	0.126	**1.000**	0.224	0.029	0.020	2.0	0.996	0.223
SINAI	0	0.115	**1.000**	0.206	0.029	0.021	**1.0**	**1.000**	0.206
SINAI	1	0.124	**1.000**	0.221	0.028	0.020	2.0	0.996	0.220
SINAI	2	0.108	**1.000**	0.195	0.03	0.022	**1.0**	**1.000**	0.195
SINAI	4	0.092	0.981	0.168	0.044	0.027	3.0	0.992	0.166
UMU [27]	1	**1.000**	0.388	0.559	0.047	0.043	94.5	0.651	0.364
UMU	0	0.086	**1.000**	0.158	0.039	0.029	2.0	0.996	0.157
UMU	2	0.048	**1.000**	0.092	0.057	0.044	2.0	0.996	0.091
UMU	3	0.593	0.311	0.408	0.048	0.045	80.0	0.701	0.286
UMU	4	0.048	**1.000**	0.091	0.053	0.045	2.0	0.996	0.090
NLP-UNED	0	0.057	0.903	0.108	0.052	0.052	**1.0**	**1.000**	0.108
NLP-UNED	1	0.053	0.845	0.099	0.064	0.063	141.0	0.502	0.050
NLP-UNED	2	0.054	0.854	0.101	0.056	0.055	**1.0**	**1.000**	0.101
NLP-UNED	3	0.055	0.874	0.103	0.056	0.056	**1.0**	**1.000**	0.103
NLP-UNED	4	0.066	0.728	0.121	0.071	0.071	142.0	0.500	0.060

Table 9. Ranking-based evaluation for Task 2

Team	Run	1 writing			100 writings			500 writings			1000 writings		
		P@10	NDCG@10	NDCG@100	P@10	NDCG@10	NDCG@100	P@10	NDCG@10	NDCG@100	P@10	NDCG@10	NDCG@100
UNSL [39]	0	1.00	1.00	0.46	1.00	1.00	0.70	1.00	1.00	0.64	1.00	1.00	0.64
UNSL	1	1.00	1.00	0.57	1.00	1.00	0.78	1.00	1.00	0.67	1.00	1.00	0.70
UNSL	2	1.00	1.00	0.55	1.00	1.00	0.75	1.00	1.00	0.69	1.00	1.00	0.69
ELiRF-UPV [20]	0	1.00	1.00	0.59	1.00	1.00	0.91	1.00	1.00	0.95	1.00	1.00	0.94
Xabi_EHU [9]	0	1.00	1.00	0.57	1.00	1.00	0.50	0.80	0.88	0.41	0.90	0.94	0.41
Xabi_EHU	1	1.00	1.00	0.56	0.90	0.94	0.49	0.70	0.76	0.38	0.80	0.88	0.40
Xabi_EHU	2	1.00	1.00	0.55	1.00	1.00	0.51	0.70	0.79	0.40	0.80	0.88	0.41
Xabi_EHU	3	1.00	1.00	0.56	0.90	0.94	0.51	0.80	0.86	0.41	0.90	0.94	0.42
Xabi_EHU	4	1.00	1.00	0.58	1.00	1.00	0.50	0.80	0.86	0.40	0.90	0.94	0.41
OBSER-MENH [21]	0	1.00	1.00	0.64	1.00	1.00	0.55	1.00	1.00	0.48	1.00	1.00	0.50
OBSER-MENH	1	1.00	1.00	0.65	1.00	1.00	0.56	1.00	1.00	0.49	1.00	1.00	0.50
OBSER-MENH	2	1.00	1.00	0.65	1.00	1.00	0.56	1.00	1.00	0.49	1.00	1.00	0.50
OBSER-MENH	3	1.00	1.00	0.64	1.00	1.00	0.56	1.00	1.00	0.48	1.00	1.00	0.50
OBSER-MENH	4	1.00	1.00	0.65	1.00	1.00	0.56	1.00	1.00	0.49	1.00	1.00	0.50
RELAI [22]	0	0.30	0.25	0.08	0.00	0.00	0.02	0.00	0.00	0.06			
RELAI	1	0.00	0.00	0.01	0.00	0.00	0.00	0.00	0.00	0.00			
RELAI	2	0.00	0.00	0.02	0.00	0.00	0.01	0.00	0.00	0.02			
RELAI	3	0.00	0.00	0.01	0.00	0.00	0.01	0.00	0.00	0.00			
RELAI	4	0.10	0.06	0.02	0.00	0.00	0.04	0.00	0.00	0.02			
NLP-UNED-2 [6]	0	1.00	1.00	0.32	1.00	1.00	0.83	1.00	1.00	0.90	1.00	1.00	0.90
NLP-UNED-2	1	1.00	1.00	0.40	1.00	1.00	0.86	1.00	1.00	0.93	1.00	1.00	0.94
NLP-UNED-2	2	1.00	1.00	0.40	1.00	1.00	0.85	1.00	1.00	0.92	1.00	1.00	0.93
NLP-UNED-2	3	1.00	1.00	0.59	1.00	1.00	0.92	1.00	1.00	0.95	1.00	1.00	0.93
NLP-UNED-2	4	1.00	1.00	0.57	1.00	1.00	0.89	1.00	1.00	0.89	1.00	1.00	0.87
NUS-eRisk	0	0.00	0.00	0.06	0.00	0.00	0.02	0.10	0.19	0.07	0.00	0.00	0.02
NUS-eRisk	1	0.00	0.00	0.03	0.00	0.00	0.02	0.10	0.19	0.06	0.00	0.00	0.01
NUS-eRisk	2	0.00	0.00	0.04	0.00	0.00	0.03	0.10	0.10	0.03	0.00	0.00	0.01
NUS-eRisk	3	0.00	0.00	0.06	0.00	0.00	0.04	0.10	0.19	0.08	0.00	0.00	0.02
NUS-eRisk	4	0.00	0.00	0.06	0.00	0.00	0.03	0.10	0.12	0.05	0.00	0.00	0.02
BioNLP-IISERB [38]	0	0.40	0.60	0.14									
BioNLP-IISERB	1	0.00	0.00	0.02									
BioNLP-IISERB	2	0.00	0.00	0.03									
BioNLP-IISERB	3	0.00	0.00	0.05									
BioNLP-IISERB	4	0.10	0.10	0.10									
SINAI [24]	0	1.00	1.00	0.72	1.00	1.00	0.88	1.00	1.00	0.85			
SINAI	1	1.00	1.00	0.73	1.00	1.00	0.90	1.00	1.00				
SINAI	2	1.00	1.00	0.71	1.00	1.00	0.87	1.00	1.00	0.84			
SINAI	3	1.00	1.00	0.72	1.00	1.00	0.89	1.00	1.00	0.86			
SINAI	4	0.80	0.86	0.53	0.90	0.94	0.56	0.70	0.80	0.47			
UMU [27]	0	1.00	1.00	0.63	0.00	0.00	0.01	0.00	0.00	0.00	0.00	0.00	0.00
UMU	1	0.00	0.00	0.03	0.00	0.00	0.01	0.00	0.00	0.00	0.00	0.00	0.00
UMU	2	0.00	0.00	0.00	0.00	0.00	0.01	0.00	0.00	0.09	0.20	0.16	0.14
UMU	3	0.00	0.00	0.03	0.30	0.31	0.12	0.40	0.36	0.20	0.50	0.50	0.23
UMU	4	0.00	0.00	0.00	0.00	0.00	0.01	0.00	0.00	0.09	0.20	0.16	0.14
NLP-UNED	0	0.00	0.00	0.03	0.00	0.00	0.02	0.00	0.00	0.02	0.00	0.00	0.02
NLP-UNED	1	0.00	0.00	0.03	0.00	0.00	0.02	0.00	0.00	0.02	0.00	0.00	0.02
NLP-UNED	2	0.00	0.00	0.03	0.00	0.00	0.02	0.00	0.00	0.02	0.00	0.00	0.02
NLP-UNED	3	0.00	0.00	0.03	0.00	0.00	0.02	0.00	0.00	0.02	0.00	0.00	0.02
NLP-UNED	4	0.00	0.00	0.03	0.00	0.00	0.02	0.00	0.00	0.02	0.00	0.00	0.02

the best $F1$. The majority of teams made quick decisions. Overall, these findings indicate that some systems achieved a relatively high level of effectiveness with only a few user submissions. Social and public health systems may use the best predictive algorithms to assist expert humans in detecting signs of pathological gambling as early as possible.

Table 9 presents the ranking-based results. Because some teams only processed a few dozens of user writings, we could only compute their user rankings for the initial rounds. For tie breaking in the scores for the users, we used the traditional *docid* criteria (subject name). SINAI (run 1) obtained the best overall values after only one writing. At the other evaluation points, ELiRF-UPV was again the best performing.

4   Task 3: Measuring the Severity of Eating Disorders

The objective of the task is to estimate the severity of various symptoms related to the diagnosis of eating disorders. Participants were provided with a thread of user submissions to work with. For each user, a history of posts and comments from Social Media was given, and participants had to estimate the user's responses to a standardised eating disorder questionnaire based on the evidence found in the history of posts/comments.

The questionnaire used in the task is derived from the Eating Disorder Examination Questionnaire (EDE-Q)[3], which is a self-reported questionnaire consisting of 28 items. It is adapted from the semi-structured interview Eating Disorder Examination (EDE)[4] [7]. For this task, we focused on questions 1–12 and 19–28 from the EDE-Q. This questionnaire is designed to assess various aspects and severity of features associated with eating disorders. It includes four subscales: Restraint, Eating Concern, Shape Concern, and Weight Concern, along with a global score. Table 10 shows a excerpt of the EDE-Q.

---

[3] https://www.corc.uk.net/media/1273/ede-q_quesionnaire.pdf.
[4] https://www.corc.uk.net/media/1951/ede_170d.pdf.

Table 10. Eating Disorder Examination Questionarie

```
Instructions:

The following questions are concerned with the past four weeks (28 days) only. Please read
each question carefully. Please answer all the questions. Thank you..

1. Have you been deliberately trying to limit the amount of food you eat to influence your
shape or weight (whether or not you have succeeded)   0. NO DAYS
   1. 1-5 DAYS
   2. 6-12 DAYS
   3. 13-15 DAYS
   4. 16-22 DAYS
   5. 23-27 DAYS
   6. EVERY DAY

2. Have you gone for long periods of time (8 waking hours or more) without eating anything at
all in order to influence your shape or weight?
   0. NO DAYS
   1. 1-5 DAYS
   2. 6-12 DAYS
   3. 13-15 DAYS
   4. 16-22 DAYS
   5. 23-27 DAYS
   6. EVERY DAY

3. Have you tried to exclude from your diet any foods that you like in order to influence
your shape or weight (whether or not you have succeeded)?
   0. NO DAYS
   1. 1-5 DAYS
   2. 6-12 DAYS
   3. 13-15 DAYS
   4. 16-22 DAYS
   5. 23-27 DAYS
   6. EVERY DAY

   ⋮

22. Has your weight influenced how you think about (judge) yourself as a person?
   0. NOT AT ALL (0)
   1. SLIGHTY (1)
   2. SLIGHTY (2)
   3. MODERATELY (3)
   4. MODERATELY (4)
   5. MARKEDLY (5)
   6. MARKEDLY (6)
```

23. Has your shape influenced how you think about (judge) yourself as a person?
    0. NOT AT ALL (0)
    1. SLIGHTY (1)
    2. SLIGHTY (2)
    3. MODERATELY (3)
    4. MODERATELY (4)
    5. MARKEDLY (5)
    6. MARKEDLY (6)

24. How much would it have upset you if you had been asked to weigh yourself once
a week (no more, or less, often) for the next four weeks?
    0. NOT AT ALL (0)
    1. SLIGHTY (1)
    2. SLIGHTY (2)
    3. MODERATELY (3)
    4. MODERATELY (4)
    5. MARKEDLY (5)
    6. MARKEDLY (6)

The primary objective of this task was to explore the possibility of automatically estimating the severity of multiple symptoms related to eating disorders. The algorithms are required to estimate the user's response to each individual question based on their writing history. To evaluate the performance of the participating systems, we collected questionnaires completed by Social Media users along with their corresponding writing history. The user-completed questionnaires serve as the ground truth against which the responses provided by the systems are evaluated.

During the training phase, participants were provided with data from 28 users from the 2022 edition. This training data included the writing history of the users as well as their responses to the EDE-Q questions. In the test phase, there were 46 new users for whom the participating systems had to generate results. The results were expected to follow the following specific file structure:

username1 answer1 answer2...answer12 answer19...answer28

username2 answer1 answer2...answer12 answer19...answer28

Each line has the username and 22 values (no answers from 13 to 18). These values correspond with the responses to the questions above (the possible values are 0, 1, 2, 3, 4, 5, 6).

4.1 Evaluation Metrics

Evaluation is based on the following effectiveness metrics:

- **Mean Zero-One Error** ($MZOE$) between the questionnaire filled by the real user and the questionnaire filled by the system (i.e. fraction of incorrect predictions).

$$MZOE(f, Q) = \frac{|\{q_i \in Q : R(q_i) \neq f(q_i)\}|}{|Q|} \qquad (1)$$

where f denotes the classification done by an automatic system, Q is the set of questions of each questionnaire, q_i is the i-th question, $R(q_i)$ is the real user's answer for the i-th question and $f(q_i)$ is the predicted answer of the system for the i-th question. Each user produces a single $MZOE$ score and the reported $MZOE$ is the average over all $MZOE$ values (mean $MZOE$ over all users).

– **Mean Absolute Error (MAE)** between the questionnaire filled by the real user and the questionnaire filled by the system (i.e. average deviation of the predicted response from the true response).

$$MAE(f,Q) = \frac{\sum_{q_i \in Q} |R(q_i) - f(q_i)|}{|Q|} \qquad (2)$$

Again, each user produces a single MAE score and the reported MAE is the average over all MAE values (mean MAE over all users).

– **Macroaveraged Mean Absolute Error (MAE_{macro})** between the questionnaire filled by the real user and the questionnaire filled by the system (see [2]).

$$MAE_{macro}(f,Q) = \frac{1}{7} \sum_{j=0}^{6} \frac{\sum_{q_i \in Q_j} |R(q_i) - f(q_i)|}{|Q_j|} \qquad (3)$$

where Q_j represents the set of questions whose true answer is j (note that j goes from 0 to 6 because those are the possible answers to each question). Again, each user produces a single MAE_{macro} score and the reported MAE_{macro} is the average over all MAE_{macro} values (mean MAE_{macro} over all users).

The following measures are based on aggregated scores obtained from the questionnaires. Further details about the EDE-Q instruments can be found elsewhere (e.g. see the scoring section of the questionnaire).

– **Restraint Subscale (RS):** Given a questionnaire, its restraint score is obtained as the mean response to the first five questions. This measure computes the RMSE between the restraint ED score obtained from the questionnaire filled by the real user and the restraint ED score obtained from the questionnaire filled by the system.

Each user u_i is associated with a real subscale ED score (referred to as $R_{RS}(u_i)$) and an estimated subscale ED score (referred to as $f_{RS}(u_i)$). This metric computes the RMSE between the real and an estimated subscale ED scores as follows:

$$RMSE(f,U) = \sqrt{\frac{\sum_{u_i \in U} (R_{RS}(u_i) - f_{RS}(u_i))^2}{|U|}} \qquad (4)$$

where U is the user set.

– **Eating Concern Subscale (ECS):** Given a questionnaire, its eating concern score is obtained as the mean response to the following questions (7, 9, 19, 21, 20). This metric computes the RMSE (Eq. 5) between the eating concern ED score obtained from the questionnaire filled by the real user and the eating concern ED score obtained from the questionnaire filled by the system.

$$RMSE(f, U) = \sqrt{\frac{\sum_{u_i \in U}(R_{ECS}(u_i) - f_{ECS}(u_i))^2}{|U|}} \qquad (5)$$

- **Shape Concern Subscale (SCS):** Given a questionnaire, its shape concern score is obtained as the mean response to the following questions (6, 8, 23, 10, 26, 27, 28, 11). This metric computes the RMSE (Eq. 6) between the shape concern ED score obtained from the questionnaire filled by the real user and the shape concern ED score obtained from the questionnaire filled by the system.

$$RMSE(f, U) = \sqrt{\frac{\sum_{u_i \in U}(R_{SCS}(u_i) - f_{SCS}(u_i))^2}{|U|}} \qquad (6)$$

- **Weight Concern Subscale (WCS):** Given a questionnaire, its weight concern score is obtained as the mean response to the following questions (22, 24, 8, 25, 12). This metric computes the RMSE (Eq. 7) between the weight concern ED score obtained from the questionnaire filled by the real user and the weight concern ED score obtained from the questionnaire filled by the system.

$$RMSE(f, U) = \sqrt{\frac{\sum_{u_i \in U}(R_{WCS}(u_i) - f_{WCS}(u_i))^2}{|U|}} \qquad (7)$$

- **Global ED (GED):** To obtain an overall or 'global' score, the four subscales scores are summed and the resulting total divided by the number of subscales (i.e. four) [7]. This metric computes the RMSE between the real and an estimated global ED scores as follows:

$$RMSE(f, U) = \sqrt{\frac{\sum_{u_i \in U}(R_{GED}(u_i) - f_{GED}(u_i))^2}{|U|}} \qquad (8)$$

4.2 Results

Table 11 reports the results obtained by the participants in this task. In order to provide some context, the table includes the performance of three baseline variants in the top block: "all 0s", "all 6s", and "average". The "all 0s" variant represents a strategy where the same response (0) is submitted for all questions. Similarly, the "all 6s" variant submits the response 6 for all questions. The "average" variant calculates the mean of the responses provided by all participants for each question and submits the response that is closest to this mean value (e.g. if the mean response provided by the participants equals 3.7 then this average approach would submit a 4).

Table 11. Task 3 Results. Participating teams and runs with corresponding scores for the metrics.

team	run ID	MAE	MZOE	MAE_{macro}	GED	RS	ECS	SCS	WCS
baseline	all0s	2.419	**0.674**	2.803	3.207	2.138	3.221	3.028	2.682
baseline	all6s	3.581	0.834	3.995	3.839	4.814	3.650	3.950	3.318
baseline	average	2.091	0.859	1.957	2.391	**1.592**	2.398	2.162	**2.002**
BFH-AMI [23]	0	2.407	0.719	2.729	3.169	2.597	2.854	2.923	2.414
GMU-FAST	0	2.529	0.902	2.012	2.498	2.585	1.948	1.950	2.221
GMU-FAST	1	2.525	0.903	1.992	2.487	2.584	**1.924**	1.917	2.219
GMU-FAST	2	2.738	0.764	2.058	2.708	2.278	2.641	2.295	2.662
GMU-FAST	3	2.671	0.833	**1.741**	**1.999**	2.740	2.053	2.083	2.401
GMU-FAST	4	2.534	0.796	1.879	2.174	2.469	2.136	2.033	2.387
RiskBusters [8]	0	2.338	0.691	1.922	2.294	1.866	2.492	1.999	2.425
RiskBusters	1	2.352	0.699	1.858	2.127	2.025	2.365	2.034	2.466
RiskBusters	2	2.396	0.704	1.861	2.178	1.859	2.484	1.957	2.468
RiskBusters	3	2.419	0.709	1.898	2.251	1.935	2.440	2.037	2.445
RiskBusters	4	2.346	0.705	1.859	2.217	1.862	2.398	**1.898**	2.395
RiskBusters	5	2.334	0.702	1.854	2.230	1.898	2.381	1.947	2.378
RiskBusters	7	2.408	0.696	1.936	2.365	2.048	2.536	1.985	2.414
RiskBusters	8	2.347	0.696	1.975	2.534	1.911	2.443	2.215	2.494
UMU [27]	0	**2.194**	0.800	2.027	2.288	1.777	2.412	2.556	2.135

The results indicate that the top-performing system in terms of Mean Absolute Error (MAE) was run 0 by UMU. However, this particular run did not outperform the naive baseline approach of submitting all 0s in terms of Mean Zero-One Error (MZOE). Among the participating systems, GMU-FAST' run 3 achieved the best performance in two metrics: MAE_{macro} and Global ED. For the Eating Concern Subscale, the best-performing system was GMU-FAST' run 1, while for the Shape subscale, RiskBusters' run 4 showed the highest performance.

5 Conclusions

This paper provided an overview of eRisk 2023, the seventh edition of the lab, which focused on three types of tasks: symptoms search (Task 1 on depression), early detection (Task 2 on pathological gambling), and severity estimations (Task 3 on eating disorders). Participants in Task 1 were given a collection of sentences and had to rank them according to their relevance to each one of the BDI-II depression symptoms. Participants in Task 2 had sequential access to social media posts and had to send alerts about individuals showing risks of gambling.

In Task 3, participants were given the full user history and had to automatically estimate the user's responses to a standard depression questionnaire.

A total of 105 runs were submitted by 20 teams for the proposed tasks. Although the effectiveness of the solutions is still modest in some of the tasks, the experimental results demonstrate the value of extracting evidence from social media, indicating that automatic or semi-automatic screening tools to detect at-risk individuals could be promising. These findings highlight the need for the development of benchmarks for text-based risk indicator screening.

Acknowledgements. This work was supported by project PLEC2021-007662 (MCIN/AEI/10.13039/501100011033, Ministerio de Ciencia e Innovación, Agencia Estatal de Investigación, Plan de Recuperación, Transformación y Resiliencia, Unión Europea-Next Generation EU). The first and second authors thank the financial support supplied by the Xunta de Galicia-Consellería de Cultura, Educación, Formación Profesional e Universidade (accreditation 2019–2022 ED431G/01, GPC ED431B 2022/33) and the European Regional Development Fund, which acknowledges the CITIC Research Center in ICT of the University of A Coruña as a Research Center of the Galician University System. The third author thanks the financial support supplied by the Xunta de Galicia-Consellería de Cultura, Educación, Formación Profesional e Universidade (accreditation 2019–2022 ED431G-2019/04, ED431C 2022/19) and the European Regional Development Fund, which acknowledges the CiTIUS-Research Center in Intelligent Technologies of the University of Santiago de Compostela as a Research Center of the Galician University System.

References

1. Abbott, M.: The epidemiology and impact of gambling disorder and other gambling-related harm. In: WHO Forum on Alcohol, Drugs and Addictive Behaviours, Geneva, Switzerland (2017)
2. Baccianella, S., Esuli, A., Sebastiani, F.: Evaluation measures for ordinal regression, pp. 283–287 (2009). https://doi.org/10.1109/ISDA.2009.230
3. Beck, A.T., Ward, C.H., Mendelson, M., Mock, J., Erbaugh, J.: An inventory for measuring depression. JAMA Psychiatry **4**(6), 561–571 (1961)
4. Bezerra, E., dos Santos, L., Nascimento, R., Lopes, R.P., Guedes, G.: NailP at eRisk 2023: search for symptoms of depression. In: Working Notes of CLEF 2023 - Conference and Labs of the Evaluation Forum, 18–21 September 2023, Thessaloniki, Greece (2023)
5. Bucur, A.M.: Utilizing ChatGPT generated data to retrieve depression symptoms from social media. In: Working Notes of CLEF 2023 - Conference and Labs of the Evaluation Forum, 18–21 September 2023, Thessaloniki, Greece (2023)
6. Fabregat, H., Duque, A., Araujo, L., Martinez-Romo, J.: NLP-UNED-2 at eRisk 2023: detecting pathological gambling in social media through dataset relabeling and neural networks. In: Working Notes of CLEF 2023 - Conference and Labs of the Evaluation Forum, 18–21 September 2023, Thessaloniki, Greece (2023)
7. Fairburn, C.G., Cooper, Z., O'Connor, M.: Eating disorder examination Edition 17.0D (2014)
8. Grigore, D.N., Pintilie, I.: Transformer-based topic modeling to measure the severity of eating disorder symptoms. In: Working Notes of CLEF 2023 - Conference and Labs of the Evaluation Forum, 18–21 September 2023, Thessaloniki, Greece (2023)

9. Larrayoz, X., Lebeña, N., Casillas, A., Pérez, A.: Representation exploration and deep learning applied to the early detection of pathological gambling risks. In: Working Notes of CLEF 2023 - Conference and Labs of the Evaluation Forum, 18–21 September 2023, Thessaloniki, Greece (2023)

10. Losada, D.E., Crestani, F.: A test collection for research on depression and language use. In: Fuhr, N., et al. (eds.) CLEF 2016. LNCS, vol. 9822, pp. 28–39. Springer, Cham (2016). https://doi.org/10.1007/978-3-319-44564-9_3

11. Losada, D.E., Crestani, F., Parapar, J.: eRisk 2017: CLEF lab on early risk prediction on the internet: Experimental foundations. In: Jones, G.J., et al. (eds.) CLEF 2017. LNCS, vol. 10456, pp. 346–360. Springer, Cham (2017). https://doi.org/10.1007/978-3-319-65813-1_30

12. Losada, D.E., Crestani, F., Parapar, J.: Overview of eRisk 2018: early risk prediction on the internet (extended lab overview). In: CEUR Proceedings of the Conference and Labs of the Evaluation Forum, CLEF 2018, Avignon, France (2018)

13. Losada, D.E., Crestani, F., Parapar, J.: Overview of eRisk: early risk prediction on the internet. In: Bellot, P., et al. (eds.) CLEF 2018. LNCS, vol. 11018, pp. 343–361. Springer, Cham (2018). https://doi.org/10.1007/978-3-319-98932-7_30

14. Losada, D.E., Crestani, F., Parapar, J.: Early detection of risks on the internet: an exploratory campaign. In: Azzopardi, L., Stein, B., Fuhr, N., Mayr, P., Hauff, C., Hiemstra, D. (eds.) ECIR 2019, Part II. LNCS, vol. 11438, pp. 259–266. Springer, Cham (2019). https://doi.org/10.1007/978-3-030-15719-7_35

15. Losada, D.E., Crestani, F., Parapar, J.: Overview of eRisk 2019: early risk prediction on the internet. In: Crestani, F., et al. (eds.) CLEF 2019. LNCS, vol. 11696, pp. 340–357. Springer, Cham (2019). https://doi.org/10.1007/978-3-030-28577-7_27

16. Losada, D.E., Crestani, F., Parapar, J.: Overview of eRisk at CLEF 2019: early risk prediction on the internet (extended overview). In: CEUR Proceedings of the Conference and Labs of the Evaluation Forum, CLEF 2019, Lugano, Switzerland (2019)

17. Losada, D.E., Crestani, F., Parapar, J.: eRisk 2020: self-harm and depression challenges. In: Jose, J., et al. (eds.) ECIR 2020, Part II. LNCS, vol. 12036, pp. 557–563. Springer, Cham (2020). https://doi.org/10.1007/978-3-030-45442-5_72

18. Losada, D.E., Crestani, F., Parapar, J.: Overview of eRisk 2020: early risk prediction on the internet. In: Experimental IR Meets Multilinguality, Multimodality, and Interaction - Proceedings of the 11th International Conference of the CLEF Association, CLEF 2020, 22–25 September 2020, Thessaloniki, Greece, pp. 272–287 (2020)

19. Losada, D.E., Crestani, F., Parapar, J.: Overview of erisk at CLEF 2020: early risk prediction on the internet (extended overview). In: Working Notes of CLEF 2020 - Conference and Labs of the Evaluation Forum, 22–25 September 2020, Thessaloniki, Greece (2020)

20. Marco, A.M., Huang, X., Hurtado, L.F., Pla, F.: ELiRF-UPV at eRisk 2023: early detection of pathological gambling using SVM. In: Working Notes of CLEF 2023 - Conference and Labs of the Evaluation Forum, 18–21 September 2023, Thessaloniki, Greece (2023)

21. Martinez-Romo, J., Araujo, L., Larrayoz, X., Oronoz, M., Pérez, A.: OBSERMENH at eRisk 2023: deep learning-based approaches for symptom detection in depression and early identification of pathological gambling indicators. In: Working Notes of CLEF 2023 - Conference and Labs of the Evaluation Forum, 18–21 September 2023, Thessaloniki, Greece (2023)

22. Maupomé, D., et al.: Lightweight methods for early risk detection. In: Working Notes of CLEF 2023 - Conference and Labs of the Evaluation Forum, 18–21 September 2023, Thessaloniki, Greece (2023)

23. Merhbene, G., Puttick, A.R., Kurpicz-Briki, M.: BFH-AMI at eRisk@CLEF 2023. In: Working Notes of CLEF 2023 - Conference and Labs of the Evaluation Forum, 18–21 September 2023, Thessaloniki, Greece (2023)

24. Mármol-Romero, A.M., del Arco, F.M.P., Montejo-Ráez, A.: NSINAI at eRisk@CLEF 2023: approaching early detection of gambling with natural language processing. In: Working Notes of CLEF 2023 - Conference and Labs of the Evaluation Forum, 18–21 September 2023, Thessaloniki, Greece (2023)

25. Otero, D., Parapar, J., Barreiro, Á.: Beaver: efficiently building test collections for novel tasks. In: Proceedings of the First Joint Conference of the Information Retrieval Communities in Europe (CIRCLE 2020), 6–9 July 2020, Samatan, Gers, France (2020)

26. Otero, D., Parapar, J., Barreiro, Á.: The wisdom of the rankers: a cost-effective method for building pooled test collections without participant systems. In: SAC 2021: The 36th ACM/SIGAPP Symposium on Applied Computing, 22–26 March 2021, Virtual Event, Republic of Korea, pp. 672–680 (2021)

27. Pan, R., Díaz, J.A.G., Valencia-Garcia, R.: UMUTeam at eRisk@CLEF 2023 shared task: transformer models for early detection of pathological gambling, depression, and eating disorder. In: Working Notes of CLEF 2023 - Conference and Labs of the Evaluation Forum, 18–21 September 2023, Thessaloniki, Greece (2023)

28. Parapar, J., Martín-Rodilla, P., Losada, D.E., Crestani, F.: erisk 2021: pathological gambling, self-harm and depression challenges. In: Hiemstra, D., Moens, M.F., Mothe, J., Perego, R., Potthast, M., Sebastiani, F. (eds.) ECIR 2021, Part II. LNCS, vol. 12657, pp. 650–656. Springer, Cham (2021). https://doi.org/10.1007/978-3-030-72240-1_76

29. Parapar, J., Martín-Rodilla, P., Losada, D.E., Crestani, F.: Overview of eRisk 2021: early risk prediction on the internet. In: Experimental IR Meets Multilinguality, Multimodality, and Interaction - Proceedings of the 12th International Conference of the CLEF Association, CLEF 2021, Virtual Event, 21–24 September 2021, pp. 324–344 (2021)

30. Parapar, J., Martín-Rodilla, P., Losada, D.E., Crestani, F.: Overview of eRisk at CLEF 2021: early risk prediction on the internet (extended overview). In: Proceedings of the Working Notes of CLEF 2021 - Conference and Labs of the Evaluation Forum, 21–24 September 2021, Bucharest, Romania, pp. 864–887 (2021)

31. Parapar, J., Martín-Rodilla, P., Losada, D.E., Crestani, F.: eRisk 2022: pathological gambling, depression, and eating disorder challenges. In: Hagen, M., et al. (eds.) ECIR 2022, Part II. LNCS, vol. 13186, pp. 436–442. Springer, Cham (2022). https://doi.org/10.1007/978-3-030-99739-7_54

32. Parapar, J., Martín-Rodilla, P., Losada, D.E., Crestani, F.: Overview of eRisk 2022: Early risk prediction on the internet. In: Barrón-Cedeño, A., et al. (eds.) CLEF 2022. LNCS, vol. 13390, pp. 233–256. Springer, Cham (2022). https://doi.org/10.1007/978-3-031-13643-6_18

33. Parapar, J., Martín-Rodilla, P., Losada, D.E., Crestani, F.: Overview of erisk at CLEF 2022: early risk prediction on the internet (extended overview). In: Proceedings of the Working Notes of CLEF 2022 - Conference and Labs of the Evaluation Forum, 5–8 September 2022, Bologna, Italy, pp. 821–850 (2022)

34. Parapar, J., Martín-Rodilla, P., Losada, D.E., Crestani, F.: eRisk 2023: depression, pathological gambling, and eating disorder challenges. In: Kamps, J., et al. (eds.) ECIR 2023, Part III. LNCS, vol. 13982, pp. 585–592. Springer, Cham (2023). https://doi.org/10.1007/978-3-031-28241-6_67

35. R, N., Bolimera, P., Gupta, Y., M, A.K.: Exploring depression symptoms through similarity methods in social media posts. In: Working Notes of CLEF 2023 - Conference and Labs of the Evaluation Forum, 18–21 September 2023, Thessaloniki, Greece (2023)

36. Sadeque, F., Xu, D., Bethard, S.: Measuring the latency of depression detection in social media. In: WSDM, pp. 495–503. ACM (2018)

37. Sakib, F.A., Choudhury, A.A., Uzuner, Ö.: MASON-NLP at eRisk 2023: deep learning-based detection of depression symptoms from social media texts. In: Working Notes of CLEF 2023 - Conference and Labs of the Evaluation Forum, 18–21 September 2023, Thessaloniki, Greece (2023)

38. Talha, A., Basu, T.: A natural language processing based risk prediction framework for pathological gambling. In: Working Notes of CLEF 2023 - Conference and Labs of the Evaluation Forum, 18–21 September 2023, Thessaloniki, Greece (2023)

39. Thompson, H., Cagnina, L., Errecalde, M.: Strategies to harness the transformers' potential: UNSL at eRisk 2023. In: Working Notes of CLEF 2023 - Conference and Labs of the Evaluation Forum, 18–21 September 2023, Thessaloniki, Greece (2023)

40. Trotzek, M., Koitka, S., Friedrich, C.: Utilizing neural networks and linguistic metadata for early detection of depression indications in text sequences. IEEE Trans. Knowl. Data Eng. **32**, 588–601 (2018)

41. Wang, Y., Inkpen, D.: uOttawa at eRisk 2023: search for symptoms of depression. In: Working Notes of CLEF 2023 - Conference and Labs of the Evaluation Forum, 18–21 September 2023, Thessaloniki, Greece (2023)

Overview of EXIST 2023 – Learning with Disagreement for Sexism Identification and Characterization

Laura Plaza[1,2](✉), Jorge Carrillo-de-Albornoz[1,2], Roser Morante[1], Enrique Amigó[1], Julio Gonzalo[1], Damiano Spina[2], and Paolo Rosso[3]

[1] Universidad Nacional de Educación a Distancia (UNED), 28040 Madrid, Spain
{lplaza,jcalbornoz,rmorant,enrique,julio}@lsi.uned.es
[2] RMIT University, 3000 Melbourne, Australia
damiano.spina@rmit.edu.au
[3] Universitat Politècnica de València (UPV), 46022 Valencia, Spain
prosso@dsic.upv.es

Abstract. In recent years, the rapid increase in the dissemination of offensive and discriminatory material aimed at women through social media platforms has emerged as a significant concern. This trend has had adverse effects on women's well-being and their ability to freely express themselves. The EXIST campaign has been promoting research in online sexism detection and categorization since 2021. The third edition of EXIST, hosted at the CLEF 2023 conference, consists of three tasks, two of which are the continuation of EXIST 2022 (*sexism identification* and *sexism categorization*), and a third and novel one is on *source intention identification*. For this edition, new test and training data are provided and the "learning with disagreement" paradigm is adopted to address disagreements in the labelling process and promote the development of equitable systems that are able to learn from different perspectives on the sexism phenomena. 28 teams participated in the three EXIST 2023 tasks, submitting 232 runs. This lab overview describes the tasks, dataset, evaluation methodology, approaches and results.

Keywords: Sexism Detection · Sexism Categorization · Data Bias · Learning with Disagreement

1 Introduction

Sexism is defined by the Oxford English Dictionary as "prejudice, stereotyping, or discrimination, typically against women, on the basis of sex". This phenomenon remains prevalent even in contemporary, developed societies, and among the younger generations who have grown up in democratic societies. Sexism continues to pose significant challenges for women in various areas of their lives, such as work, family life, and personal growth, acting as a barrier that impedes their progress.

Sexism manifests in many different forms and can be categorized according to different dimensions. Regarding the facet of the women that is attacked, we can find attitudes such as sexual objectification, stereotyping or patriarchy. Depending on the level of society at which discrimination occurs, we can find institutional sexism, interpersonal sexism and individual sexism. Depending on the expression and underlying motivation, sexism can be categorized into two main types: hostile sexism and benevolent sexism. While hostile sexism involves openly negative and antagonistic attitudes towards women, benevolent sexism appears positive but patronizing, involving protective attitudes towards women while still reinforcing restrictive gender roles.

The persistence of gender inequality and discrimination against women in society is now being replicated and amplified in the online realm, as highlighted by Azmina et al. [8]. The Internet not only perpetuates these gender differences but also normalizes sexist attitudes, as noted by Burgos [4]. Specially concerning is the spread of sexist messages through social networks. Social networks have allowed women to rise their voices to report abuses, discrimination and sexist experiences, but the anonymity that they provide has also facilitated the transmission of hateful behaviours against women. With the increase of social media use by children and adolescents, detecting and fighting against online sexism becomes a priority. Previous studies [12] have shown that media content influences how social realities are perceived, so that the exposure to sexist and even misogynous content may contribute to develop sexist attitudes or to perceive them as natural or acceptable. Moreover, social media platforms are not acting efficiently to remove or avoid sexist and hateful content.

EXIST 2023[1] at CLEF is the third edition of the EXIST (sEXism Identification in Social neTworks) challenge that aims at combating sexism on social media. In 2021 and 2022, the EXIST shared tasks were proposed at the IberLEF forum [30,31]. These editions were the first in proposing tasks focusing on identifying and classifying online sexism in a broad sense, from explicit and/or hostile to other subtle or even benevolent expressions that involve implicit sexist behaviours. The 2021 and 2022 EXIST editions more than 50 teams participated from research institutions and companies from all around the world. While the two previous editions only focused on classifying sexist messages according to the facet of the women that was being undermined, the 2023 edition tackles an additional task that aims to determine the intention of the author of a sexist message. Since social networks are usually used to report and criticize sexist situations, it is important to distinguish the messages that are sexist by themselves from those that report experiences with the aim of raising awareness against sexism.

Additionally, the main novelty of the 2023 EXIST edition, and what makes it different to other recent initiatives such as the SemEval-2023 Shared Task 10: "Explainable Detection of Online Sexism" [17], is the adoption of the "Learning with Disagreement" (LwD) paradigm [36] for the development of the dataset and for the evaluation of the systems. Our previous work showed that the perception

---

[1] http://nlp.uned.es/exist2023/. Accessed 14 June 2023.

of sexism is strongly dependent on the demographic and cultural background of the subject, so that when identifying sexist attitudes and expressions, and even when classifying them in different sexist categories, people sometimes disagree. In the LwD paradigm, instead of relying on a single "correct" label for each example, the model is trained to handle and learn from conflicting or diverse annotations. In this way, that different annotators' perspectives, biases, or interpretations may be taken into account by the systems and the learning process becomes more fair.

In addition to adopting the LwD paradigm, we have made an effort to control bias in the annotations (see Sect. 3) and have developed new evaluation metrics that take into account the disagreement. This will allow us to assess whether including the different views and sensibilities of the annotators contributes to the development of more accurate and equitable NLP systems.

In the following sections, we provide comprehensive information about the tasks, the dataset, the evaluation methodology, the results and the different approaches of the systems that participated in the EXIST 2023 Lab. The competition features three distinct tasks: (i) sexism identification, (ii) source intention classification, and (iii) sexism categorization. A total of 103 teams from 29 different countries registered to participate. Ultimately, we received 232 results from 28 teams, and 24 of them completed the process by submitting the working notes. Interestingly, a significant number of teams leveraged the diverse labels representing various demographic groups and provided soft labels as the outputs of their systems. Their results showcase the effectiveness and advantages of employing the LwD paradigm in our specific domain: sexism in social networks.

2 Tasks

The two first editions of EXIST focused on detecting sexist messages in two social networks, Twitter and Gab,[2] as well as on categorizing these messages according to the type of sexist behaviour they enclose. For the 2023 edition, we focus on Twitter only and we address an additional task, namely "source intention classification". The three tasks are described below.

2.1 Task 1: Sexism Identification

The first task is a binary classification where systems must decide whether or not a given tweet expresses ideas related to sexism in any of the three forms: it is sexist itself, it describes a sexist situation in which discrimination towards women occurs, or criticizes a sexist behaviour. The following statements show examples of sexist and not sexist messages, respectively.

(1) **Sexist:** *It's less of #adaywithoutwomen and more of a day without feminists, which, to be quite honest, sounds lovely.*

---

[2] https://gab.com/. Accessed 14 June 2023.

(2) **Not sexist:** *Just saw a woman wearing a mask outside spank her very tightly leashed dog and I gotta say I love learning absolutely everything about a stranger in a single instant.*

2.2 Task 2: Source Intention

This task aims to categorize the message according to the intention of the author. We propose a ternary classification of tweets: (i) direct sexist message, (ii) reported sexist message, and (iii) judgemental message. This distinction will allow us to differentiate sexism that is actually taking place online from sexism which is being suffered by women in other situations but that is being reported in social networks with the aim of complaining and fighting against sexism. The three categories are defined as:

- **Direct** sexist message. The intention was to write a message that is sexist by itself or incites to be sexist, as in:
  (3) *A woman needs love, to fill the fridge, if a man can give this to her in return for her services (housework, cooking, etc.), I don't see what else she needs.*
- **Reported** sexist message. The intention was to report and share a sexist situation suffered by a woman or women in first or third person, as in:
  (4) *I doze in the subway, I open my eyes feeling something weird: the hand of the man sat next to me on my leg #SquealOnYourPig.*
- **Judgemental** message. The intention was judgmental, since the tweet describes sexist situations or behaviours with the aim of condemning them.
  (5) *As usual, the woman was the one quitting her job for the family's welfare...*

2.3 Task 3: Sexism Categorization

Many facets of a woman's life may be the focus of sexist attitudes including domestic and parenting roles, career opportunities, sexual image, and life expectations, to name a few. Automatically detecting which of these facets of women are being more frequently attacked in social networks will facilitate the development of policies to fight against sexism. According to this, each sexist tweet must be assigned one or more of the following categories:

- **Ideological and inequality**. This category includes messages that discredit the feminist movement in order to devalue, belittle and defame the struggle of women in any aspect of their lives. It also includes messages that reject inequality between men and women, or present men as victims of gender-based oppression.
  (6) *#Feminism is a war on men, but it's also a war on women. It's a war on female nature, a war on femininity.*

- **Stereotyping and dominance**. This category includes messages that express false ideas about women that suggest they are more suitable or inappropriate for certain tasks. It also includes any claim that implies that men are somehow superior to women.

  (7) *Most women no longer have the desire or the knowledge to develop a high quality character, even if they wanted to.*

- **Objectification**. This category includes messages where women are presented as objects apart from their dignity and personal aspects. We also include messages that assume or describe certain physical qualities that women must have in order to fulfill traditional gender roles, for example, ideas that suggest that women should maintain a standard and ideal of beauty or attacks on a woman's physique.

  (8) *No offense but I've never seen an attractive African American hooker. Not a single one.*

- **Sexual violence**. This category includes messages where sexual suggestions, requests or harassment of a sexual nature (rape or sexual assault) are made.

  (9) *I wanna touch your tits..you can't imagine what I can do on your body.*

- **Misogyny and non sexual violence**. This category includes expressions of hatred and violence towards women.

  (10) *Domestic abuse is never okay.... Unless your wife is a bitch.*

3   Dataset: Managing Bias in Data and Providing Pre-aggregated Annotations

An important problem in NLP that has gained attention in the recent years in parallel to the growing protagonism of large language models is bias, both in the data that are used to train and test systems, and in the way algorithms learn, mainly due to the bias in the data [32]. In EXIST 2023 we tackle one aspect of this problem, i.e., the data bias that may be introduced both during the data selection and during the labeling process.

3.1   Data Sampling

To gather the data, we follow the methodology used in previous EXIST editions [30,31]. We first collected different popular expressions and terms, both in English and Spanish, commonly used to underestimate the role of women in our society. These expressions have been extracted from different sources: (a) previous works in the area; (b) Twitter accounts (journalist, teenagers, etc.) or hashtags used to report sexist situations; (c) expressions extracted from The Every Day Sexism Project;[3] and d) a compendium of feminist dictionaries. These expressions were later used as seeds to retrieve Twitter data. To mitigate the **seed bias**, we have also gathered other common hashtags and expressions less

---

[3] https://everydaysexism.com/. Accessed 14 June 2023.

frequently used in sexist contexts to ensure a balanced distribution between sex-ist/not sexist expressions. This first set of seeds contains more than 400 expressions.

The set of seeds was then used to extract tweets in English and Spanish (more than 8,000,000 tweets were downloaded). The crawling was performed during the period from the September 1, 2021 till September 30, 2022. 100 tweets were downloaded for each seed per day (no retweets and promotional tweets were included). To ensure an appropriate balance between seeds, we removed those with less than 60 tweets. The final set of seeds contains 183 seeds for Spanish and 163 seeds for English.

To mitigate the **terminology and temporal bias**, the final sets of tweets were selected as follows: for each seed, approximately 20 tweets were randomly selected within the period from 1st September 1, February 28, 2022 for the training set, taking into account a representative temporal distribution between tweets of the same seed. Similarly, 3 tweets per seed were selected for the development set within the period from 1st to 31st May of 2022, and 6 tweets per seed within the period from August 1, 2022 to September, 30 2022 were selected for the test set. Only one tweet per author was included in the final selection to avoid **author bias**. Finally, tweets containing less than 5 words were removed. As a result, we have more than 3,200 tweets per language for the training set, around 500 per language for the development set, and nearly 1,000 tweets per language for the test set.

3.2 Labeling Process

Before starting the annotation process we considered possible sources of **label bias** [14]. Label bias may be introduced by socio-demographic differences of the persons that participate in the annotation process.

The labeling of the data was carried out by crowd-workers,[4] selected according to their different demographic characteristics in order to minimize the label bias. We consider gender (male/female) and age (18–22 y.o./23–45 y.o./+46 y.o). Each tweet was annotated by 6 crowdsourcing annotators selected through Prolific.[5] The Prolific crowdsourcing platform was specifically selected because of the features it provides to define participant criteria in the recruiting process – in our case, gender, age, and fluency in the different languages.

Different quality control mechanisms were employed, including small/medium size batches to ensure that the data is labeled by a significant/diverse amount of annotators, control of the time employed to perform the task, outlier analysis, and the use of attention mechanisms and ground truth data. We communicated frequently with the workers to solve doubts and to correct errors was kept.

---

[4] No personally identifiable information about the crowd-workers was collected. Workers were informed that the tweets could contain offensive information and were allowed to withdraw voluntarily at any time. Full consent was obtained.

[5] https://www.prolific.co/. Accessed 14 June 2023.

The annotators were provided with annotation guidelines that included a detailed description of the different tasks along with numerous examples, both positive and negative, for all different categories/labels. The guidelines were developed by two experts in gender issues.

3.3 Pre-aggregated Annotations: The Learning with Disagreement Paradigm

As stated by Uma et al. [36], the assumption that natural language expressions have a single and clearly identifiable interpretation in a given context is a convenient idealization, but far from reality. To deal with this, Uma et al. [36] have proposed the Learning With Disagreement (LwD) paradigm, which consists mainly of letting systems learn from datasets with information about the annotations from all annotators, in an attempt to gather the diversity of views. In the case of sexism identification, this is particularly relevant, since the perception of a situation as sexist or not can be subjective and may depend on the gender, age and cultural background of the person who is judging it. Following methods proposed for training directly from the data with disagreements, instead of using an aggregated label [24,29,34], the EXIST 2023 dataset provides multiple annotations per example. The LwD paradigm may also help to mitigate bias and produce equitable NLP systems. The selection of annotators for the development of the EXIST 2023 dataset took into account the heterogeneity necessary to avoid gender and age biases.

4   Evaluation Methodology and Metrics

As in SemEval 2021, we have carried out a **"soft evaluation"** and a **"hard evaluation"**. The soft evaluation corresponds to the LwD paradigm and is intended to measure the ability of the model to capture disagreements, by considering the probability distribution of labels in the output as a soft label and comparing it with the probability distribution of the annotations. The hard evaluation is the most standard evaluation paradigm and assumes that a single label is provided by the systems for every instance in the dataset.

From the point of view of evaluation metrics, the tasks can be described as follows:

- Task 1 (sexism identification): binary classification, monolabel.
- Task 2 (source intention): multiclass hierarchical classification, monolabel. The hierarchy of classes has a first level with two categories, sexist/not sexist, and a second level for the sexist category with three mutually-exclusive subcategories: direct/reported/judgemental. A suitable evaluation metric must reflect the fact that a confusion between not sexist and a sexist category is more severe than a confusion between two sexist subcategories.
- Task 3 (sexism categorization): multiclass hierarchical classification, multilabel. Again the first level is a binary distinction between sexist/not sexist, and

there is a second level for the sexist category that includes five subcategories: ideological and inequality, stereotyping and dominance, objectification, sexual violence, and misogyny and non-sexual violence. These classes are not mutually exclusive: a tweet may belong to several subcategories at the same time.

The LwD paradigm can be considered in both sides of the evaluation process:

- **The ground truth.** In a "hard" setting, the variability in the human annotations is reduced by selecting one and only one gold category per instance, the hard label. In a "soft" setting, the gold standard label for one instance is the set of all the human annotations existing for that instance. Therefore, the evaluation metric incorporates the proportion of human annotators that have selected each category (soft labels). Note that in Tasks 1 and 2, which are monolabel problems, the sum of probabilities of each class must be one. But in Task 3, which is multilabel, each annotator may select more than one category for a single instance. Therefore, the sum of probabilities of each class may be larger than one.
- **The system output.** In a "hard", traditional setting, the system predicts one or more categories for each instance. In a "soft" setting, the system predicts a probability for each category, for each instance. The evaluation score is maximized when the probabilities predicted match the actual probabilities in a soft ground truth.

In EXIST 2023, for each of the tasks, three types of evaluation have been performed:

1. **Soft-soft evaluation.** For systems that provide probabilities for each category, we provide a soft-soft evaluation that compares the probabilities assigned by the system with the probabilities assigned by the set of human annotators. The probabilities of the classes for each instance are calculated according to the distribution of labels and the number of annotators for that instance. We use a modification of the original ICM metric (Information Contrast Measure [1]), ICM-Soft (see details below), as the official evaluation metric in this variant and we also provide results for the normalized version of ICM-Soft (ICM-Soft Norm). It is important to note that ICM is a measure that quantifies information, and its upper and lower bounds are $+\infty$ and $-\infty$, respectively. To normalize the results, we used the gold standard score as the upper bound and minority class baseline as the lower bound. We also provide results for Cross Entropy.
2. **Hard-hard evaluation.** For systems that provide a hard, conventional output, we provide a hard-hard evaluation. To derive the hard labels in the ground truth from the different annotators' labels, we use a probabilistic threshold computed for each task. As a result, for Task 1, the class annotated by more than 3 annotators is selected; for Task 2, the class annotated by more than 2 annotators is selected; and for Task 3 (multilabel), the classes annotated by more than 1 annotator are selected. The instances for which there is

no majority class (i.e., no class receives more probability than the threshold) are removed from this evaluation scheme. The official metric for this task is the original ICM, as defined by Amigó and Delgado [1]. We also report a normalized version of ICM (ICM Norm) and F1. In Task 1, we use F1 for the positive class. In Tasks 2 and 3, we use the average of F1 for all classes. Note, however, that F1 is not ideal in our experimental setting: although it can handle multilabel situations, it does not take into account the relationships between classes. In particular, a confusion between not sexist and any of the sexist subclasses, and a confusion between two of the sexist subclasses, are penalized equally.

3. **Hard-soft evaluation.** For systems that provide a hard output, we will also provide a hard-soft evaluation comparing the categories assigned by the system with the probabilities assigned to each category in the ground truth. As in the previous case, we use ICM-Soft as the official evaluation metric in this variant. In this evaluation, the hard outputs are transformed into soft outputs by assigning a probability of 1.0 to the selected class and 0.0 to the other classes.

ICM is a similarity function that generalizes Pointwise Mutual Information (PMI), and can be used to evaluate system outputs in classification problems by computing their similarity to the ground truth. The general definition of ICM is:

$$\text{ICM}(A, B) = \alpha_1 IC(A) + \alpha_2 IC(B) - \beta IC(A \cup B)$$

Where $IC(A)$ is the Information Content of the instance represented by the set of features A. ICM maps into PMI when all parameters take a value of 1. The general definition of ICM by [1] is applied to cases where categories have a hierarchical structure and instances may belong to more than one category. The resulting evaluation metric is proved to be analytically superior to the alternatives in the state of the art. The definition of ICM in this context is:

$$\text{ICM}(s(d), g(d)) = 2IC(s(d)) + 2IC(g(d)) - 3IC(s(d) \cup g(d))$$

Where $IC()$ stands for Information Content, $s(d)$ is the set of categories assigned to document d by system s, and $g(d)$ the set of categories assigned to document d in the gold standard.

As there is not, to the best of our knowledge, any current metric that fits hierarchical multilabel classification problems in a learning with disagreement scenario, we have defined an extension of ICM (ICM-soft) that accepts both soft system outputs and soft ground truth assignments. ICM-soft works as follows: first, we define the Information Content of a single assignment of a category c with an agreement v to a given instance:

$$I(\{\langle c, v \rangle\}) = -\log_2(P(\{d \in D : g_c(d) \geq v\}))$$

Note that the information content of assigning a category c with an agreement v grows inversely with the probability of finding an instance that receives category c with agreement equal or larger than v. To this end, we compute the mean and deviation of the agreement levels for each class across instances, and applying the cumulative probability over the inferred normal distribution.[6]

The system output and the gold standards are sets of assignments. Therefore, in order to estimate their information content, we apply a recursive function similar to the one described by Amigó and Delgado [1].

$$IC\left(\bigcup_{i=1}^{n}\{\langle c_i, v_i\rangle\}\right) = IC(\langle c_1, v_1\rangle) + IC\left(\bigcup_{i=2}^{n}\{\langle c_i, v_i\rangle\}\right)$$
$$- IC\left(\bigcup_{i=2}^{n}\{\langle \mathrm{lca}(c_1, c_i), min(v_1, v_i)\rangle\}\right) \quad (1)$$

where $\mathrm{lca}(a, b)$ is the lowest common ancestor of categories a and b.

5   Overview of Approaches

Although 103 teams from 29 different countries registered for participation, the number of participants who finally submitted results were 28, submitting 232 runs. Teams were allowed to participate in any of the three tasks and submit hard and/or soft outputs. Table 1 summarizes the participation in the different tasks and evaluation contexts.

Table 1. Runs submitted per task and evaluation scenario.

	Task 1		Task 2		Task 3	
	Hard	Soft	Hard	Soft	Hard	Soft
# runs	67	54	33	25	30	23
# teams	28		15		14	

The evaluation campaign started on February 13, 2023 with the release of the training set. The development set was released on March 27, and the test set was made available on April 10. The participant teams were provided with the official evaluation script. Runs had to be submitted by May 15. Each team could submit up to three runs per task, that may contain soft and/or hard outputs.

For a comprehensive description of the systems submitted by the participants, please refer to extended overview [26] and the participants' working notes [2,3, 5–7,9–11,13,15,16,18–23,25,27,28,33,35,37,38]. Here we summarize the main approaches.

Approximately 90% of the systems submitted utilized large language models, both monolingual and multilingual. Some teams employed ensembles of multiple language models to enhance the overall performance. Only two teams utilized

---

[6] In the case of zero variance, we must consider that the probability for values equals or below the mean is 1 (zero IC) and the probability for values above the mean must be smoothed.

deep learning architectures such as BiLSTM, CNN, and RNN, while two others opted for traditional machine learning methods, including perceptrons, Naive Bayes, and SVM.

Data augmentation techniques were used by several teams, involving the translation of tweets, the utilization of data from similar tasks (such as previous EXIST editions), and the duplication of instances within the training set. Additionally, Twitter-specific models and transfer learning techniques from domains like hate speech, toxicity, and sentiment analysis were also utilized.

Most participants made use of the soft labels and applied the LwD paradigm, rather than opting for a traditional approach and providing only hard labels as outputs.

For each of the three tasks, the organization also provided different baseline runs:

- **Majority class:** non-informative baseline that classifies all instances as the majority class.
- **Minority class:** non-informative baseline that classifies all instances as the minority class.
- **Oracle most voted:** hard approach that selects the most voted label following the same procedure as the one used to generate the gold hard. Note that this baseline is only employed in the hard-soft evaluation.

6   Results

In the next subsections, we report the results of the participants and the baseline systems for each task.

6.1   Task 1 Sexism Identification

We first report and analyze the results for Task 1, which focuses on sexism identification. This task involves a binary classification. As discussed in Sect. 4, we report three sets of evaluation results.

Soft-Soft Evaluation. Table 2 presents the results for this evaluation context. 54 runs were submitted. 46 runs outperformed the non-informative majority class baseline (all instances are labeled as "NO"), while 51 runs surpassed the non-informative minority class baseline (all instances are labeled as "YES").

Looking at the results in Table 2, we observe a notable variation in the performance of the runs, ranging from an ICM-Soft score of 0.903 (equivalent to a 64% ICM-Soft Norm) to −5.6659 (lower than the empirically determined lower bound). However, it is worth mentioning that the best run only reached a 64% ICM-Soft Norm. This suggests that there is still room for improvement when it comes to capturing the appropriate distribution that represents real data.

These findings highlight the complexity of modeling the distribution of disagreements in a subjective task such as sexism identification.

Table 2. Systems' results for Task 1 in the Soft-soft evaluation.

Run	Rank	ICM-Soft	ICM-Soft Norm	Cross Entropy
Gold_soft	0	3.1182	1	0.5472
SINAI_3	1	0.9030	0.6421	0.7960
CLassifiers_3	2	0.9027	0.6421	0.9754
CLassifiers_2	3	0.8698	0.6368	0.9823
CLassifiers_1	4	0.8172	0.6283	0.9672
CIC-SDS.KN_2	5	0.7960	0.6248	0.7770
CIC-SDS.KN_3	6	0.7555	0.6183	0.7620
AI-UPV_2	7	0.7343	0.6149	1.3607
CIC-SDS.KN_1	8	0.7200	0.6126	0.7846
IUEXIST_1	9	0.7115	0.6112	1.1537
Tlatlamiztli_1	10	0.6879	0.6074	1.0538
UMUTeam_1	11	0.6818	0.6064	0.8707
JPM_UNED_1	12	0.6779	0.6058	0.8023
AI-UPV_3	13	0.6772	0.6056	1.6400
Mario_1	14	0.6696	0.6044	1.9247
Mario_2	15	0.6629	0.6033	1.8536
Mario_3	16	0.6603	0.6029	1.9503
IUEXIST_2	17	0.6141	0.5955	1.8418
JPM_UNED_2	18	0.5972	0.5927	0.8852
AIT_FHSTP_3	19	0.5955	0.5924	0.9392
AIT_FHSTP_1	20	0.5648	0.5875	1.1491
AI-UPV_1	21	0.5448	0.5843	1.5543
DRIM_1	22	0.5433	0.5840	0.8932
UMUTeam_2	23	0.4969	0.5765	0.8100
SINAI_1	24	0.4863	0.5748	1.5759
Alex_P_UPB_1	25	0.3214	0.5482	1.1709
SMS_1	26	0.3142	0.5470	1.6650
SMS_2	27	0.3142	0.5470	1.6650
M&S_NLP_1	28	0.2990	0.5445	0.8496
JPM_UNED_3	29	0.2467	0.5361	2.2342
M&S_NLP_2	30	0.1802	0.5254	0.9724
IU-NLP-JeDi_3	31	0.1244	0.5163	1.0878
InsightX_2	32	−0.0357	0.4905	0.9122
Awakened_2	33	−0.1496	0.4721	0.8706
IU-NLP-JeDi_2	34	−0.1499	0.4720	0.9097
InsightX_1	35	−0.2351	0.4583	0.9145

(*continued*)

Table 2. (*continued*)

Run	Rank	ICM-Soft	ICM-Soft Norm	Cross Entropy
UMUTeam_3	36	−0.3460	0.4403	0.9318
Awakened_3	37	−0.4369	0.4257	0.9282
IU-Percival_2	38	−0.4435	0.4246	3.1682
IU-Percival_1	39	−0.4612	0.4217	3.1777
IU-Percival_3	40	−0.5491	0.4075	3.2560
CNLP-NITS-PP_2	41	−0.5775	0.4029	2.0155
Awakened_1	42	−0.5931	0.4004	0.9625
InsightX_3	43	−0.6356	0.3936	0.9402
iimasGIL_NLP_3	44	−1.9161	0.1867	1.8619
iimasGIL_NLP_2	45	−1.9438	0.1822	1.8151
iimasGIL_NLP_1	46	−2.1678	0.1460	2.2546
Majority_class	47	−2.3585	0.1152	4.6115
M&S_NLP_3	48	−2.4596	0.0989	3.1670
roh-neil_2	49	−2.8848	0.0302	1.5472
roh-neil_1	50	−2.8851	0.0301	7.3091
roh-neil_3	51	−2.8851	0.0301	6.7608
Minority_class	52	−3.0717	0	5.3572
I2C-Huelva-1_3	53	−4.0609	−0.1598	2.5776
I2C-Huelva-1_1	54	−4.1626	−0.1762	2.4439
I2C-Huelva-1_2	55	−4.3175	−0.2013	2.9025
SINAI_2	56	−5.6559	−0.4175	7.3080

We next analyze the performance of the top ten systems. As shown in Table 2, the top five systems achieve similar results in terms of ICM-Soft, with a difference of less than 2% percentage points in terms of ICM-Soft Norm. The difference among the top ten systems was also less than 4% percentage points.

Among the top ten systems, the top nine utilized multilingual learning models, while the tenth system used a monolingual Spanish model. The variations between the systems primarily stemmed from the use of data augmentation techniques, such as in the fourth and tenth ranked systems, as well as the incorporation of domain-specific Twitter models and ensembles.

Hard-Hard Evaluation. Table 3 presents the results for the Hard-hard evaluation. In this scenario, the annotations from the six annotators are combined into a single label using the majority vote, resulting in the loss of information about the different perspectives provided by each annotator. Out of the 67 systems submitted for this task, 65 ranked above the majority class baseline (all instances labeled as "NO"). All systems surpassed the minority class baseline (all instances labeled as "YES").

Similar to the Soft-soft evaluation, the results vary considerably for the ICM-Hard metric, ranging from 0.6575 (78.50% ICM-Hard Norm) to -0.5335 (2.59%). However, the impact of this variation is not as prominent when considering the F1 score, indicating that ICM-Hard penalizes systems that predominantly suggest only one class for all instances more severely. For example, systems like "shm2023_1" labeled 2063 out of 2076 instances as "NO", while "M&S_NLP_3" labeled 1594 out of 2076 instances as "YES", thus getting very poor results in the ICM metrics.

Furthermore, the comparison between the ICM-Hard and F1 scores, as reflected in the ranking, shows a similar distribution, particularly at the top of the table. A strong correlation between the two metrics has been observed. However, in contrast to the Soft-soft evaluation, the behavior of the best systems in the Hard-hard evaluation aligns more closely with the gold standard. This highlights the higher complexity of the soft scenario and the inherent differences between the soft and hard evaluation contexts.

Table 3. Systems' results for Task 1 in the Hard-hard evaluation.

Run	Rank	ICM-Hard	ICM-Hard Norm	F1
Gold_hard	0	0.9948	1	1
Mario_3	1	0.6575	0.7850	0.8109
Mario_1	2	0.6540	0.7828	0.8058
Mario_2	3	0.6120	0.7560	0.8029
roh-neil_1	4	0.5795	0.7353	0.7840
roh-neil_2	5	0.5795	0.7353	0.7840
CIC-SDS.KN_2	6	0.5715	0.7302	0.7775
CIC-SDS.KN_3	7	0.5647	0.7259	0.7721
SINAI_1	8	0.5584	0.7219	0.7804
SINAI_2	9	0.5543	0.7192	0.7719
roh-neil_3	10	0.5474	0.7149	0.7754
SINAI_3	11	0.5440	0.7127	0.7715
CLassifiers_2	12	0.5390	0.7095	0.7702
UniBo_1	13	0.5381	0.7089	0.7716
UniBo_2	14	0.5381	0.7089	0.7716
IUEXIST_2	15	0.5341	0.7064	0.7717
IUEXIST_1	16	0.5313	0.7046	0.7734
CIC-SDS.KN_1	17	0.5303	0.7040	0.7677
CLassifiers_3	18	0.5282	0.7026	0.7642

(*continued*)

Table 3. (*continued*)

Run	Rank	ICM-Hard	ICM-Hard Norm	F1
JPM_UNED_3	19	0.5223	0.6989	0.7623
AI-UPV_3	20	0.5119	0.6922	0.7574
CLassifiers_1	21	0.5113	0.6918	0.7615
AI-UPV_2	22	0.5106	0.6914	0.7589
AIT_FHSTP_2	23	0.5086	0.6901	0.7571
UMUTeam_2	24	0.5083	0.6899	0.7604
I2C-Huelva-1_1	25	0.5075	0.6894	0.7611
I2C-Huelva-1_2	26	0.5075	0.6894	0.7611
I2C-Huelva-1_3	27	0.5075	0.6894	0.7611
JPM_UNED_1	28	0.5057	0.6883	0.7560
UMUTeam_1	29	0.5053	0.6880	0.7611
iimasGIL_NLP_3	30	0.5024	0.6862	0.7568
Tlatlamiztli_1	31	0.5013	0.6855	0.7535
MART_1	32	0.4937	0.6806	0.7587
JPM_UNED_2	33	0.4863	0.6759	0.7533
AIT_FHSTP_1	34	0.4850	0.6751	0.7550
AIT_FHSTP_3	35	0.4832	0.6739	0.7544
iimasGIL_NLP_1	36	0.4751	0.6688	0.7484
AI-UPV_1	37	0.4700	0.6655	0.7445
MART_2	38	0.4672	0.6637	0.7490
iimasGIL_NLP_2	39	0.4626	0.6608	0.7477
Alex_P_UPB_1	40	0.4021	0.6222	0.7302
M&S_NLP_1	41	0.3975	0.6193	0.7202
ZaRa-IU-NLP_2	42	0.3914	0.6154	0.7305
Awakened_2	43	0.3623	0.5969	0.7222
M&S_NLP_2	44	0.3057	0.5608	0.682
IU-Percival_1	45	0.3024	0.5587	0.6971
IU-Percival_2	46	0.2964	0.5549	0.6981
ZaRa-IU-NLP_1	47	0.2842	0.5471	0.6955
IU-NLP-JeDi_3	48	0.2753	0.5414	0.6909
InsightX_2	49	0.2689	0.5373	0.6883
IU-NLP-JeDi_1	50	0.2676	0.5365	0.6872
IU-Percival_3	51	0.2675	0.5365	0.6907
SMS_3	52	0.2469	0.5233	0.6717
InsightX_1	53	0.2458	0.5226	0.6809
SMS_1	54	0.2369	0.5170	0.6787

(*continued*)

Table 3. (*continued*)

Run	Rank	ICM-Hard	ICM-Hard Norm	F1
SMS_2	55	0.2369	0.5170	0.6787
UMUTeam_3	56	0.1882	0.4859	0.6639
IU-NLP-JeDi_2	57	0.1851	0.4839	0.6485
Awakened_3	58	0.1457	0.4588	0.6400
CNLP-NITS-PP_1	59	0.1093	0.4356	0.6409
CNLP-NITS-PP_2	60	0.1093	0.4356	0.6409
IIIT SURAT_1	61	0.1042	0.4324	0.6355
Awakened_1	62	0.0723	0.4120	0.6322
InsightX_3	63	−0.0403	0.3403	0.4804
shm2023_2	64	−0.1470	0.2723	0.4638
KUCST_2	65	−0.3578	0.1379	0.4062
Majority_class	66	−0.4413	0.0847	0
shm2023_1	67	−0.4473	0.0809	0.0071
M&S_NLP_3	68	−0.5335	0.0259	0.5312
Minority_class	69	−0.5742	0	0.5698

As shown in Table 3, the performance of the top ten runs is more remarkable in the hard context evaluation, with a difference of 7% percentage points in terms of ICM-Hard Norm. This difference is primarily due to the good performance of the top three systems ("Mario" team) which outperform the 4th ranked system by 5%. The "Mario" team utilizes a cascade model consisting of two fine-tuned monolingual GPT-based models trained on EXIST 2023 data and on data from other related hate speech tasks. Interestingly, the efficiency of the "Mario" team's runs in the soft context is considerably lower, as they are ranked 14th, 15th, and 16th. Among the other top ten teams, all of them employ multilingual approaches, with some using Twitter-specific models and none utilizing data augmentation techniques. It is also worth noting the performance of the "CIS-SDK.KN" team, which achieves similar rankings in both evaluations (5th and 6th in the soft context, and 6th and 7th in hard context).

Hard-Soft Evaluation. The Hard-soft evaluation context aims to assess the impact of information loss when using the majority vote strategy. Although the results of the Soft-soft and hard-soft evaluations are not strictly comparable, they provide insights into the performance of hard systems compared to soft systems in a real-world context.

Due to length restrictions, a detailed discussion of the hard-soft evaluation is provided in the extended version of the overview [26]. Here, we highlight the main findings. The evaluation included 67 systems, 65 outperformed the majority class baseline, and all surpassed the minority class baseline. Notably, the loss of information in the hard-soft context was more significant than expected when compared to the Soft-soft context. This had a clear impact on the behaviour of the systems, as evidenced by the first-ranked system achieving 64.21% in terms of ICM-Soft in the Soft-soft evaluation context compared to the 57.25% achieved in the Hard-soft context. This phenomenon is also reflected in the performance of the "EXIST2023_oracle_most_voted" baseline, which considers the most voted label by the annotators, and achieves a score of 1.1977 in the hard-soft context and a score of 3.1182 in the Soft-soft context.

6.2   Task 2 Source Intention

The second task is a hierarchical multiclass classification problem where systems must determine if a tweet is sexist or not, and categorize the sexist tweets according to the source intention in: "JUDGEMENTAL", "REPORTED", and "DIRECT".

Soft-Soft Evaluation. Table 4 presents the results for the Soft-soft evaluation of Task 2. The table shows that 25 runs were submitted. Among them, 21 runs achieved better results compared to the majority class baseline (where all instances are labeled as "NO"). Furthermore, all of the submitted runs outperformed the minority class baseline (where all instances are labeled as "REPORTED"). The differences between the scores achieved by the gold standard and the worst-performing system, the non-informative minority class baseline, are higher compared to Task 1. This can be attributed to the hierarchical and multiclass nature of Task 2, which allows for a wider range of values in the quantification of information by the ICM-Soft metric. It is also worth noting the correlation between the ICM-Soft and Cross-Entropy measures. The results indicate a strong correlation between the two metrics, but some differences can still be observed. Our initial analysis suggests that ICM-Soft is better at capturing the hierarchical nature of the task, as evidenced by a preliminary analysis of runs "JPM_UNED_2" and "SINAI_3". This observation is further supported by the fact that "JPM_UNED_2" utilizes a cascade model to determine whether a tweet is sexist or not sexist as a first step. However, further analysis is required to confirm this observation.

Table 4. Systems' results for Task 2 in the Soft-soft evaluation.

Run	Rank	ICM-Soft	ICM-Soft Norm	Cross Entropy
Gold_soft	0	6.2057	1	0.9128
DRIM_1	1	−1.3443	0.8072	1.7833
AIT_FHSTP_2	2	−1.4350	0.8049	1.6486
JPM_UNED_2	3	−1.6750	0.7988	2.5549
AI-UPV_2	4	−1.6836	0.7985	2.1697
JPM_UNED_3	5	−1.6888	0.7984	2.5561
AI-UPV_3	6	−1.7691	0.7964	2.5424
AIT_FHSTP_3	7	−2.1619	0.7863	2.0897
SINAI_3	8	−2.2900	0.7831	1.6753
UMUTeam_3	9	−2.5405	0.7767	2.2271
UMUTeam_1	10	−2.5674	0.7760	1.8102
Alex_P_UPB_1	11	−3.1765	0.7604	3.2050
Awakened_2	12	−3.1954	0.7599	1.6668
iimasGIL_NLP_3	13	−3.5072	0.7520	1.8860
Mario_2	14	−3.5509	0.7509	3.0061
iimasGIL_NLP_1	15	−3.5570	0.7507	1.9067
iimasGIL_NLP_2	16	−3.6387	0.7486	1.9149
Awakened_1	17	−3.9152	0.7416	1.7741
UMUTeam_2	18	−4.0482	0.7382	4.0452
SINAI_1	19	−4.2437	0.7332	2.3710
Awakened_3	20	−4.3598	0.7302	1.8594
AI-UPV_1	21	−4.3632	0.7301	3.0172
Majority_class	22	−5.4460	0.7025	4.6233
roh-neil_1	23	−5.7590	0.6945	3.7519
roh-neil_2	24	−5.7592	0.6945	3.2828
SINAI_2	25	−10.9851	0.5610	4.6237
M&S_NLP_1	26	−12.5531	0.5210	7.5639
Minority_class	27	−32.9552	0	8.8517

The top-performing run achieved a score of −1.3443 in terms of ICM-Soft, which is quite similar to the scores obtained by the other top nine systems. The tenth system achieved a score of −2.5674, resulting in a difference of only 3.1% in terms of ICM-Soft normalized. The first system ("DRIM_1") proposed an ensemble model consisting of three monolingual models, each trained on one of the three EXIST tasks. These models were fine-tuned using the ICM-Soft measure and optimized with manual features and annotator distribution information to calculate similarities.

Regarding the other nine systems, most of them utilized multilingual models, with the exception of "JPM_UNED". The systems employed a combination of techniques such as data augmentation, transfer learning, and ensembles, as in the previous task. One notable difference among the top ten systems is the use of socio-demographic knowledge by the "JPM_UNED" team, where six models were trained for each population's cohort to calculate the final distribution of probabilities.

Finally, upon comparing the performance of the top ten systems in Task 1 and Task 2 in the Soft-soft context, we find that the "AI-UPV_2" and "SINAI_3" teams consistently excel in accurately identifying sexism content and its properties in both tasks.

Hard-Hard Evaluation. Table 5 presents the results of the Hard-hard evaluation for Task 2, where 33 systems were assessed against the hard gold standard. 28 runs outperform the majority class baseline (all instances labeled as "NO"), while all systems demonstrate superior performance compared to the minority class baseline (all instances labeled as "REPORTED"). Similar to the Soft-soft evaluation context, the discrepancies between the gold standard and the worst-performing system (minority class baseline) are higher in Task 2 than in Task 1. The correlation between ICM-Hard and F1 is generally high, with slight variations observed among the top-ranked systems, and higher variability towards the end of the table. This discrepancy can be attributed to the inability of F1 to capture the hierarchical nature of the task, unlike ICM-Hard, which penalizes misclassifications between different levels of the hierarchy more severely.

Table 5. Systems' results for Task 2 in the Hard-hard evaluation.

Run	Rank	ICM-Hard	ICM-Hard Norm	F1
Gold_hard	0	1.5378	1	1
Mario_2	1	0.4887	0.7764	0.5715
roh-neil_1	2	0.3883	0.7550	0.5480
roh-neil_2	3	0.3883	0.7550	0.5480
UniBo_2	4	0.2786	0.7316	0.5283
UniBo_1	5	0.2439	0.7243	0.5217
AIT_FHSTP_1	6	0.2229	0.7198	0.5029
AI-UPV_2	7	0.1951	0.7139	0.4962
AI-UPV_3	8	0.1870	0.7121	0.4993
JPM_UNED_2	9	0.1862	0.7120	0.5054
JPM_UNED_3	10	0.1806	0.7108	0.5092
JPM_UNED_1	11	0.1673	0.7079	0.5032
AIT_FHSTP_3	12	0.1662	0.7077	0.4911

(*continued*)

Table 5. (*continued*)

Run	Rank	ICM-Hard	ICM-Hard Norm	F1
AIT_FHSTP_2	13	0.1475	0.7037	0.4759
UMUTeam_1	14	0.1409	0.7023	0.5013
AI-UPV_1	15	0.1217	0.6982	0.4897
Awakened_2	16	0.0088	0.6741	0.4606
UMUTeam_2	17	−0.0453	0.6626	0.4495
SINAI_2	18	−0.0496	0.6617	0.4924
SMS_1	19	−0.0892	0.6533	0.3654
SMS_3	20	−0.1226	0.6461	0.3504
UMUTeam_3	21	−0.1349	0.6435	0.4300
Alex_P_UPB_1	22	−0.1481	0.6407	0.4278
SMS_2	23	−0.2571	0.6175	0.3246
CNLP-NITS-PP_1	24	−0.3601	0.5955	0.3663
SINAI_1	25	−0.5959	0.5453	0.2562
Awakened_1	26	−0.7515	0.5121	0.2910
Awakened_3	27	−0.9048	0.4794	0.3087
KUCST_2	28	−0.9333	0.4734	0.2383
Majority_class	29	−0.9504	0.4697	0.1603
SINAI_3	30	−0.9646	0.4667	0.2544
iimasGIL_NLP_3	31	−0.9925	0.4608	0.2910
iimasGIL_NLP_1	32	−1.0631	0.4457	0.2505
iimasGIL_NLP_2	33	−1.0778	0.4426	0.2629
M&S_NLP_1	34	−2.9687	0.0396	0.0765
Minority_class	35	−3.1545	0	0.0280

Regarding the comparison between the top ten runs in terms of the ICM-Hard, the first system ("Mario_2") achieves a score of 0.4887 and the tenth system ("JPM_UNED_3") achieves a score of 0.1806 (only a 6.6% point difference). Unlike the previous task and context evaluation, where multilingual models were predominant, this task comprises an equal presence of multilingual and monolingual approaches among the top systems. Specifically, the first-ranked system is a monolingual GPT-based model that utilizes transfer learning and a cascade approach to identify and filter sexist tweets. The other top systems employ similar techniques, including transfer learning, data augmentation, and Twitter domain-specific models. Once again, the "JPM_UNED" team stands out as the only top-performing system that leverages the demographic features of the annotators, training six models, one for each cohort. It is interesting to note the comparison with Task 1 in the Hard-hard context, where the systems from the "roh-neil" and "Mario" teams are also among the top 5 in the ranking.

Hard-Soft Evaluation. The Hard-soft evaluation for Task 2 is discussed in the extended version of the overview [26]. However, we provide here a brief summary of the main findings. In this evaluation, 35 hard approaches were evaluated against the soft gold standard. Only 2 systems outperformed the majority class baseline, while all systems showed significant improvement over the minority class baseline. This behaviour is attributed to the fact that ICM-Soft penalizes errors in the minority classes to a greater extent, as they provide more informative signals than the majority classes. When comparing the results with the Soft-soft evaluation, similar trends were observed as in Task 1. The system that performed the best in the Soft-soft evaluation achieved an ICM-Soft Norm score of 80.72%, while the top-performing system in the hard-soft evaluation scored 71.08%.

6.3 Task 3 Sexism Categorization

The third task is the most challenging one since is a hierarchical multiclass and multilabel classification problem, where systems must determine if a tweet is sexist or not, and categorize the sexist tweets according to the five categories of sexism defined in Sect. 2.

Soft-Soft Evaluation. Table 6 displays the results of the Soft-soft evaluation for Task 3. A total of 23 runs were submitted, with 14 runs surpassing the majority class baseline (all instances labeled as "NO"), and all systems outperforming the minority class baseline (all instances labeled as "SEXUAL-VIOLENCE"). This highlights the complexity of categorizing sexism in social networks. The comparison among the system scores and the gold standard further emphasizes this difficulty, with a notable difference of more than 10 points in ICM-Soft scores: 9.4686 for the gold standard and −2.3183 for the best system ("AI-UPV_3"). Additionally, the differences between the best and the worst systems are significantly higher than in any of the previous tasks. However, despite the complexity, the results of the ICM-Soft Norm metric indicate that the systems are still able to correctly capture relevant information concerning the different types of sexism.

The best system in this task utilizes an ensemble approach with two multilingual models that have been optimized with the number of annotators to calculate probability distributions, achieving an ICM-Soft score of −2.3183. The differences between the top ten systems in this task are higher compared to previous tasks, with a nearly 9% difference in terms of ICM-Soft Norm between the 1st and 10th system. In terms of the techniques employed by the top ten systems, the prevalent architecture is multilingual, often combined with data augmentation techniques, domain-specific models for Twitter, and ensembles. It is noteworthy that the "AI-UPV_2" and "SINAI_3" teams consistently perform well in all three Soft-soft evaluation tasks, securing a position in the top ten of the ranking.

Table 6. Systems' results for Task 3 in the Soft-soft evaluation.

Run	Rank	ICM-Soft	ICM-Soft Norm
Gold_soft	0	9.4686	1
AI-UPV_3	1	−2.3183	0.7879
AI-UPV_2	2	−2.5616	0.7835
AI-UPV_1	3	−3.3437	0.7695
DRIM_1	4	−3.6842	0.7633
Alex_P_UPB_1	5	−4.2139	0.7538
CLassifiers_1	6	−6.4072	0.7143
roh-neil_1	7	−6.6622	0.7098
roh-neil_2	8	−6.6622	0.7098
roh-neil_3	9	−6.6622	0.7098
SINAI_3	10	−7.1306	0.7013
iimasGIL_NLP_3	11	−7.7704	0.6898
iimasGIL_NLP_2	12	−7.8073	0.6892
iimasGIL_NLP_1	13	−7.8867	0.6877
M&S_NLP_1	14	−8.3574	0.6793
Majority_class	15	−8.7089	0.6729
Mario_1	16	−11.4241	0.6241
Mario_2	17	−11.4241	0.6241
Mario_3	18	−11.4241	0.6241
SINAI_2	19	−13.5493	0.5858
CLassifiers_2	20	−14.7828	0.5636
Awakened_1	21	−20.0399	0.4690
Awakened_2	22	−23.6389	0.4043
Awakened_3	23	−25.9233	0.3632
SINAI_1	24	−34.9362	0.201
Minority_class	25	−46.108	0

Hard-Hard Evaluation. In the Hard-hard evaluation context for the last task, a total of 30 systems were submitted. As shown in Table 7, 26 systems outperformed the majority class baseline (all instances labeled as "NO"), while all systems achieved better results than the minority class baseline (all instances labeled as "SEXUAL-VIOLENCE"). Similar to Task 2, the discrepancies between the gold standard and the worst-performing system (minority class baseline) are higher in Task 3 due to its more complex nature of being hierarchical, multiclass, and multilabel.

The variation in results among different runs follows a similar distribution to that observed in Task 2, except for the last four systems which obtained substantially lower results due to that a high number of not sexist instances have been incorrectly assigned to different sexist subclasses, resulting in a strong penalization by the ICM-Hard metric that considers the class hierarchy.

The correlation between the F1 and the ICM-Hard measure is not as strong as in Task 2. This can be attributed to the fact that the ICM-Hard measure takes into account the hierarchy and penalizes errors between hierarchy levels more severely.

Finally, comparing the behaviour of the different tasks in a hard-hard context, the efficiency of the systems in this task, in terms of ICM-Hard Norm, is substantially lower than in previous tasks, further highlighting the complexity of categorizing sexism.

Table 7. Systems' results for Task 3 in the Hard-hard evaluation.

Run	Rank	ICM-Hard	ICM-Hard Norm	F1
EXIST2023_test_gold_hard	0	2.1533	1	1
roh-neil_1	1	0.4433	0.6763	0.6296
roh-neil_2	2	0.4433	0.6763	0.6296
AIT_FHSTP_1	3	0.2366	0.6372	0.5842
UniBo_2	4	0.2263	0.6352	0.5909
UniBo_1	5	0.1776	0.6260	0.5850
Mario_3	6	0.1700	0.6246	0.5323
SINAI_2	7	0.1472	0.6203	0.5822
Mario_2	8	0.1228	0.6156	0.5145
Mario_1	9	0.0896	0.6094	0.5011
SINAI_3	10	0.0249	0.5971	0.5033
AI-UPV_1	11	−0.1862	0.5571	0.4732
Alex_P_UPB_1	12	−0.1948	0.5555	0.4817
AI-UPV_2	13	−0.2516	0.5448	0.4757
SINAI_1	14	−0.3020	0.5352	0.5306
Awakened_2	15	−0.4276	0.5115	0.4027
UMUTeam_3	16	−0.5121	0.4955	0.5130
AI-UPV_3	17	−0.5788	0.4828	0.4195
UMUTeam_1	18	−0.5963	0.4795	0.5108
iimasGIL_NLP_3	19	−0.6510	0.4692	0.4482
Awakened_3	20	−0.6731	0.4650	0.3794
iimasGIL_NLP_1	21	−0.6859	0.4626	0.4406
iimasGIL_NLP_2	22	−0.7786	0.4450	0.4255
CNLP-NITS-PP_1	23	−0.8412	0.4332	0.3199
Awakened_1	24	−0.9507	0.4124	0.3283
roh-neil_3	25	−0.9626	0.4102	0.3139
UMUTeam_2	26	−0.9727	0.4083	0.4630
EXIST2023_test_majority_class	27	−1.5984	0.2898	0.1069
KUCST_2	28	−1.7934	0.2529	0.2889
CLassifiers_2	29	−1.8664	0.2391	0.3047
CLassifiers_1	30	−1.8852	0.2355	0.3126
M&S_NLP_1	31	−2.1587	0.1838	0.0017
EXIST2023_test_minority_class	32	−3.1295	0	0.0288

As in the Soft-soft evaluation of Task 3, the differences among the top ten systems are higher than in other tasks, up to 8% points between the 1st ("roh-neil_1") and 10th ("SINAI_3") systems. In this scenario, the best system proposed a multilingual domain specific model trained on Twitter data, and uses parameter optimization using the Optuna framework. Roughly an equal number of systems utilized monolingual and multilingual approaches. The majority of these systems employed a combination of techniques, including data augmentation, domain-specific models, and transfer learning. Interestingly, the only team that exploit the demographic knowledge is the "SINAI" team. Upon analyzing the performance of the top ten hard approaches across all tasks, it is evident that the "Mario" and "roh-neil" teams consistently achieve favorable results, ranking within the top 10.

Hard-Soft Evaluation. Finally, for completeness we provide a brief summary of the main findings for the hard-soft evaluation, while a detailed description can be found in the extended overview paper [26]. In this evaluation, a total of 30 systems were evaluated, and none of them outperformed the majority class baseline, while all of them improved upon the minority class baseline. As mentioned in the hard-soft evaluation for Task 2, the ICM-Soft measure heavily penalizes errors in the minority classes, while errors in the majority class are not as heavily penalized. This behaviour is particularly significant in the hard-soft evaluation due to the automatic assignment of higher probabilities to the hard labels. When comparing the results with the Soft-soft evaluation, we observe that the best performance system in the Soft-soft evaluation achieved an ICM-Soft Norm score of 78.79%, while the top-performing system in the hard-soft evaluation scored 66.52%.

7   Conclusions

The objective of the EXIST challenge is to encourage research on the automated detection and modeling of sexism in online environments, with a specific focus on social networks. The EXIST Lab held in 2023 as part of CLEF garnered significant interest, attracting over 100 registered teams for participation. We received a total of 232 submissions. Participants adopted a wide range of approaches including data augmentation through automatic translation, data duplication and utilization of past EXIST editions' data, multilingual language models, Twitter-specific language models, and transfer learning techniques from domains like hate speech, toxicity, and sentiment analysis. Fortunately, most participants chose to leverage the multiple annotations available and embrace the learning with disagreements paradigm, rather than opting for a traditional approach and providing only hard labels as outputs.

In addition to obtaining new insights into the detection and categorization of sexist messages in social networks, the Lab has also contributed to raise awareness about the importance of addressing annotator disagreements and label bias by selecting annotators from diverse population groups. This is particularly true for tasks where subjectivity and moral considerations come into play, and interpretations may vary across cultures and over time. Moreover, the EXIST lab has provided a valuable dataset that can be utilized for future experimentation and benchmarking purposes, further contributing to the advancement of research in this field.

For future editions of EXIST, we plan to extend our study to other languages, communication channels and media, as well as studying sexism in particular scenarios and population groups.

Acknowledgments. This work has been financed by the European Union (NextGenerationEU funds) through the "Plan de Recuperación, Transformación y Resiliencia", by the Ministry of Economic Affairs and Digital Transformation and by the UNED University. It has also been financed by the Spanish Ministry of Science and Innovation (project FairTransNLP (PID2021-124361OB-C31 and PID2021-124361OB-C32)) funded by MCIN/AEI/10.13039/501100011033 and by ERDF, EU A way of making Europe, and by the Australian Research Council (DE200100064 and CE200100005).

References

1. Amigó, E., Delgado, A.: Evaluating extreme hierarchical multi-label classification. In: Proceedings of the 60th Annual Meeting of the Association for Computational Linguistics. vol. 1: Long Papers, pp. 5809–5819. ACL, Dublin, Ireland (2022)
2. Angel, J., Aroyehun, S., Gelbukh, A.: Multilingual Sexism Identification using contrastive learning. In: Working Notes of CLEF 2023 - Conference and Labs of the Evaluation Forum (2023)
3. Asnani, H., Davis, A., Rajanala, A., Kübler, S.: Tlatlamiztli: fine-tuned RoBERTuito for sexism detection. In: Working Notes of CLEF 2023 - Conference and Labs of the Evaluation Forum (2023)
4. Burgos, A., et al.: Violencias de género 2.0. (2014)
5. Buzzell, M., Dickinson, J., Singh, N., Kübler, S.: IU-NLP-JeDi: investigating sexism detection in English and Spanish. In: Working Notes of CLEF 2023 - Conference and Labs of the Evaluation Forum (2023)
6. Böck, J., et al.: AIT_FHSTP at EXIST 2023 benchmark: sexism detection by transfer learning, sentiment and toxicity embeddings and hand-crafted features. In: Working Notes of CLEF 2023 - Conference and Labs of the Evaluation Forum (2023)
7. Chaudhary, A., Kumar, R.: Sexism identification in social networks. In: Working Notes of CLEF 2023 - Conference and Labs of the Evaluation Forum (2023)
8. Dhrodia, A.: Social media and the silencing effect: why misogyny online is a human rights issue (2017). https://www.newstatesman.com/culture/social-media/2017/11/social-media-and-silencing-effect-why-misogyny-online-human-rights-issue
9. Díaz, J.A.G., Pan, R., Valencia-Garcia, R.: UMUTeam at EXIST 2023: sexism identification and categorisation fine-tuning multilingual large language models. In: Working Notes of CLEF 2023 - Conference and Labs of the Evaluation Forum (2023)
10. Erbani, J., Egyed-Zsigmond, E., Nurbakova, D., Portier, P.E.: When multiple perspectives and an optimization process lead to better performance, an automatic sexism identification on social media with pretrained transformers in a soft label

context. In: Working Notes of CLEF 2023 - Conference and Labs of the Evaluation Forum (2023)

11. Gabel, E., Redman, H., Swanson, D., Kübler, S.: IU-Percival: linear models for sexism detection. In: Working Notes of CLEF 2023 - Conference and Labs of the Evaluation Forum (2023)

12. Gerbner, G., Gross, L., Morgan, M., Signorielli, N.: Living with television: the dynamics of the cultivation process. Perspect. Media Effects **1986**, 17–40 (1986)

13. Hatekar, Y., Abdo, M., Khanna, S., Kübler, S.: IUEXIST: multilingual pre-trained language models for sexism detection on twitter in EXIST2023. In: Working Notes of CLEF 2023 - Conference and Labs of the Evaluation Forum (2023)

14. Hovy, D., Prabhumoye, S.: Five sources of bias in natural language processing. Lang. Linguist. Compass **15**(8), e12432 (2021)

15. Jhakal, C., Singal, K., Suri, M., Chaudhary, D., Gorton, I., Kumar, B.: Detection of sexism on social media with multiple simple transformers. In: Working Notes of CLEF 2023 - Conference and Labs of the Evaluation Forum (2023)

16. Kelkar, S., Ravi, S., Madasamy, A.K.: LSTM-attention architecture for online bilingual sexism detection. In: Working Notes of CLEF 2023 - Conference and Labs of the Evaluation Forum (2023)

17. Kirk, H.R., Yin, W., Vidgen, B., Röttger, P.: SemEval-2023 task 10: explainable detection of online sexism. In: Proceedings of the 17th International Workshop on Semantic Evaluation (SemEval) (2023)

18. Koonireddy, R., Adel, N.: ROH_NEIL@EXIST2023: detecting sexism in tweets using multilingual language models. In: Working Notes of CLEF 2023 - Conference and Labs of the Evaluation Forum (2023)

19. Mohammadi, H., Giachanou, A., Bagheri, A.: Towards robust online sexism detection: a multi-model approach with BERT, XLM-RoBERTa, and DistilBERT for EXIST 2023 Tasks. In: Working Notes of CLEF 2023 - Conference and Labs of the Evaluation Forum (2023)

20. Muti, A., Mancini, E.: Enriching hate-tuned transformer-based embeddings with emotions for the categorization of sexism. In: Working Notes of CLEF 2023 - Conference and Labs of the Evaluation Forum (2023)

21. Cordón, P., Jacinto Mata, V.P., Domínguez, J.L.: I2C at CLEF-2023 EXIST task: leveraging ensembling language models to detect multilingual sexism in social media. In: Working Notes of CLEF 2023 - Conference and Labs of the Evaluation Forum (2023)

22. de Paula, A.F.M., Rizzi, G., Fersini, E., Spina, D.: AI-UPV at EXIST 2023 - sexism characterization using large language models under the learning with disagreement regime. In: Working Notes of CLEF 2023 - Conference and Labs of the Evaluation Forum (2023)

23. Pedrosa-Marín, J., de Albornoz, J.C., Plaza, L.: Combining large language models with socio-demographic information for improving sexism detection in social media. In: Working Notes of CLEF 2023 - Conference and Labs of the Evaluation Forum (2023)

24. Peterson, J., Battleday, R., Griffiths, T., Russakovsky, O.: Human uncertainty makes classification more robust. In: 2019 IEEE/CVF International Conference on Computer Vision (ICCV), pp. 9616–9625. IEEE Computer Society, Los Alamitos, CA, USA (2019)

25. Petrescu, A.: Leveraging MiniLMv2 pipelines for EXIST2023. In: Working Notes of CLEF 2023 - Conference and Labs of the Evaluation Forum (2023)

26. Plaza, L., et al.: Overview of EXIST 2023 - learning with disagreement for sexism identification and characterization (extended overview). In: Working Notes of CLEF 2023 - Conference and Labs of the Evaluation Forum (2023)

27. Radler, G., Ersoy, B.I., Carpentieri, S.: CLassifiers at EXIST 2023: detecting sexism in spanish and english tweets with XLM-T. In: Working Notes of CLEF 2023 - Conference and Labs of the Evaluation Forum (2023)

28. Regulagedda, R.M., Leech, Z., Kübler, S.: Specialized or generalized? Sexism detection for EXIST 2023 at CLEF. In: Working Notes of CLEF 2023 - Conference and Labs of the Evaluation Forum (2023)

29. Rodrigues, F., Pereira, F.C.: Deep learning from crowds. In: Proceedings of the Thirty-Second AAAI Conference on Artificial Intelligence and Thirtieth Innovative Applications of Artificial Intelligence Conference and Eighth AAAI Symposium on Educational Advances in Artificial Intelligence. AAAI 2018/IAAI 2018/EAAI 2018, AAAI Press (2018)

30. Rodríguez-Sánchez, F., et al.: Overview of EXIST 2021: sexism identification in social networks. Procesamiento del Lenguaje Nat. **67**, 195–207 (2021)

31. Rodríguez-Sánchez, F., et al.: Overview of EXIST 2022: sexism identification in social networks. Procesamiento del Lenguaje Nat. **69**, 229–240 (2022)

32. Roselli, D., Matthews, J., Talagala, N.: Managing bias in AI. In: Companion Proceedings of The 2019 World Wide Web Conference, pp. 539–544. WWW 2019, Association for Computing Machinery, New York, NY, USA (2019)

33. Sanchez-Urbina, A., Gómez-Adorno, H., Bel-Enguix, G., Rodríguez-Figueroa, V., Monge-Barrera, A.: iimasGIL_NLP@Exist2023: unveiling sexism on twitter with fine-tuned transformers. In: Working Notes of CLEF 2023 - Conference and Labs of the Evaluation Forum (2023)

34. Sheng, V.S., Provost, F., Ipeirotis, P.G.: Get another label? improving data quality and data mining using multiple, noisy labelers. In: Proceedings of the 14th ACM SIGKDD International Conference on Knowledge Discovery and Data Mining, pp. 614–622. KDD 2008, Association for Computing Machinery, New York, NY, USA (2008)

35. Tian, L., Huang, N., Zhang, X.: Efficient multilingual sexism detection via large language models cascades. In: Working Notes of CLEF 2023 - Conference and Labs of the Evaluation Forum (2023)

36. Uma, A., et al.: SemEval-2021 task 12: learning with disagreements. In: Proceedings of the 15th International Workshop on Semantic Evaluation (SemEval-2021), pp. 338–347. Association for Computational Linguistics, Online (2021)

37. Vallecillo-Rodríguez, M.E., del Arco, F.M.P., Ureña-López, L.A., Martín-Valdivia, M.T., Montejo-Ráez, A.: Integrating annotator information in transformer fine-tuning for sexism detection. In: Working Notes of CLEF 2023 - Conference and Labs of the Evaluation Forum (2023)

38. Vetagiri, A., Adhikary, P.K., Pakray, P., Das, A.: Leveraging GPT-2 for automated classification of online sexist content. In: Working Notes of CLEF 2023 - Conference and Labs of the Evaluation Forum (2023)

Intelligent Disease Progression Prediction: Overview of iDPP@CLEF 2023

Guglielmo Faggioli[1(✉)], Alessandro Guazzo[1], Stefano Marchesin[1],
Laura Menotti[1], Isotta Trescato[1], Helena Aidos[2], Roberto Bergamaschi[3],
Giovanni Birolo[4], Paola Cavalla[5], Adriano Chiò[4], Arianna Dagliati[3],
Mamede de Carvalho[2], Giorgio Maria Di Nunzio[1], Piero Fariselli[4],
Jose Manuel García Dominguez[6], Marta Gromicho[2], Enrico Longato[1],
Sara C. Madeira[2], Umberto Manera[4], Gianmaria Silvello[1], Eleonora Tavazzi[7],
Erica Tavazzi[1], Martina Vettoretti[1], Barbara Di Camillo[1], and Nicola Ferro[1]

[1] University of Padua, Padua, Italy
{guglielmo.faggioli,stefano.marchesin,laura.menotti,giorgiomaria.dinunzio,
enrico.longato,gianmaria.silvello,erica.tavazzi,martina.vettoretti,
barbara.dicamillo,nicola.ferro}@unipd.it,
{alessandro.guazzo,isotta.trescato}@phd.unipd.it
[2] University of Lisbon, Lisbon, Portugal
{haidos,sacmadeira}@fc.ul.pt, mamedemg@mail.telepac.pt,
mgromichosilva@medicina.ulisboa.pt
[3] University of Pavia, Pavia, Italy
roberto.bergamaschi@mondino.it, arianna.dagliati@unipv.it
[4] University of Turin, Turin, Italy
{giovanni.birolo,adriano.chio,piero.fariselli,umberto.manera}@unito.it
[5] "Città della Salute e della Scienza", Turin, Italy
paola.cavalla@unito.it
[6] Gregorio Marañon Hospital in Madrid, Madrid, Spain
jgarciadominguez@salud.madrid.org
[7] IRCCS Foundation C. Mondino in Pavia, Pavia, Italy

Abstract. Amyotrophic Lateral Sclerosis (ALS) and Multiple Sclerosis (MS) are chronic diseases that cause progressive or alternating neurological impairments in motor, sensory, visual, and cognitive functions. Affected patients must manage hospital stays and home care while facing uncertainty and significant psychological and economic burdens that also affect their caregivers. To ease these challenges, clinicians need automatic tools to support them in all phases of patient treatment, suggest personalized therapeutic paths, and preemptively indicate urgent interventions.

iDPP@CLEF aims at developing an evaluation infrastructure for AI algorithms to describe ALS and MS mechanisms, stratify patients based on their phenotype, and predict disease progression in a probabilistic,

G. Faggioli, A. Guazzo, S. Marchesin, L. Menotti and I. Trescato—These authors contributed equally.

A. Arampatzis et al. (Eds.): CLEF 2023, LNCS 14163, pp. 343–369, 2023.
https://doi.org/10.1007/978-3-031-42448-9_24

time-dependent manner.

iDPP@CLEF 2022 ran as a pilot lab in CLEF 2022, with tasks related to predicting ALS progression and explainable AI algorithms for prediction. iDPP@CLEF 2023 will continue in CLEF 2023, with a focus on predicting MS progression and exploring whether pollution and environmental data can improve the prediction of ALS progression.

1   Introduction

Amyotrophic Lateral Sclerosis (ALS) and *Multiple Sclerosis (MS)* are severe chronic diseases that cause progressive neurological impairment. They exhibit high heterogeneity in terms of symptoms and disease progression, leading to differing needs for patients. The heterogeneity of these diseases partly explains the lack of effective prognostic tools and the current lack of therapies that can effectively slow or reverse their course. This poses challenges for caregivers and clinicians alike. Furthermore, the timing of worsening or significant events – such as the need for *Non-Invasive Ventilation (NIV)* or *Percutaneous Endoscopic Gastrostomy (PEG)* in the case of ALS – is uncertain and hard to predict. Being able to preemptively recognize the need for specific medical treatments would have significant implications for the quality of life of patients. Therefore, it would be of uttermost importance to devise automatic tools that could aid clinicians in their decision-making in all phases of disease progression and facilitate personalized therapeutic choices.

To address these challenges and develop *Artificial Intelligence (AI)* predictive algorithms researchers need a framework to design and evaluate approaches to:

- stratify patients according to their phenotype all over the disease evolution;
- predict the progression of the disease in a probabilistic, time-dependent way;
- describe better and in an explainable fashion the mechanisms underlying MS and ALS diseases.

In this context, it is crucial to develop shared approaches, promote common benchmarks, and foster experiment comparability and replicability, even if not yet so common. The *Intelligent Disease Progression Prediction at CLEF (iDPP@CLEF)* lab[1] aims to provide an evaluation infrastructure for the development of such AI algorithms. Unlike previous challenges in the field, iDPP@CLEF systematically addresses issues related to the application of AI in clinical practice for ALS and MS. Apart from defining risk scores based on the probability of events occurring in the short or long term, iDPP@CLEF also deals with providing clinicians with structured and understandable data.

The paper is organized as follows: Sect. 2 presents related challenges; Sect. 3 describes its tasks; Sect. 4 discusses the developed dataset; Sect. 5 explains the setup of the lab and introduces the participants; Sect. 6 introduces the evaluation measures adopted to score the runs; Sect. 7 analyzes the experimental results for the different tasks; finally, Sect. 8 draws some conclusions and outlooks some future work.

---

[1] https://brainteaser.health/open-evaluation-challenges/.

2   Related Challenges

Within CLEF, there have been no other labs on this or similar topics before the start of iDPP@CLEF. iDPP@CLEF 2022, whose details are summarized below, was the first iteration of the lab and the current is the second one.

Outside CLEF, there have been a recent challenge on Kaggle[2] in 2021 and some older ones, the DREAM 7 ALS Prediction challenge[3] in 2012 and the DREAM ALS Stratification challenge[4] in 2015. The Kaggle challenge used a mix of clinical and genomic data to seek insights about the mechanisms of ALS and the difference between people with ALS who progress faster versus those who develop it more slowly. The DREAM 7 ALS Prediction challenge [15] asked to use 3 months of ALS clinical trial information (months 0–3) to predict the future progression of the disease (months 3–12), expressed as the slope of change in *ALS Functional Rating Scale Revisited (ALSFRS-R)* [5], a functional scale that ranges between 0 and 40. The DREAM ALS Stratification challenge asked participants to stratify ALS patients into meaningful subgroups, to enable better understanding of patient profiles and application of personalized ALS treatments. Differently from these previous challenges, iDPP@CLEF focuses on explainable AI and on temporal progression of the disease.

Finally, when it comes to *Multiple Sclerosis (MS)*, studies are mostly conducted on closed and proprietary datasets and iDPP@CLEF represents one of the first attempts to create a public and shared dataset.

2.1   iDPP@CLEF 2022

iDPP@CLEF 2022 ran as a pilot lab for the first time in CLEF 2022[5] [7,8] and focused on activities aimed at ALS progression prediction as well as at an understanding of the challenges and limitations to refine and tune the labs itself for future iterations. iDPP@CLEF 2022 consisted of the following tasks:

- **Pilot Task 1 - Ranking Risk of Impairment**: it focused on ranking patients based on the risk of impairment. We used the ALSFRS-R scale [5] to monitor speech, swallowing, handwriting, dressing/hygiene, walking and respiratory ability in time and asked participants to rank patients based on the time-to-event risk of experiencing impairment in each specific domain.
- **Pilot Task 2 - Predicting Time of Impairment**: it refined Task 1 by asking participants to predict when specific impairments will occur (i.e. in the correct time-window). In this regard, we assessed model calibration in terms of the ability of the proposed algorithms to estimate a probability of an event close to the true probability within a specified time-window.
- **Position Paper Task 3 - Explainability of AI algorithms**: we evaluated proposals of different frameworks able to explain the multivariate nature of the data and the model predictions.

---

[2] https://www.kaggle.com/alsgroup/end-als.

[3] https://dreamchallenges.org/dream-7-phil-bowen-als-prediction-prize4life/.

[4] https://dx.doi.org/10.7303/syn2873386.

[5] https://brainteaser.health/open-evaluation-challenges/idpp-2022/.

iDPP@CLEF 2022 created 3 datasets, for the prediction of specific events related to ALS, consisting of fully anonymized data from 2,250 real patients from medical institutions in Turin, Italy, and Lisbon, Portugal. The datasets contain both static data about patients, e.g. age, onset date, gender, . . . and event data, i.e. 18,512 ALSFRS-R questionnaires and 4,015 spyrometries. 6 groups participated in iDPP@CLEF 2022 and submitted a total of 120 runs.

3   Tasks

iDPP@CLEF 2023 is the second iteration of the lab, expanding its scope to include both ALS and MS in the study of disease progression. The activities in iDPP@CLEF 2023 focus on two objectives: exploring the prediction of MS worsening and conducting a more in-depth analysis of ALS compared to iDPP@CLEF 2022, with the addition of environmental data.

Following iDPP@CLEF 2022, iDPP@CLEF 2023 targets three tasks:

- Pilot tasks (Task 1 and Task 2) on predicting the progression of the MS, focusing on its worsening;
- Position papers (Task 3) on the impact that environmental data can have on the progression of the ALS.

In the remainder of this section, we describe each task more in detail.

3.1   Task 1: Predicting Risk of Disease Worsening (MS)

Task 1 focuses on MS and requires ranking subjects based on the risk of worsening, setting the problem as a survival analysis task. More specifically the risk of worsening predicted by the algorithm should reflect how early a patient experiences the "worsening" event and should range between 0 and 1.

Worsening is defined on the basis of the *Expanded Disability Status Scale (EDSS)* [16], according to clinical standards. In particular, we consider two different definitions of worsening corresponding to two different sub-tasks:

- Task1a: the patient crosses the threshold EDSS ≥ 3 at least twice within a one-year interval;
- Task1b: the second definition of worsening depends on the first recorded value, according to current clinical protocols:
  - if the baseline is EDSS < 1, then the worsening event occurs when an increase of EDSS by 1.5 points is first observed;
  - if the baseline is $1 \leq$ EDSS < 5.5, then the worsening event occurs when an increase of EDSS by 1 point is first observed;
  - if the baseline is EDSS ≥ 5.5, then the worsening event occurs when an increase of EDSS by 0.5 points is first observed.

For each sub-task, participants are given a dataset containing 2.5 years of visits, with the occurrence of the worsening event and the time of occurrence pre-computed by the challenge organizers.

3.2   Task 2: Predicting Cumulative Probability of Worsening (MS)

Task 2 refines Task 1 by asking participants to explicitly assign the cumulative probability of worsening at different time windows, i.e., between years 0 and 2, 0 and 4, 0 and 6, 0 and 8, 0 and 10. In particular, as in Task 1, we consider two different definitions of worsening corresponding to two different sub-tasks:

- Task2a: the patient crosses the threshold EDSS ≥ 3 at least twice within a one-year interval;
- Task2b: the second definition of worsening depends on the first recorded value, according to current clinical protocols:
  - if the baseline is EDSS < 1, then the worsening event occurs when an increase of EDSS by 1.5 points is first observed;
  - if the baseline is $1 \leq$ EDSS < 5.5, then the worsening event occurs when an increase of EDSS by 1 point is first observed;
  - if the baseline is EDSS ≥ 5.5, then worsening event occurs when an increase of EDSS by 0.5 points is first observed.

For each sub-task, participants are given a dataset containing 2.5 years of visits, with the occurrence of the worsening event and the time of occurrence pre-computed by the challenge organizers.

3.3   Task 3: Position Papers on the Impact of Exposition to Pollutants (ALS)

Participants in Task 3 are required to propose approaches to assess if exposure to different pollutants is a useful variable to predict time to PEG, NIV, and death in ALS patients. This task is based on the same design as Task 1 in iDPP@CLEF 2022 and employs the same data as well. Therefore, both training and test data are available immediately. Compared to iDPP@CLEF 2022, the dataset is complemented with environmental data to investigate the impact of exposition to pollutants on the prediction of disease progression. The task consists in ranking subjects based on the risk of early occurrence of:

- Task3a: NIV or (competing event) death, whichever occurs first;
- Task3b: PEG or (competing event) Death, whichever occurs first;
- Task3c: Death.

Since test data were already released at the end of iDPP@CLEF 2022 it is impossible to produce a fair leaderboard. Therefore, participants are required to produce position papers in which they describe their approaches and findings concerning the link between environmental factors and ALS progression.

4   Dataset

For iDPP@CLEF 2023, we provided 5 datasets, two for MS and three for ALS, using data from three clinical institutions in Turin and Pavia, Italy, and Lisbon,

Portugal. The datasets are fully anonymized: identifiers and pseudo-identifiers, e.g. place of birth or city of residence, have been removed; dates are reported as relative spans in days with respect to a `Time` 0, i.e., a reference moment in time that depends on the considered disease. For MS, `Time` 0 denotes the first visit to assess EDSS after the patient has received a confirmed diagnosis of MS. In the context of ALS, `Time` 0 represents the date of the first ALSFRS-R questionnaire.

4.1   Task 1 and Task 2: MS Datasets

Tasks 1 and 2 share the same datasets – each MS dataset corresponds to a specific sub-task (a and b). As training features, we provide:

- Static data, containing information on patient's demographics, diagnostic delay, and symptoms at the onset;
- Dynamic data (2.5 years), containing information on: relapses, EDSS scores, evoked potentials, MRIs, and MS course.

The following data are available as ground-truth:

- The worsening occurrence, as defined in Sect. 3, expressed as a Boolean variable with 0 meaning "not occurred" and 1 meaning "occurred".
- The time-of-occurrence, expressed as relative delta with respect to `Time` 0 in years (also fractions).

Each of dataset contains the following groups of variables:

- `static vars.`, representing static variables associated with a patient. The complete list of available static variables is available at http://brainteaser. dei.unipd.it/challenges/idpp2023/assets/other/ms/static-vars.txt.
- `MS type`, containing information about the MS type and the (relative) date when the MS type has been observed.
- `relapses` consisting of the (relative) initial dates of relapses.
- `EDSS`, containing EDSS scores and the (relative) date when they were recorded.
- `evoked potentials`, reporting the results of evoked potential tests. The complete list of variables for each evoked potential test is available at http://brainteaser.dei.unipd.it/challenges/idpp2023/assets/other/ ms/evoked-potentials.txt.
- `MRI`, containing the data involving MRIs; e.g., the area on which MRIs have been performed and the observed lesions. The complete list of variables about MRIs is available at http://brainteaser.dei.unipd.it/challenges/ idpp2023/assets/other/ms/mri.txt.
- `outcomes`, detailing the patients' worsening occurrence, together with the time of occurrence. More in detail, `outcomes` contain one record for each patient where:
  - The first column is the patient ID;
  - The second column indicates if the worsening occurred (1) or not (0).

Table 1. Training and test datasets for MS tasks.

Training

Sub-task	Patients	Relapses	EDSS	Evoked Potentials	MRIs	MS courses
Sub-task a	440	480	2,660	1,210	960	310
Sub-task b	510	552	3,068	1,521	965	324

Test

Sub-task	Patients	Relapses	EDSS	Evoked Potentials	MRIs	MS courses
Sub-task a	110	94	674	277	236	68
Sub-task b	128	124	812	298	265	74

- The third column is the time of occurrence, defined as a floating point number in the range [0,15].

Table 1 reports the number of records for each group of variables for training and test sets for each sub-task.

Creation of the Datasets. To obtain the iDPP@CLEF 2023 MS datasets, we processed two datasets coming from Turin and Pavia research centres. The source datasets contained approximately 1,800 records linked to patients, with approximately 6,700 records for relapses, 28,600 records on EDSS, 6,200 on evoked potentials, 10,300 on MRIs, and 3,700 on MS courses. To remove minor inconsistencies and typos present in the original data, we first processed the data removing records that were likely wrong or did not provide enough information for AI methods to perform predictions. We removed patients' records without:

- onset date;
- first visit date;
- functional systems scores and corresponding EDSS scores.

Other records associated with such patients (e.g., EDSS or MRIs) have been discarded as well. As for relapses, we removed those records where no information about the relapse was given. We removed MRI records not reporting information about T1 and T2 lesions. After cleaning, to generate the challenge datasets, we restricted visits data to a 2.5 years window prior to Time 0.

Split into Training and Test. Each of the two MS datasets underwent a division into a training set and a test set, with proportions of 80% and 20% respectively. In order to ensure a well-stratified distribution of variables across the datasets and to avoid any biases during the splitting process, the data were randomly partitioned 100 times using 100 different random seeds. To assess the appropriateness of the stratification, a comparison of variable distributions was conducted for each training/test pair. Statistical tests were performed on each variable based on its type: the Kruskal-Wallis test [18] was applied to continuous variables, while the Chi-squared test [22] was employed for categorical and ordinal variables. A variable was considered well-stratified depending on the test result. For each split, the percentage of well-stratified variables was calculated using Eq. 1.

$$perc_{well-stratified} = \frac{number\ of\ positive\ tests}{total\ number\ of\ variables} * 100 \tag{1}$$

To identify the split that achieved the best stratification between those that achieved the highest percentage, equal to 97%, a visual inspection was then conducted. Density plots were used for continuous variables, bar plots for categorical and ordinal variables, and Kaplan-Meier curves [20] for the outcome time in the survival setting. A careful examination of the outcome occurrence and time was performed to ensure that the models' performance would not be influenced by the data splitting. Furthermore, special attention was given to sparsely observed levels in categorical variables. The splitting process allowed for the possibility that a rare level might only appear in the training set, but not vice versa. Table 2 and 3 report the comparison of the variables' distributions in the training and test sets for sub-task a[6]. Since the distributions are similar, we concluded that the training/test split provided to the participants met best-practice quality standards.

4.2   Task 3: ALS Dataset

The datasets used for Task 3 in iDPP@CLEF 2023 have the same structure and most of the records as the one used in iDPP@CLEF 2022. There are three datasets concerning patients affected by ALS, Dataset ALSa, Dataset ALSb, and Dataset ALSc. Each dataset concerns a specific type of event that might to patients affected by ALS. Datasets ALSa and ALSb regard respectively the moment in which a patient undergoes NIV or PEG. While dataset ALSc concerns the death of the patient. For a detailed description of the data, cleaning procedures, and additional statistics, please refer to [7,8].

iDPP@CLEF 2023 dataset extends the previous version by providing participants with environmental data. Furthermore, due to its release at the end of iDPP@CLEF 2022, the ground truth is available to the challenge participants since the beginning of the challenge.

Updates over iDPP@CLEF 2022. In the 2023 version of the dataset, a small subset of patients (less than 50) has been removed from the dataset used for iDPP@CLEF 2022. Indeed, such patients were characterized by the absence of relevant events (i.e., NIV, PEG or death), but did not receive further ALSFRS-R

---

[6] A more complete and detailed comparison, including the information for the other sub-task, is shown in the extended overview [6].

Table 2. Sub-task a, comparison between training and test populations. Continuous variables are presented as *mean (sd)*; categorical variables as *count (percentage on available data)*, for each level. Demographic and onset-related features.

	Variable	Level	Levels train	Levels test
static vars.	sex	Female	305 (69.32%)	76 (69.09%)
		Male	135 (30.68%)	34 (30.91%)
	residence_classification	Cities	120 (27.27%)	32 (29.09%)
		Rural Area	100 (22.73%)	18 (16.36%)
		Towns	208 (47.27%)	54 (49.09%)
		NA	12 (2.73%)	6 (5.45%)
	ethnicity	Caucasian	424 (96.36%)	99 (90.00%)
		Hispanic	-	4 (3.64%)
		Black_African	-	2 (1.82%)
		NA	16 (3.64%)	5(4.55%)
	ms_in_pediatric_age	FALSE	410 (93.18%)	103 (93.64%)
		TRUE	30 (6.82%)	7 (6.36%)
	age_at_onset	mean (sd)	31 (9.427)	30 (8.775)
	diagnostic_delay	mean (sd)	1029 (1727.8)	967 (1447.6)
		NA	12 (2.73%)	1 (0.91%)
	spinal_cord_symptom	FALSE	348 (79.09%)	83 (75.45%)
		TRUE	92 (20.91%)	27 (24.55%)
	brainstem_symptom	FALSE	305 (69.32%)	79 (71.82%)
		TRUE	135 (30.68%)	31 (28.18%)
	eye_symptom	FALSE	318 (72.27%)	82 (74.55%)
		TRUE	122 (27.73%)	28 (25.45%)
	supratentorial_symptom	FALSE	301 (68.41%)	74 (67.27%)
		TRUE	139 (31.59%)	36 (32.73%)
	other_symptoms	False	431 (97.95%)	107 (97.27%)
		RM+	3 (0.68%)	2 (1.82%)
		Sensory	4 (0.91%)	1 (0.91%)
		Epilepsy	2 (0.45%)	0 (—)
	time_since_onset	mean (sd)	2524 (2448.3)	2446 (2235.9)
MS type	multiple_sclerosis_type	CIS	99 (32.04%)	18 (26.87%)
		RR	210 (67.96%)	49 (73.13%)
	delta_observation_time0	mean (sd)	−718 (210.2)	−715 (237.6)

assessments after the first. Therefore, such patients were annotated with the censoring event happening at time 0 making it impossible to provide a sensible prediction. Such patients were removed from the 2023 version of the iDPP@CLEF ALS dataset. Table 4 reports the number of removed patients compared to the original iDPP@CLEF ALS dataset. Notice that, by construction, all the removed patients were labelled with event NONE. Spyrometries and ALSFRS-R questionnaires associated with dropped patients have been removed as well.

Table 3. Table 2 contd. Dynamical assessments and outcome features.

	Variable	Level	Levels train	Levels test
edss	edss_as_evaluated_by_clinician	mean (sd)	2 (0.716)	2 (0.655)
		NA	37 (1.39%)	3 (0.45%)
	delta_edss_time0	mean (sd)	−499 (251.6)	−499 (254.4)
evoked potentials	altered_potential	Auditory	280 (23.14%)	58 (20.94%)
		Motor	101 (8.35%)	19 (6.86%)
		Somatosensory	482 (39.83%)	111 (40.07%)
		Visual	347 (28.68%)	89 (32.13%)
	potential_value	mean (sd)	0 (0.401)	0 (0.415)
	location	left	311 (25.70%)	73 (26.35%)
		lower left	126 (10.41%)	29 (10.47%)
		lower right	136 (11.24%)	31 (11.19%)
		right	316 (26.12%)	74 (26.71%)
		upper left	156 (12.89%)	34 (12.27%)
		upper right	165 (13.64%)	36 (13.00%)
	delta_evoked_potential_time0	mean (sd)	−712 (206.3)	−731 (213.3)
relapses	delta_relapse_time0	mean (sd)	−561 (286.1)	−551 (286.5)
magnetic resonance image	mri_area_label	Brain Stem	681 (71.01%)	164 (69.79%)
		Cervical	62 (6.47%)	25 (10.64%)
		Spinal Cord	201 (20.96%)	36 (15.32%)
		Spinal Cord	15 (1.56%)	10 (4.26%)
		Thoracic Spinal Cord		
	lesions_T1	FALSE	175 (18.25%)	45 (19.15%)
		TRUE	149 (15.54%)	29 (12.34%)
		NA	635 (66.21%)	161 (68.51%)
	lesions_T1_gadolinium	FALSE	575 (59.96%)	145 (61.70%)
		TRUE	247 (25.76%)	51 (21.70%)
		NA	137 (14.29%)	39 (16.1%)
	number_of_lesions_T1_gadolinium	mean (sd)	0 (1.0)	0 (1.0)
		NA	187 (19.5%)	48 (20.43%)
	new_or_enlarged_lesions_T2	FALSE	377 (39.31%)	107 (45.53%)
		TRUE	240 (25.03%)	52 (22.13%)
		NA	342 (35.66%)	76 (32.34%)
	number_of_new_or_enlarged_lesions_T2	mean (sd)	1 (1.486)	1 (1.401)
		NA	349 (36.39%)	76 (32.34%)
	lesions_T2	FALSE	55 (5.74%)	10 (4.26%)
		TRUE	275 (28.68%)	62 (26.38%)
		NA	629 (65.59%)	163 (69.36%)
	number_of_total_lesions_T2	0	55 (7.74%)	10 (4.26%)
		1-2	66 (6.88%)	14 (5.96%)
		>=3	70 (7.30%)	14 (5.96%)
		>=9	139 (14.49%)	24 (14.47%)
		NA	629 (65.59%)	163 (69.36%)
	delta_mri_time0	mean (sd)	-512 (282.0)	-534 (275.5)
outcomes	outcome_occurred	0	367 (83.41%)	93 (84.55%)
		1	73 (16.59%)	17 (15.45%)
	outcome_time	mean (sd)	5 (4.4)	5 (4.1)

Table 4. Patients removed from the iDPP@CLEF ALS dataset 2023 due to having an unrealistic censoring event time. Between parentheses the original number of patients available in the dataset.

	Train	Test	Total
Dataset ALSa	22 (orig. 1454)	4 (orig. 350)	26 (orig. 1806)
Dataset ALSb	36 (orig. 1715)	8 (orig. 430)	44 (orig. 2145)
Dataset ALSc	40 (orig. 1756)	8 (orig. 494)	48 (orig. 2250)

Environmental Data. One of the primary objectives of iDPP@CLEF 2023 is to promote research on the influence of environmental factors on the progression of ALS disease. Task 3, which specifically focuses on this aspect, requires participants to submit position papers investigating the impact of exposure to pollutants.

To address this objective, the iDPP@CLEF 2022 datasets have been expanded to include information about patients' exposure to environmental agents. This includes various environmental factors such as daily mean, minimum, and maximum temperatures, daily precipitation, daily averaged sea level pressure and relative humidity, daily mean wind speed, and daily mean global radiation. Additionally, the iDPP@CLEF 2023 ALS datasets also provide information on the concentration of seven pollutants: PM10, PM25, O_3, C_6H_6, CO, SO_2, and NO_2. For each environmental parameter, both the raw observations collected each day and the calibrated version of the observations, following best practices [10,23], are made available.

It is important to note that not all patients have the same amount of environmental information due to varying diagnosis times and data availability. Several patients could not be associated with environmental data, as their disease progression occurred before public environmental data repositories were established. Approximately 20% of the iDPP@CLEF 2023 ALS datasets, corresponding to 434 to 574 patients, are linked to environmental data.

Considering that the impact of environmental factors may occur well before the diagnosis, we include the maximum amount of available information before Time 0 for all patients with historical records. Depending on the patient, this corresponds to a maximum of 4 to 6 years of data. However, no more than 6 months of data after Time 0 are considered. If a patient has more than 180 days of information after the first ALSFRS-R assessment, the subsequent days are excluded from the released dataset.

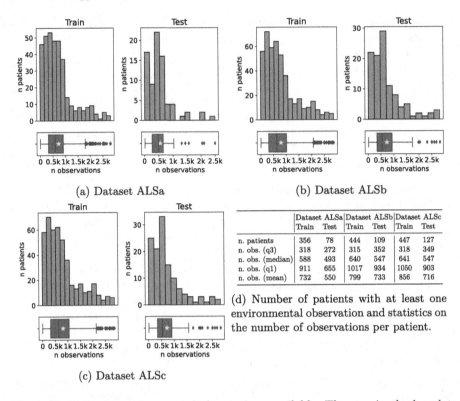

(a) Dataset ALSa   (b) Dataset ALSb

(c) Dataset ALSc

(d) Number of patients with at least one environmental observation and statistics on the number of observations per patient.

	Dataset ALSa		Dataset ALSb		Dataset ALSc	
	Train	Test	Train	Test	Train	Test
n. patients	356	78	444	109	447	127
n. obs. (q3)	318	272	315	352	318	349
n. obs. (median)	588	493	640	547	641	547
n. obs. (q1)	911	655	1017	934	1050	903
n. obs. (mean)	732	550	799	733	856	716

Fig. 1. Statistics on environmental observations available. The star in the boxplots indicates the mean.

Figure 1 reports the number of patients associated with environmental data as well as the number of records of environmental observations available. It is possible to observe that on average, on the training set, there are 732, 799 and 856 days of observations in the case of Datasets ALSa ALSb, and ALSc respectively. Patients within the test set contain slightly lower numbers of records.

Figure 2 shows the proportion of patients (among those with environmental data) having observations for a given day in (their) history. For example, it is possible to observe that roughly 80% of the patients have a record of their Time 0, this number grows to approximately 95% if we consider the Time 180, the last day for which we release information. Going back in time, we observe that, for roughly 40% of the patients, we have at least 2 years (Time -730) of information before their Time 0.

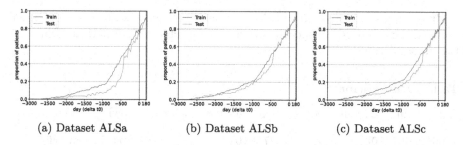

(a) Dataset ALSa   (b) Dataset ALSb   (c) Dataset ALSc

Fig. 2. Proportion of patients having environmental observations on a given day in (their relative) time.

5 Lab Setup and Participation

In the remainder of this section, we detail the guidelines the participants had to comply with to submit their runs and the submissions received by iDPP@CLEF. In the remainder, we describe the guidelines provided to participating teams.

5.1 Guidelines

- The runs should be submitted in the textual format described below;
- Each group can submit a maximum of 10 runs for each subtask, thus amounting to maximum 20 runs for each of Task 1 and Task 2 and 30 runs for Task 3.

Task 1 Run Format

Runs should be submitted as a text file (.txt) with the following format:

```
10061925618906738677048445096063212421 0.897 upd_T1a_survRF
10160033396142711512526634552182640753 0.773 upd_T1a_survRF
10287479530859953246187859713708391150 0.773 upd_T1a_survRF
12398828804459792215818261570544715022 0.615 upd_T1a_survRF
10038199677222038202107097495517621823 0.317 upd_T1a_survRF
. . .
```

where:

- Columns are separated by a white space;
- The first column is the `patient ID`, an hashed version of the original patient ID (should be considered just as a string);
- The second column is the risk score. It is expected to be a floating point number in the range [0, 1];
- The third column is the run identifier, according to the format described below. It must uniquely identify the participating team and the submitted run.

It is important to include all the columns and have a white space delimiter between the columns. No specific ordering is expected among patients (rows) in the submission file.

Task 2 Run Format

Runs should be submitted as a text file (.txt) with the following format:

```
10061925618906... 0.221 0.437 0.515 0.817 0.916 upd_T2b_survRF
10160033396142... 0.213 0.617 0.713 0.799 0.822 upd_T2b_survRF
10287479530859... 0.205 0.312 0.418 0.781 0.856 upd_T2b_survRF
12398828804459... 0.197 0.517 0.617 0.921 0.978 upd_T2b_survRF
10038199677222... 0.184 0.197 0.315 0.763 0.901 upd_T2b_survRF
...
```

where:

- Columns are separated by a white space;
- The first column is the patient ID, a hashed version of the original patient ID (should be considered just as a string);
- The second column is the cumulative probability of worsening between years 0 and 2. It is expected to be a floating point number in the range [0, 1].
- The third column is the cumulative probability of worsening between years 0 and 4. It is expected to be a floating point number in the range [0, 1].
- The fourth column is the cumulative probability of worsening between years 0 and 6. It is expected to be a floating point number in the range [0, 1].
- The fifth column is the cumulative probability of worsening between years 0 and 8. It is expected to be a floating point number in the range [0, 1].
- The sixth column is the cumulative probability of worsening between years 0 and 10. It is expected to be a floating point number in the range [0, 1].
- The seventh column is the run identifier, according to the format described below. It must uniquely identify the participating team and the submitted run.

It is important to include all the columns and have a white space delimiter between the columns. No specific ordering is expected among patients (rows) in the submission file.

Task 3 Run Format

Runs should be submitted as a text file (.txt) with the following format:

```
0x4bed50627d141453da7499a7f6ae84ab 0.897 upd_T3a_EW6_survRF
0x4d0e8370abe97d0fdedbded6787ebcfc 0.773 upd_T3a_EW6_survRF
0x5bbf2927feefd8617b58b5005f75fc0d 0.773 upd_T3a_EW6_survRF
0x814ec836b32264453c04bb989f7825d4 0.615 upd_T3a_EW6_survRF
0x71dabb094f55fab5fc719e348dffc85x 0.317 upd_T3a_EW6_survRF
...
```

where:

- Columns are separated by a white space;
- The first column is the `patient` ID, a 128 bit hex number (should be considered just as a string);
- The second column is the risk score. It is expected to be a floating point number in the range $[0, 1]$;
- The third column is the run identifier, according to the format described below. It must uniquely identify the participating team and the submitted run.

It is important to include all the columns and have a white space delimiter between the columns. No specific ordering is expected among patients (rows) in the submission file. Since different time windows may be considered, participants are allowed to submit predictions for a variable number of patients. We encourage participants to submit predictions for as many patients as possible. To avoid favoring runs that consider only a few patients, submitted runs will be evaluated based on their correctness as well as the number of patients included. The number of patients included is also reported in the output of the evaluation scripts.

Submission Upload

Runs should be uploaded in the repository provided by the organizers. Following the repository structure discussed above, for example, a run submitted for the first task should be included in `submission/task1`.

Runs should be uploaded using the following name convention for their identifiers:

`<teamname>_T<1|2|3><a|b|c>_[type_]<freefield>`

where:

- `teamname` is the name of the participating team;
- `T<1|2><a|b|c>` is the identifier of the task the run is submitted to, e.g. `T1b` for Task 1, subtask b;
  - `type` describes the type of run only in the case of Task 3 (it can be omitted for Task 1 and 2). It should be one among:
  - `base` for a baseline run;
  - `EW6` when using environmental data in a time window of 6 months before and after `Time 0`;
  - `EWP` when using environmental data in a time windows chosen by the participant; in this case it is suggested to use `freefield` to provide information about the adopted time window;
- `freefield` is a free field that participants can use as they prefer to further distinguish among their runs. Please, keep it short and informative.

For example, a complete run identifier may look like:

`upd_T3a_EW6_survRF`

where:

- upd is the University of Padua team;
- T3a means that the run is submitted for Task 3, subtask a;
- EW6 means that environmental data in a time window of 6 months before and after Time 0 have been used;
- survRF suggests that participants have used survival random forests as a prediction method.

The name of the text file containing the run must be the identifier of the run followed by the .txt extension. In the above example:

upd_T3a_EW6_survRF.txt

Run Scores

Performance scores for the submitted runs will be returned by the organizers in the score folder, which follows the same structure as the submission folder.

For each submitted run, participants will find a file named

<teamname>_T<1|2|3><a|b|c>_[type_]<freefield>.score.txt

where <teamname>_T<1|2|3><a|b|c>_[type_]<freefield> matches the corresponding run. The file will contain performance scores for each of the evaluation measures described below. In the above example:

upd_T3a_EW6_survRF.score.txt

5.2 Participants

Overall, 45 teams registered for participating in iDPP@CLEF but only 10 of them actually managed to submit runs for at least one of the offered tasks. Table 5 reports the details about the participating teams.

Table 6 provides breakdown of the number of runs submitted by each participant for each task and sub-task. Overall, we have received 163 runs with a prevalence of submissions for Task 1 (76 runs), followed by Task 2 (48 runs), and lastly, Task 3 (49 runs).

6 Evaluation Measures

iDPP@CLEF adopted several state-of-the-art evaluation measures to assess the performance of the prediction algorithms, among which:

- *Area Under the ROC Curve (AUC)* [11] to show the trade-off between clinical sensitivity and specificity for every possible cut-off of the risk scores;
- *Harrel's Concordance Index (C-index)* [13] to summarize how well a predicted risk score describes an observed sequence of events.
- *O/E ratio* to assess whether or not the observed event rates match expected event rates in subgroups of the model population.

Table 5. Teams participating in iDPP@CLEF 2023.

Team Name	Description	Country	Repository	Paper
CompBioMed	Department of Medical Sciences, University of Turin	Italy	https://bitbucket.org/ brainteaser-health/ idpp2023-compbiomed	Rossi et al. [21]
FCOOL	Faculty of Sciences of the University of Lisbon	Portugal	https://bitbucket.org/ brainteaser-health/ idpp2023-fcool	Branco et al. [2,3]
HULAT-UC3M	Polytechnic School Universidad Carlos III de Madrid	Spain	https://bitbucket.org/ brainteaser-health/ idpp2023-hulat-u3m	Ramos et al. [19]
NeuroTN	Independent Researcher, Sfax	Tunisia	https://bitbucket.org/ brainteaser-health/ idpp2023-neurotn	Karray [14]
Onto-Med	Ontomed	Bulgaria	https://bitbucket.org/ brainteaser-health/ idpp2023-onto-med	Asamov et al. [1]
SBB	University of Padua	Italy	https://bitbucket.org/ brainteaser-health/ idpp2023-sbb	Guazzo et al. [9]
Stefagroup	University of Pavia, BMI lab "Mario Stefanelli"	Italy	https://bitbucket.org/ brainteaser-health/ idpp2023-stefagroup	Buonocore et al. [4]
SisInfLab_AIBio	Polytechnic University of Bari	Italy	https://bitbucket.org/ brainteaser-health/ idpp2023-sisinfo-aibio	Lombardi et al. [17]
UHU-ETSI-1	Universidad de Huelva	Spain	https://bitbucket.org/ brainteaser-health/ idpp2023-uhu-etsi	Not Submitted
UWB	University of West Bohemia	Czech Republic	https://bitbucket.org/ brainteaser-health/ idpp2023-uwb	Hanzl and Picek [12]

To ease the computation and reproducibility of the results, scripts for computing the measures are available in the following repository: https://bitbucket.org/brainteaser-health/idpp2023-performance-computation.

6.1 Task 1: Measures to Evaluate the Prediction of the Risk of Disease Worsening (MS)

For Task 1, the effectiveness of the submitted runs is evaluated using Harrell's Concordance Index (C-index) [13]. This score quantifies the model's ability in ranking pairs of observations based on their predicted outcomes. A C-index value of 1 indicates perfect concordance, meaning the model can accurately distinguish between higher and lower-risk individuals. Conversely, a value of 0.5 suggests random guessing, while values below 0.5 indicate a counter-correlation.

Table 6. Break-down of the runs submitted by participants for each task and sub-task. Participation in Task 3 does not involve submission of runs and it is marked just with a tick.

Team	Task 1		Task 2		Task 3			Total
	a	b	a	b	a	b	c	
CompBioMed	3	3	3	2	—	—	—	11
FCOOL	5	5	—	—	9	9	9	37
HULAT-UC3M	2	1	2	1	—	—	—	6
NeuroTN	—	—	—	—	4	4	4	12
Onto-Med	5	4	5	4	—	—	—	18
SBB	3	3	3	3	—	—	—	12
SisInfLab_AIBio	5	4	5	4	—	—	—	18
Stefagroup	2	—	—	—	—	—	—	2
UHU-ETSI-1	6	7	3	3	—	—	—	19
UWB	9	9	5	5	—	—	—	28
Total	40	36	26	22	13	13	13	163

6.2   Task 2: Measures to Evaluate the Prediction of the Cumulative Probability of Worsening (MS)

The effectiveness of the submitted runs is evaluated with the following measures:

– *Area Under the ROC curve (AUROC)* at each of the time intervals (0–2, 0–4, 0–6, 0–8, 0–10 years);
– O/E Ratio: the ratio of observed to expected events at each of the time intervals (0–2, 0–4, 0–6, 0–8, 0–10 years).

The *Receiver Operating Characteristic (ROC)* curve is a graphical representation of the model's true positive rate (sensitivity) against the false positive rate (1 - specificity) at different classification thresholds. The AUROC ranges from 0 to 1, where a value of 1 indicates a perfect model that can accurately distinguish between individuals who will experience worsening and those who will not. An AUROC value of 0.5 suggests a model that performs no better than random chance. Therefore, a higher AUROC reflects a better ability of the model to discriminate between different outcomes.

The O/E (Observed-to-Expected) ratio provides a measure of calibration for the model's predictions. It compares the actual number of observed worsening events to the number of events expected based on the model's predictions. Ideally, the O/E ratio should be close to 1, indicating good calibration and alignment between predicted and observed outcomes. A ratio significantly above 1 suggests an overestimation of the number of worsening events, while a ratio below 1 indicates an underestimation. Monitoring the O/E ratio at each time interval allows for assessing the model's calibration performance over time.

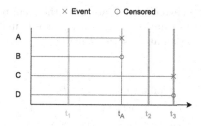

Fig. 3. The set of possible outcomes and censoring time scenarios.

To compute the AUROC and O/E Ratio, we applied censoring to the ground truth values using the following schema. Let A, B, C, and D be four subjects, where:

– A experienced the outcome at t_A;
– B was censored at t_A;
– C experienced the outcome at t_3;
– D was censored at t_3.

The scenario is represented in Fig. 3.

Table 7 reports the outcome occurrence label and outcome time for each possible scenario of censoring time, which we refer to as t_1, t_2, and t_3. When t_1 is considered as censoring time, all four example subjects have yet to experience the event or be censored, as a result, their outcome occurrence label at this time is set to 0 as shown in the first column of Table 7. When t_2 is considered to perform censoring (second column of Table 7), instead, only subjects C and D have yet to experience either the even or the censoring, and their outcome label is then set to 0. In this scenario, subject A had the event before t_2 and its outcome label is then set to 1. Subject B was censored before t_2 and, as its outcome at this time is unknown, it must be excluded from performance evaluation. Finally, when t_3 is considered to perform censoring (third column of Table 7), outcome labels of subjects A and B are equal to those considered for t_2 since their situation at this time is unchanged compared to the previous one. However, subject C experienced the vent at t_3 and now its outcome label must be set to 1 and subject D was censored at t_3 and its outcome label is then set to 0.

6.3 Task 3: Measures to Evaluate the Impact of Exposition to Pollutants (ALS)

The effectiveness of the submitted runs is evaluated with the following measures:

– AUROC: the area under the receiver operating characteristic curve at each of the time intervals (6, 12, 18, 24, 30, 36 months);
– C-index.

Table 7. Outcome time/occurrence annotation for the example in Fig. 3. * indicates that being the outcome of the subject at censoring time t_i unknown, the subject can not be considered for evaluation at censoring time t_i.

		t_1	t_2	t_3
A	outcome time	t_1	t_A	t_A
	outcome occurred	0	1	1
B	outcome time	t_1	NA	NA
	outcome occurred	0	NA*	NA*
C	outcome time	t_1	t_2	t_3
	outcome occurred	0	0	1
D	outcome time	t_1	t_2	t_3
	outcome occurred	0	0	0

7   Results

For each task, we report the analysis of the performance of the runs submitted by the Lab's participants according to the measures described in Sect. 6.

7.1   Task 1: Predicting Risk of Disease Worsening (Multiple Sclerosis)

Figure 4 shows the C-index with its 95% confidence intervals computed for all runs submitted for Task 1 sub-task a and for the random classifier (last row)[7]. Discrimination performance varies across the different submitted runs ranging from 0.4 to above 0.8. Runs submitted by the UWB team [12] lead the pack (C-index >0.8), followed by CompBioMed (CBMUnitTO) [21], and FCOOL [3]. The best-performing approach for UWB and FCOOL and SisInfLab_AIBio [17] are Survival Random Forests. CompBioMed [21], HULAT [19], and SBB [9] achieve the best performance with Cox regression and CoxNets.

7.2   Task 2: Predicting Cumulative Probability of Worsening (Multiple Sclerosis)

Table 8 presents the AUROC and the O/E ratios, with their 95% confidence intervals computed for all runs submitted for task 2 sub-task a. To avoid cluttering, we report the performance obtained for the two-year time window; complete results for subtask a and the results for sub-task b, are shown in the extended overview [6]. As highlighted in Table 8, the approach obtaining the best result in terms of AUCROC corresponds to the run uwb_T2a_survRFmri, while the best results for O/E ratio are shown by uwb_T2a_survGB_minVal. In general, survival Gradient Boosting approaches proposed by UWB achieve good performance in AUROC, with a good O/E as well.

---

[7] Results for sub-task b are available on the extended overview [6].

Table 8. AUROC and OE ratio for all the submitted runs for task 2 subtask a, with a two-year time window. We report the measure as well as the 95% confidence interval.

identifier	AUROC	O/E ratio
CBMUniTO_T2a_coxnet	0.890 (0.739, 1.000)	0.443 (−0.018, 0.904)
CBMUniTO_T2a_cwgbsa	0.841 (0.618, 1.000)	0.467 (−0.007, 0.940)
CBMUniTO_T2a_evilcox	0.854 (0.655, 1.000)	0.449 (−0.015, 0.913)
HULATUC3M_T2a_survcoxnet	0.864 (0.770, 0.958)	0.437 (−0.021, 0.895)
HULATUC3M_T2a_survRF	0.840 (0.710, 0.969)	0.451 (−0.014, 0.917)
onto-med_T2a_0.01.1.0e-5.10000.100.adj	0.731 (0.482, 0.980)	0.133 (−0.120, 0.386)
onto-med_T2a_0.2.1.0e-5.10000.100	0.696 (0.440, 0.951)	0.269 (−0.090, 0.628)
onto-med_T2a_0.2.1.0e-5.10000.200	0.716 (0.446, 0.987)	0.234 (−0.101, 0.570)
onto-med_T2a_0.2.1.0e-5.5000.100	0.647 (0.399, 0.896)	0.380 (−0.047, 0.807)
onto-med_T2a_0.2.1.0e-5.5000.200	0.590 (0.337, 0.842)	0.358 (−0.057, 0.772)
sbb_T2a_Cox	0.708 (0.491, 0.926)	0.389 (−0.043, 0.821)
sbb_T2a_RSF	0.604 (0.386, 0.822)	0.385 (−0.045, 0.815)
sbb_T2a_SSVM	0.624 (0.461, 0.787)	0.358 (−0.057, 0.772)
sisinflab-aibio_T2a_GB1	0.677 (0.462, 0.893)	0.000 (0.000, 0.000)
sisinflab-aibio_T2a_GB2	0.782 (0.618, 0.945)	0.000 (0.000, 0.000)
sisinflab-aibio_T2a_GB3	0.481 (0.259, 0.703)	0.000 (−0.002, 0.002)
sisinflab-aibio_T2a_RF1	0.754 (0.537, 0.970)	0.017 (−0.073, 0.107)
sisinflab-aibio_T2a_RF2	0.569 (0.347, 0.791)	0.010 (−0.060, 0.081)
uhu-etsi-1_T2a_03	0.769 (0.621, 0.916)	0.678 (0.107, 1.248)
uhu-etsi-1_T2a_04	0.812 (0.690, 0.933)	0.713 (0.128, 1.298)
uhu-etsi-1_T2a_05	0.774 (0.636, 0.912)	0.697 (0.119, 1.276)
uwb_T2a_CGBSA	0.862 (0.731, 0.993)	3.106 (1.885, 4.327)
uwb_T2a_survGB	0.877 (0.745, 1.000)	0.919 (0.255, 1.583)
uwb_T2a_survGB_minVal	0.894 (0.787, 1.000)	**0.946** (0.272, 1.620)
uwb_T2a_survRF	0.914 (0.784, 1.000)	1.811 (0.879, 2.744)
uwb_T2a_survRFmri	**0.924** (0.800, 1.000)	1.889 (0.937, 2.842)

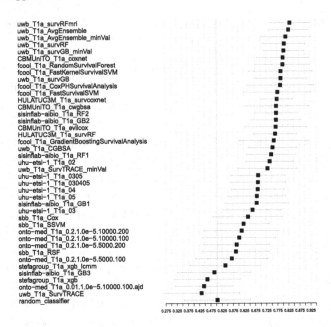

uwb_T1a_survRFmri
uwb_T1a_AvgEnsemble
uwb_T1a_AvgEnsemble_minVal
uwb_T1a_survRF
uwb_T1a_survGB_minVal
CBMUniTO_T1a_coxnet
fcool_T1a_RandomSurvivalForest
fcool_T1a_FastKernelSurvivalSVM
uwb_T1a_survGB
fcool_T1a_CoxPHSurvivalAnalysis
fcool_T1a_FastSurvivalSVM
HULATUC3M_T1a_survcoxnet
CBMUniTO_T1a_cwgbsa
sisinflab-aibio_T1a_RF2
sisinflab-aibio_T1a_GB2
CBMUniTO_T1a_evilcox
HULATUC3M_T1a_survRF
fcool_T1a_GradientBoostingSurvivalAnalysis
uwb_T1a_CGBSA
sisinflab-aibio_T1a_RF1
uhu-etsi-1_T1a_02
uwb_T1a_SurvTRACE_minVal
uhu-etsi-1_T1a_0305
uhu-etsi-1_T1a_030405
uhu-etsi-1_T1a_04
uhu-etsi-1_T1a_05
sisinflab-aibio_T1a_GB1
uhu-etsi-1_T1a_03
sbb_T1a_Cox
sbb_T1a_SSVM
onto-med_T1a_0.2.1.0e-5.10000.200
onto-med_T1a_0.2.1.0e-5.10000.100
onto-med_T1a_0.2.1.0e-5.5000.200
sbb_T1a_RSF
onto-med_T1a_0.2.1.0e-5.5000.100
stefagroup_T1a_xgb_lcmm
sisinflab-aibio_T1a_GB3
stefagroup_T1a_xgb
onto-med_T1a_0.01.1.0e-5.10000.100.ajd
uwb_T1a_SurvTRACE
random_classifier

0.275 0.325 0.375 0.425 0.475 0.525 0.575 0.625 0.675 0.725 0.775 0.825 0.875 0.925

Fig. 4. C-index (with 95% confidence interval) achieved by runs submitted to Task 1a.

7.3   Task 3: Position Papers on Impact of Exposition to Pollutants (Amyotrophic Lateral Sclerosis)

Figure 5 shows the C-index and 95% confidence intervals achieved on Task 1 sub-task a[8] by the submitted runs and for the random classifier (last row). As observed by Karray [14] and Branco et al. [2] runs including environmental data (runs tagged with EWP and EW6) tend to perform worse than their counterpart that does not rely on the environmental data. The best-performing approach is provided by the NeuroTN team [14] and corresponds to the classifier ensemble (see Subsect. 7.4).

7.4   Approaches

In this section, we provide a short summary of the approaches adopted by participants in iDPP@CLEF. There are two separate sub-sections, one for Task 1 and 2 – focused on MS worsening prediction – and one for Task 3 – which concerns the impact of exposition to pollutants on the ALS progression.

Tasks 1 and 2. CompBioMed [21] experiments with CoxNet, Component-wise Gradient Boosting Survival Analysis (CWGBSA), and a hybrid method where the most important features selected by CWGBSA are used to build a CoxNet

---

[8] Results for sub-tasks b and c are available on the extended overview [6].

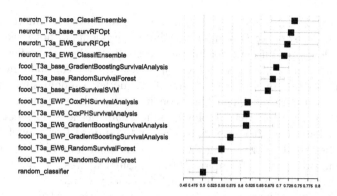

Fig. 5. C-index (with 95% confidence interval) achieved by runs submitted to Task 3a.

model (EvilCox). They also test non-linear methods such as Random Survival Forest and Gradient Boosting Survival Analysis, observing a tendency to overfit the training data. To assess the importance of the features, Rossi et al. [21] perform Permutation-based Feature Importance Analysis. In general, they observe that Coxnet is the best-performing approach for all tasks and subtasks. Nevertheless, they also observed that CWGBSA is resistant to over-fitting and aggressive in eliminating features. CWGBSA cross-validated performance is almost on par with that of CoxNet, despite using a smaller set of features.

FCOOL [3] explores several survival prediction methods to rank MS patients according to the risk of worsening. The considered methods are Random Survival Forest, Gradient Boosting, Fast Survival SVM, Fast Kernel Survival SVM, and the Cox Proportional-Hazards model. A data preprocessing phase is conducted prior to training to manage the temporal nature of patient data by choosing relevant features and by computing additional ones – which capture the temporal progression of the disease. Overall, Random Survival Forest performs best on subtask 1a, whereas Fast Kernel Survival SVM on subtask 1b. Subtask 1b was found to be more complex because of the different definition of the worsening event.

HULAT [19] investigates the effectiveness of Random Survival Forest and Cox regression with Elastic Net regularization (CoxNet) methods on MS worsening prediction. As well as other groups, Ramos et al. [19] perform a data preprocessing phase involving data cleaning, format transformation, normalization, and outliers removal. In particular, the preprocessing step removes all the dynamic features containing a high number of missing values.

Onto-Med [1] develop a Maximum Likelihood Estimation approach to predict MS progression. The proposed method relies on patients' covariates and employs a multi-layer perceptron to approximate the optimal distribution parameters. To handle both tasks, Asamov et al. [1] used the whole training data to build a model and estimate a maximum likelihood distribution for each patient given their features. The method uses a cumulative probability estimate instead of coherent risk measures to accommodate the requirements of bot tasks.

SBB [9] develops different machine-learning approaches to predict a worsening in patient disability caused by MS. Specifically, they consider the following well-known survival analysis approaches: Cox model, random survival forests, and survival support machine. They conclude that these approaches achieve modest performance and that employing non-linear methods does not lead to a discernible advantage with respect to the gold standard Cox model. Nonetheless, they observe that improving data pre-processing may be a key operation to perform in order to obtain more relevant input features and augment model discrimination with the aim of obtaining satisfactory results.

Stefagroup [4] explores two post-hoc model-agnostic XAI methods, namely SHAP and AraucanaXAI, to provide insights about the most predictive factors of worsening in MS patients. Buonocore et al. [4] evaluate the proposed XAI approaches using commonly adopted measures in XAI for healthcare such as identity, fidelity, separability and time. By leveraging SHAP and AraucanaXAI, the authors gained a deeper understanding of the shortcomings and limitations of their classifiers through feature importance and navigable decision trees.

SisInfLab_AIBio [17] uses Random Survival Forests, an extension of random forests specifically designed for survival analysis, and Boosting Machines for time-to-event analysis. To assess the importance of features for both ML models, the permutation feature importance is computed as well. Lombardi et al. [17] observe that, if the definition of worsening is more complex and condition-dependent (tasks 1b and 2b) significantly lower their approach performs worse than with a simpler definition of worsening (tasks 1a and 2a).

UWB [12] evaluates various ML methods – such as Random Forest and Gradient Boosting – for survival analysis, as well as a Deep Learning survival analysis method based on the Transformer architecture: SurfTRACE. Among the different methods, the authors report top performance with Random Forest. Hanzl and Picek [12] observe that three aspects are instrumental to achieving good performance: (i) data preprocessing, (ii) hyper-parameter tuning, and (iii) validation.

Task 3. FCOOL [2] investigates four models to assess the importance of environmental data in predicting the risk of early occurrence of NIV, PEG or death: Cox Proportional-Hazards, Random Survival Forest, Survival SVM, and Gradient Boosting. Without the introduction of environmental data, the models perform reasonably well. Nevertheless, Branco et al. [2] observe an evident degradation in performance when providing the model with environmental and clinical data in all three tasks. For task A, they observe an even larger degradation when unconstrained amounts of environmental data are provided, compared to what was observed with only 6 months of data. This pattern does not hold for Tasks B and C, where the amount of data does not harm the results, which are, in any case, lower than what was observed without environmental data.

NeuroTN [14] Proposes an approach to stratify patients relying on the disease progression patterns according to features extracted from applying staging systems on visits data. Clusters of patients are then profiled to determine their common characteristics: clinical, demographic and environmental. A second clustering procedure is carried on to detect clusters of patients with similar exposure concentrations to 3 different air pollutants. Then, Karray [14] performs risk prediction on each cluster separately and combines the predictions. In particular Karray [14] relies on two ensembles of classifiers trained on a different data representation (data with Environmental Features and data without Environmental Features). Furthermore, they explored also Survival Random Forests. As for Branco et al. [2], the introduction of environmental features does not seem to benefit both models and causes performance deterioration.

8 Conclusions and Future Work

The second iteration of iDPP@CLEF focuses on predicting the temporal progression of MS and ALS. In particular, iDPP@CLEF 2023 comprises three tasks. The first two tasks concern MS and participants were provided clinical data and had the objective of predicting the risk of worsening. The third task centres around ALS and builds upon the foundation laid by iDPP@CLEF 2022. This task follows a similar design, involving the prediction of NIV, PEG, or death, but with the addition of environmental data to explore the impact of pollutant exposure on the progression of ALS.

We developed 5 datasets, two for MS and three for ALS, based on the anonymized data provided by three medical institutions in Turin, Lisbon, and Pavia. Out of 45 registered participants, 10 managed to submit a total of 163 runs with a prevalence of submissions for Tasks 1 and 2. Participants adopted a range of approaches, such as Survival Random Forests and Coxnets.

The next iteration of iDPP@CLEF will maintain its dual focus on both ALS and MS. We will extend the amount of available information, by considering also time-series concerning patients' vital parameters produced by wearable devices.

Acknowledgments. The work reported in this paper has been partially supported by the BRAINTEASER (https://brainteaser.health/) project (contract n. GA101017598), as a part of the European Union's Horizon 2020 research and innovation programme.

References

1. Asamov, T., Aksenova, A., Ivanov, P., Boytcheva, S., Taskov, D.: Maximum likelihood estimation with deep learning for multiple sclerosis progression prediction. In: Aliannejadi, M., Faggioli, G., Ferro, N., Vlachos, M. (eds.) CLEF 2023 Working Notes (2023)
2. Branco, R., et al.: Investigating the impact of environmental data on ALS prognosis with survival analysis. In: Aliannejadi, M., Faggioli, G., Ferro, N., Vlachos, M. (eds.) CLEF 2023 Working Notes (2023)

3. Branco, R., et al.: Survival analysis for multiple sclerosis: predicting risk of disease worsening. In: Aliannejadi, M., Faggioli, G., Ferro, N., Vlachos, M. (eds.) CLEF 2023 Working Notes (2023)
4. Buonocore, T., et al.: Predicting and explaining risk of disease worsening using temporal features in multiple sclerosis notebook for the iDPP lab on intelligent disease progression prediction at clef 2023. In: CLEF 2023 Working Notes (2023)
5. Cedarbaum, J.M., et al.: The ALSFRS-R: a revised ALS functional rating scale that incorporates assessments of respiratory function. J. Neurol. Sci. **169**(1–2), 13–21 (1999)
6. Faggioli, G., et al.: Overview of iDPP@CLEF 2023: the intelligent disease progression prediction challenge. In: Aliannejadi, M., Faggioli, G., Ferro, N., Vlachos, M. (eds.) CLEF 2023 Working Notes, CEUR Workshop Proceedings (CEUR-WS.org) (2023). ISSN 1613–0073
7. Guazzo, A., et al.: Intelligent disease progression prediction: overview of iDPP@CLEF 2022. In: Barron-Cedeno, A., et al. (eds.) Experimental IR Meets Multilinguality, Multimodality, and Interaction. CLEF 2022. LNCS, vol. 13390, pp. 395–422. Springer, Cham (2022). https://doi.org/10.1007/978-3-031-13643-6_25
8. Guazzo, A., et al.: Overview of iDPP@CLEF 2022: the intelligent disease progression prediction challenge. In: Faggioli, G., Ferro, N., Hanbury, A., Potthast, M. (eds.) CLEF 2022 Working Notes, pp. 1130–1210, CEUR Workshop Proceedings (CEUR-WS.org) (2022). ISSN 1613–0073. http://ceur-ws.org/Vol-3180/
9. Guazzo, A., Trescato, I., Longato, E., Tavazzi, E., Vettoretti, M., Camillo, B.: Baseline machine learning approaches to predict multiple sclerosis disease progression. In: Aliannejadi, M., Faggioli, G., Ferro, N., Vlachos, M. (eds.) CLEF 2023 Working Notes (2023)
10. Hagan, D.H., et al.: Calibration and assessment of electrochemical air quality sensors by co-location with regulatory-grade instruments. Atmos. Meas. Tech. **11**(1), 315–328 (2018). ISSN 1867–8548, https://doi.org/10.5194/amt-11-315-2018
11. Hanley, J.A., McNeil, B.J.: The meaning and use of the area under a receiver operating characteristic (ROC) curve. Radiology **143**(1), 29–36 (1982). pMID: 7063747
12. Hanzl, M., Picek, L.: Predicting risk of multiple sclerosis worsening. In: Aliannejadi, M., Faggioli, G., Ferro, N., Vlachos, M. (eds.) CLEF 2023 Working Notes (2023)
13. Harrell, F.E.J., Califf, R.M., Pryor, D.B., Lee, K.L., Rosati, R.A.: Evaluating the yield of medical tests. JAMA **247**(18), 2543–2546 (1982). ISSN 0098–7484
14. Karray, M.: Air pollution profiling through patient stratification: study of ALS staging systems usefulness in facilitating data-driven disease subtyping and discovery of hazardous ambient air pollutants. In: Aliannejadi, M., Faggioli, G., Ferro, N., Vlachos, M. (eds.) CLEF 2023 Working Notes (2023)
15. Küffner, R., et al.: Crowdsourced analysis of clinical trial data to predict amyotrophic lateral sclerosis progression. Nat. Biotechnol. **33**(1), 51–57 (2015)
16. Kurtzke, J.F.: Rating neurologic impairment in multiple sclerosis. Neurology **33**(11), 1444–1444 (1983). ISSN 0028–3878, https://doi.org/10.1212/WNL.33.11.1444, https://n.neurology.org/content/33/11/1444
17. Lombardi, A., et al.: Time-to-event interpretable machine learning for multiple sclerosis worsening prediction: results from iDPP@CLEF 2023. In: Aliannejadi, M., Faggioli, G., Ferro, N., Vlachos, M. (eds.) CLEF 2023 Working Notes (2023)
18. McKight, P.E., Najab, J.: Kruskal-wallis test. The corsini encyclopedia of psychology, p. 1 (2010)
19. Ramos, A., Martínez, P., González-Carrasco, I.: Hulat@iddp clef 2023: intelligent prediction of disease progression in multiple sclerosis patients. In: Aliannejadi, M., Faggioli, G., Ferro, N., Vlachos, M. (eds.) CLEF 2023 Working Notes (2023)

20. Rich, J.T., Neely, J.G., Paniello, R.C., Voelker, C.C., Nussenbaum, B., Wang, E.W.: A practical guide to understanding Kaplan-Meier curves. Otolaryngol.-Head Neck Surg. **143**(3), 331–336 (2010)
21. Rossi, I., Birolo, G., Fariselli, P.: iDPP@CLEF 2023 results from dsm-compbio unito. In: Aliannejadi, M., Faggioli, G., Ferro, N., Vlachos, M. (eds.) CLEF 2023 Working Notes (2023)
22. Tallarida, R.J., Murray, R.B., Tallarida, R.J., Murray, R.B.: Chi-Square Test. In: Manual of Pharmacologic Calculations: With Computer Programs, pp. 140–142. Springer, New York, NY (1987). https://doi.org/10.1007/978-1-4612-4974-0_43
23. Vogt, M., Schneider, P., Castell, N., Hamer, P.: Assessment of low-cost particulate matter sensor systems against optical and gravimetric methods in a field co-location in Norway. Atmosphere **12**(8), 961 (2021). ISSN 2073–4433, https://doi.org/10.3390/atmos12080961

Overview of the ImageCLEF 2023: Multimedia Retrieval in Medical, Social Media and Internet Applications

Bogdan Ionescu[1], Henning Müller[2], Ana-Maria Drăgulinescu[1(✉)],
Wen-Wai Yim[3], Asma Ben Abacha[3], Neal Snider[4], Griffin Adams[5],
Meliha Yetisgen[6], Johannes Rückert[7], Alba García Seco de Herrera[8],
Christoph M. Friedrich[7], Louise Bloch[7], Raphael Brüngel[7],
Ahmad Idrissi-Yaghir[7], Henning Schäfer[7], Steven A. Hicks[9],
Michael A. Riegler[9], Vajira Thambawita[9], Andrea M. Storås[9], Pål Halvorsen[9],
Nikolaos Papachrysos[10], Johanna Schöler[10], Debesh Jha[9,11],
Alexandra-Georgiana Andrei[1], Ioan Coman[1], Vassili Kovalev[12,13],
Ahmedkhan Radzhabov[13], Yuri Prokopchuk[13], Liviu-Daniel Ştefan[1],
Mihai-Gabriel Constantin[1], Mihai Dogariu[1], Jérôme Deshayes[14],
and Adrian Popescu[14]

[1] Politehnica University of Bucharest, Bucharest, Romania
{bogdan.ionescu,ana.dragulinescu}@upb.ro
[2] University of Applied Sciences Western Switzerland (HES-SO), Sierre, Switzerland
henning.mueller@hevs.ch
[3] Microsoft, Redmond, USA
yimwenwai@microsoft.com
[4] Microsoft/Nuance, Redmond, USA
[5] Columbia University, New York, USA
[6] University of Washington, Seattle, USA
[7] University of Applied Sciences and Arts Dortmund, Dortmund, Germany
[8] University of Essex, Colchester, UK
[9] SimulaMet, Oslo, Norway
[10] Sahlgrenska University Hospital, Gothenburg, Sweden
[11] Northwestern University, Chicago, USA
[12] Belarus State University, Minsk, Belarus
[13] Belarusian National Academy of Sciences, Minsk, Belarus
[14] CEA LIST, Palaiseau, France

Abstract. This paper presents an overview of the ImageCLEF 2023 lab, which was organized in the frame of the Conference and Labs of the Evaluation Forum – CLEF Labs 2023. ImageCLEF is an ongoing evaluation event that started in 2003 and that encourage the evaluation of the technologies for annotation, indexing and retrieval of multimodal data with the goal of providing information access to large collections of data in various usage scenarios and domains. In 2023, the 21st edition of ImageCLEF runs three main tasks: (i) a *medical* task which included the sequel of the caption analysis task and three new tasks, namely, GANs for medical images, Visual Question Answering for colonoscopy images, and

A. Arampatzis et al. (Eds.): CLEF 2023, LNCS 14163, pp. 370–396, 2023.
https://doi.org/10.1007/978-3-031-42448-9_25

medical dialogue summarization; (ii) a sequel of the *fusion* task addressing the design of late fusion schemes for boosting the performance, with
two real-world applications: image search diversification (retrieval) and
prediction of visual interestingness (regression); and (iii) a sequel of the
social media aware task on potential real-life effects awareness of online
image sharing. The benchmark campaign was a real success and received
the participation of over 45 groups submitting more than 240 runs.

Keywords: Medical text summarization · medical image caption
analysis · visual question answering · Generative Adversarial Networks
(GANs) · late fusion for search diversification and interestingness
prediction · prediction of effects of online image sharing · ImageCLEF
lab

1   Introduction

Started in 2003 with only four participants [14], ImageCLEF[1] is the image
retrieval and classification lab of the CLEF (Conference and Labs of the Evaluation Forum) conference and it rapidly increased its impact when the medical tasks
were included in 2004 [13]. Then, over 20 participants were attracted. Its growing trend lead to more than 200 participants in 2019 and even more than 110 in
2020 during the COVID-19 pandemic. Even though the tasks were added, changed
or discontinued, the general objective remained the same, i.e., *to combine multi-
modal data to retrieve and classify visual information.* Tasks have evolved along
the time from more general object classification and retrieval to specific application domains, e.g., medical, Internet and social media, nature, and even security.
In [31], one presents a thorough analysis of several tasks and the creation of the
data sets. ImageCLEF impact over the years was assessed in [44,45].

Starting with 2018, ImageCLEF used the crowdAI platform, which migrated
to AIcrowd[2] from 2020, to distribute the data sets and receive the submitted
runs. The system allowed the assignment of an online leader board and gave
the opportunity to keep the data sets accessible beyond competition, including
a continuous submission of runs and addition to the leader board. In 2023, the
ImageCLEF team developed its own system, as migrating to the AI4Media[3]
benchmarking platform (based on Codalab[4]). Over the years, ImageCLEF and
also CLEF have shown a strong scholarly impact that was assessed in [44,45]. For
instance, the term "ImageCLEF" returns on Google Scholar[5] over 6,850 article
results (search on June 26th, 2023). This underlines the importance of the evaluation campaigns for disseminating best scientific practices. We introduce here
the three tasks that were run in the 2023 edition[6], namely: ImageCLEFmedical,
ImageCLEFfusion, and ImageCLEFaware (Fig. 1).

---

[1] http://www.imageclef.org/.
[2] https://www.aicrowd.com/.
[3] https://www.ai4media.eu/.
[4] https://codalab.org/.
[5] https://scholar.google.com/.
[6] https://www.imageclef.org/2023/.

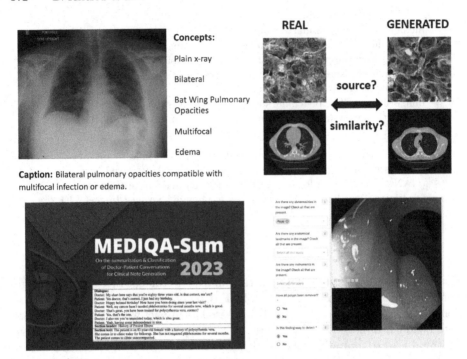

Fig. 1. Sample images from (left to right, top to bottom): ImageCLEFmedical-caption with an image and its corresponding CUIs and captions, ImageCLEFmedical-GAN with an example of real and generated images, ImageCLEFmedical-Mediqa with an example of doctor-patient conversation, and ImageCLEFmedical-VQA with examples of questions and answers in the area of colonoscopy.

2 Overview of Tasks and Participation

ImageCLEF 2023 consists of three main tasks with the objective of covering a *diverse range* of multimedia retrieval applications, namely: *medicine, social media,* and *Internet* applications. It followed the 2019 tradition [26] of diversifying the use cases [1,23,37,40,43,49]. The 2023 tasks are presented as follows:

- **ImageCLEFmedical.** Since 2004, in the frame of ImageCLEF benchmarking, medical tasks were organised. Despite the fact that in 2018, for example, all but one task were medical, one could remark little interaction between the medical tasks. Consequently, starting with 2019, the medical tasks were focused towards one specific problem but combined as a single task with several subtasks. In this way, one could allow synergies between the domains:
  - *MEDIQA-Sum*: This is the fourth edition of the MEDIQA tasks and its first edition in the text format. The 2019 MEDIQA task featured several medical natural language semantic retrieval-related tasks, including natural language inference (NLI) classification of MIMIC-III clinical note sentences, recognizing question entailment (RQE) in consumer health ques-

tions, and reranking retrieved answers to consumer health questions [7]. Continuing in 2021, the next MEDIQA task resumed hosting one clinical subtask and two consumer-health question-answer related subtasks [5]. Different from the 2019 subtasks, MEDIQA 2021 focused on summarization; summarization of clinical radiology note findings, consumer health questions, and consumer health answers. This year's MEDIQA tasks include clinical dialogue section header classification, short-dialogue note summarization, and full-encounter generation. This task is introduced as part of the ImageCLEF challenges as an experimental precursor to a multimodal image and dialogue summarization task [49]. An overlapping dataset with an additional dialogue generation task was part of the ACL 2023 Clinical NLP MEDIQA-CHAT challenge [8].

- *Caption*: This is the 7th edition of the task in this format, however, it is based on previous medical tasks. The task is once again running with both the "concept detection" and "caption prediction" subtasks [40], after the former was brought back in 2021 due to participants' demands [18,20,21,33–35,39]. The "caption prediction" subtask focuses on composing coherent captions for the entirety of a radiology image, while the "concept detection" subtask focuses on identifying the presence of relevant concepts in the same corpus of radiology images. After a smaller data set of manually annotated radiology images was used in 2021, the 2023 edition once again uses a larger dataset based on ROCO data [36], which was already used in 2019, 2020, and 2022.

- *GANs*: This is the first edition of the task [1]. The objective of the task is to investigate the hypothesis that generative models generate medical images that exhibit resemblances to the images employed during their training. This addresses concerns surrounding the privacy and security of personal medical image data in the context of generating and utilizing artificial images in various real-world scenarios. The task aims to identify distinctive features or "fingerprints" within synthetic biomedical image data, allowing us to determine which real images were used during the training process to generate the synthetic images.

- *MEDVQA-GI*: Analysis of gastrointestinal images and videos is a very popular topic in both the medical and computer science community. Usually, research and methods focus on images as a single modality. The MEDVQA-GI [23] introduces the task of visual question answering (VQA) [3,4,6,19] in the field of GI endoscopy extending the modalities with text. The idea is that through the combination of text and image data, the output of the analysis gets easier to use by medical experts. For the task, a new dataset based on previously published open datasets [12,28,29] was developed. The extended dataset has additional data corresponding to questions regarding the type of examinations, anomaly location, number of findings, colors of the findings, to name a few.

- **ImageCLEFfusion**. This is the 2nd edition of the task [42,43]. The main objective for this task is the development of late fusion or ensembling approaches, that are able to use prediction results from pre-computed inducers in order to generate better, improved prediction outputs. The present iteration of this task encompasses three distinct challenges: the continuation of the previous year's regression challenge utilizing media interestingness data, the continuation of the retrieval challenge involving image search result diversification data, and the addition of a new multi-label classification task focused on concepts detection in medical data. Notably, the tasks employ inducers that have been developed by actual users, ensuring their real-world applicability.

– **ImageCLEFaware**. This was the 3rd edition of the task and it focuses on personal data disclosure-awareness as users' data can be reused in other contexts when they share it for specific purposes. Consequently, the feedback to the users is very important when dealing with the effects of personal data sharing. The objective of the task resided in automatically providing a rating of a visual user profile in different real-life situations. The dataset created specifically for the 2021 edition of the task was expanded in order to make the evaluation more robust. Data were sampled from YFCC100 dataset and were further anonymized in order to comply with GPDR.

Table 1. Key figures regarding participation in ImageCLEF 2023.

Task	Groups that submitted results	Submitted runs	Submitted working notes
Caption	13	116	12
Mediqa	12	48	12
GANs	8	40	9
MedVQA	12	14	4
Fusion	2	23	2
Aware	0	0	0
Overall	47	241	39

In order to participate in the evaluation campaign, the research groups had to register by following the instructions on the ImageCLEF 2023 web page[7]. In 2022, the challenge was organized through the AIcrowd platform[8] to ease the overall management of the campaign, but in 2023 we setup our own registration and submission system and next year we will use the AI4Media platform based on codalab[9] to manage the benchmarking campaign. As in previous year, to actually get access to the data sets, the participants were required to submit a signed End User Agreement (EUA). Table 1 summarizes the participation in

---

[7] https://www.imageclef.org/2023/.

[8] https://www.aicrowd.com/.

[9] https://github.com/AIMultimediaLab/AI4Media-EaaS-prototype-Py2-public.

ImageCLEF 2023, indicated the statistics both per task and for the overall lab. The table also shows the number of groups that submitted runs and the ones that submitted a working notes paper describing the techniques used. Teams were allowed to register for several tasks.

After a decrease in participation in 2016, the participation increased in 2017 and 2018, and increased again in 2019. In 2018, 31 teams completed the tasks and 28 working notes papers were received. In 2019, 63 teams completed the tasks and 50 working notes papers were retrieved. In 2020, 40 teams completed the tasks and submitted working notes papers. In 2021, 42 teams completed the tasks and we received 30 working notes papers. In 2022, 28 teams completed the tasks and we received 26 working notes papers. In 2023, 47 teams submitted the results and we received 39 working notes, thus experiencing the revival of the campaign. Also, visual question answering, not organized in 2022, was retaken this year focusing on the text modality. Nevertheless, the number of submitted runs dropped compared to 2021 and 2022 with more teams involved 258 (2021) and 256 (2022) vs 241 (2023). This could be due to the fact that the teams were focused on finding higher-quality solutions at the expense of the numer of the runs. Thus, ImageCLEF continues to provide a strong evaluation benchmark for the community.

In the following sections, we present the tasks. Only a short overview is reported, including general objectives, description of the tasks and data sets, and a short summary of the results. A detailed review of the received submissions for each task is provided with the task overview working notes: Caption [39], Mediqa [49], GAN [1], MedVQA [23], and Fusion [43].

3   The Caption Task

The caption task was first proposed as part of the ImageCLEFmedical [21] in 2016 aiming to extract the most relevant information from medical images. Hence, the task was created to condense visual information into textual descriptions. In 2017 and 2018 [18,20], the ImageCLEFcaption task comprised two subtasks: concept detection and caption prediction. In 2019 [34] and 2020 [35], the task concentrated on the the concept detection task, extracting Unified Medical Language System® (UMLS) Concept Unique Identifiers (CUIs) [11] from radiology images. In 2021 [33], both subtasks, concept detection and caption prediction, were running again due to participants' demands. The focus in 2021 was on making the task more realistic by using fewer images which were all manually annotated by medical doctors. For the 2022 ImageCLEFmedical Caption task [39], both subtasks were continued albeit with an extended version of the ROCO data set used for both subtasks, which was already used in 2020 and 2019. The 2023 edition of ImageCLEFmedical caption [40] continues in the same vein, once again using a ROCO-based data set for both subtasks, but switching from BLEU [32] to BERTScore [50] as the primary evaluation metric for caption prediction.

3.1 Task Setup

The ImageCLEFmedical Caption 2023 [40] follows the format of the previous ImageCLEFmedical caption tasks. In 2023, the overall task comprises two sub-tasks: "Concept Detection" and "Caption Prediction". The concept detection sub-task focuses on predicting Unified Medical Language System® (UMLS) Concept Unique Identifiers (CUIs) [11] based on the visual image representation in a given image. The caption prediction subtask focuses on composing coherent captions for the entirety of the images.

The detected concepts are evaluated using the balanced precision and recall trade-off in terms of F1-scores, as in previous years. Like last year, a secondary F1-score is computed using a subset of concepts that was manually curated and only contains x-ray anatomy, directionality, and image modality concepts. For the first time this year, BERTScore was used as the primary metric for the evaluation of the caption prediction subtask, replacing the BLEU score, which had been used in previous years. BERTScore evaluates the semantic similarity of the predicted captions, whereas BLEU focuses more on n-gram overlap. In addition to the BERTScore, a secondary ROUGE score, which measures the overlap of content between the predicted captions and reference captions, was provided. After the submission period ended, a number of additional scores were calculated and published: METEOR [2], CIDEr [47], CLIPScore [22], BLEU and BLEURT [41].

3.2 Data Set

In 2023, an extended subset of the ROCO [36] data set is used for both subtasks. The ROCO data set originates from biomedical articles of the PMC Open Access Subset[10] [38] and was extended with new images added since the last time the data set was updated. For this year, only CC BY and CC BY-NC licensed images are included. From the captions, UMLS® concepts were extracted, and concepts regarding anatomy and image modality were manually validated for all images. New for this year was the addition of manually validated x-ray directionality concepts.

Following this approach, we provided new training, validation, and test sets for both tasks:

- *Training set* including 60,918 radiology images and associated captions and concepts.
- *Validation set* including 10,437 radiology images and associated captions and concepts.
- *Test set* including 10,473 radiology images.

---

[10] https://www.ncbi.nlm.nih.gov/pmc/tools/openftlist/.

Table 2. Performance of the participating teams in the ImageCLEFmedical 2023 concept detection subtask. The best run per team is selected. Teams with previous participation in 2022 are marked with an asterisk.

Team	Institution	F1-Score
AUEB-NLP-Group*	Department of Informatics, Athens University of Economics and Business, Athens, Greece	0.5223
KDE-Lab_Med*	KDE Laboratory, Department of Computer Science and Engineering, Toyohashi University of Technology, Aichi, Japan	0.5074
VCMI*	University of Porto, Porto, Portugal and INESC TEC, Porto, Portugal	0.4998
IUST_NLPLAB*	School of Computer Engineering, Iran University of Science and Technology, Tehran, Islamic Republic Of Iran	0.4959
Clef-CSE-GAN-Team	SSN College Of Engineering, Chennai, India	0.4957
CS_Morgan*	Computer Science Department, Morgan State University, Baltimore, Maryland	0.4834
SSNSheerinKavitha	Department of CSE, Sri Sivasubramaniya Nadar College of Engineering, India	0.4649
closeAI2023	Baidu Intelligent Health Unit, Beijing, China and Peng Cheng Laboratory, Shenzhen, China	0.0900
SSN_MLRG	Department of CSE, Sri Sivasubramaniya Nadar College of Engineering, India	0.0173

3.3   Participating Groups and Submitted Runs

In the seventh edition of the ImageCLEFmedical Caption task, 27 teams registered and signed the End-User-Agreement that is needed to download the development data. 13 teams submitted 116 graded runs (12 teams submitted working notes) attracting similar attention to 2022. Each of the groups was allowed a maximum of 10 graded runs per subtask. Unlike last year, participants did not have access to their own scores until after the submission period was over. 9 teams participated in the concept detection subtask this year, 6 of those teams also participated in 2022. 13 teams submitted runs to the caption prediction subtask, 7 of those teams also participated in 2022. Overall, 9 teams participated in both subtasks, and four teams participated only in the caption prediction subtask. Unlike in 2022, no teams participated only in the concept detection subtask.

In the concept detection subtasks, the groups used primarily multi-label classification systems, with image retrieval systems consistently performing worse for teams who experimented with them. One team successfully used an image retrieval system as a fallback when the multi-label classification system did not predict any concepts. Last year's winners once again achieved the top scores by increasing their ensemble from two to three models.

In the caption prediction subtask, most teams experimented with encoder-decoder frameworks with different backbones and LSTM [24] decoders. Unsur-

Table 3. Performance of the participating teams in the ImageCLEFmedical 2023 caption prediction subtask. The best run per team is selected. Teams with previous participation in 2022 are marked with an asterisk.

Team	Institution	BERTScore
CSIRO*	Australian e-Health Research Centre, Commonwealth Scientific and Industrial Research Organisation, Herston, Queensland, Australia and CSIRO Data61, Imaging and Computer Vision Group, Pullenvale, Queensland, Australia and Queensland University of Technology, Brisbane, Queensland, Australia	0.6413
closeAI2023	Baidu Intelligent Health Unit, Beijing, China and Peng Cheng Laboratory, Shenzhen, China	0.6281
AUEB-NLP-Group*	Department of Informatics, Athens University of Economics and Business, Athens, Greece	0.6170
PCLmed	Peng Cheng Laboratory, Shenzhen, China and ADSPLAB, School of Electronic and Computer Engineering, Peking University, Shenzhen, China	0.6152
VCMI*	University of Porto, Porto, Portugal and INESC TEC, Porto, Portugal	0.6147
KDE-Lab_Med*	KDE Laboratory, Department of Computer Science and Engineering, Toyohashi University of Technology, Aichi, Japan	0.6145
SSN_MLRG	Department of CSE, Sri Sivasubramaniya Nadar College of Engineering, India	0.6019
DLNU_CCSE	Unknown	0.6005
CS_Morgan*	Computer Science Department, Morgan State University, Baltimore, Maryland	0.5819
Clef-CSE-GAN-Team	SSN College Of Engineering, Chennai, India	0.5816
Bluefield-2023	Toyohashi University of Technology, Aichi, Japan and Toyohashi Heart Center, Aichi, Japan	0.5780
IUST_NLPLAB*	School of Computer Engineering, Iran University of Science and Technology, Tehran, Islamic Republic Of Iran	0.5669
SSNSheerinKavitha*	Department of CSE, Sri Sivasubramaniya Nadar College of Engineering, India	0.5441

prisingly, teams increasingly used Large Language Models (LLMs) in the decoding step and to help generate or refine captions. BLIP-2 [30] was used for the first time and achieved good results (second and fourth place). One novelty was the use of reinforcement learning to refine and improve upon last year's best solution in terms of BERTScore, which ended up winning this year's competition after the change of primary scores from BLEU to BERTScore.

To get a better overview of the submitted runs, the primary scores of the best results for each team are presented in Tables 2 and 3.

3.4   Results

For the concept detection subtask, the overall F1 scores increased compared to last year, which is not surprising considering the reduced number of concepts for this year's edition of the challenge.

While one team experimented with a novel autoregressive multi-label classification system that tries to model relationships between concepts and another team tried training separate models for the different modalities, these experiments did not yield better results compared to the winning approach.

BERTScore and ROUGE scores were used to predict captions. Unlike last year's edition, BERTScore replaced BLEU as the primary score for a more nuanced evaluation of captions. The adoption of BERTScore reflects the intent to prioritize semantic alignment and information preservation in the generated captions and not focus on the frequency of n-gram matches, which is the basis of BLEU.

The aforementioned change of evaluation metrics had a big effect on the outcome of the caption prediction challenge, with last year's winner placing second to last according to the BERTScore evaluation while still winning in terms of the ROUGE, BLEU and METEOR scores with a similar approach as last year. An in-depth analysis is presented in [39].

3.5   Lessons Learned and Next Steps

This year's caption task of ImageCLEFmedical once again ran with both subtasks, concept detection and caption prediction. Like last year, it used a ROCO-based data set for both challenges after a smaller, manually annotated data set was used in 2021. Manually validated concepts for X-ray directionality information was added for this year's dataset and caption pre-processing was kept minimal. It attracted 13 teams who submitted a total of 116 graded runs, a similar level of participation to last year. Some changes were introduced for the scores, with a switch from BLEU to BERTScore as the primary evaluation metric for the caption prediction. As mentioned before, this switch had a large impact on the results, and we will continue to evaluate and explore different possible metrics or combination of metrics, but the evaluation of generated captions remains difficult.

Like last year, most teams were more successful in training multi-label classification models compared to image retrieval models for the concept detection. For the caption prediction, most teams used Transformer-based models [46], with LLMs making an appearance as part of some of the approaches.

For next year's ImageCLEFmedical Caption challenge, some possible improvements include an improved caption prediction evaluation metric which is specific to medical texts, and improving manually validated concept quality with the help of a medical professional. It will also be important to make sure that no models are used that were pre-trained on PubMedCentral data, since these models will already have seen the original captions.

4   The MEDIQA-Sum Task

The MEDIQA tasks aim to pose natural language problems related to medical language and semantics [7]. The first edition hosted the challenges of clinical note sentence NLI, as well as consumer health RQE, and answer retrieval re-ranking. The focus of the last edition, in 2021, involved summarization tasks in the areas of clinical radiology note findings, consumer health question summarization, and multiple answer summarization [5]. In 2023, two editions were hosted. The 2023 ACL Clinical NLP MEDIQA-CHAT challenge included three subtasks including short-dialogue section header and note generation, full-encounter dialogue-to-note generation, and full-encounter note-to-dialogue generation [8]. In the 2023 ImageCLEF edition, the MEDIQA-SUM subtasks included short dialogue-to-topic classification, short dialogue and topic- to note summarization, and full encounter dialogue-to-note summarization [49].

4.1   Task Setup

The MEDIQA-SUM 2023 overall task comprises three sub-tasks:

- (A) dialogue2topic (section header) classification
- (B) dialogue2note summarization given the target section header
- (C) full-encounter dialogue2note summarization.

Subtask A topic classification was evaluated using accuracy. The subtask B snippet summarization was evaluated using the mean of BLEURT, BERTscore, and ROUGE-1; metrics found to be correlated to human evaluation in several independent health summarization datasets [9]. Full-encounter summarization in Subtask C used two metrics: (1) a full-note ROUGE-1 score and (2) an equally weighted division-based (subjective, objective_exam, objective_results, assessment_and_plan) aggregate score of the BLEURT, BERTscore, and ROUGE-1 metric.

Subtask A and B use the same test set. After Subtask A was closed, the gold standard section header was released so that it would be available as input to Subtask B. Code submissions were required at submission. The organizers checked output of code against submitted runs and documented each team's code replicability status.

4.2   Data Set

The 2023 MEDIQA-SUM challenge includes data from two collections: MTS-Dialog [10] and ACI-BENCH [48]. Subtasks A and B consist of 1,201 pairs of conversations and associated section headers and contents; 100 examples in validation, and 200 pairs in test. Subtask C includes full encounters with 67 examples in training, 20 in validation, and 40 in test.

Table 4. Performance of the participating teams in the MEDIQA-Sum 2023 Subtask A on topic classification. The best run per team is selected.

Team	Institution	Accuracy
Cadence	Cadence Solutions, USA	0.820
HuskyScribe	University of Washington, USA	0.815
Tredence	Tredence Inc, India	0.800
StellEllaStars	University of Michigan School of Information, USA	0.765
SSNSheerinKavitha	Sri Sivasubramaniya Nadar College of Engineering, India	0.740
SuryaKiran	University of Mumbai, India	0.735
SSNdhanyadivyakavitha	Sri Sivasubramaniya Nadar College of Engineering, India	0.720
ds4dh	University of Geneva, Switzerland	0.710
uetcorn	University of Engineering and Technology, Vietnam National University, Hanoi, Vietnam	0.710
SKKU-DSAIL	Department of Applied Artificial Intelligence, Sungkyunkwan University, South Korea	0.700
MLRG-JBTTM	Sri Sivasubramaniya Nadar College of Engineering, India	0.665

Table 5. Performance of the participating teams in the MEDIQA-Sum 2023 Subtask B on dialogue2note summarization. The best run per team is selected.

Team	Institution	Aggregated Score
SuryaKiran	University of Mumbai, India	0.573
PULSAR	ASUS AICS/University of Manchester, Singapore/UK	0.569
Tredence	Tredence Inc, India	0.559
HuskyScribe	University of Washington, USA	0.529
uetcorn	University of Engineering and Technology, Vietnam National University, Hanoi, Vietnam	0.481
SKKU-DSAIL	Department of Applied Artificial Intelligence, Sungkyunkwan University, South Korea	0.461
SSNSheerinKavitha	Sri Sivasubramaniya Nadar College of Engineering, India	0.419

Table 6. Performance of the participating teams in the MEDIQA-Sum 2023 Subtask C on full-encounter dialogue2note summarization, ranked by ROUGE-1. The best run per team is selected.

Team	Institution	ROUGE-1
Tredence	Tredence Inc, USA	0.500
uetcorn	University of Engineering and Technology, Vietnam National University, Hanoi, Vietnam	0.498
HuskyScribe	University of Washington, USA	0.470
PULSAR	ASUS AICS/University of Manchester, Singapore/UK	0.294

Table 7. Performance of the participating teams in the MEDIQA-Sum 2023 Subtask C on full-encounter dialogue2note summarization, ranked by the aggregated score. The best run per team is selected.

Team	Institution	Aggregated Score
Tredence	Tredence Inc, USA	0.455
uetcorn	University of Engineering and Technology, Vietnam National University, Hanoi, Vietnam	0.441
HuskyScribe	University of Washington, USA	0.413
PULSAR	ASUS AICS/University of Manchester, Singapore/UK	0.247

4.3   Participating Groups and Submitted Runs

Overall 12 teams participated with a total of 48 runs. Subtask A included 23 valid submissions among 11 teams. Subtask B included 16 submissions among 7 teams. Subtask C included 9 submissions among 4 teams. At most three runs were allowed per team in each subtask. With the exception of 1 team, all teams participated in Subtask A. Four teams participated in two subtasks. Three teams participated in all three subtasks.

4.4   Results

The best teams achieved 0.8 Accuracy on Subtask A topic classification (Table 4) and an aggregate score of 0.43 for Subtask B (Table 5). The top two systems for Subtask C achieved ROUGE-1 at 0.49 F1 (Table 6) and aggregated scores at 0.44 (Table 7).

Subtask A submissions included classic machine learning algorithms such as SVM, KNN, Random Forest, with some pre-processing such as TF-IDF, lemmatization. This task also featured the use of pre-trained models such as GPT3.5, clinical-BERT, clinical T5, and their low-rank adaptation (LoRA). Eight out of 23 submissions either used additional training data or adjusted data sampling.

Subtask B primarily consisted of pre-trained sequence-to-sequence models such as llama, bart, flan T5, biobart, and their LoRA versions, fine-tuned on the training and validation sets. Eight out of 16 submissions used the gold standard section headers released from Subtask A.

Subtask C submissions had a diverse set of systems that used creative means to circumvent a low-resource generation problem. Specifically, Uetcorn, HuskyScribe, and Tredence all divided the problem into multiple parts. Firstly, relevant parts of the dialogue were grouped together as related to particular sections. Each team used a different method to achieve this; the UETCorn team identified relevant parts of dialogue for specific note section key points (e.g. "chief complaint" or "medications"), using a similarity function between dialogue sentences and a hand-crafted section-specific description; afterwards, several note generation strategies were used for each key point. HuskyScribe built a model classifying smaller dialogue exchanges into the same categories, while Tredence classified dialogues chunked by various window sizes. In the second step, grouped dialogue chunks were sent through a text generator to produce parts of the note. The use of pre-trained models such as BART/BioBART and flan T5 for the generation was typical. The Uetcorn and Tredence team included some section/keypoint specific questions as part of the generation input, e.g. (e.g. input: "question: {question} context: {conversation}", output: summary). The Uetcorn team also experimented with a reading comprehension answer extraction based on specially designed key point query (e.g. "names of medication used") and post-processing. The HuskyScribe team additionally used Subtask A data to generate additional synthetic data for training. Finally, the completed note was assembled through concatenation and post-processing. Unlike the other three groups, the PULSAR team employed an end-to-end approach, experimenting with flan T5 and llama models with additional data created using MTSamples data processed through GPT3.5.

With the exception of two runs in subtask A and one team's runs in Subtask B, all submissions were reproducible based on participants' submitted code. A more detailed account can be found in the MEDIQA-Sum overview paper [49].

4.5   Lessons Learned and Next Steps

This year's MEDIQA tasks hosted similar problems on an overlapping dataset with the 2023 ACL ClinicalNLP MEDIQA-Chat Shared Tasks [8]. A striking difference between the participants in this edition was that there were no GPT4 submissions. As GPT4 access requires a subscription, we can view the solutions from this evaluation lab as a whole to be solutions constrained to only using open-source or free models and data. A more detailed comparison can be found in the MEDIQA-Sum overview paper [49].

The requirement of code submissions in this year's MEDIQA challenges was successful and ensured that final submissions would be of high quality; it also encouraged the release of open-source code into the community beyond the challenge. In this year's edition, the code was run manually by the organizers. In future editions, we will explore the use of platforms, e.g. https://codalab.org/, that will provide a standard management and packaging pipeline allowing submissions to be more easily and quickly evaluated.

Natural language evaluation is a challenging and active area of research. Evaluation for long documents is even less explored. While we used several metrics associated with human-labeled facts for our dataset (ROUGE-1 and an aggregated BLEURT, BERTScore, ROUGE1 metric), new metrics can be further explored for future challenges.

In our next future edition, we plan on running a multi-modal medical dialogue summarization task - we will use the lessons learned in this edition.

5   The GANs Task

The development of generative models in the area of artificial intelligence in recent years has generated a great deal of attention and creativity, altering many industries and the way we approach different tasks. Task offered an environment for investigating GANs' effects on the creation of synthetic medical images by providing a benchmark to explore the impact of GANs on artificial biomedical image generation. Medical image generation is essential for patient care improvement, healthcare professional education, and medical research. While obtaining genuine patient data can be expensive, insufficient, or ethically problematic, the ability to generate artificial yet realistic biological images can fill these gaps and provide researchers, doctors, and educators more leverage. As a result, generative models have shown to be remarkably effective at producing high-quality images that closely resemble the traits and patterns of real data.

5.1 Task Setup

This is the first edition of the task and consists of one challenge. The task aims to identify distinctive features or "fingerprints" within synthetic biomedical image data, allowing us to determine which real images were used during the training process to generate the synthetic images.

5.2 Data Set

A data set containing axial chest CT scans of lung tuberculosis patients was provided for the task. This means that some of them may appear pretty "normal" whereas the others may contain certain lung lesions including the severe ones. These images are stored in the form of 8 bit/pixel PNG images with dimensions of 256×256 pixels. The artificial slice images are 256×256 pixels in size. All of them were generated using Diffuse Neural Networks.

- *Development (Train) dataset*: consists of 500 artificial images and 160 real images annotated according to their use in the training of the generative network. Out of the real images, 80 were used during training.
- *Test (Evaluation) dataset* was created in similar way. The only difference is that the two subsets of real images are mixed and no proportion of non-used and used ones has been disclosed. Thus, a total of 10,000 generated and 200 real images are provided.

5.3 Participating Groups and Submitted Runs

Overall, 23 teams registered to the task, 8 of them finalizing the task and submitting runs. A total of 40 runs were received.

5.4 Results

An analysis of the proposed methods shows a great diversity among them, ranging from texture analysis, similarity-based approaches that join inducer predictions like SVM or KNN, to deep learning approaches and even multi-stage transfer learning. More detailed results, including methods presentation and other performance measures, are presented in the overview article [1]. The task was evaluated as a binary-class classification problem and the evaluation was carried out by measuring the F1-score, the official evaluation metric of this year's edition. The results are presented in Table 8.

5.5   Lessons Learned and Next Steps

The first edition of the ImageCLEF medical GANs task attracted a total of 8 teams that submitted runs, with all of them completing their submissions by creating a working notes papers. A prediction-based task was proposed to the participants. The best result for the task is an F1-score of 0.802 obtained by VCMI team followed by PicusLabMed with an F1-score of 0.666 and AIMulti-mediaLab with an F1-score of 0.626. We are pleased to report a high level of diversity in the identification strategies put forth by the participants. Future iterations of this task will diversify various elements, such as datasets and generation techniques, and broaden the study fields of synthetic medical data. We also intend to add more tasks based on various aspects of the security and privacy of the created data.

6   The MedVQA-GI Task

Identifying lesions in colonoscopy images is one of the most popular applications of artificial intelligence in medicine. Until now, the research has focused on single-image or video analysis. With this task, we aim to bring a new aspect to the field by adding multiple modalities to the picture. The main focus of the task will be on visual question answering (VQA) and visual question generation (VQG). The goal is that through the combination of text and image data, the output of the analysis gets easier to use by medical experts. The task has three sub-tasks.

For the VQA subtask, the participants need to combine images and text answers to answer the questions. In the VQG subtask, the participants are asked to generate text questions from a given image and answer. Example questions for both VQA and VQG: How many polyps are in the image? Are there any polyps in the image? What disease is visible in the image? The third subtask is the visual location question answering (VLQA), where the participants get an image and a question and are required to answer it by providing a segmentation mask for the object in the question. Example questions are: Where exactly in the image is the polyp? Where exactly in the image is the instrument?

6.1   Task Setup

The task had three sub-tasks that the participants could work on. There was no requirement on which task should be finished or not. For the first sub-task (VQA), participants were asked to generate text answers given a text question and image pair. For subtask 2 (VQG), the task was to generate questions based

Table 8. Summary on the participant submissions and their results for GAN task.

Group rank	Group name	Submission #	F1-score
#1	VCMI	submission 2	0.802
#2	VCMI	submission 1	0.731
#3	VCMI	submission 3	0.707
#4	PicusLabMed	submission 8	0.666
#5	VCMI	submission 4	0.654
#6	AIMultimediaLab	submission 1	0.626
#7	PicusLabMed	submission 6	0.624
#8	VCMI	submission 5	0.621
#9	Clef-CSE-GAN-Team	submission 1	0.614
#10	VCMI	submission 7	0.613
#11	VCMI	submission 6	0.605
#12	VCMI	submission 10	0.594
#13	AIMultimediaLab	submission 2	0.585
#14	one five one zero	submission 2	0.563
#15	PicusLabMed	submission 9	0.562
#16	PicusLabMed	submission 4	0.552
#17	KDE lab	submission 5	0.548
#18	one five one zero	submission 3	0.522
#19	Clef-CSE-GAN-Team	submission 2	0.521
#20	VCMI	submission 9	0.514
#21	one five one zero	submission 1	0.507
#22	GAN-ISI	submission 5	0.502
#23	GAN-ISI	submission 2	0.489
#24	PicusLabMed	submission 10	0.487
#25	GAN-ISI	submission 3	0.486
#26	GAN-ISI	submission 4	0.483
#27	DMK	submission 1	0.480
#28	PicusLabMed	submission 2	0.470
#29	KDE lab	submission 2	0.469
#30	GAN-ISI	submission 1	0.469
#31	KDE lab	submission 1	0.465
#32	KDE lab	submission 4	0.457
#33	DMK	submission 2	0.449
#34	VCMI	submission 8	0.448
#35	PicusLabMed	submission 1	0.434
#36	Clef-CSE-GAN-Team	submission 3	0.431
#37	PicusLabMed	submission 3	0.419
#38	PicusLabMed	submission 5	0.417
#39	KDE lab	submission 3	0.407
#40	PicusLabMed	submission 7	0.093

on a given text answer and image pair. The final subtask (VLQA) asked the participants to segment parts of an image given a text question and image pair. For the different tasks, we used different metrics to evaluate the performance. More details on the tasks and evaluation metrics can be found in the task overview paper [23].

6.2   Data Set

The dataset consisted of images from the GI tract and ground truth regarding specific questions and answers related to the images, and was based on open GI data sets previously published by the organizers [12,28,29]. The data set was developed with medical experts having several years of experience working in GI endoscopy. Moreover, segmentation masks were included for subtask 3, since the subtask asked for segmentation masks as answers to input pairs of images and textual questions. For the challenge, the dataset was split in two, a development dataset and a testing dataset. The development dataset contained 2,000 samples (imaged and question-answer pairs), and the testing dataset consisted of 1,949 samples. The participants were only provided with the ground truth for the development dataset. The data and evaluation scripts will be made publicly available after the competition of the challenge.

6.3   Results

In total, 16 valid runs were submitted to the task from 8 different teams. One team did not submit their task description paper. Overall, the teams achieved reasonably good results ranging from an accuracy of around 0.21 to 0.82 for subtask 1. For subtask 3, four teams submitted a solution, and there we observed a large performance difference with IoU ranging from 0.234 to 0.666. For subtask 2, teams only submitted an inverse of subtask 1, which was not a meaningful way to approach the task. In future iterations of the task, we will consider this and create a totally separate ground truth in addition to more strict task requirements. Table 9 provides an overview of all teams and their metrics for the different subtasks.

Table 9. An overview of the results for each task available at MedVQA-GI.

Team Name	Task 1 (Accuracy)	Task 2	Task 3 (IoU)
wsq4747	0.740	–	0.234
BITM	0.819	–	–
SSNSheerinKavitha	0.441	–	–
SSN_KDC	0.820	–	–
utk	0.471	–	–
VisionQAries	0.548	–	0.666
DLNU_CCSE	0.213	–	–
UIT-Saviors	0.752	–	–

6.4   Lessons Learned and Next Steps

Overall, we observe quite some interest in the task, with many teams signing up. We also experienced that the task was somehow perceived as difficult due to the different modalities. One important lesson we learned is that subtask 2 could have worked better, and teams only submitted an inverse of subtask 1, which was difficult to evaluate in a meaningful way. In conclusion, there was great interest in the task, and it was shown that the problem is complex but not impossible. We plan to extend the ground truth and refine some of the tasks for future iterations.

7   The Fusion Task

The generalization ability and performance of machine learning models show signs of reaching a plateau in many domains, where the performance improvements over the years are not significant. Therefore, exploring the performance and optimizing the efficiency of machine learning methods is important for real-world applications as they can only use limited, noisy data. In this context, fusion methods are gaining popularity by harnessing the complementary knowledge of multiple base models to build more robust and accurate models compared with single models.

Several challenges must be explored by the participants in this task, such as *diversity*, which refers to a set of classifiers that, given the same instance, output different predictions; *voting mechanism*, which regulate how individual outputs from the base models are used during prediction; *dependency*, which refers to the way a base model affects the construction of the next model in the fusion chain; *cardinality*, which refers to the number of individual base models that form the ensemble – one needs to find a balance, as diversity may be reduced if too many models are incorporated in the fusion; the *learning mode* of the base models, which is the property that balance the classifiers' ability to adapt properly to new, previously unseen, data while at the same time retaining the previously learned knowledge.

7.1   Task Setup

This second edition of the ImageCLEFfusion task [43] consists of three challenges: a regression challenge involving media interestingness (ImageCLEFfusion –int) for which we provide output data from 29 inducers, a retrieval challenge involving result diversification (ImageCLEFfusion-div) for which we provide outputs data from 56 inducers, and a multi-label classification task involving concepts detection in medical data (ImageCLEFfusion–cap) for which we provided 84 inducers. Participants were required to devise late fusion learning strategies based on the outputs of the inducers associated with the media samples for each of the subtasks. The evaluation of the participants' submissions was conducted using the Mean Average Precision at 10 (mAP@10) metric for the

ImageCLEFfusion–int task, F1 at 20 (F1@20) and Cluster Recall at 20 (Cluster Recall@20) metrics for the ImageCLEFfusion-div task, and the F1 metric for the ImageCLEFfusion–cap task. Participants were encouraged to submit their solutions for all three tasks.

7.2 Data Set

The three tasks in ImageCLEFfusion make use of different datasets and associated challenges. The ImageCLEFfusion–int task focuses on the Interestingness10k dataset [16], specifically utilizing the image-based prediction data from the 2017 MediaEval Predicting Media Interestingness task [17]. In this task, we provide prediction outputs from 29 systems that were submitted during the benchmarking task. To facilitate training and testing, the available data is divided into 1877 samples for training the fusion systems and 558 samples for testing.

On the other hand, the ImageCLEFfusion–div task relies on the Retrieving Diverse Social Images dataset [27], specifically targeting the DIV150Multi challenge [25]. For this task, we provide retrieval outputs from 56 systems, which are further divided into 60 queries for the training data and 63 queries for the testing set.

Lastly, the ImageCLEFfusion–cap task is derived from the ImageCLEF Medical Caption Task [39]. This task involves the extraction of multi-label outputs from 84 inducers. The data used for this task consists of 6101 images for the development set and 1500 images for the testing set.

In the training sets of all three tasks, we provide participants with the inducer outputs, along with the requisite scripts for metric computation. Additionally, the performance of each inducer is disclosed based on the official metrics, and ground truth data is made available. However, for the testing sets, only the inducer outputs are provided. It is crucial to emphasize that participants were strictly prohibited from utilizing external inducers. They were solely permitted to employ the inducers we provided. This constraint ensures a fair assessment of the performance of the late fusion approach, without introducing any alterations to the inducer set.

Table 10. Participation in the ImageCLEF-int 2023 task: the best score from all runs for each team. We also included a baseline that consists of the average performance of all the provided inducers.

Team	#Runs	mAP@10
Gnana	10	0.1331
baseline	–	0.0946

Table 11. Participation in the ImageCLEF-div 2023 task: the best score from all runs for each team. We also included a baseline that consists of the average performance of all the provided inducers.

Team	#Runs	F1@20	CR@20
Gnana	10	0.5708	0.449
baseline	–	0.5313	0.414

7.3   Participating Groups and Submitted Runs

A single team participated in both the ImageCLEFfusion-int and ImageCLEF fusion-div tasks, submitting a total of 20 runs, with 10 runs for each task. However, no runs were recorded for the ImageCLEFfusion–cap task.

7.4   Results

The results are presented in Table 10 for the interestingness task, and Table 11 for the diversification task. The participating team employed a diverse range of techniques for the tasks. For the result diversification task, they explored various machine learning algorithms including Elastic Net, Gradient Boosting Regressor, and Decision Tree. In addition, for the image interestingness task, they utilized XGBoost Classifier, k-Nearest Neighbors Classifier, and Decision Tree. A voting classifier and an ensemble learning model based on StackingClassifier were also tested that combined the three base models for each task. The results demonstrate the superiority of the ensemble learning approach over the other tested methods in both subtasks. For the diversification task, the ensemble learning approach achieved an F1 score of 0.5708 and a Cluster Recall (CR) score of 0.4295. In the interestingness task, the ensemble learning approach achieved a mean Average Precision at 10 (mAP@10) score of 0.1331.

7.5   Lessons Learned and Next Steps

Despite the reduced number of participants compared to the previous year, with only one team submitting runs for both the ImageCLEFfusion–int and ImageCLEFfusion–div tasks, the participating team achieved a performance that surpassed the majority of the participants in the previous year, but still under the state-of-the-art result of the last year achieved by [15].

For the next edition of this task, we believe it is very important to continue with these three datasets, as this will allow us to study the year-to-year improvement of the proposed fusion techniques.

8   The Aware Task

Social networks engage the users to share their personal data in order to interact with other users. The context of the sharing is chosen by the users but they do not have control on further data use. These data are automatically aggregated into profiles which are exploited by social networks to propose personalized advertising/services to users. Depending on their visibility, data can be also consulted by other entities to make decisions which have a high impact on the user's life. It is thus important to give users feedback about the potential real-life effects of their personal data sharing.

We designed a task focused on the automatic rating of visual user profile in four impactful situations. Each profile includes 100 photos and its appeal is manually evaluated via crowdsourcing. Participants are asked to provide automatic visual profile ratings obtained by using a training set which includes visual- and situation-related information. These ratings are then ranked and compared to manual ones in order to assess the feasibility of providing automatic feedback related to the effects of personal photos sharing.

Six teams registered for the task this year, but, unfortunately, none of them submitted runs. Given the low interest for the task, there will be no next edition. However, the datasets and evaluation scripts will be kept available in case other research teams will be interested in working with them later.

9   Conclusion

This paper presents a global picture of the tasks and outcomes of the Image-CLEF 2023 benchmarking campaign. Three main tasks were organised, covering challenges in the medical domain (caption analysis, visual question answering, medical dialogue summarisation, GANs for medical image generation) and social networks and Internet (analysis of the real-life effects of personal data sharing, fusion techniques for retrieval and interestingness prediction). With respect to the previous year, we experienced a 67% increase in the number of teams completing the tasks (28 in 2022 vs. 47 in 2023). They successfully submitted 241 runs and 39 working notes papers.

As in the previous year, almost all solutions provided by the participants were based on machine learning and deep learning techniques. In ImageCLEF-caption, multi-label classification systems were used, as well as image retrieval systems, the latter performing worse. Mediqa task determined the participants to use classic machine learning algorithms such as SVM, KNN, Random Forest, with pre-processing methods such as TF-IDF, lemmatization. In addition, the participants used pre-trained models such as GPT3.5, clinical-BERT, clinical T5, and their low-rank adaptation (LoRA). For ImageCLEF-GAN task, the participants explored a large variety of methods as texture analysis, similarity-based approaches that join inducer predictions like SVM or KNN, and even deep learning approaches and multi-stage transfer learning. For ImageCLEF-MedVQA, the participants employed transformer-based pre-trained models. In ImageCLEFfusion, being at the 2nd edition, the participants explored machine learning algorithms as Elastic Net, Gradient Boosting Regressor, Decision Tree, XGBoost Classifier, and k-Nearest Neighbors Classifier. In ImageCLEFaware, the participation decreased even more and no run was submitted. ImageCLEF

2023 provided to the participants and to the community an interesting symbiosis of tasks and approaches and we are looking forward to participating at the CLEF 2023 workshop and to present the current achievements and the future plans.

Acknowledgements. The lab is supported under the H2020 AI4Media "A European Excellence Centre for Media, Society and Democracy" project, contract #951911, as well as the ImageCLEFaware, ImageCLEFfusion tasks. The work of Louise Bloch and Raphael Brüngel was partially funded by a PhD grant from the University of Applied Sciences and Arts Dortmund (FH Dortmund), Germany. The work of Ahmad Idrissi-Yaghir and Henning Schäfer was funded by a PhD grant from the DFG Research Training Group 2535 Knowledge- and data-based personalisation of medicine at the point of care (WisPerMed).

References

1. Andrei, A., Radzhabov, A., Coman, I., Kovalev, V., Ionescu, B., Müller, H.: Overview of ImageCLEFmedical GANs 2023 task - identifying training data "Fingerprints" in synthetic biomedical images generated by GANs for medical image security. In: CLEF2023 Working Notes. CEUR Workshop Proceedings, CEUR-WS.org, Thessaloniki, Greece, 18–21 September 2023
2. Banerjee, S., Lavie, A.: Meteor: an automatic metric for MT evaluation with improved correlation with human judgments. In: Proceedings of the ACL Workshop on Intrinsic and Extrinsic Evaluation Measures for Machine Translation and/or Summarization, pp. 65–72. Association for Computational Linguistics, Ann Arbor, Michigan, June 2005. http://aclanthology.org/W05-0909
3. Ben Abacha, A., Datla, V.V., Hasan, S.A., Demner-Fushman, D., Müller, H.: Overview of the VQA-med task at ImageCLEF 2020: visual question answering and generation in the medical domain. In: CLEF 2020 Working Notes. CEUR Workshop Proceedings, CEUR-WS.org, Thessaloniki, Greece, 22–25 September 2020
4. Ben Abacha, A., Hasan, S.A., Datla, V.V., Liu, J., Demner-Fushman, D., Müller, H.: VQA-Med: overview of the medical visual question answering task at ImageCLEF 2019. In: CLEF2019 Working Notes. CEUR Workshop Proceedings, CEUR-WS.org, Lugano, Switzerland, 09–12 September 2019. http://ceur-ws.org
5. Ben Abacha, A., Mrabet, Y., Zhang, Y., Shivade, C., Langlotz, C.P., Demner-Fushman, D.: Overview of the MEDIQA 2021 shared task on summarization in the medical domain. In: Proceedings of the 20th Workshop on Biomedical Language Processing, BioNLP@NAACL-HLT 2021, Online, 11 June 2021, pp. 74–85. Association for Computational Linguistics (2021). http://doi.org/10.18653/v1/2021.bionlp-1.8
6. Ben Abacha, A., Sarrouti, M., Demner-Fushman, D., Hasan, S.A., Müller, H.: Overview of the VQA-med task at ImageCLEF 2021: visual question answering and generation in the medical domain. In: CLEF 2021 Working Notes. CEUR Workshop Proceedings, CEUR-WS.org, Bucharest, Romania, 21–24 September 2021
7. Ben Abacha, A., Shivade, C., Demner-Fushman, D.: Overview of the MEDIQA 2019 shared task on textual inference, question entailment and question answering. In: Proceedings of the 18th BioNLP Workshop and Shared Task, BioNLP@ACL 2019, Florence, Italy, 1 August 2019, pp. 370–379. Association for Computational Linguistics (2019). http://doi.org/10.18653/v1/w19-5039

8. Ben Abacha, A., Yim, W.W., Adams, G., Snider, N., Yetisgen, M.: Overview of the MEDIQA-Chat 2023 shared tasks on the summarization and generation of doctor-patient conversations. In: ACL-ClinicalNLP 2023 (2023)

9. Ben Abacha, A., Yim, W.W., Michalopoulos, G., Lin, T.: An investigation of evaluation metrics for automated medical note generation (2023)

10. Ben Abacha, A., Yim, W.W., Fan, Y., Lin, T.: An empirical study of clinical note generation from doctor-patient encounters. In: Proceedings of the 17th Conference of the European Chapter of the Association for Computational Linguistics, pp. 2291–2302. Association for Computational Linguistics, Dubrovnik, Croatia, May 2023. http://aclanthology.org/2023.eacl-main.168

11. Bodenreider, O.: The unified medical language system (UMLS): integrating biomedical terminology. Nucleic Acids Res. **32**(Database-Issue), 267–270 (2004). https://doi.org/10.1093/nar/gkh061

12. Borgli, H., et al.: Hyperkvasir, a comprehensive multi-class image and video dataset for gastrointestinal endoscopy. Sci. Data **7**(1) (2020). https://doi.org/10.1038/s41597-020-00622-y

13. Clough, P., Müller, H., Sanderson, M.: The CLEF 2004 cross-language image retrieval track. In: Peters, C., Clough, P., Gonzalo, J., Jones, G.J.F., Kluck, M., Magnini, B. (eds.) CLEF 2004. LNCS, vol. 3491, pp. 597–613. Springer, Heidelberg (2005). https://doi.org/10.1007/11519645_59

14. Clough, P., Sanderson, M.: The CLEF 2003 cross language image retrieval task. In: Proceedings of the Cross Language Evaluation Forum (CLEF 2003) (2004)

15. Constantin, M.G., Ştefan, L.D., Dogariu, M., Ionescu, B.: AI multimedia lab at imagecleffusion 2022: deepfusion methods for ensembling in diverse scenarios. In: CLEF2022 Working Notes, CEUR Workshop Proceedings, CEUR-WS. org, Bologna, Italy (2022)

16. Constantin, M.G., Ştefan, L.D., Ionescu, B., Duong, N.Q., Demarty, C.H., Sjöberg, M.: Visual interestingness prediction: a benchmark framework and literature review. Int. J. Comput. Vis. **129**(5), 1526–1550 (2021)

17. Demarty, C.H., Sjöberg, M., Ionescu, B., Do, T.T., Gygli, M., Duong, N.: Mediaeval 2017 predicting media interestingness task. In: MediaEval workshop (2017)

18. Dicente Cid, Y., Kalinovsky, A., Liauchuk, V., Kovalev, V., Müller, H.: Overview of ImageCLEFtuberculosis 2017 - predicting tuberculosis type and drug resistances. In: CLEF2017 Working Notes. CEUR Workshop Proceedings, CEUR-WS.org, Dublin, Ireland, 11–14 September 2017. http://ceur-ws.org

19. Hasan, S.A., Ling, Y., Farri, O., Liu, J., Lungren, M., Müller, H.: Overview of the ImageCLEF 2018 medical domain visual question answering task. In: CLEF2018 Working Notes. CEUR Workshop Proceedings, CEUR-WS.org, Avignon, France, 10–14 September 2018. http://ceur-ws.org

20. García Seco de Herrera, A., Eickhoff, C., Andrearczyk, V., Müller, H.: Overview of the ImageCLEF 2018 caption prediction tasks. In: CLEF2018 Working Notes. CEUR Workshop Proceedings, CEUR-WS.org, Avignon, France, 10–14 September 2018. http://ceur-ws.org

21. García Seco de Herrera, A., Schaer, R., Bromuri, S., Müller, H.: Overview of the ImageCLEF 2016 medical task. In: Working Notes of CLEF 2016 (Cross Language Evaluation Forum), September 2016

22. Hessel, J., Holtzman, A., Forbes, M., Bras, R.L., Choi, Y.: Clipscore: a reference-free evaluation metric for image captioning. In: Moens, M., Huang, X., Specia, L., Yih, S.W. (eds.) Proceedings of the 2021 Conference on Empirical Methods in

Natural Language Processing, EMNLP 2021, Virtual Event/Punta Cana, Dominican Republic, 7–11 November 2021, pp. 7514–7528. Association for Computational Linguistics (2021). https://doi.org/10.18653/v1/2021.emnlp-main.595

23. Hicks, S.A., Storås, A., Halvorsen, P., de Lange, T., Riegler, M.A., Thambawita, V.: Overview of imageclefmedical 2023 - medical visual question answering for gastrointestinal tract. In: CLEF2023 Working Notes. CEUR Workshop Proceedings, CEUR-WS.org, Thessaloniki, Greece, September 2023

24. Hochreiter, S., Schmidhuber, J.: Long short-term memory. Neural Comput. **9**(8), 1735–1780 (1997). https://doi.org/10.1162/neco.1997.9.8.1735

25. Ionescu, B., Gînscă, A.L., Boteanu, B., Lupu, M., Popescu, A., Müller, H.: Div150multi: a social image retrieval result diversification dataset with multi-topic queries. In: Proceedings of the 7th International Conference on Multimedia Systems, pp. 1–6 (2016)

26. Ionescu, B., et al.: ImageCLEF 2019: multimedia retrieval in medicine, lifelogging, security and nature. In: Crestani, F., et al. (eds.) CLEF 2019. LNCS, vol. 11696, pp. 358–386. Springer, Cham (2019). https://doi.org/10.1007/978-3-030-28577-7_28

27. Ionescu, B., Rohm, M., Boteanu, B., Gînscă, A.L., Lupu, M., Müller, H.: Benchmarking image retrieval diversification techniques for social media. IEEE Trans. Multimed. **23**, 677–691 (2020)

28. Jha, D., et al.: Kvasir-Instrument: diagnostic and therapeutic tool segmentation dataset in gastrointestinal endoscopy. In: Proceedings of the International Conference on MultiMedia Modeling (MMM). pp. 218–229 (2021). http://doi.org/10.1007/978-3-030-67835-7_19

29. Jha, D., et al.: Kvasir-SEG: a segmented polyp dataset. In: Proceeding of the International Conference on Multimedia Modeling (MMM), vol. 11962, pp. 451–462 (2020). http://doi.org/10.1007/978-3-030-37734-2_37

30. Li, J., Li, D., Savarese, S., Hoi, S.C.H.: BLIP-2: bootstrapping language-image pre-training with frozen image encoders and large language models. CoRR abs/2301.12597 (2023). 10.48550/arXiv. 2301.12597, http://doi.org/10.48550/arXiv.2301.12597

31. Müller, H., Kalpathy-Cramer, J.: The ImageCLEF medical retrieval task at ICPR 2010 — information fusion to combine visual and textual information. In: Ünay, D., Çataltepe, Z., Aksoy, S. (eds.) ICPR 2010. LNCS, vol. 6388, pp. 99–108. Springer, Heidelberg (2010). https://doi.org/10.1007/978-3-642-17711-8_11

32. Papineni, K., Roukos, S., Ward, T., Zhu, W.J.: Bleu: a method for automatic evaluation of machine translation. In: Proceedings of the 40th Annual Meeting of the Association for Computational Linguistics, pp. 311–318. Association for Computational Linguistics, Philadelphia, Pennsylvania, USA, July 2002. https://doi.org/10.3115/1073083.1073135, http://aclanthology.org/P02-1040

33. Pelka, O., Ben Abacha, A., García Seco de Herrera, A., Jacutprakart, J., Friedrich, C.M., Müller, H.: Overview of the ImageCLEFmed 2021 concept & caption prediction task. In: CLEF2021 Working Notes, pp. 1101–1112. CEUR Workshop Proceedings, CEUR-WS.org, Bucharest, Romania, 21–24 September 2021

34. Pelka, O., Friedrich, C.M., García Seco de Herrera, A., Müller, H.: Overview of the ImageCLEFmed 2019 concept prediction task. In: CLEF2019 Working Notes. CEUR Workshop Proceedings, CEUR-WS.org, Lugano, Switzerland, 09–12 September 2019. http://ceur-ws.org

35. Pelka, O., Friedrich, C.M., García Seco de Herrera, A., Müller, H.: Overview of the ImageCLEFmed 2020 concept prediction task: medical image understanding. In: CLEF2020 Working Notes. CEUR Workshop Proceedings, CEUR-WS.org, Thessaloniki, Greece, 22–25 September 2020

36. Pelka, O., Koitka, S., Rückert, J., Nensa, F., Friedrich, C.M.: Radiology objects in context (ROCO): a multimodal image dataset. In: Stoyanov, D., et al. (eds.) LABELS/CVII/STENT -2018. LNCS, vol. 11043, pp. 180–189. Springer, Cham (2018). https://doi.org/10.1007/978-3-030-01364-6_20

37. Popescu, A., Deshayes-Chossart, J., Schindler, H., Ionescu, B.: Overview of the ImageCLEF 2022 aware task. In: Proceedings of the Working Notes of CLEF 2022 - Conference and Labs of the Evaluation Forum, CEUR Workshop Proceedings, Bologna, Italy, 5–8 September 2022, vol. 3180, pp. 1329–1338 (2022)

38. Roberts, R.J.: PubMed central: the GenBank of the published literature. Proc. Natl. Acad. Sci. U.S.A. **98**(2), 381–382 (2001). https://doi.org/10.1073/pnas.98.2.381

39. Rückert, J., et al.: Overview of ImageCLEFmedical 2022 - caption prediction and concept detection. In: CLEF2022 Working Notes. CEUR Workshop Proceedings, CEUR-WS.org, Bologna, Italy, 5–8 September 2022

40. Rückert, J., et al.: Overview of ImageCLEFmedical 2023 - caption prediction and concept detection. In: CLEF2023 Working Notes. CEUR Workshop Proceedings, CEUR-WS.org, Thessaloniki, Greece, 18–21 September 2023

41. Sellam, T., Das, D., Parikh, A.P.: BLEURT: learning robust metrics for text generation. In: Jurafsky, D., Chai, J., Schluter, N., Tetreault, J.R. (eds.) Proceedings of the 58th Annual Meeting of the Association for Computational Linguistics, ACL 2020, Online, 5–10 July 2020, pp. 7881–7892. Association for Computational Linguistics (2020). https://doi.org/10.18653/v1/2020.acl-main.704

42. Ştefan, L.D., Constantin, M.G., Dogariu, M., Ionescu, B.: Overview of image-cleffusion 2022 task-ensembling methods for media interestingness prediction and result diversification. In: CLEF2022 Working Notes, CEUR Workshop Proceedings, CEUR-WS. org, Bologna, Italy (2022)

43. Ştefan, L.D., Constantin, M.G., Dogariu, M., Ionescu, B.: Overview of imagecleffusion 2023 task - testing ensembling methods in diverse scenarios. In: Experimental IR Meets Multilinguality, Multimodality, and Interaction. CEUR Workshop Proceedings, CEUR-WS.org, Thessaloniki, Greece, 18–21 September 2023

44. Tsikrika, T., de Herrera, A.G.S., Müller, H.: Assessing the scholarly impact of ImageCLEF. In: Forner, P., Gonzalo, J., Kekäläinen, J., Lalmas, M., de Rijke, M. (eds.) CLEF 2011. LNCS, vol. 6941, pp. 95–106. Springer, Heidelberg (2011). https://doi.org/10.1007/978-3-642-23708-9_12

45. Tsikrika, T., Larsen, B., Müller, H., Endrullis, S., Rahm, E.: The scholarly impact of CLEF (2000–2009). In: Forner, P., Müller, H., Paredes, R., Rosso, P., Stein, B. (eds.) CLEF 2013. LNCS, vol. 8138, pp. 1–12. Springer, Heidelberg (2013). https://doi.org/10.1007/978-3-642-40802-1_1

46. Vaswani, A., et al.: Attention is all you need. In: Guyon, I., et al., Garnett, R. (eds.) Advances in Neural Information Processing Systems 30: Annual Conference on Neural Information Processing Systems 2017, 4–9 December 2017, Long Beach, CA, USA, pp. 5998–6008 (2017). http://proceedings.neurips.cc/paper/2017/hash/3f5ee243547dee91fbd053c1c4a845aa-Abstract.html

47. Vedantam, R., Zitnick, C.L., Parikh, D.: Cider: consensus-based image description evaluation. In: IEEE Conference on Computer Vision and Pattern Recognition, CVPR 2015, Boston, MA, USA, 7–12 June 2015, pp. 4566–4575. IEEE Computer Society (2015). https://doi.org/10.1109/CVPR.2015.7299087

48. Yim, W.W., Fu, Y., Abacha, A.B., Snider, N., Lin, T., Yetisgen, M.: ACI-BENCH: a novel ambient clinical intelligence dataset for benchmarking automatic visit note generation (2023)

49. Yim, W., Ben Abacha, A., Snider, N., Adams, G., Yetisgen, M.: Overview of the MEDIQA-Sum task at ImageCLEF 2023: summarization and classification of doctor-patient conversations. In: CLEF 2023 Working Notes. CEUR Workshop Proceedings, CEUR-WS.org, Thessaloniki, Greece, 18–21 September 2023
50. Zhang, T., Kishore, V., Wu, F., Weinberger, K.Q., Artzi, Y.: BERTScore: evaluating text generation with BERT. In: 8th International Conference on Learning Representations, ICLR 2020, Addis Ababa, Ethiopia, 26–30 April 2020. OpenReview.net (2020). http://openreview.net/forum?id=SkeHuCVFDr

Overview of JOKER – CLEF-2023 Track on Automatic Wordplay Analysis

Liana Ermakova[1], Tristan Miller[2](✉), Anne-Gwenn Bosser[3],
Victor Manuel Palma Preciado[1,4], Grigori Sidorov[4], and Adam Jatowt[5]

[1] Université de Bretagne Occidentale, HCTI, Brest, France
liana.ermakova@univ-brest.fr
[2] Austrian Research Institute for Artificial Intelligence (OFAI), Vienna, Austria
tristan.miller@ofai.at
[3] École Nationale d'Ingénieurs de Brest, Lab-STICC CNRS UMR 6285,
Plouzané, France
[4] Instituto Politécnico Nacional (IPN), Centro de Investigación en
Computación (CIC), Mexico City, Mexico
[5] University of Innsbruck, Innsbruck, Austria

Abstract. The goal of the JOKER track series is to bring together linguists, translators, and computer scientists to foster progress on the automatic interpretation, generation, and translation of wordplay. Being clearly important for various applications, these tasks are still extremely challenging despite significant recent progress in AI in information retrieval and natural language processing. Building on the lessons learned from last year's edition, JOKER-2023 held three shared tasks aligned with human approaches to the translation of wordplay, or more specifically of puns in English, French, and Spanish: detection, location and interpretation, and finally translation. In this paper, we define these three tasks and describe our approaches to corpus creation and evaluation. We then present an overview of the participating systems, including the summaries of their approaches and a comparison of their performance. As in JOKER-2022, this year's track also solicited contributions making further use of our data (an "unshared task"), which we also report on.

Keywords: Wordplay · Puns · Humour · Wordplay interpretation · Wordplay detection · Wordplay generation · Machine translation

1 Introduction

Intercultural communication relies heavily on translation. It is therefore vitally important that semantics-oriented language technology be capable of detecting, interpreting, and appropriately dealing with non-literal expressions such as wordplay. However, wordplay remains one of the most elusive aspects of translation, as it requires an attuned understanding of implicit cultural knowledge, and a keen grasp of language form to understand how to bend it to the desired effect.

A. Arampatzis et al. (Eds.): CLEF 2023, LNCS 14163, pp. 397–415, 2023.
https://doi.org/10.1007/978-3-031-42448-9_26

Furthermore, wordplay appears in all languages and is present in most discourse types. It is used by novelists, poets, playwrights, scriptwriters, and copywriters. It is often employed in titles, headlines, or slogans for its salience and its playful or subversive character. But while modern translation is heavily aided by technological tools, there is little support for humour and wordplay, and even the most current language models struggle to imitate human humour [15].

If the objective of an AI-based translation tool able to deal with wordplay is to be attained, we will almost certainly need to rely on a multilingual parallel corpus: such a tool would necessarily require training on a sizeable quantity of data. This is essentially what the JOKER track at the Conference and Labs of the Evaluation Forum (CLEF) provides, together with tasks designed to establish and advance the current state of the art for wordplay processing.

While humour and wordplay are widely studied in the humanities and social sciences, they have been largely ignored in information retrieval, including dedicated neural net-based retrieval methods and large language models [9]. This is partly because modern AI tools tend to require quality and quantity of training data that has historically been lacking for humour and wordplay. Wordplay detection is useful for information retrieval, digital humanities, conversational agents, and other humour-aware text processing applications. Wordplay location is a prerequisite for the retrieval of jokes containing a specified punning word.

Building on insights gained at the 2022 edition of the JOKER lab [12], we have organized four shared tasks based on our newly expanded, multilingual, parallel corpus of wordplay in English, French, and (new in this year's edition) Spanish [10]:

Task 1 Pun detection in English, French, and Spanish.
Task 2 Pun location and interpretation in English, French, and Spanish.
Task 3 Pun translation from English to French and from English to Spanish.
Open Task We encouraged the use of our data for other tasks related to computational wordplay and humour. These could take the form of, for example, experiments on humour perception, humour evaluation, wordplay generation, or user studies.

Fifty teams registered for our JOKER track at CLEF 2023; of these, thirteen teams participated in the tasks, submitting a total of 176 runs for the numbered tasks. The statistics for these runs are presented in Table 1. In addition, we received three submissions for the open task, covering various areas: an attempt at automated sentiment analysis on the corpus, a pipeline for pun generation in English, and a user study evaluating how well non-native English speakers of varying proficiency levels and countries of origin did on the shared tasks that we had aimed at machines.

2   Task 1: Pun Detection in English, French, and Spanish

A *pun* is a form of wordplay in which a word or phrase evokes the meaning of another word or phrase with a similar or identical pronunciation [16]. *Pun*

Table 1. Statistics on submitted runs by task

Team	Task 1: Detection			Task 2.1: Location			Task 2.2: Interpret.		Task 3: Translation		Total
	EN	FR	ES	EN	FR	ES	EN	FR	EN→FR	EN→ES	
Croland	1	1	1	1	1	1	1		1	1	9
LJGG	3	3	3						4	5	18
Les_miserables	3	3	3	3	3	3	1				19
MiCroGerk	6			6			4			7	23
Smroltra	7	7	7	4	4	4	6		6	6	41
TeamCAU	6			3					3		12
TheLangVerse	1								1	1	3
ThePunDetectives	6			5					2	2	15
UBO	1	1	1	1	1	1	1		3	3	13
UBO-RT							1	1			2
Akranlu	2	2	2	2	2	2	1	1			14
Innsbruck	3										3
NPalma	1		1							2	4
Total	40	17	18	25	11	11	9	2	20	27	176

detection is a binary classification task where the goal is to distinguish between texts containing a pun (the *positive examples*) and texts not containing any pun (the *negative examples*) [20]. Performance on this task is evaluated using the standard precision, recall, accuracy, and F-score metrics from text classification and information retrieval [19, Ch. 8.3].

2.1 Data

Most of the English- and French-language data used for our tasks is described in detail in a resource paper published at SIGIR 2023 [9]. Below we briefly describe the overall data collection process and then discuss in detail the way in which the Spanish-language data was created. For Task 1, the relevant portions of these subcorpora consist of positive and negative texts that are not otherwise annotated or marked up in any way. The positive examples are all short jokes (one-liners), each containing a single pun. In contrast to previously published punning datasets, our negative examples are generated by the data augmentation techniques of manually or semi-automatically editing positive examples in such a way that the wordplay is lost but most of the rest of the meaning still remains. More specifically, in each positive text we made some minimal edits – generally substituting a single word, which may or may not have been the word forming the pun. We adopted this approach in order to minimize the differences in length, vocabulary, style, etc. that manifested across the positive and negative subsets of previous pun detection datasets and on which classifiers could rely on, inadvertently or otherwise, to distinguish those subsets. For the French subcorpus, additional negative examples were sourced through machine translation of the English positive examples, a process through which the wordplay is almost always lost.

The Spanish data was collected primarily via two methods. The first of these was to scrape a manually seeded set of web pages known to collect jokes, and then to manually filter out non-puns and other inappropriate texts. Our data source was Twitter, for which we used Twarc[1] to extract some 195K tweets with the hashtags #humor, #juegodepalabras, and #chiste (meaning "humour", "pun", and "joke", respectively). Here, too, we manually filtered out non-punning examples or those containing extraneous information (images, URLs, emoticons, extra hashtags, etc.). All told, we were able to collect about a thousand examples, about a quarter of which were from web pages and the remainder from Twitter. Negative examples were then generated using essentially the same data augmentation technique used for the English and French data.

The data for each language was split into test and training sets and provided to Task 1 participants in simple JSON and delimited text formats with fields giving a unique ID, the text to classify, and (for training data) a boolean value indicating whether or not the text contains a pun. Participants could choose which language(s) they wished to submit classification runs for. The expected output format was a similarly simple, delimited text file with fields for the run ID, the text ID, the boolean classification result, and a boolean flag indicating whether the classification was made manually or automatically.

Table 2 provides statistics on the size of the dataset, broken down by language and task. Statistics specific to Task 1 are presented in Table 3.

Table 2. Overall dataset statistics

Language	Task 1		Task 2		Task 3			
	Train	Test	Train	Test	Train		Test	
					target	source	target	source
English	5,292	3,183	2,315	1,205	—	—	—	—
French	3,999	12,873	2,000	4,655	5,838	1,405	6,590	1,197
Spanish	1,994	2,241	876	960	644	217	5,727	544

Table 3. Task 1 data statistics

Language	Train			Test		
	Positive	Negative	Total	Positive	Negative	Total
English	3,085	2,207	5,292	809	2,374	3,183
French	1,998	2,001	3,999	5,308	7,565	12,873
Spanish	855	1,139	1,994	952	1,289	2,241

---

[1] https://github.com/DocNow/twarc/.

2.2 Participants' Approaches

The Akranlu team [7] described two methods for pun detection which are based on sentence embeddings with a binary classifier and a sequence classification using XLM-Roberta. A six-layer neural network with a classifier head for the three languages was used.

The NLPalma team [24] experimented with models based on the multilingual BERT architecture. The authors concluded that this approach is promising, but indicated that more fine-tuning of models should lead to better results.

The MiCroGerk participants [25] used six different runs to classify sentences, with systems based on FastText, T5 (based on SimpleT5 library), BLOOM alone, MLP (multilayer perceptron), Naive Bayes and Ridge along with the TF–IDF vectorizer and Count vectorizer. T5 obtained the highest score compared to the other methods.

TheLangverse team used a combination of FastText and an MLP as a classifier layer, achieving somewhat good results compared to what they found in their surveys.

ThePunDetectives team [21] used several models, including Random, Fast-Text, Ridge, Naive Bayes, T5 (SimpleTransformersT5), and RoBERTa (SimpleTransformersRoBERTa) for the classification task, with RoBERTa (barely) achieving the best results among their models.

The participants from the UBO team [8] used T5 (SimpleT5) to solve Task 1, achieving mixed results across languages, with French having the highest success rate.

TeamCAU [2] used different models including Large Language Models (LLMs), FastText, and TF–IDF. In the case of LLMs, they used BLOOM, Jurassic-2 through AI21's inference API. and T5 (SimpleT5). The authors reported that, among their runs, they obtained the best results using LLMs.

The Smroltra team [23] experimented extensively with different classification methods: Random Forest, FastText, Naive Bayes, Logistic Regresion, TF–IDF, MLP, and finally a T5 (SimpleT5) transformer which is already commonly used in the tasks concerning humour. They obtained quite similar results for the Spanish and French datasets. Their approaches were not as reliable for detecting English puns, despite the fact that most of the methods tend to be used mainly for English.

The Croland team [17] tackled Task 1 using OpenAI's GPT-3, under the assumption that LLMs should possess a good understanding of humour.

The Innsbruck team [26] experimented with different data augmentation (DA) techniques, including synonym replacement, back-translation, shortening, in order to improve humour recognition.

The LJGG team experimented with different ways of training a T5 (SimpleT5) model.

Finally, the Les_miserables team (who did not submit a system description paper) submitted SimpleT5- and FastText-based predictions, as well as random baseline results.

2.3   Results

Tables 4, 5, and 6 report participants' results for wordplay detection in English, French, and Spanish, respectively. As some participants submitted only partial runs, we provide separate precision, recall, F_1, and accuracy scores for the total number of instances in the test set (P, R, F_1, A) and for only the number of instances $(\#)$ where a classification was attempted (P^*, R^*, F_1^*, A^*). Our results suggest that wordplay detection is still challenging for all models and all three languages. The improvement of the best runs over the random results are less than 15 points according to F_1 score for all three languages.

The best results according to the F_1 metric for English are achieved with T5 model by two teams: LJGG and UBO. Still the results are lower than 60%. We also observe that the results of the same methods depend heavily on implementation, fine-tuning, and/or used prompts.

For French, the best results were also achieved by the teams applying T5 with F_1 going up to 66.45. This improvement over English might be explained by higher similarity between train and test data in French as this data is coming from the translation of the overlapping sets English puns, while the test set in English contains different puns without semantic or vocabulary similarity. Surprisingly, Logistic Regression and TF–IDF classifier demonstrated comparable results. These results might suggest that efficient training of lighter models could help to achieve results comparable to ones from large pre-trained models which are very expensive and resource-consuming.

For Spanish, the best results were achieved again by T5 and the Akranlu team who applied sentence embeddings with a binary classifier and a sequence classification using XLM-Roberta ($F_1 = 59.64$). Note that Smroltra's random prediction obtained $F_1 = 51.92$ on the same data.

3   Task 2: Pun Location and Interpretation in English, French, and Spanish

Pun location (Task 2.1) is a finer-grained version of pun detection, where the goal is to identify which words carry the double meaning in a text known *a priori* to contain a pun. For example, the first of the following sentences contains a pun where the word *propane* evokes the similar-sounding word *profane*, and the second sentence contains a pun exploiting two distinct meanings of the word *interest*:

- (1) When the church bought gas for their annual barbecue, proceeds went from the sacred to the propane.
- (2) I used to be a banker but I lost interest.

While for the pun detection task, the correct answer for these two instances would be "true", for the pun location task, the correct answers are respectively "propane" and "interest". System performance is reported in terms of accuracy.

Table 4. Results for Task 1 (pun detection) in English

run ID	#	P	R	F_1	A	P^*	R^*	F_1^*	A^*
Croland_EN_GPT3	3183	100.00	0.86	1.71	74.80	100.00	0.86	1.71	74.80
LJGG_t5_large_easy_en	3183	42.73	71.94	**53.61**	68.36	42.73	71.94	53.61	68.36
LJGG_t5_large_label_en	3183	25.41	100.00	40.53	25.41	25.41	100.00	40.53	25.41
LJGG_t5_large_no_label_en	3183	25.41	100.00	40.53	25.41	25.41	100.00	40.53	25.41
Les_miserables_fasttext	3183	25.78	80.96	39.11	35.94	25.78	80.96	39.11	35.94
Les_miserables_random	3183	26.43	51.29	34.88	51.33	26.43	51.29	34.88	51.33
Les_miserables_simplet5	3183	28.13	88.75	42.72	39.52	28.13	88.75	42.72	39.52
MiCroGerk_EN_BLOOM	13.00	8.33	0.12	0.24	74.26	8.33	100.00	15.38	15.38
MiCroGerk_EN_FastText	3183	25.87	82.94	39.44	35.28	25.87	82.94	39.44	35.28
MiCroGerk_EN_MLP	3183	29.04	72.92	41.54	47.84	29.04	72.92	41.54	47.84
MiCroGerk_EN_NB	3183	25.98	95.42	40.84	29.75	25.98	95.42	40.84	29.75
MiCroGerk_EN_Ridge	3183	26.74	85.16	40.70	36.94	26.74	85.16	40.70	36.94
MiCroGerk_EN_SimpleT5	3183	30.75	83.06	44.88	48.16	30.75	83.06	44.88	48.16
Smroltra_EN_FastText	3183	25.62	80.34	38.85	35.72	25.62	80.34	38.85	35.72
Smroltra_EN_Logistic-Regression	3183	26.14	86.15	40.11	34.62	26.14	86.15	40.11	34.62
Smroltra_EN_MLP	3183	27.78	72.43	40.16	45.14	27.78	72.43	40.16	45.14
Smroltra_EN_NBC	3183	26.12	95.55	41.02	30.19	26.12	95.55	41.02	30.19
Smroltra_EN_Random	3183	25.54	66.99	36.98	41.97	25.54	66.99	36.98	41.97
Smroltra_EN_SimpleT5	3183	31.97	83.68	46.27	50.61	31.97	83.68	46.27	50.61
Smroltra_EN_TFIDF	3183	26.90	84.05	40.76	37.92	26.90	84.05	40.76	37.92
TeamCAU_EN_AI21	40	27.58	0.98	1.90	74.17	27.58	80.00	41.02	0.43
TeamCAU_EN_BLOOM	40	30.00	0.37	0.73	74.45	30.00	30.00	30.00	65.00
TeamCAU_EN_FastText	3183	25.71	80.84	39.02	35.78	25.71	80.84	39.02	35.78
TeamCAU_EN_RandomForestWithTFidfEncoding	3183	25.69	83.43	39.28	34.46	25.69	83.43	39.28	34.46
TeamCAU_EN_ST5	3183	26.99	93.32	41.87	34.15	26.99	93.32	41.87	34.15
TeamCAU_EN_TFidfRidge	3183	26.74	85.16	40.70	36.94	26.74	85.16	40.70	36.94
TheLangVerse_fasttext-MLP	3183	26.31	75.40	39.01	40.08	26.31	75.40	39.01	40.08
ThePunDetectives_Fasttext	3183	26.07	80.22	39.35	37.16	26.07	80.22	39.35	37.16
ThePunDetectives_NaiveBayes	3183	25.43	99.62	40.52	25.66	25.43	99.62	40.52	25.66
ThePunDetectives_Random	3183	25.96	50.55	34.31	50.80	25.96	50.55	34.31	50.80
ThePunDetectives_Ridge	3183	27.44	88.75	41.92	37.51	27.44	88.75	41.92	37.51
ThePunDetectives_Roberta	3183	26.11	91.96	40.67	31.82	26.11	91.96	40.67	31.82
ThePunDetectives_SimpleT5	3183	29.21	93.20	44.48	40.87	29.21	93.20	44.48	40.87
UBO_SimpleT5	3183	36.51	85.53	51.18	58.52	36.51	85.53	51.18	58.52
Akranlu_sentemb	3183	26.29	86.40	40.32	34.99	26.29	86.40	40.32	34.99
Akranlu_seqclassification	3183	25.41	100.00	40.53	25.41	25.41	100.00	40.53	25.41
Innsbruck_DS_backtranslation	3183	27.35	84.91	41.38	38.86	27.35	84.91	41.38	38.86
Innsbruck_DS_r1	3183	27.32	86.89	41.57	37.92	27.32	86.89	41.57	37.92
Innsbruck_DS_synonym	3183	27.15	86.89	41.37	37.41	27.15	86.89	41.37	37.41

Table 5. Results for Task 1 (pun detection) in French

run ID	#	P	R	F_1	A	P^*	R^*	F_1^*	A^*
Croland_FR_GPT3	12873	100.00	01.14	02.27	59.24	100.00	01.14	02.27	59.24
LJGG_t5_large_easy_fr	12873	55.13	64.29	59.36	63.70	55.13	64.29	59.36	63.70
LJGG_t5_large_label_fr	12873	41.23	100.00	58.39	41.23	41.23	100.00	58.39	41.23
LJGG_t5_large_no_label_fr	12873	41.23	100.00	58.39	41.23	41.23	100.00	58.39	41.23
Les_miserables_fasttext	12873	58.57	19.76	29.55	61.15	58.57	19.76	29.55	61.15
Les_miserables_random	12873	41.14	49.81	45.06	49.92	41.14	49.81	45.06	49.92
Les_miserables_simplet5	12873	59.72	74.88	**66.45**	68.82	59.72	74.88	66.45	68.82
Smroltra_FR_FastText	12873	55.24	25.00	34.42	60.72	55.24	25.00	34.42	60.72
Smroltra_FR_Logistic-Regression	12873	58.43	60.39	59.40	65.95	58.43	60.39	59.40	65.95
Smroltra_FR_MLP	12873	56.49	62.88	59.52	64.73	56.49	62.88	59.52	64.73
Smroltra_FR_NBC	12873	56.73	63.18	59.78	64.94	56.73	63.18	59.78	64.94
Smroltra_FR_Random	12873	42.14	67.70	51.95	48.36	42.14	67.70	51.95	48.36
Smroltra_FR_SimpleT5	12873	61.21	67.69	64.29	68.99	61.21	67.69	64.29	68.99
Smroltra_FR_TFIDF	12873	58.77	62.09	60.38	66.41	58.77	62.09	60.38	66.41
UBO_SimpleT5	12871	67.80	58.76	62.95	71.49	67.80	58.76	62.95	71.48
Akranlu_sentemb	12873	41.18	73.88	52.88	45.71	41.18	73.88	52.88	45.71
Akranlu_seqclassification	12873	41.23	100.00	58.39	41.23	41.23	100.00	58.39	41.23

Table 6. Results for Task 1 (pun detection) in Spanish

run ID	#	P	R	F_1	A	P^*	R^*	F_1^*	A^*
Croland_ES_GPT3	2241	98.07	05.35	10.15	59.75	98.07	05.35	10.15	59.75
LJGG_t5_large_easy_es	2230	50.34	54.09	52.15	57.83	50.34	54.26	52.23	57.75
LJGG_t5_large_label_es	2230	42.55	99.68	**59.64**	42.70	42.55	100.00	59.70	42.55
LJGG_t5_large_no_label_es	2230	42.55	99.68	**59.64**	42.70	42.55	100.00	59.70	42.55
Les_miserables_fasttext	2230	0.00	0.00	0.00	57.51	0.00	0.00	0.00	57.44
Les_miserables_random	2230	43.43	51.78	47.24	50.87	43.43	51.94	47.31	50.76
Les_miserables_simplet5	2230	51.10	17.01	25.53	57.83	51.10	17.07	25.59	57.75
NLPalma_BERT	2230	55.94	40.54	47.01	61.17	55.94	40.67	47.10	61.12
Smroltra_ES_FastText	2238	40.75	0.625	49.33	45.47	40.75	0.625	49.33	45.39
Smroltra_ES_Logistic-Regression	2238	0.50	49.05	49.52	57.51	0.50	49.05	49.52	57.46
Smroltra_ES_MLP	2238	55.45	44.32	49.27	61.22	55.45	44.32	49.27	61.17
Smroltra_ES_NBC	2238	47.69	56.40	51.68	55.19	47.69	56.40	51.68	55.13
Smroltra_ES_Random	2241	42.05	67.85	51.92	46.63	42.05	67.85	51.92	46.63
Smroltra_ES_SimpleT5	2238	44.31	46.21	45.24	52.47	44.31	46.21	45.24	52.41
Smroltra_ES_TFIDF	2238	53.34	46.11	49.46	59.97	53.34	46.11	49.46	59.91
UBO_SimpleT5	2230	51.28	62.92	56.50	58.85	51.28	63.11	56.58	58.78
Akranlu_sentemb	2230	41.39	72.26	52.63	44.75	41.39	72.49	52.70	44.61
Akranlu_seqclassification	2230	42.55	99.68	**59.64**	42.70	42.55	100.00	59.70	42.55

In pun interpretation (Task 2.2), systems must indicate the two meanings of the pun. In JOKER-2023, semantic annotations are in the form of a pair of lemmatized word sets. Following the practice used in lexical substitution datasets, these word sets contain the synonyms (or if absent, then hypernyms) of the two words involved in the pun, except for any synonyms/hypernyms that happen to share the same spelling with the pun as written.

For example, for the punning joke introduced in Example 1 above, the word sets are {*gas, fuel*} and {*profane*}, and for Example 2, the word sets are {*involvement*} and {*fixed charge, fixed cost, fixed costs*}.

Task 2.2 is evaluated with the precision, recall, and F-score metrics as used in word sense disambiguation [22], except that each instance is scored as the average score for every of its senses. Systems need to guess only one word for each sense of the pun; a guess is considered correct if it matches any of the words in the gold-standard set. For example, a system guessing {*fuel*}, {*profane*} would receive a score of 1 for Example 1, and a system guessing {*fuel*}, {*prophet*} would receive a score of $1/2$.

3.1   Data

The pun location data is drawn from the positive examples of Task 1, with each text being accompanied by an annotation that reproduces the word being punned upon, as described above.

For the English pun interpretation data, we manually annotated each pun according to its senses in WordNet 3.1 and then automatically extracted the synonyms (or if there were none, the hypernyms) of those words to form the two word sets. In some cases, one or both of the senses of the pun was not present in WordNet, or WordNet did contain neither synonyms nor hypernyms for the annotated senses. (This was particularly the case with adjectives and adverbs, which WordNet does not arrange into a hypernymic hierarchy.) In these cases, we sourced the synonym/hypernym sets from human annotators. For French data, we used a simplified version of the annotation made in JOKER-2022 [11].

As in Task 1, the data was split into test and training sets and provided to participants as JSON or delimited text files with fields containing the text of the punning joke and a unique ID. For training data for the pun location task, there is an additional field reproducing the pun word; for training data for the pun interpretation task, there is an additional field giving the two synonym/hypernym sets. System output is expected as a JSON or delimited text file with fields for the run ID, text ID, the pun word (for pun location) or its synonym/hypernym sets (for pun interpretation), and a boolean flag indicating whether the run is manual or automatic.

3.2   Participants' Approaches

The Akranlu team participants [7] employ the token classification method with a tagging schema that relies on assigning a tag of 1 to every pun word and 0 to every word that is not a punning word. For pun interpretation, the results from the pun location subtask were used to disambiguate the appropriate senses of the pun word based on the sentence content and find two synonyms for those senses, sourced from WordNet, that were most similar to sentence embedding.

The MiCroGerk team [25] chose an LLM approach for Task 2, using T5 (SimpleT5), BLOOM, and models from OpenAI and AI21. They also submitted a baseline that uses last word in the sentence as a prediction, as well as a random baseline. It is noteworthy that the BLOOM model presented the worst results compared to the others.

The Smroltra team [23] observed that models based on GPT-3, SpaCy, T5, and BLOOM showed very good performance when it came to Spanish and English, while for French the results were worse. This was particularly the case for SpaCy, which is believed to be not as developed for French as for English. On the other hand, for interpretation, the other methods except for BLOOM were not as effective as expected, including even GPT-3 and various combinations of the methods used with WordNet for location prediction.

TeamCAU [2] used various LLMs. T5 showed good results in comparison to BLOOM and models from AI21 (albeit for partial runs only).

FastText, Ridge, Naive Bayes, SimpleT5, and SimpleTransformersT5 were used by the participants of ThePunDetectives team [21]. They found the best results to be produced by the pre-trained models. In particular, T5 achieved good performance, as predicted by the authors.

For the location and interpretation tasks, the UBO team [8] opted to use T5 (SimpleT5).

The UBO-RT team [4] approached pun location and interpretation in English and French using post-edited output of ChatGPT. A zero-shot strategy was used in their approach and the analysis of the results reveals quite poor capabilities of ChatGPT in interpreting puns, especially those involving homophonic components.

The Croland team [17] used GPT-3.

The Les_miserables team (who did not submit a system description paper) submitted two baseline runs, one where the system selects the final word of the sentence as the pun location, and another run that randomly predicts words; they also submitted a run using the T5 (SimpleT5) model.

3.3   Results

Table 7 reports the participants' results for wordplay location in English, French, and Spanish. As some participants submitted only partial runs, we provide two sets of accuracy scores: those labelled A are based on the total number of instances in the test set, while those labelled A^* are based on the actual number of attempted instances (#).

Accuracy scores for pun location in English and Spanish (A ≈ 80) are twice as good as those for French (A ≈ 40). This could be explained by the fact that participants used large language models that might have included in their training data some of the same puns found in our corpus. By contrast, the French wordplay data was largely constructed by us and not previously published online.

Table 7. Results for Task 2.1 (pun location)

run ID	EN			FR			ES		
	#	A	A*	#	A	A*	#	A	A*
Croland_GPT3	19	0.41	26.31	61	0.20	18.03	51	1.77	33.33
Les_miserables_random	1205	8.87	8.87	4655	4.37	4.98	960	6.14	6.14
Les_miserables_simplet5	1205	76.18	76.18	4655	39.92	45.49	960	55.41	55.41
Les_miserables_word	1205	49.54	49.54	4655	28.67	32.67	960	51.56	51.56
Smroltra_BLOOM	32	1.74	65.62	65	0.41	33.84	57	2.60	43.85
Smroltra_GPT3	32	2.15	81.25	65	0.56	46.15	57	5.20	87.71
Smroltra_SimpleT5	1205	79.50	79.50	4655	39.86	45.43	960	**82.81**	82.81
Smroltra_SpaCy	1205	44.48	44.48	4655	0.00	0.00	960	24.16	24.16
UBO_SimpleT5	1205	77.67	77.67	4655	40.39	46.03	960	57.70	57.70
Akranlu_tokenclassification_x	1205	77.51	77.51	4655	40.56	46.22	960	54.27	54.27
Akranlu_tokenclassification_y	1205	79.17	79.17	4655	**41.35**	47.13	960	56.14	56.14
TeamCAU_AI21	32	1.16	43.75						
TeamCAU_BLOOM	32	1.24	46.87						
TeamCAU_ST5	1205	80.66	80.66						
ThePunDetectives_Fasttext	1205	5.06	5.06						
ThePunDetectives_NaiveBayes	1205	2.07	2.07						
ThePunDetectives_Ridge	1205	50.20	50.20						
ThePunDetectives_SimpleT5	1205	80.41	80.41						
ThePunDetectives_SimpleTransformersT5	1205	**83.15**	83.15						
MiCroGerk_AI21	17	1.32	94.11						
MiCroGerk_BLOOM	17	0.99	70.58						
MiCroGerk_OpenAI	17	1.24	88.23						
MiCroGerk_SimpleT5	1205	79.91	79.91						
MiCroGerk_lastWord	1205	54.43	54.43						
MiCroGerk_random	1205	13.94	13.94						

Owing to various scheduling and technical issues, the pun interpretation results were not ready at the time the manuscript for this paper was submitted. We will provide them in a future article, to be published either in the CLEF CEUR proceedings [1] or on a public preprint server such as arXiv. A link to this article will be provided on the JOKER website at http://www.joker-project.com/.

4   Task 3: Translation of Puns from English to French and Spanish

In Task 3, participating systems attempt to translate English punning jokes into French and Spanish. The translations should aim to preserve, to the extent possible, both the form and meaning of the original wordplay – that is, to implement the pun→pun strategy described in Delabastita's typology of pun translation strategies [5,6]. For example, Example 2 above ("I used to be a banker but I lost interest") might be rendered into French as "*J'ai été banquier mais j'en ai perdu tout l'intérêt"*. This fairly straightforward translation preserves the pun, since *interest* and *intérêt* share the same ambiguity.

4.1 Data

Our French training data contains 5,838 translations of 1,405 distinct puns in English as in Tasks 1 and 2. These translations come from translation contests and the JOKER-2022 track [11,12]. A detailed description of the corpus can be found in our SIGIR 2023 paper [9]. For the test set, we provided participants with 4,290 distinct puns in English to be translated into French and Spanish. Then, we manually evaluated 6,590 French translations of 1,197 distinct puns in English pooled from the participants' runs used as the final test data.

We also provide new sets of English–Spanish translations of punning jokes, similar to English–French datasets we produced for JOKER-2022. These translations were sourced via a translation contest in which professional translators were asked to translate 400 English puns. In total, they produced 2,459 pairs of translated puns. These translations underwent an expert review to ensure compliance with the data set's criteria of preserving both wordplay and the general meaning. We kept 644 translations of 217 distinct English puns for training data. We manually evaluated 5,727 translations of 544 distinct English puns.

The training and test data was provided in JSON and in delimited text formats with fields containing the text of the punning joke and a unique ID; for training there were one or two additional fields containing gold-standard translations of the text into French and/or Spanish. Systems were expected to output a JSON or delimited text file containing the run ID, text ID, the text of the translation(s) into French and/or Spanish, and a boolean flag indicating whether the run was manual or automatic.

4.2 Evaluation

As we have previously argued [11,12], vocabulary overlap metrics such as BLEU are unsuitable for evaluating wordplay translations. We therefore continue JOKER-2022's practice of having trained experts manually evaluate system translations according to features such as lexical field preservation, sense preservation, wordplay form preservation, style shift, humorousness shift, etc. and the presence or absence of errors in syntax, word choice, etc.

Participants' runs were subject to whitespace trimming, lower-casing, and were pooled together. We then filtered out French and Spanish translations identical to the original wordplay in English, as we considered these wordplay instances to be untranslated. The runs are ranked according to the number of successful translations – i.e., translations preserving, to the extent possible, both the form and sense of the original wordplay.

4.3 Participants' Approaches

The LJGG team submitted runs for translation from English to French and Spanish. Their model is a three-stage architecture based on T5 (SimpleT5). The two stages calculate the information necessary to concatenate the English sentence, which forms an input for the third neural network. For training the models, they enlarged Task 3's dataset with the data prepared for Task 1. They also used the DeepL translator to compare their results and found that the DeepL translations are better.

The NLPalma team [24] approached the translation of wordplay from English to Spanish using BLOOMZ & mT5, which is an improved version of BLOOM.

The MiCroGerk team [25] used SimpleT5-, BLOOM-, OpenAI-, and AI21-based models and the models from the EasyNMT package (Opus-MT, mBART50_m2m, and M2M_10) for the English–Spanish translation task. The OpenAI- and AI21-based models proved to be the best, with the lowest-ranked models being SimpleT5. According to the authors, however, there is still plenty of room for improvement.

The UBO team [8] used the models from the EasyNMT package – namely, Opus-MT, mBART50_m2m, and M2M_100.

The TheLangVerse team made use of the j2-grande model from the AI21 platform. They also combined the datasets to provide more content for fine-tuning, obtaining results comparable to those obtained from their surveys.

Opus-MT and M2M_100 from the EasyNMT package were selected by participants of ThePunDetectives team [21]. The authors found that M2M_100 made translations that diverged from the original senses at the expense of precision. In contrast, Opus-MT presented a slightly better translation capability, being able to comprehend some types of humour.

The solution of the Smroltra team [23] was to use the GPT-3, BLOOM, Opus-MT, and mBART50_m2m models from EasyNMT; SimpleT5; and the Google Translate service for both English–Spanish and English–French translations. The best results were obtained using GPT-3, while the worst came from T5, which produced incoherent sentences. GPT-3 and BLOOM obtained the highest scores on both datasets, although according to the authors, the translation of the datasets requires more data and time.

Finally, the Croland team [17] approached the task using GPT-3.

4.4 Results

Tables 8 and 9 present the scores for participants' runs submitted for translations into French and Spanish, respectively. We report the following scores:

#**E** number of manually evaluated translations
#**T** number of submitted translations used for evaluation
#**M** number of translations preserving the meaning of the source puns
%**M** percentage of translations preserving the meaning of the source puns
#**W** number of translations containing wordplay

%W percentage of translations containing wordplay

#S number of translations containing wordplay and preserving the meaning of the source puns

%S percentage of translations containing wordplay and preserving the meaning of the source puns

%R percentage of translations containing wordplay and preserving the meaning of the source puns over the total test set

We rank the runs according to **#S**. For French, the best results were obtained by the Jurassic-2 model and T5. (Note that participants trained the T5 model on the training set while other LLMs were used in a few-shot setup.) For Spanish, the best results were obtained by systems using Google Translate or T5. As in 2022 [11, 12], we observe that the success rate of wordplay translation is extremely low even in the case of LLMs, with the maximum value of 6% over the total evaluated test set for French. This score goes up to 18% for Spanish.

Table 8. Results for Task 3 (pun translation, English to French)

run ID	#E	#T	#M	%M	#W	%W	#S	%S	%R
Croland_task_3_EN_FR_GPT3	16	28	4	25	0	0	0	0	0
LJGG_Google_Translator_EN_FR_auto	1,076	1,197	580	53	67	6	63	5	5
LJGG_task3_fr_mt5_base_auto	2	1,197	2	100	1	50	1	50	0
LJGG_task3_fr_mt5_base_no_label_auto	1	1,197	1	100	0	0	0	0	0
LJGG_task3_fr_t5_large_auto	90	1,197	24	26	2	2	2	2	0
LJGG_task3_fr_t5_large_no_label_auto	140	1,197	80	57	15	10	15	10	1
Smroltra_task_3_EN-FR_BLOOM	31	32	8	25	0	0	0	0	0
Smroltra_task_3_EN-FR_EasyNMT-Opus	786	1,197	427	54	58	7	56	7	4
Smroltra_task_3_EN-FR_EasyNMT-mbart	1139	1,197	613	53	68	5	64	5	5
Smroltra_task_3_EN-FR_GPT3	30	32	8	26	0	0	0	0	0
Smroltra_task_3_EN-FR_GoogleTranslation	1109	1,197	602	54	71	6	67	6	5
Smroltra_task_3_EN-FR_SimpleT5	1043	1,197	562	53	66	6	65	6	5
TeamCAU_task_3_EN-FR_AI21	30	32	8	26	0	0	0	0	0
TeamCAU_task_3_EN-FR_BLOOM	32	32	8	25	0	0	0	0	0
TeamCAU_task_3_EN-FR_ST5	1090	1,197	577	52	71	6	69	6	5
TheLangVerse_task_3_j2-grande-finetuned	1176	1,197	636	54	76	6	**72**	6	6
ThePunDetectives_task_1,3_EN-FR_M2M100	13	340	9	69	2	15	2	15	0
ThePunDetectives_task_1,3_EN-FR_OpusMT	183	340	92	50	19	10	17	9	1
UBO_task_3_SimpleT5	73	1,195	47	64	5	6	5	6	0
UBO_task_3_SimpleT5_x	1148	1,195	616	53	71	6	67	5	5
UBO_task_3_SimpleT5_y	791	1,194	429	54	61	7	59	7	5

Table 9. Results for Task 3 (pun translation, English to Spanish)

run ID	#E	#T	#M	%M	#W	%W	#S	%S	%R
Croland_task_3_ENESGPT3	45	47	9	20.00	3	6.66	3	6.66	0
LJGG_task3_es_mt5_base_auto	34	544	16	47.05	5	14.70	5	14.70	0
LJGG_task3_es_mt5_base_no_label_auto	34	544	16	47.05	5	14.70	5	14.70	0
LJGG_task3_es_t5_large_auto	34	544	16	47.05	5	14.70	5	14.70	0
LJGG_task3_es_t5_large_no_label_auto	34	544	16	47.05	5	14.70	5	14.70	0
LJGG_task_3_GoogleTranslatorENESauto	544	544	274	50.36	106	19.48	99	18.19	18
NLPalma_task_3_BLOOMZ_x	359	359	215	59.88	85	23.67	80	22.28	14
NLPalma_task_3_BLOOMZ_y	359	359	215	59.88	85	23.67	80	22.28	14
Smroltra_task_3_EN-ES_EasyNMT-Opus	529	544	263	49.71	100	18.90	93	17.58	17
Smroltra_task_3_EN-ES_EasyNMT-Opus_x	529	544	263	49.71	100	18.90	93	17.58	17
Smroltra_task_3_EN-ES_EasyNMT-Opus_y	529	544	263	49.71	100	18.90	93	17.58	17
Smroltra_task_3_EN-ES_GoogleTranslation	532	544	267	50.18	103	19.36	96	18.04	17
Smroltra_task_3_EN-ES_SimpleT5	531	544	265	49.90	101	19.02	94	17.70	17
Smroltra_task_3_ENESBLOOM	45	47	8	17.77	2	4.44	2	4.44	0
TheLangVerse_task_3_j2-grande-finetuned	415	544	200	48.19	70	16.86	65	15.66	11
ThePunDetectives_task_1.3_EN-ES_M2M100	33	430	16	48.48	7	21.21	7	21.21	1
ThePunDetectives_task_1.3_ENESOpusMT	428	430	208	48.59	71	16.58	66	15.42	12
MiCroGerk_task_3_EN-ES_OpenAI	6	17	3	0.5	1	16.66	1	16.66	0
MiCroGerk_task_3_EN-ES_mbart50_m2m_x	543	544	274	50.46	106	19.52	99	18.23	18
MiCroGerk_task_3_EN-ES_AI21_x	1	17	1	1	0	0	0	0	0
MiCroGerk_task_3_EN-ES_mbart50_m2m_y	543	544	274	50.46	106	19.52	99	18.23	18
MiCroGerk_task_3_EN-ES_m2m_100_418M	43	544	23	53.48	11	25.58	11	25.58	2
MiCroGerk_task_3_EN-ES_SimpleT5	5	544	4	0.8	3	0.6	3	0.6	0

5   Open Task

We received three submissions for the Open Task, raising different challenges.

5.1   Pun Generation for Text Transformation and Conversational Systems

Glemarec & Charles [14] proposed an experiment for wordplay generation. Their motivation was to integrate similar techniques into interactive systems (narratives, virtual agents) to favor engagement. This work is an update and expansion of a paper presented at the previous JOKER lab at CLEF 2022 [13]. In the latter work, the authors had proposed a pipeline for pun generation in French, using Jurassic [18] as a Large Language Model and substituting a word in a source sentence containing a homophonic word, to provide a context appropriate for creating a new punning sentence. In this year's submission, they used the GPT-3 API in addition to libraries for recognizing paronyms in English, hence not requiring the services of a phonetic lexicon. Several examples of generated outputs are presented in their paper. Although there is no quantitative or qualitative evaluation provided, and despite that the target language being different (which makes it difficult to compare to the last year's results), the curated examples provided seem successful at providing new humorous puns.

5.2   Sentiment Analysis for Wordplay

Thomas-Young & Ermakova [27] presented a sentiment analysis of the corpora, using the Microsoft Azure Service. They deem their results inconclusive: the use of sentiment analysis at the word level as well as for context analysis seems to require specially designed models.

5.3   Comparison of Machine and Human Performances

Große-Bolting et al. considered the performance not just of machine learning algorithms but also humans in JOKER tasks using the English corpus.

For the evaluation of human competence on JOKER tasks, a survey was conducted in four countries where English is not a mother tongue: Poland, France, Spain and Germany. The survey used ten randomly selected punning sentences from a curation of 100 in our corpus. In addition to questions for estimating the English proficiency of respondents, questions were asked for determining if respondents could locate the pun, understand the pun, and provide a translation of the pun in their native languages for ten random entries of the JOKER corpus. The answers allowed the authors to check how well the participants performed on the location, interpretation, and translation tasks.

Standard metrics such as recall, precision and F_1 were used to compare respective performances. Participants scored $0.74\,F_1$ for classification, $0.2\,F_1$ for location, albeit with low inter-rater reliability. The authors noted that the low score humans achieved in pun location can be beaten by a simple system which always selects the last word of the punning sentence (which, the authors claim, achieves an F_1 of 0.35).

Their results echo the work of Bell [3], who noted that being able to understand humour in a foreign language is a particular challenge for learners.

6   Conclusion

In this paper, we described the JOKER track at CLEF 2023, consisting of three interconnected shared tasks on automatic wordplay analysis and translation, as well as an open task. These tasks aim to advance the automation of creative-language translation by developing the requisite parallel data and evaluation metrics for detecting, locating, interpreting, and translating wordplay. Thirteen teams submitted 176 runs for the shared tasks. We received many partial runs due to token/time constraints of LLMs.

Our results in general suggest that wordplay detection and location are still a challenge for LLMs despite their recent significant advances. For the pun detection task for all three languages, the improvement of the best runs over the random results are less than 15% points according to F_1 score. We also observe that the results of the same methods depend heavily on implementation, fine-tuning, and/or prompts used. For French, we can see a slight improvement over English, which might be explained by higher similarity between training and test data in French; this data comes from the translation of overlapping sets English puns, while the test set in English contains different puns without semantic or vocabulary similarity. Surprisingly, Logistic Regression and the TF–IDF classifier demonstrated comparable results. These results might suggest that efficient training of lighter models could help to achieve results comparable to large pre-trained models, which are very expensive and resource-consuming.

Accuracy scores for pun location in English and Spanish are twice as high as those for French. This could be explained by the fact that participants used large language models that might have included in their training data some of the puns found in our corpus, which were sourced from the web directly or indirectly by applying LLMs. By contrast, the French wordplay data was largely constructed by us and not previously published online.

We observe that the success rate of wordplay translation is extremely low even in the case of LLMs, with the maximum value of 6% over the total evaluated test set for French and 18% for Spanish.

Further details on the shared tasks and the submitted runs can be found in the CLEF CEUR proceedings [1]. Additional information on the track is available on the JOKER website: http://www.joker-project.com/.

Acknowledgements. This project has received a government grant managed by the National Research Agency under the program *"Investissements d'avenir"* integrated into France 2030, with the Reference ANR-19-GURE-0001. JOKER is supported by *La Maison des sciences de l'homme en Bretagne*. We thank Carolina Palma Preciado, Leopoldo Jesús Gutierrez Galeano, Khatima El Krirh, Nathalie Narváez Bruneau, and Rachel Kinlay for their help and support in the first Spanish pun translation contest. We also thank Quentin Dubreuil, Keith Salina, Constance Germann, Océane Brunelière, Aurianne Damoy, Angelique Robert, and all other colleagues and students who participated in data construction, the translation contests, and the CLEF JOKER track.

References

1. Aliannejadi, M., Faggioli, G., Ferro, N., Vlachos, M. (eds.): Proceedings of the Working Notes of CLEF 2023: Conference and Labs of the Evaluation Forum. CEUR Workshop Proceedings, CEUR-WS.org (2023)
2. Anjum, A., Lieberum, N.: Exploring humor in natural language processing: a comprehensive review of JOKER tasks at CLEF symposium 2023. In: Proceedings of the Working Notes of CLEF 2023: Conference and Labs of the Evaluation Forum [1] (2023)
3. Bell, N.D.: Learning about and through humor in the second language classroom. Lang. Teach. Res. **13**(3), 241–258 (2009). https://doi.org/10.1177/1362168809104697
4. Brunelière, O., Germann, C., Salina, K.: CLEF 2023 JOKER Task 2: using Chat GPT for pun location and interpretation. In: Proceedings of the Working Notes of CLEF 2023: Conference and Labs of the Evaluation Forum [1] (2023)

5. Delabastita, D.: There's a Double Tongue: an Investigation into the Translation of Shakespeare's Wordplay, with Special Reference to Hamlet. Rodopi, Amsterdam (1993)
6. Delabastita, D.: Wordplay as a translation problem: a linguistic perspective. In: Ein internationales Handbuch zur Übersetzungsforschung, vol. 1, pp. 600–606. De Gruyter Mouton, June 2008. https://doi.org/10.1515/9783110137088.1.6.600
7. Dsilva, R.R.: AKRaNLU @ CLEF JOKER 2023: using sentence embeddings and multilingual models to detect and interpret wordplay. In: Proceedings of the Working Notes of CLEF 2023: Conference and Labs of the Evaluation Forum [1] (2023)
8. Dubreuil, Q.: UBO team @ CLEF JOKER 2023 track for task 1, 2 and 3 - applying AI models in regards to pun translation. In: Proceedings of the Working Notes of CLEF 2023: Conference and Labs of the Evaluation Forum [1] (2023)
9. Ermakova, L., Gwenn-Bosser, A., Jatowt, A., Miller, T.: The JOKER corpus: English-French parallel data for multilingual wordplay recognition. In: SIGIR '23: Proceedings of the 46th International ACM SIGIR Conference on Research and Development in Information Retrieval. Association for Computing Machinery, New York, NY (2023). https://doi.org/10.1145/3539618.3591885
10. Ermakova, L., Miller, T., Bosser, A.G., Preciado, V.M.P., Sidorov, G., Jatowt, A.: Science for fun: the CLEF 2023 JOKER track on automatic wordplay analysis. In: Kamps, J., et al. (eds.) Advances in Information Retrieval. ECIR 2023. LNCS, vol. 13982, pp. 546–556. Springer, Cham (2023). https://doi.org/10.1007/978-3-031-28241-6_63
11. Ermakova, L., et al.: Overview of JOKER@CLEF 2022: automatic wordplay and humour translation workshop. In: Barron-Cedeno, A., et al. (eds.) Experimental IR Meets Multilinguality, Multimodality, and Interaction. CLEF 2022. LNCS, vol. 13390, pp. 447–469. Springer, Cham (2022). https://doi.org/10.1007/978-3-031-13643-6_27
12. Ermakova, L., et al.: Overview of the CLEF 2022 JOKER task 3: pun translation from English into French. In: Faggioli, G., Ferro, N., Hanbury, A., Potthast, M. (eds.) Proceedings of the Working Notes of CLEF 2022 - Conference and Labs of the Evaluation Forum, Bologna, Italy, 5th to 8th September 2022. CEUR Workshop Proceedings, vol. 3180, pp. 1681–1700, August 2022
13. Glemarec, L., Bosser, A., Boccou, J., Ermakova, L.: Humorous wordplay generation in French. In: Faggioli, G., Ferro, N., Hanbury, A., Potthast, M. (eds.) Proceedings of the Working Notes of CLEF 2022 - Conference and Labs of the Evaluation Forum, Bologna, Italy, 5th to 8th September 2022. CEUR Workshop Proceedings, vol. 3180, pp. 1793–1806. CEUR-WS.org (2022)
14. Glemarec, L., Charles, F.: BU-Pier team @ CLEF JOKER 2023 open task: slip of the tongue generation to improve social interaction with virtual agents. In: Proceedings of the Working Notes of CLEF 2023: Conference and Labs of the Evaluation Forum [1] (2023)
15. Jentzsch, S., Kersting, K.: ChatGPT is fun, but it is not funny! Humor is still challenging large language models (2023)
16. Kolb, W., Miller, T.: Human-computer interaction in pun translation. In: Hadley, J.L., Taivalkoski-Shilov, K., Teixeira, C.S.C., Toral, A. (eds.) Using Technologies for Creative-Text Translation, pp. 66–88. Routledge (2022). https://doi.org/10.4324/9781003094159-4
17. Komorowska, J., Čatipović, I., Vujica, D.: CLEF2023' JOKER working notes. In: Proceedings of the Working Notes of CLEF 2023: Conference and Labs of the Evaluation Forum [1] (2023)

18. Lieber, O., Sharir, O., Lentz, B., Shoham, Y.: Jurassic-1: technical details and evaluation. White paper, AI21 Labs, August 2021. https://uploads-ssl.webflow.com/60fd4503684b466578c0d307/61138924626a6981ee09caf6_jurassic_tech_paper.pdf
19. Manning, C.D., Raghavan, P., Schütze, H.: Introduction to Information Retrieval. Cambridge University Press, Cambridge (2008)
20. Miller, T., Hempelmann, C.F., Gurevych, I.: SemEval-2017 Task 7: detection and interpretation of English puns. In: Proceedings of the 11th International Workshop on Semantic Evaluation, pp. 58–68, August 2017. https://doi.org/10.18653/v1/S17-2005
21. Ohnesorge, F., Gutiérrez, M.Á., Plichta, J.: CLEF 2023 JOKER tasks 2 and 3: using NLP models for pun location, interpretation and translation. In: Proceedings of the Working Notes of CLEF 2023: Conference and Labs of the Evaluation Forum [1] (2023)
22. Palmer, M., Ng, H.T., Dang, H.T.: Evaluation of WSD systems. In: Agirre, E., Edmonds, P. (eds.) Word Sense Disambiguation. Text, Speech and Language Technology, vol. 33, pp. 75–106. Springer, Dordrecht (2007). https://doi.org/10.1007/978-1-4020-4809-8_4
23. Popova, O., Dadić, P.: Does AI have a sense of humor? CLEF 2023 JOKER tasks 1, 2 and 3: using BLOOM, GPT, SimpleT5, and more for pun detection, location, interpretation and translation. In: Proceedings of the Working Notes of CLEF 2023: Conference and Labs of the Evaluation Forum [1] (2023)
24. Preciado, V.M.P., Preciado, C.P., Sidorov, G.: NLPalma @ CLEF 2023 JOKER: a BLOOMZ and BERT approach for wordplay detection and translation. In: Proceedings of the Working Notes of CLEF 2023: Conference and Labs of the Evaluation Forum [1] (2023)
25. Prnjak, A., Davari, D.R., Schmitt, K.: CLEF 2023 JOKER task 1, 2, 3: pun detection, pun interpretation, and pun translation. In: Proceedings of the Working Notes of CLEF 2023: Conference and Labs of the Evaluation Forum [1] (2023)
26. Reicho, S., Jatowt, A.: Innsbruck @ CLEF JOKER 2023 track's task 1: data augmentation techniques for humor recognition in text. In: Proceedings of the Working Notes of CLEF 2023: Conference and Labs of the Evaluation Forum [1] (2023)
27. Thomas-Young, T.: Why sentiment analysis is a joke with JOKER data? word-level and interpretation analysis (CLEF 2023 JOKER Task 2). In: Proceedings of the Working Notes of CLEF 2023: Conference and Labs of the Evaluation Forum [1] (2023)

Overview of LifeCLEF 2023: Evaluation of AI Models for the Identification and Prediction of Birds, Plants, Snakes and Fungi

Alexis Joly[1], Christophe Botella[1], Lukáš Picek[8], Stefan Kahl[6,11],
Hervé Goëau[2(✉)], Benjamin Deneu[1], Diego Marcos[1],
Joaquim Estopinan[1], Cesar Leblanc[1], Théo Larcher[1], Rail Chamidullin[8],
Milan Šulc[10], Marek Hrúz[8], Maximilien Servajean[7], Hervé Glotin[3],
Robert Planqué[4], Willem-Pier Vellinga[4], Holger Klinck[6], Tom Denton[9],
Ivan Eggel[5], Pierre Bonnet[2], and Henning Müller[5]

[1] Inria, LIRMM, Univ Montpellier, CNRS, Montpellier, France
[2] CIRAD, UMR AMAP, Montpellier, Occitanie, France
herve.goeau@cirad.fr
[3] Univ. Toulon, Aix Marseille Univ., CNRS, LIS, DYNI team, Marseille, France
[4] Xeno-canto Foundation, Leiden, The Netherlands
[5] Informatics Insitute, HES-SO Valais, Sierre, Switzerland
[6] K. Lisa Yang Center for Conservation Bioacoustics, Cornell Lab of Ornithology,
Cornell University, Ithaca, USA
[7] LIRMM, AMIS, Univ Paul Valéry Montpellier, Univ Montpellier, CNRS,
Montpellier, France
[8] Department of Cybernetics, FAV, University of West Bohemia, Pilsen, Czechia
[9] Google Research, San Francisco, USA
[10] Second Foundation, Prague, Czech Republic
[11] Chemnitz University of Technology, Chemnitz, Germany

Abstract. Biodiversity monitoring through AI approaches is essential, as it enables the efficient analysis of vast amounts of data, providing comprehensive insights into species distribution and ecosystem health and aiding in informed conservation decisions. Species identification based on images and sounds, in particular, is invaluable for facilitating biodiversity monitoring efforts and enabling prompt conservation actions to protect threatened and endangered species. The LifeCLEF virtual lab has been promoting and evaluating advances in this domain since 2011. The 2023 edition proposes five data-oriented challenges related to the identification and prediction of biodiversity: (i) BirdCLEF: bird species recognition in long-term audio recordings (soundscapes), (ii) SnakeCLEF: snake identification in medically important scenarios, (iii) PlantCLEF: very large-scale plant identification, (iv) FungiCLEF: fungi recognition beyond 0–1 cost, and (v) GeoLifeCLEF: remote sensing-based prediction of species. This paper overviews the motivation, methodology, and main outcomes of that five challenges.

1 LifeCLEF Lab Overview

Accurately identifying organisms observed in the wild is an essential step in eco-logical studies. It forms the foundation for understanding species interactions, population dynamics, and ecological processes, allowing researchers to accurately assess biodiversity, track changes over time, and make informed management and conservation decisions. However, observing and identifying living organisms requires high levels of expertise. For instance, vascular plants alone account for more than 300,000 different species and the distinctions between them can be quite subtle. The worldwide shortage of trained taxonomists and curators capa-ble of identifying organisms has come to be known as the *taxonomic impediment*. Since the Rio Conference of 1992, it has been recognized as one of the major obstacles to the global implementation of the Convention on Biological Diver-sity[1]. In 2004, Gaston and O'Neill [10] discussed the potential of automated approaches for species identification. They suggested that if the scientific com-munity were able to (i) produce large training datasets, (ii) precisely evaluate error rates, (iii) scale-up automated approaches, and (iv) detect novel species, then it would be possible to develop a generic automated species identification system that would open up new vistas for research in biology and related fields.

Since the publication of [10], automated species identification has been stud-ied in many contexts [6,12,13,20,31,49,50,58]. This area continues to expand rapidly, particularly due to advances in deep learning [2,11,32,34,35,52–54,56]. Biodiversity monitoring through AI approaches is now recognized as a key solu-tion to collect and analyze vast amounts of data from various sources, enabling us to gain a comprehensive understanding of species distribution, abundance, and ecosystem health [3]. This information is essential for making informed con-servation decisions and identifying areas in need of protection.

To measure progress in a sustainable and repeatable way, the LifeCLEF[2] virtual lab was created in 2014 as a continuation and extension of the plant identification task that had been run within the ImageCLEF lab[3] since 2011 [15–17]. Since 2014, LifeCLEF has expanded the challenge by considering ani-mals and fungi in addition to plants and including audio and video content in addition to images [21–29]. Nearly a thousand researchers and data scientists register yearly to LifeCLEF to download the data, subscribe to the mailing list, benefit from the shared evaluation tools, etc. The number of participants who finally crossed the finish line by submitting runs was respectively: 22 in 2014, 18 in 2015, 17 in 2016, 18 in 2017, 13 in 2018, 16 in 2019, 16 in 2020, $1,022$ in 2021 and 1146 in 2022. LifeCLEF 2023 consists of five challenges (BirdCLEF, SnakeCLEF, PlantCLEF, FungiCLEF, GeoLifeCLEF) whose methodology and main outcomes are described in this paper. Table 1 provides an overview of the data and tasks of the five challenges.

---

[1] https://www.cbd.int/.
[2] http://www.lifeclef.org/.
[3] http://www.imageclef.org/.

Table 1. Overview of the data and tasks of the five LifeCLEF challenges

	Modality	#species	#items	Task	Metric
BirdCLEF	audio	264	16,900	Multi-Label Classification	cmAP
SnakeCLEF	images metadata	1,500	150–200K	Classification	ad-hoc metric
FungiCLEF	images metadata	1,600	300K	Classification	ad-hoc metric
PlantCLEF	images	80,000	4.0M	Classification	Macro-Average MRR
GeoLifeCLEF	images time-series tabular	10,040	5.3M	Multi-Label Classification	Micro-Average F1

The systems used to run the challenges (registration, submission, leaderboard, etc.) were the Kaggle platform for the BirdCLEF and GeoLifeCLEF challenges, the Hugging Face competition platform for SnakeCLEF and FungiCLEF challenges, and the AICrowd platform for the PlantCLEF challenge. Three of the challenges (GeoLifeCLEF, SnakeCLEF, and FungiCLEF) were organized jointly with FGVC 10, an annual workshop dedicated to Fine-Grained Visual Categorization organized in the context of the CVPR international conference on computer vision and pattern recognition.

In total, $1,226$ people/teams participated to LifeCLEF 2023 edition by submitting runs to at least one of the five challenges ($1,189$ only for the BirdCLEF challenge). Only some of them managed to get the results right, and 17 of them went all the way through the CLEF process by writing and submitting a *working note* describing their approach and results (for publication in CEUR-WS proceedings. In the following sections, we provide a synthesis of the methodology and main outcomes of each of the five challenges. More details can be found in the extended overview reports of each challenge and in the individual working notes of the participants (references provided below).

2   BirdCLEF Challenge: Bird Call Identification in Soundscapes

A detailed description of the challenge and a more complete discussion of the results can be found in the dedicated working note [30].

2.1   Objective

Recognizing bird sounds in complex soundscapes is an important sampling tool that often helps reduce the limitations of point counts. In the future, archives of recorded soundscapes will become increasingly valuable as the habitats in which they were recorded will be lost. In the past few years, deep learning approaches have transformed the field of automated soundscape analysis. Yet, when training data is sparse, detection systems struggle to recognize bird species reliably. The goal of this competition was to establish training and test datasets that can serve as real-world applicable evaluation scenarios for endangered habitats and help the scientific community to advance their conservation efforts through automated bird sound recognition.

2.2   Dataset

We built on the experience from previous editions and adjusted the overall task to encourage participants to focus on task-specific model designs. We selected training and test data to suit this demand. As in previous iterations, Xeno-canto was the primary source for training data, and expertly annotated soundscape recordings were used for testing. We focused on bird species which are usually underrepresented in large bird sound collections, but we also included common species so that participants were able to train good recognition systems. In search of suitable test data, we considered different data sources with varying complexity (call density, chorus, signal-to-noise ratio, man-made sounds, etc.) and quality (mono and stereo recordings). We also wanted to focus on very specific real-world use cases (e.g., conservation efforts in Africa) and framed the competition based on the demand of the particular use case.

2.3   Evaluation Protocol

The challenge was held on Kaggle, and the evaluation mode resembled the test mode of previous iterations, i.e., hidden test data, code competition, etc. We used the class-wise mean average precision (cmAP) as a metric, which allowed organizers to assess system performance independent of fine-tuned confidence thresholds. Participants were asked to return a list of species for short audio segments extracted from labeled soundscape data. We used 5-s segments, which reflect a good compromise between typical signal length and sufficiently long context windows. Again, we kept the dataset size reasonably small (<50 GB) and easy to process, and we also provided introductory code repositories and write-ups to lower the entry-level of the competition.

2.4   Participants and Results

1,397 participants across 1,189 teams participated in the BirdCLEF 2023 challenge and submitted a total of 21,519 runs. In Fig. 1 we report the performance achieved by the top 25 collected runs. The private leaderboard score is the primary metric and was revealed to participants after the submission deadline to avoid probing the hidden test data. Public leaderboard scores were visible to participants over the course of the entire challenge.

The baseline cmAP-score in this year's edition was 0.602 (public 0.717) with random confidence scores for all birds for all segments, and 1,165 teams managed

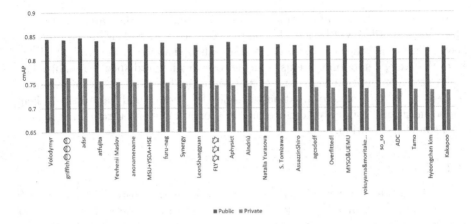

Fig. 1. BirdCLEF 2023 results of the top 25 teams.

to score above this threshold. The best submission achieved a cmAP-score of 0.7639 (public 0.8444) and the top 10 best performing systems were within only 1.5% difference in score. The vast majority of approaches were based on convolutional neural network ensembles and mostly differed in pre- and post-processing and neural network backbone. Interestingly, few-shot learning techniques were vastly underrepresented despite the fact that some target species only had a handful of training samples. Some teams utilized embeddings of pre-trained bird recognition models (such as BirdNET or Google Perch, both were provided as supporting models) to train on high-level features, which somewhat mitigated the need for extensive training data. Due to the limited CPU runtime for submissions, participants focused on accelerating model inference and efficient architectures, with EfficientNet backbones being the most common choice. Interestingly, participants also experimented with ONNX and openVINO to improve model inference speed.

3 SnakeCLEF Challenge: Snake Identification in Medically Important Scenarios

A detailed description of the challenge and a more complete discussion of the results can be found in the dedicated overview paper [37].

3.1 Motivation

Developing a robust system for identifying species of snakes from photographs is an important goal in biodiversity but also for human health. With over half a million victims of death & disability from venomous snakebite annually, understanding the global distribution of the >4,000 species of snakes and differentiating species from images (particularly images of low quality) will significantly improve epidemiology data and treatment outcomes. We have learned from previous editions that *"machines"* can accurately recognize ($F_1^C \approx 90\%$ and Top1 Accuracy $\approx 90\%$) even in scenarios with long-tailed distributions and $\approx 1,600$ species. Thus, testing over real Medically Important Scenarios and specific countries (India and Central America) and integrating the medical importance of species is the next step that should provide a more reliable machine prediction.

3.2   Objective

The main objective of this competition is to create a machine learning model that can accurately predict snake species for given observation data, i.e., images and location, and: (i) fits limits for memory footprint (max size of $1\,GB$), (ii) minimizes the danger to human life, i.e., the venomous $\longleftrightarrow$ harmless confusion, (iii) generalize to all countries and geographic regions.

3.3   Dataset

The dataset was constructed from observations submitted to the citizen science platforms – iNaturalist and HerpMapper – and combined roughly 110,000 reals snake specimen observations with community-verified species labels. The number of species was extended up to $\approx 1,800$ snake species from around the world. Apart from image data, we have provided information about medical importance (i.e., how venomous the species is), and country-species relevance was provided for each species. We list the dataset statistics in Table 2.

Table 2. SnakeCLEF 2023 dataset statistics for each subset.

Subset	#Species	#Countries	#Images	#Observations
Training	1,784	212	168,144	95,588
↪ *iNaturalist*	*1,784*	*210*	*154,301*	*85,843*
↪ *HerpMapper*	*889*	*119*	*13,843*	*9,745*
Validation	1,599	177	14,117	7,816
Public Test	1,784	191	28,274	15,632
Private Test	182	8	8,080	3,765
↪ *India*	*76*	*1*	*2,892*	*2,395*
↪ *Central America*	*107*	*4*	*5,188*	*1,370*

Geographical Bias: There is a lack of data from remote parts of developing countries that tend to lack herpetological expertise and have high snake diversity, and snakebites are common (i.e., Asia, Africa, and Central/South America).

3.4   Evaluation Protocol

To motivate research in recognition scenarios with uneven costs for different errors, such as mistaking a venomous snake for a harmless one, this year's challenge goes beyond the 0–1 loss common in classification. We make some assumptions to reduce the complexity of the evaluation. We consider that there exists a universal antivenom that is applicable to all venomous snake bites. Furthermore, such antivenom is not lethal or seriously harmful when applied to a healthy human. Hence, we will penalize the misclassification of a venomous species with a harmless one more than the other way around. Although this solution is not perfect, it is a first step into a more complex evaluation of snake bites. We specify two metrics (T_1, T_2) reflecting these different scenarios.

$$T_1 = \frac{w_1 F_1 + w_2 C_{h \longrightarrow h} + w_3 C_{h \longrightarrow v} + w_4 C_{v \longrightarrow v} + w_5 C_{v \longrightarrow h}}{w_1 + w_2 + w_3 + w_4 + w_5}, \qquad (1)$$

where C is equal to 1–ratio of misclassified samples, confusing h-armless and v-enomous species. This metric has a lower bound of 0% and an upper bound of 100%. The lower bound is achieved when all species are misclassified, including misclassifications of harmless species as venomous and vice versa. On the other hand, if the F1-score reaches 100%, indicating the correct classification of all species, each C value must be zero, leading to an overall score of 100%.

$$T_2 = \sum_i L(y_i, \hat{y}_i), \qquad L(y, \hat{y}) = \begin{cases} 0 & \text{if } y = \hat{y} \\ 1 & \text{if } y \neq \hat{y} \text{ and } p(y) = 0 \text{ and } p(\hat{y}) = 0 \\ 2 & \text{if } y \neq \hat{y} \text{ and } p(y) = 0 \text{ and } p(\hat{y}) = 1 \\ 2 & \text{if } y \neq \hat{y} \text{ and } p(y) = 1 \text{ and } p(\hat{y}) = 1 \\ 5 & \text{if } y \neq \hat{y} \text{ and } p(y) = 1 \text{ and } p(\hat{y}) = 0 \end{cases}, \qquad (2)$$

where the function p returns 0 if y is a harmless species and 1 if it is venomous.

3.5   Participants and Results

This year a total of 16 teams participated in the SnakeCLEF. However, just five teams submitted their models for private evaluation together with the working notes. Details of the best methods and systems used are synthesized in the competition overview paper [1].

In Fig. 2 and Fig. 3 we report the public and private leaderboard performance achieved by individual teams using: (i) Track 1 Metric (T_1), (ii) Track 2 Metric (T_2) and (iii) the macro F_1 score. The main outcomes we can derive from the achieved results are as follows:

NLP Model Encoded Metadata Might be the Next Big Thing. Same as in previous years, most of the teams used the provided metadata and showed that by doing so the competition metric improves. CLIP [44] – a strong multimodal descriptor, was used for the first time in this competition to encode the metadata. This trend may lead to the utilization of bigger NLP models.

Transformers for the Win. But Do Not Rule Out the CNNs Yet. On the vision part, convolutional models (ResNet [18], EfficientNet [48], ConvNext [57]) and Transformer models (MetaFormer [8], Swin [33], VOLO [59]) were used to extract the visual features. When teams compared the architectures side-by-side, most of the times the Transformer architecture performed better. However, the winning team used ConvNextv2. Due to the lack of a fair and exhaustive ablation study, it is not clear how a Transformer model would fare.

Task-Tailored Losses and Self-supervision are the Key to Learning. Traditionally, Seesaw loss [55] and SimCLR [7] were used to cope with the long-tailed data. Some teams introduced a weighted version of the loss functions tackling the different penalization for different errors. Multi-Instance Learning [19] was applied to make use of more images per observation.

Medically Important Scenarios Might be on to Something. The final results on the private dataset show an interesting behavior of the models. The best team (named *word2vector*) achieved macro F_1 score of 53.58% with the competition score of 91.31%. The runner-up (BBracke) actually achieved a much better F_1 score of 61.39% but had a lower competition score of 90.19%. We hypothesize that this was possible due to the post-processing step of team *word2vector*. When they observed that the top-5 results contained a venomous species, the observation was classified as such.

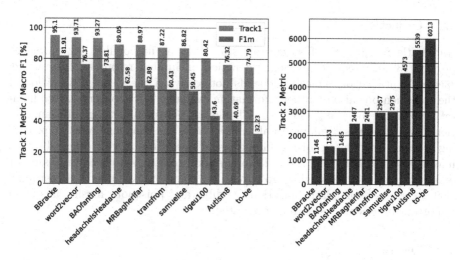

Fig. 2. Public Leaderboard – SnakeCLEF 2023 competition – Top10 teams.

424   A. Joly et al.

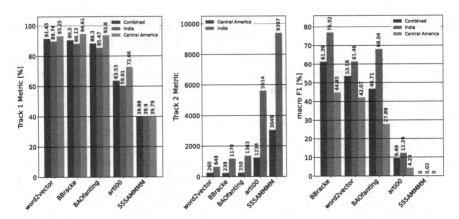

Fig. 3. Private Leaderboard – SnakeCLEF 2023 competition – 5 teams.

4   FungiCLEF Challenge: Fungi Recognition Beyond 0–1 Cost

A detailed description of the challenge and a more complete discussion of the results can be found in the dedicated working note [42].

4.1   Objective

Automatic recognition of species at scale, such as in popular citizen-science projects [39,47], requires efficient prediction on limited resources. In practice, species identification typically depends not solely on the visual observation of the specimen but also on other information available to the observer, e.g., habitat, substrate, location, and time. Thanks to rich metadata, precise annotations, and baselines available to all competitors, the challenge aims at providing a major benchmark for combining visual observations with other observed information. Additionally, the 2023 competition considers decision processes for different usage scenarios, which go beyond the commonly assumed 0/1 cost function – e.g., cost for misclassification of edible and poisonous mushrooms is an important practical aspect to be evaluated.

4.2   Dataset

The challenge builds upon the Danish Fungi 2020 dataset [40], which comes from a citizen science project, the Atlas of Danish Fungi, where all samples went through an expert validation process, guaranteeing a high quality of labels. Rich metadata (Habitat, Substrate, Timestamp, GPS, EXIF etc.) are provided for most samples. The training set will be the union of the training and public-test set (without out-of-scope samples) from the 2022 challenge [41] – i.e., 295,938 training images belonging to 1,604 species observed mostly in Denmark.

The validation and test sets include all the expert validates observations with species labels collected in 2021 and 2022. respectively. Both the validation and test set cover roughly 3,000 fungi species and include a high number of observations with *"unknown"* species. The test set was further split (50/50 ratio) to provide different data for a public and private evaluation. We list the dataset statistics in Table 3.

Table 3. FungiCLEF 2023 dataset statistics for each subset.

Subset	Species	$\rightarrow$ Known/Unknown	Images	Observations
Training	1,604	1,604 / –	295,938	177,170
Validation	2,713	1,084 / 1,629	60,832	30,131
Public Test	2,650	1,085 / 1,565	60,225	30,130
Private Test	3,299	1,116 / 2,183	91,231	45,021

4.3 Evaluation Protocol

Given the set of real fungi species observations and corresponding metadata, the goal of the task is to create a classification model that predicts a species for each given observation. The classification model must fit limits for memory footprint (max size of 1 GB) and should have to consider and minimize the danger to human life, i.e., the confusion between poisonous and edible species.

FungiCLEF 2023 considered five different decision scenarios, minimizing the empirical loss $L = \sum_i W(y_i, q(x_i))$ for decisions $q(x)$ over observations x and true labels y, given a cost function $W(y, q(x))$. Five cost functions were given for the following scenarios:

- Track 1: Standard classification with "unknown" category;
- Track 2: Cost for confusing edible species for poisonous and vice versa;
- Track 3: An application user-focused loss composed of both the classification error (e.g., accuracy) and the poisonous $\longleftrightarrow$ edible confusion;
- Track 4: Cost for missing *"unknown"* species is higher; misclassifying for *"unknown"* is cheaper than confusing species;

Baseline procedures of how metadata can help the classification, pre-trained baseline classifiers, and code submission example were provided to all participants as part of the task description.

4.4 Participants and Results

Twelve teams participated in the FungiCLEF 2023 challenge; four provided their models for a private evaluation, and three submitted working notes. Details of the best methods and systems used are synthesized in the overview working note paper of the task [38] and further developed in the individual working notes of participants (see references in [38]). In Fig. 4 and Fig. 5, we report the performance achieved by the participants. Interestingly, none of the teams that submitted working notes optimized decision-making for each of the five tasks.

The best-performing team – *meng18* – combined visual information with metadata using MetaFormer [8], tackled class imbalance with the Seesaw loss [55], proposed an entropy-guided recognition of unknown species, and introduced an additional poisonous-classification loss.

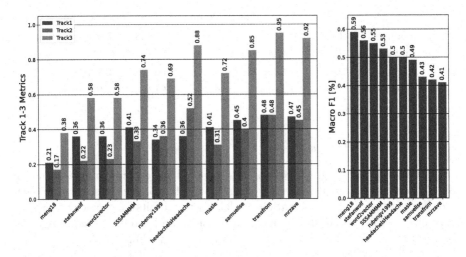

Fig. 4. Public Leaderboard – FungiCLEF 2023 competition – Top10 teams.

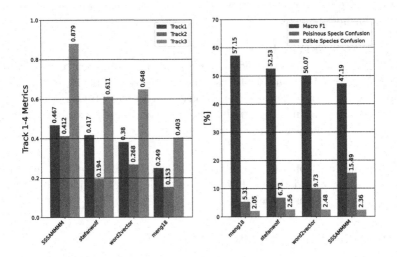

Fig. 5. Private Leaderboard – FungiCLEF 2023 competition – 4 teams.

5   PlantCLEF Challenge: Identify the World's Flora

A detailed description of the challenge and a more complete discussion of the results can be found in the dedicated working note [14].

5.1   Objective

Advancements in deep learning and the growing abundance of field photographs have significantly enhanced the automated identification of plants. A notable milestone was achieved during LifeCLEF 2018, where a top-1 classification accuracy of up to 90% was attained for over 10k species. This demonstrated that automated systems had made remarkable progress and are approaching human expertise in this domain [21]. However, it is crucial to recognize that such impressive performance levels are still a long way off from encompassing the vastness of the world's flora. Presently, science has identified approximately 391,000 vascular plant species, with new discoveries and descriptions being made each year. The significance of this plant diversity extends beyond the mere existence of species; it plays a pivotal role in ecosystem functioning and the advancement of human civilization. Regrettably, the majority of these species remain poorly understood, and there is an acute scarcity of training images available for the vast majority of them [43].

The objective of the PlantCLEF challenges in 2022 and 2023 was to advance the field of plant identification on a global scale. To achieve this, a training dataset was curated, encompassing a remarkable 80,000 species and comprising 4 million images. This expansive dataset was made accessible to the community through a challenge hosted on the AIcrowd platform[4], providing an opportunity for researchers and enthusiasts to contribute to the development of plant recognition.

5.2   Dataset

The training set consists of two distinct subsets. The first subset, referred to as the trusted training dataset, is derived from the GBIF (Global Biodiversity Information Facility) portal[5], which is the largest biodiversity data portal globally. This subset comprises over 2.9 million images encompassing 80,000 plant species. These images have been shared and collected primarily through GBIF, with some contributions from the Encyclopedia of Life[6] (EOL). The sources of these images include academic institutions such as museums, universities, and national institutions, as well as collaborative platforms like iNaturalist and Pl@ntNet, implying a fairly high certainty of determination quality (collaborative platforms only share their highest quality data qualified as "research graded"). To maintain a manageable training set size and address class imbalance, the number of images per

---

[4] https://www.aicrowd.com/challenges/lifeclef-2022-23-plant/.
[5] https://gbif.org/.
[6] https://eol.org/.

species was restricted to approximately 100. Additionally, the selection process favored specific views that are conducive to plant identification, such as close-ups of flowers, fruits, leaves, trunks, and other relevant features. This approach ensures that the training dataset comprises informative and relevant images for accurate plant recognition.

In contrast, a second "web" training dataset comprises images obtained from commercial search engines like Google and Bing. This dataset comes with its own set of challenges. The raw downloaded data from these search engines contains a notable number of species identification errors and a substantial presence of (near)-duplicates and images that are not well-suited for plant identification purposes. For instance, the dataset includes images of herbarium sheets, landscapes, microscopic views, and various other non-relevant visuals. Moreover, the web dataset contains a significant amount of unrelated images, such as portraits of botanists, maps, graphs, images from other kingdoms of living organisms, and even manufactured objects. To address these issues, a semi-automatic filtering approach was adopted. This process involved multiple iterations of training Convolutional Neural Networks (CNNs), conducting inference, and human labeling. Through this iterative process, the raw data was as best as possible cleaned up, leading to a drastic reduction in the number of irrelevant pictures. Furthermore, the image quality was improved by prioritizing close-ups of flowers, fruits, leaves, trunks, and other relevant plant features. As a result of this filtering process, the web dataset consists of approximately 1.1 million images, covering approximately 57k plant species.

Participants were allowed to use complementary training data (e.g. for pre-training purposes) but at the condition that (i) the experiment is entirely reproducible, i.e. that the used external resource is clearly referenced and accessible to any other research group in the world, (ii) the use of external training data or not is mentioned for each run, and (iii) the additional resource does not contain any of the test observations. External training data was allowed but participants had to provide at least one submission that used only the provided data.

The test set used in the PlantCLEF challenge was constructed using multi-image plant observations obtained from the Pl@ntNet platform during the year 2021. These observations had not been shared through GBIF, meaning they were not present in the training set. Only observations that received a very high confidence score in the Pl@ntNet collaborative review process were selected for the challenge to ensure the highest possible quality of determination. This process involves people with a wide range of skills (from beginners to world-leading experts), but these have different weights in the decision algorithms. Finally, the test set contains about 27k plant observations related to about 55k images (a plant can be associated with several images) covering about 7.3k species.

5.3   Evaluation Protocol

The evaluation of the task in the PlantCLEF challenge primarily relies on the Mean Reciprocal Rank (MRR) metric. MRR is a statistical measure used to assess processes that generate a list of potential responses to a set of queries, ordered by the probability of correctness. It quantifies the performance of a system by considering the reciprocal rank of the first correct answer for each query. The reciprocal rank of a query response is calculated as the multiplicative inverse of the rank of the first correct answer. In other words, if the correct answer is ranked first, the reciprocal rank is 1. If it is ranked second, the reciprocal rank is 1/2, and so on. To determine the MRR for the entire test set, the reciprocal ranks for all the queries are averaged together:

$$MRR = \frac{1}{Q} \sum_{q=1}^{Q} \frac{1}{\text{rank}_q} \tag{3}$$

where Q is the total number of query occurrences (plant observations) in the test set. However, the macro-average version of the MRR (average MRR per species in the test set - MA-MRR) was used because of the long tail of the data distribution to re-balance the results between under- and over-represented species in the test set.

5.4 Participants and Results

Although over a hundred participants signed up for the challenge, in the end only 3 participants from 3 countries participated to the PlantCLEF 2023 challenge and submitted a total of 22 runs. Details of the best methods and systems used are synthesized in the overview working notes paper of the task [14]. In Fig. 6 we report the performance achieved by the different runs of the participants.

The main outcomes we can derive from that results are the following:

- The most impressive outcomes were achieved by vision transformer-based approaches, particularly the vision-centric foundation model EVA [9], that was the state-of-the-art position during the challenge in the first quarter of 2023. While CNN-based approaches also produced respectable results, with a maximum MA-MRR of 0.618 (Neuon AI Run 9), they still fell notably short of the highest score attained by an EVA approach. The best EVA approach, Mingle Xu Run 8, achieved a remarkable MA-MRR of 0.674.
- Utilizing the complete PlantCLEF training dataset, comprising both the trusted and web datasets, proved advantageous, despite the added training time and the residual noise inherent in the web dataset. The inclusion of the web training dataset resulted in a noticeable improvement, with the MA-MRR reaching 0.674, compared to a maximum of 0.65 without it.
- The reduction of the training set by removing the classes with the fewest images (Mingle Xu Run 1-4-2-6 vs Run 5) implies a significant drop in performance. This demonstrates that there might not always be a direct connection between the training data and the test data, emphasizing the importance of considering all classes, including those linked to uncommon species, when addressing the task of monitoring plant biodiversity

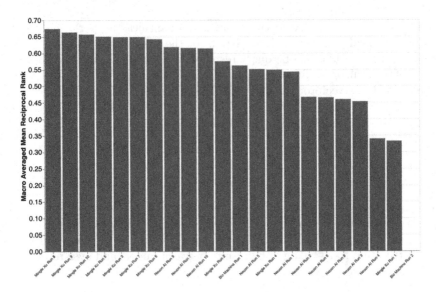

Fig. 6. PlantCLEF 2023 results

6   GeoLifeCLEF Challenge: Species Composition Prediction with High Spatial Resolution at Continental Scale Using Remote Sensing

A detailed description of the challenge and a more complete discussion of the results can be found in the dedicated working note [5]. A graphical abstract of the challenge is provided in Fig. 7.

6.1   Objective

Predicting which species are present in a given area through Species Distribution Models (SDM) is a central problem in ecology and a crucial issue for biodiversity conservation. Such predictions are a fundamental element of many decision-making processes, whether for land use planning, the definition of protected areas, or the implementation of more ecological agricultural practices. Classical SDMs are well-established but have the drawback of covering only a limited number of species at spatial resolutions often coarse in the order of kilometers, or hundreds of meters at best. In addition, while the use of the massive presence-only data arising from large citizen science platforms has grown, the SDM built from such data are affected by many sampling biases, as, for instance, species detection bias or species set size bias. Developing scalable methods suited to account and correct for these biases is a necessary step to update regularly species distributions maps by capitalizing on the massive flow of citizen science data. The objective of GeoLifeCLEF is to evaluate models with orders of magnitude hitherto unseen, whether in terms of the number of species covered

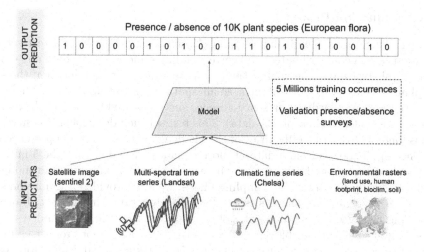

Fig. 7. GeoLifeCLEF 2023 graphical abstract

(thousands), spatial resolution (on the order of 10 m), or the number of occurrences used as training data (several million). These models have the potential to greatly improve biodiversity management processes, especially at the local level (e.g. municipalities), where the need for spatial and taxonomic precision is greatest.

6.2 Training Dataset

A brand new dataset was built for the 2023 edition of GeoLifeCLEF in the framework of a large-scale European project on biodiversity monitoring (MAMBO, Horizon EU program). It contains about 5 million plant species presence-only records (single positive labels, hereafter PO) covering 10 thousand species extracted from thirteen selected datasets of the Global Biodiversity Information Facility (GBIF) and covers the whole EU territory (38 countries including E.U. members). We also provided the participants with a validation set of 5 thousand standardized presence-absence (hereafter PA) surveys of small spatial plots (multi-label) to help calibrate the models, and specifically to correct for sampling biases. For the explanatory variables (to be used as inputs of the models), the dataset contains both high-resolution remote sensing data (10m resolution Sentinel-2 satellite images and Landsat multi-spectral time-series at each data location, along with elevation) and coarser resolution environmental raster data (land cover, human footprint, bioclimatic and soil variables). The geo-coordinates and date of the species observations are also provided and can also be used as one modality. Participants are free to use one, several, or all available modalities in their models. The detailed description of the GeoLifeCLEF 2023 dataset is provided in [4].

6.3   Evaluation Protocol

The challenge is as a multi-label (/set) classification task. Given a test set of
locations (i.e., geo-coordinates) and corresponding remote sensing data and environmental covariates, the goal of the task is to return for each location the set
of plant species truly present in a small spatial plot (of area 10–400 m^2) as
reported in a standardized presence-absence survey carried by botanical experts
(same type as the validation PA data). This test set includes 22,404 PA surveys.
Thus, one of the major difficulties of the challenge is to predict the presence or
absence of all species from a dataset mostly made of PO data (i.e. the 5 million
GIF records). As noted earlier, to enable participants to calibrate their models,
and specifically to correct for sampling biases, we also provided a validation set
of only 5 thousand PA surveys, spatially separated from the test set. Indeed,
following the recommendations of [46], the split of the validation and test set
was done using a spatial blocking strategy that enables a more robust estimation of the model's performance (based on a 50 × 50 km spatial grid). Moreover,
we have excluded the PO records located less than 500 m from the test plots
to avoid the risk that some may have originated from these plots. The detailed
protocol is described in [4].

The evaluation metric is the F1 score. It measures the precision and recall
score for each test plot x and computes their harmonic mean:

$$F1(x) = \frac{2}{\frac{1}{Precision(x)} + \frac{1}{Recall(x)}}$$

It is equivalent to the Sørensen-Dice coefficient defined as the size of the
intersection between the predicted and true set of species, divided by the mean
of their respective size.

The final global metric is calculated by averaging the F1 score of all plots in
the test set.

6.4   Participants and Results

Six participants from four countries participated in the GeoLifeCLEF 2023 challenge and submitted a total of 121 entries (i.e. /textitruns). Details of the best
methods and systems used are synthesized in the overview paper of the task [5]
and the winning team methodology is explained in details in their working note
([51]). In Table 4 we report the performance achieved by the best performing
methods of the participants as well as the baseline methods developed by the
organizers. Hereafter, we briefly describe those different methods:

Participant's Methods

- **KDDI research**: This team trained various convolutional neural networks,
  all based on the ResNet backbone (ResNet34 and 50). One of the CNN was
  trained solely on the 19 bioclimatic rasters while others were multi-modal

networks with a late fusion layer to merge the different modalities used (see Table 4). The best performing run was an ensemble of the best models based on a simple average of their output. The best models were trained in three steps, firstly on the PA plots with a binary cross-entropy loss, then fine-tuned on the PO records with a cross-entropy loss, and finally fine-tuned again on the PA with the binary-cross-entropy loss. This team carried an ablation showing the importance of these three steps.

– **Jiexun Xu**: This researcher focused on the tabular environmental data only, i.e. he didn't use the spatial structure of the environmental co-variates nor the remotes sensing images and times series. The model used is XGBoost and it was trained on the PA plots. He also added the one-hot encoded species presences in GBIF in a 1km radius of these plots as input variables.
– **Lucas Morin**: This researcher optimized a K-Nearest Neighbor predictor using only the spatial coordinates and the PA plots.
– **QuantMetry**: This team trained various models on the PA data, and their best scoring model was a ResNet50 using only the Sentinel2 satellite images (RGB+NIR) as input. The model was pre-trained on the satellite images in a prior work ([60]) and fine-tuned to the PA data in the challenge.
– **Nina van Tiel**: This researcher used a small CNN, with two convolutional layers and two fully connected layers on the RGB images, along the bioclimatic, soil and land-cover rasters, trained on the PA plots.
– **Ousmane Youme**: This researcher focused solely on the Landsat time series data at the location of the PA plots. He used a Conv1D neural network model with a binary-cross entropy loss. A common probability threshold was used to convert the predicted species-wise presence probabilities into a set of predicted species.

Organizer's Baselines

– **MAXENT**: the MAXENT method is a modeling approach widely used in ecology to predict the distribution of a given species based on tabular environmental variables. It is not adapted to handle complex input data such as the Sentinel images or Landsat time series. The model creates a pre-defined set of non-linear transformations of the input environmental variables consistent with the theoretical ecological response of species to environmental gradients (e.g. quadratic and threshold responses, see [36]). The statistical model is equivalent to a Poisson regression modeling the count of a species per location ([45]). We fitted one Maxent model per species present in the PA plots. The species count was set to one when present or zero otherwise. The environmental input variables included were the climate, soil, land cover and human footprint variables, but only a subset of these variables were included for species with a smal number of observations. One random subset of the PA plots was used to train all species models while the other was used to assess the predictive accuracy of each species model. We thus determined that it was optimal to keep only the 391 most trustable species, in terms of validation score, for the final prediction, the left-out species being always predicted absent. A run including on all species models was also submitted, achieving a much lower performance due to an over-prediction of rare species in extrapolation (see [5] for details).

- **Environmental Random Forest**: Random forests are also widely used in ecology to predict the distribution of species based on a set of environmental variables. As for Maxent, the Env. Random Forest models were trained only on the environmental tabular variables at the location of the PA plots. One Random Forest was trained per species in the PA plots and its hyperparameters were optimized through a cross-validation grid search.
- **Spatial Random Forest**: Contrary to the two previous baseline, this Random Forest were trained solely on the spatial coordinates of the PA plots, regardless the environmental variables.
- **Species co-occurrence**: Conditionally to the presence of each species, we computed the proportion of presences of all other species among the PA plots. Then, for each test location, we combined the species probabilities conditionally to the species observed in the PO data in a 1km-radius into a predicted species set through a weighted average. Therefore, this method doesn't use any input variable except the spatial coordinates.
- **Constant predictor**: this baseline always predict the same set of species, i.e. the ones that are the K most frequent in the PA plots, where K maximizes the F1-micro score over these PA plots ($K = 25$ species).

Outcomes. The main outcomes we can derive from the challenge are the following:

- The problem remains very difficult and the best model only achieves a F1-score of 0.27
- The MAXENT method remains a strong baseline when considering only the tabular environmental data, regardless the spatial structure of the environment or the more complex data such as remote sensing images.
- Training a model on the PO data (with a cross-entropy loss) and fine-tuning it on PA (with a binary cross-entropy loss) resulted in a considerable performance gain. This shows the wealth of information that can be mobilised in the PO data, provided that the learning strategy avoids sampling biases.
- The best model was based on a Convolutional Neural Network which confirms that this kind of model is relevant for the task. It allows capturing complex patterns in the input data while allowing elaborated training strategy such as transfer learning.
- Making use exclusively of PO data remains a major hurdle, and all the methods that did so had a very low performance. Most participants used only the PA validation data in the training of their models, and the best method succeeded by combining both. Much work lies ahead to extract the information from PO without complementary standardized data, if that is even possible.

Table 4. Overview of the results of GeoLifeCLEF 2023 challenge - the acronyms PA and PO respectively stand for (PA) Presence/Absence: meaning that the plots in the validation set were used to fit the model, and (PO) Presence-Only which means that the GBIF occurrences of the training set were used to fit the model.

Team/scientist	Method	Used data	Used modalities	Score
KDDI research	Ensemble of CNN models Multimodal & bioclim	PO & PA	Sentinel-2 RGB-NIR, soil, bio-climatic, human footprint	0.270
KDDI research	Multi-modal CNN (3 x ResNet-50)	PO & PA	Sentinel-2 RGB-NIR, soil, bio-climatic, human footprint	0.249
KDDI research	Bioclim CNN (ResNet-50 w/ 19 channels)	PO & PA	bio-climatic	0.239
Organizer (baselines)	MAXENT (391 most confident species)	PA	soil, bio-climatic, human footprint	0.224
Jiexun Xu	XGBoost	PO & PA	soil, bio-climatic, human footprint	0.223
Lucas Morin	K-Nearest Neighbors (K=500)	PA	lat. / long.	0.208
Quantmetry	ResNet50	PA	Sentinel-2 RGB-NIR	0.206
Organizer (baselines)	Spatial Random Forest	PA	lat. / long.	0.191
Organizer (baselines)	Env. Random Forest	PA	soil, bio-climatic, human footprint	0.188
Organizer (baselines)	Species co-occurrence	PA/PO	lat./long.	0.167
Organizer (baselines)	Constant predictor	PA	NONE	0.160
Nina van Tiel	Small CNN	PA	Sentinel-2 RGB, soil, bio-climatic, human footprint	0.158
Ousmane Youm	Conv1d CNN	PA	Landsat time series	0.134

7 Conclusions and Perspectives

The main outcome of this collaborative evaluation is a new snapshot of the performance of state-of-the-art computer vision, bioacoustic, and machine learning techniques toward building real-world biodiversity monitoring systems. Overall, this study shows that the field continues to progress year after year, and that, although the challenges that are most closely related to common tasks, such as multi-class classification based on images, are able to profit from the most recent advances in computer vision, certain problems are still wide open, such as the prediction of species as a function of location (as part of the GeoLifeCLEF challenge). In terms of the methods used, the results show that convolutional neural networks are still a very powerful method for image and sound processing. In 4 of the 5 challenges, the best results were obtained using CNNs. Only the PlantCLEF challenge obtained much better results (for the identification of plants from images) with the use of foundation vision transformer models such as EVA [9]. The best submission to FungiCLEF was based on MetaFormer [8], utilizing both a convolutional backbone and a transformer to fuse visual and meta information. Complementary to vision-based models, NLP models were also used successfully, in particular hybrid models such as CLIP [44] that efficiently learn visual concepts from natural language supervision. We believe that this principle of combining different modalities in the training of deep learning models will be a key to future progress in AI for biodiversity.

Acknowledgements. The research described in this paper was partly funded by the European Commission via the GUARDEN and MAMBO projects, which have received funding from the European Union's Horizon Europe research and innovation program under grant agreements 101060693 and 101060639. The opinions expressed in this work are those of the authors and are not necessarily those of the GUARDEN or MAMBO partners or the European Commission.

References

1. Overview of SnakeCLEF 2023: Snake identification in medically important scenarios. In: Working Notes of CLEF 2023 - Conference and Labs of the Evaluation Forum (2023)
2. Banan, A., Nasiri, A., Taheri-Garavand, A.: Deep learning-based appearance features extraction for automated carp species identification. Aquacult. Eng. **89**, 102053 (2020)
3. Besson, M., et al.: Towards the fully automated monitoring of ecological communities. Ecol. Lett. **25**(12), 2753–2775 (2022)
4. Botella, C., et al.: The GeoLifeCLEF 2023 dataset to evaluate plant species distribution models at high spatial resolution across Europe. XXXX (2023)
5. Botella, C., et al.: Overview of GeoLifeCLEF 2023: Species composition prediction with high spatial resolution at continental scale using remote sensing. In: Working Notes of CLEF 2023 - Conference and Labs of the Evaluation Forum (2023)
6. Cai, J., Ee, D., Pham, B., Roe, P., Zhang, J.: Sensor network for the monitoring of ecosystem: Bird species recognition. In: Intelligent Sensors, Sensor Networks and Information, 2007. ISSNIP 2007. 3rd International Conference on (2007). https://doi.org/10.1109/ISSNIP.2007.4496859
7. Chen, T., Kornblith, S., Norouzi, M., Hinton, G.: A simple framework for contrastive learning of visual representations. In: International Conference on Machine Learning, pp. 1597–1607. PMLR (2020)
8. Diao, Q., Jiang, Y., Wen, B., Sun, J., Yuan, Z.: MetaFormer: a unified meta framework for fine-grained recognition. arXiv preprint arXiv:2203.02751 (2022)
9. Fang, Y., et al.: EVA: exploring the limits of masked visual representation learning at scale (2022)
10. Gaston, K.J., O'Neill, M.A.: Automated species identification: why not? Philos. Trans. Royal Soc. Lond. B: Biol. Sci. **359**(1444), 655–667 (2004)
11. Ghazi, M.M., Yanikoglu, B., Aptoula, E.: Plant identification using deep neural networks via optimization of transfer learning parameters. Neurocomputing **235**, 228–235 (2017)

12. Glotin, H., Clark, C., LeCun, Y., Dugan, P., Halkias, X., Sueur, J.: Proceedings of the 1st workshop on Machine Learning for Bioacoustics - ICML4B. ICML, Atlanta USA (2013). https://sabiod.org/ICML4B2013_book.pdf

13. Glotin, H., LeCun, Y., Artières, T., Mallat, S., Tchernichovski, O., Halkias, X.: Neural Information Processing Scaled for Bioacoustics, from Neurons to Big Data. Proceedings of the NIPS International Conference (2013). https://sabiod.org/nips4b

14. Goëau, H., Bonnet, P., Joly, A.: Overview of PlantCLEF 2023: image-based plant identification at global scale. In: Working Notes of CLEF 2023 - Conference and Labs of the Evaluation Forum (2023)

15. Goëau, H., et al.: The imageCLEF 2013 plant identification task. In: CLEF task overview 2013, CLEF: Conference and Labs of the Evaluation Forum, September 2013, Valencia, Spain. Valencia (2013)

16. Goëau, H., et al.: The imageCLEF 2011 plant images classification task. In: CLEF Task Overview 2011, CLEF: Conference and Labs of the Evaluation Forum, September 2011, Amsterdam, Netherlands (2011)

17. Goëau, H., et al.: ImageCLEF 2012 plant images identification task. In: CLEF Task Overview 2012, CLEF: Conference and Labs of the Evaluation Forum, September 2012, Rome, Italy. Rome (2012)

18. He, K., Zhang, X., Ren, S., Sun, J.: Deep residual learning for image recognition. In: Proceedings of the IEEE Conference on Computer Vision and Pattern Recognition, pp. 770–778 (2016)

19. Ilse, M., Tomczak, J., Welling, M.: Attention-based deep multiple instance learning. In: International conference on machine learning, pp. 2127–2136. PMLR (2018)

20. Joly, A., et al.: Interactive plant identification based on social image data. Eco. Inform. **23**, 22–34 (2014)

21. Joly, A., et al.: Overview of LifeCLEF 2018: a large-scale evaluation of species identification and recommendation algorithms in the era of AI. In: Bellot, P., et al. (eds.) CLEF 2018. LNCS, vol. 11018, pp. 247–266. Springer, Cham (2018). https://doi.org/10.1007/978-3-319-98932-7_24

22. Joly, A., et al.: Overview of LifeCLEF 2019: identification of Amazonian plants, South & North American birds, and Niche prediction. In: Crestani, F., et al. (eds.) CLEF 2019. LNCS, vol. 11696, pp. 387–401. Springer, Cham (2019). https://doi.org/10.1007/978-3-030-28577-7_29. https://hal.umontpellier.fr/hal-02281455

23. Joly, A., et al.: LifeCLEF 2016: multimedia life species identification challenges. In: Fuhr, N., et al. (eds.) CLEF 2016. LNCS, vol. 9822, pp. 286–310. Springer, Cham (2016). https://doi.org/10.1007/978-3-319-44564-9_26. https://hal.archives-ouvertes.fr/hal-01373781

24. Joly, A., et al.: LifeCLEF 2017 lab overview: multimedia species identification challenges. In: Jones, G.J.F., et al. (eds.) CLEF 2017. LNCS, vol. 10456, pp. 255–274. Springer, Cham (2017). https://doi.org/10.1007/978-3-319-65813-1_24. https://hal.archives-ouvertes.fr/hal-01629191

25. Joly, A., et al.: LifeCLEF 2014: multimedia life species identification challenges. In: Kanoulas, E., et al. (eds.) CLEF 2014. LNCS, vol. 8685, pp. 229–249. Springer, Cham (2014). https://doi.org/10.1007/978-3-319-11382-1_20. https://hal.inria.fr/hal-01075770

26. Joly, A., et al.: LifeCLEF 2015: multimedia life species identification challenges. In: Mothe, J., et al. (eds.) CLEF 2015. LNCS, vol. 9283, pp. 462–483. Springer, Cham (2015). https://doi.org/10.1007/978-3-319-24027-5_46

27. Joly, A., et al.: Overview of LifeCLEF 2020: a system-oriented evaluation of auto-mated species identification and species distribution prediction. In: Arampatzis, A., et al. (eds.) CLEF 2020. LNCS, vol. 12260, pp. 342–363. Springer, Cham (2020). https://doi.org/10.1007/978-3-030-58219-7_23

28. Joly, A., et al.: Overview of lifeCLEF 2022: an evaluation of machine-learning based species identification and species distribution prediction. In: Barrón-Cedeño, A., et al. (eds.) CLEF 2022. LNCS, vol. 13390, pp. 257–285. Springer, Cham (2022). https://doi.org/10.1007/978-3-031-13643-6_19

29. Joly, A., et al.: Overview of LifeCLEF 2021: an evaluation of machine-learning based species identification and species distribution prediction. In: Candan, K.S., et al. (eds.) CLEF 2021. LNCS, vol. 12880, pp. 371–393. Springer, Cham (2021). https://doi.org/10.1007/978-3-030-85251-1_24

30. Kahl, S., et al.: Overview of BirdCLEF 2023: automated bird species identification in Eastern Africa. In: Working Notes of CLEF 2023 - Conference and Labs of the Evaluation Forum (2023)

31. Lee, D.J., Schoenberger, R.B., Shiozawa, D., Xu, X., Zhan, P.: Contour matching for a fish recognition and migration-monitoring system. In: Optics East, pp. 37–48. International Society for Optics and Photonics (2004)

32. Lee, S.H., Chan, C.S., Remagnino, P.: Multi-organ plant classification based on convolutional and recurrent neural networks. IEEE Trans. Image Process. **27**(9), 4287–4301 (2018)

33. Liu, Z., et al.: Swin transformer: hierarchical vision transformer using shifted win-dows. In: Proceedings of the IEEE/CVF International Conference on Computer Vision, pp. 10012–10022 (2021)

34. Norouzzadeh, M.S., Morris, D., Beery, S., Joshi, N., Jojic, N., Clune, J.: A deep active learning system for species identification and counting in camera trap images. Meth. Ecol. Evol. **12**(1), 150–161 (2021)

35. Ovalle, J.C., Vilas, C., Antelo, L.T.: On the use of deep learning for fish species recognition and quantification on board fishing vessels. Mar. Policy **139**, 105015 (2022)

36. Phillips, S.J., Dudík, M.: Modeling of species distributions with maxent: new exten-sions and a comprehensive evaluation. Ecography **31**(2), 161–175 (2008)

37. Picek, L., Šulc, M., Chamidullin, R., Durso, A.M.: Overview of snakeCLEF 2023: snake identification in medically important scenarios. In: CLEF 2023-Conference and Labs of the Evaluation Forum (2023)

38. Picek, L., Šulc, M., Chamidullin, R., Matas, J.: Overview of fungiCLEF 2023: fungi recognition beyond 1/0 cost. In: CLEF 2023-Conference and Labs of the Evaluation Forum (2023)

39. Picek, L., Šulc, M., Matas, J., Heilmann-Clausen, J., Jeppesen, T.S., Lind, E.: Automatic fungi recognition: deep learning meets mycology. Sensors **22**(2), 633 (2022)

40. Picek, L., et al.: Danish fungi 2020-not just another image recognition dataset. In: Proceedings of the IEEE/CVF Winter Conference on Applications of Computer Vision, pp. 1525–1535 (2022)

41. Picek, L., Šulc, M., Heilmann-Clausen, J., Matas, J.: Overview of FungiCLEF 2022: fungi recognition as an open set classification problem. In: Working Notes of CLEF 2022 - Conference and Labs of the Evaluation Forum (2022)

42. Picek, L., Šulc, M., Heilmann-Clausen, J., Matas, J.: Overview of FungiCLEF 2023: fungi recognition beyond 0–1 cost. In: Working Notes of CLEF 2023 - Conference and Labs of the Evaluation Forum (2023)

43. Pitman, N.C., et al.: Identifying gaps in the photographic record of the vascular plant flora of the Americas. Nat. Plants **7**(8), 1010–1014 (2021)

44. Radford, A., et al.: Learning transferable visual models from natural language supervision. In: International Conference on Machine Learning, pp. 8748–8763. PMLR (2021)

45. Renner, I.W., Warton, D.I.: Equivalence of maxent and Poisson point process models for species distribution modeling in ecology. Biometrics **69**(1), 274–281 (2013)

46. Roberts, D.R., et al.: Cross-validation strategies for data with temporal, spatial, hierarchical, or phylogenetic structure. Ecography **40**(8), 913–929 (2017)

47. Sulc, M., Picek, L., Matas, J., Jeppesen, T., Heilmann-Clausen, J.: Fungi recognition: a practical use case. In: Proceedings of the IEEE/CVF Winter Conference on Applications of Computer Vision, pp. 2316–2324 (2020)

48. Tan, M., Le, Q.: EfficientNet: rethinking model scaling for convolutional neural networks. In: International Conference on Machine Learning, pp. 6105–6114. PMLR (2019)

49. Towsey, M., Planitz, B., Nantes, A., Wimmer, J., Roe, P.: A toolbox for animal call recognition. Bioacoustics **21**(2), 107–125 (2012)

50. Trifa, V.M., Kirschel, A.N., Taylor, C.E., Vallejo, E.E.: Automated species recognition of antbirds in a Mexican rainforest using hidden Markov models. J. Acoust. Soc. Am **123**, 2424 (2008)

51. Ung, H.Q., Kojima, R., Wada, S.: Leverage samples with single positive labels to train CNN-based models for multi-label plant species prediction. In: Working Notes of CLEF 2023 - Conference and Labs of the Evaluation Forum (2023)

52. Van Horn, G., et al.: The iNaturalist species classification and detection dataset. In: CVPR (2018)

53. Villon, S., Mouillot, D., Chaumont, M., Subsol, G., Claverie, T., Villéger, S.: A new method to control error rates in automated species identification with deep learning algorithms. Sci. Rep. **10**(1), 1–13 (2020)

54. Wäldchen, J., Rzanny, M., Seeland, M., Mäder, P.: Automated plant species identification-trends and future directions. PLoS Comput. Biol. **14**(4), e1005993 (2018)

55. Wang, J., et al.: Seesaw loss for long-tailed instance segmentation. In: Proceedings of the IEEE/CVF Conference on Computer Vision and Pattern Recognition, pp. 9695–9704 (2021)

56. Wang, Y., Zhang, Y., Feng, Y., Shang, Y.: Deep learning methods for animal counting in camera trap images. In: 2022 IEEE 34th International Conference on Tools with Artificial Intelligence (ICTAI), pp. 939–943. IEEE (2022)

57. Woo, S., et al.: ConvNeXt v2: Co-designing and scaling convnets with masked autoencoders. arXiv preprint arXiv:2301.00808 (2023)

58. Yu, X., Wang, J., Kays, R., Jansen, P.A., Wang, T., Huang, T.: Automated identification of animal species in camera trap images. EURASIP J. Image Video Process. **2013**, 1–10 (2013)

59. Yuan, L., Hou, Q., Jiang, Z., Feng, J., Yan, S.: VOLO: vision outlooker for visual recognition. IEEE Trans. Pattern Anal. Mach. Intell. **45**, 6575–6586 (2022)

60. Zheng, Z., Ma, A., Zhang, L., Zhong, Y.: Change is everywhere: single-temporal supervised object change detection in remote sensing imagery. In: Proceedings of the IEEE/CVF International Conference on Computer Vision, pp. 15193–15202 (2021)

Overview of the CLEF-2023 LongEval Lab on Longitudinal Evaluation of Model Performance

Rabab Alkhalifa[1,2], Iman Bilal[3], Hsuvas Borkakoty[4],
Jose Camacho-Collados[4], Romain Deveaud[6], Alaa El-Ebshihy[9],
Luis Espinosa-Anke[4,12], Gabriela Gonzalez-Saez[7], Petra Galuščáková[7],
Lorraine Goeuriot[7], Elena Kochkina[1,5], Maria Liakata[1,3,5],
Daniel Loureiro[4], Philippe Mulhem[7(✉)], Florina Piroi[9],
Martin Popel[10], Christophe Servan[6,11], Harish Tayyar Madabushi[8],
and Arkaitz Zubiaga[1]

[1] Queen Mary University of London, London , UK
[2] Imam Abdulrahman Bin Faisal University, Dammam, Saudi Arabia
[3] University of Warwick, Coventry, UK
[4] Cardiff University, Cardiff, UK
[5] Alan Turing Institute, London, UK
[6] Qwant, Paris, France
[7] Univ. Grenoble Alpes, CNRS, Grenoble INP
(Institute of Engineering Univ. Grenoble Alpes.), LIG, Grenoble, France
`Philippe.Mulhem@imag.fr`
[8] University of Bath, Bath, UK
[9] Research Studios Austria, Data Science Studio, Vienna, Austria
[10] Charles University, Prague, Czech Republic
[11] Paris-Saclay University, CNRS, LISN, Orsay, France
[12] AMPLYFI, Cardiff, UK

Abstract. We describe the first edition of the LongEval CLEF 2023 shared task. This lab evaluates the temporal persistence of Information Retrieval (IR) systems and Text Classifiers. Task 1 requires IR systems to run on corpora acquired at several timestamps, and evaluates the drop in system quality (NDCG) along these timestamps. Task 2 tackles binary sentiment classification at different points in time, and evaluates the performance drop for different temporal gaps. Overall, 37 teams registered for Task 1 and 25 for Task 2. Ultimately, 14 and 4 teams participated in Task 1 and Task 2, respectively.

Keywords: Evaluation · Temporal Persistence · Temporal Generalisability

1 Introduction

Datasets collected across different time periods can vary in several aspects, including the language used, the data format, as well as other structural changes.

A. Arampatzis et al. (Eds.): CLEF 2023, LNCS 14163, pp. 440–458, 2023.
https://doi.org/10.1007/978-3-031-42448-9_28

Time is however a dimension that is often overlooked when conducting experiments with static datasets. As recent research has demonstrated, however, models trained on data pertaining to a particular time period struggle to keep their performance levels when applied on test data that is distant in time. This has been shown to be the case for information retrieval (IR) systems as well as for text classification models [3].

With the aim of tackling this challenge of making models persistent over time, the objective of the LongEval lab is twofold: (i) to explore the extent to which the evolution of evaluation datasets deteriorates performance of information retrieval and classification systems, and (ii) to propose improved methods that mitigate performance drop by making models more robust over time.

The LongEval lab took place as part of the Conference and Labs of the Evaluation Forum (CLEF) 2023, and consisted in two separate tasks: (i) Task 1, focused on information retrieval, and (ii) Task 2, focused on text classification for sentiment analysis. Both tasks provided labeled datasets enabling analysis and evaluation of models over longitudinally evolving data.

In what follows, we describe the datasets, experiment settings as well as final results for each of these two tasks.

2 Task 1 - Retrieval

The goal of the retrieval task is to explore the effect of changes in datasets on retrieval of text documents. More specifically, we focus on a setup in which the datasets are evolving. This means, that one dataset can be acquired from another by adding, removing (and replacing) a limited number of documents and queries. We explore two main scenarios and the setup of the task thus reflects the details of these two problems.

A Single System in An Evolving Setup
We explore how one selected system behaves if we evaluate it using several collections, which evolve across the time. Specifically, we explore the effect of changes in datasets on retrieval performances in a **Web search** domain. In this domain, the documents, queries and also the perception of relevance naturally continuously evolves and Web search engines need to deal with this situation. The evaluation in this scenario is thus very specific and should take into account the evolving nature of the data. Evaluation should ideally reflect the changes in the collection and especially signal substantial changes that could lead to performance drop. This would allow to re-train the search engine model, exactly when it is really needed, and enable much more efficient overall training.

This problem emerges also with the popularity of neural networks. The stability of the performance of the neural networks seems to be lower than in the case of the statistical model. Moreover, the performance strongly depends on the data used for training the neural model. One objective of the task is to explore the behavior of the neural system in the evolving data scenario.

Comparison of Multiple Systems in An Evolving Setup
While in the first point, we explore a single system, comparison of this systems

with multiple systems across evolving collections, should provide more information about systems stability and robustness.

2.1 Description of the Task

The task datasets were created over sequential time periods, which allows doing observations at different time stamps t, and most importantly, comparing the performance across different time stamps t and t'. Two sub-tasks are organized as follows:

A) Short-term (ST) Persistence task that aim to assess the performance difference between t and t' when t' occurs right after or shortly after t
B) Long-term (LT) Persistence task that aim to examine the performance difference between two t and t'', when t'' occurs several months after t (and thus $|t'' - t| > |t' - t|$).

In addition to this, we provide Within-time (WT) dataset, which contains the same documents (but different queries) as the training data. This data are used as a control group and applied to measure a change against the training data.

2.2 Dataset

Data for this task were provided by the French search engine Qwant. They consist of the queries issued by the users of this search engine, cleaned Web documents, which were 1) selected to correspond to the queries, and 2) to add additional noise, and relevance judgments, which were created using a click model. The dataset is fully described in [5]. We provided training data, which included 672 train queries, with corresponding 9,656 assessments and 1,570,734 Web pages. In addition to this, the training data included the 98 heldout WT queries. All training and heldout data were collected during June 2022. Test data were split into two collections, each corresponding to a single sub-task. The data for the short-term persistence sub-task was collected over July 2022 and this dataset contains 1,593,376 documents and 882 queries. The data for the long-term persistence sub-task was collected over September 2022 and this dataset consists of 1,081,334 documents and 923 queries. All the datasets are freely available at Lindat/Clarin. As the data were initially collected by French search engine and are all in French, we also provide automatic English translations of both queries and documents.

Though online evaluation is more frequent in Web search scenarios, we focus on offline evaluation, which allows us to make the collection re-usable. However, we use two different relevance judgments: the judgments acquired by the click model, based on the raw clicks of the users; and manual relevance judgment on a pooled subset. This allows us to interconnect the advantages of offline and online evaluation approaches. As the manual evaluations are ongoing, in this paper we only report the relevance judgments acquired from the click model.

For evaluating both subtasks, we use the NDCG measure (calculated for each dataset), as well as the drop between the ST and LT collection against the training data (WT collection).

2.3   Submissions

In total 14 teams submitted their systems to the Retrieval task. 12 of these teams submitted the results into both Short-term and Long-term retrieval sub-tasks, two teams only submitted the results for the Short-term retrieval sub-tasks. As per the requirements, all participating teams needed to submit their systems also on the within-time dataset, which was created at the same dataframe as the training data, which allows measuring relative drop between the datasets. All teams, except one, which submitted 4 systems, decided to submit 5 systems. Together, with 4 baseline runs provided by the Université Grenoble Alpes (marked as UGA), this creates a pool of 73 systems available on the within-time (WT, corresponding to the Heldout queries runs on the Train corpus) and short-term (ST) collections and 63 systems available on the long-term collection.

2.4   Absolute Scores

The overview of NDCG and MAP scores for each submitted run on different datasets (WT, ST, LT) is presented in Table 1. In this table, one column indicates, for each run, which language was used (English, French, or both), whether any neural approach (yes/no) was involved, and whether a combination of several approaches (yes/no) or a single approach was used.

From Table 1, we see that the systems which are the best for the WT data are also among the top for the ST and LT datasets. For instance, the best system in the WT according to the NDCG measure (FADERIC_Fr-BM25-S50-LS-S-F-SC-R20W6), is ranked best also for ST, and considering the systems that did get non-zero evaluation for the two tasks, the best system for NDCG in WT, SQUID_SEARCHERAI, is also the best on ST and LT datasets. This finding does not hold for the MAP measures: considering the systems that participated to the two tasks, the best system for MAP in WT, CLOSE_SBERT_BM25, is the second best on the ST dataset and the fourth best on the LT dataset. An explanation may come from the fact that the NDCG emphasizes on the top ranked documents of the runs.

We describe now the methods used in the top-3 runs, according to the NDCG evaluation measure, for each WT, ST and LT. For the WT Dataset Heldout queries, the top systems are:

1. CLOSE_SBERT_BM25 from the CLOSE team: The system uses query variant generated from GPT using dedicated prompts, and applies sentence BERT to rerank the initial BM25 results.
2. gwca_lightstem-phrase-qexp from de GWCA team: this systems uses a French stoplist and stemmer, a query expression is composed of the original text, phrases extracted from the query, and text generated using GPT 3.5.
3. SQUID_SEARCHERAI from the Squid team: this systems relies on Lucene indexing and searcher on French documents and queries. It uses several fields for the documents (title/url/body) with different boost values, and expands the queries with synonyms from GPT 3.5.

For the ST Dataset, the top-3 systems are:

1. FADERIC_Fr-BM25-S50-LS-S-F-SC-R20W6 from the FADERIC team. The matching is based on BM25, fine-tuned on the training set. The query processing use the Lucene *fuzzy* matching, able to allow partial match of words, and integrate synomyms expansion. A reranking fuses linearly the BM25 scores and BERT for the 20 top BM25 documents. Though the runs from the FADERIC team achieve the highest NDCG scores on the ST collection, unfortunately the scores achieved on the LT collection is zero, presumably due to an error.

2. FADERIC_Fr-BM25-S50-LS-S-F-R30 from the FADERIC team. This run is similar to the one above, the differences rely on the number of document reranked (here 30) and a different weight of BM25 score in the linear combination.

3. SQUID_SEARCHERAI from the Squid team, already described above.

For the LT Dataset, the top-3 systems are:

1. CLOSE_SBERT_BM25 from the CLOSE team, already described;

2. SQUID_W2V from the Squid team: this system relies on Lucene indexing and searcher on french documents and queries. It uses several fields for the documents (title/url/body) with different boost values, and expands the queries with word2Vec similar terms.

3. SQUID_SEARCHERAI from the Squid team, already described above.

Thus, the best approaches all rely to some extent on query expansion techniques, and integrate at one point or another embeddings or Large Language Models. The best results use French documents and queries. The effect of the translation provided by the lab has a clear impact. This remark is exemplified by the *UGA* baselines: the UGA_BM25_French outperforms the UGA_BM25_English default, and similarly the reranking using T5 French run (UGA_T5_French) outperforms its English counterpart (UGA_T5_English).

2.5   Changes in the Scores

The main part of the task is to see the changes in the scores between the collections. All collections were created using the same approach and procedure and have a high overlap in terms of both queries and documents. In Table 2, we thus provide the relative drops between the collections ST and WT and between the collections LT and WT. The definition of the value "WT-ST" NDCG change is defined, for a run r as:

$$\frac{\mathrm{NDCG}_{WT}(r) - \mathrm{NDCG}_{ST}(r)}{\mathrm{NDCG}_{WT}(r)}$$

Table 1. NDCG and MAP scores for three test datasets (WT, ST, LT). Results are sorted according to the NDCG scores achieved on the ST dataset.

System	Neural	Comb.	Language	NDCG WT	NDCG ST	NDCG LT	MAP WT	MAP ST	MAP LT
FADERIC_Fr-BM25-S50-LS-S-F-SC-R20W6	yes	no	French	0.4169	0.4239	0	0.2474	0.2665	0
FADERIC_Fr-BM25-S50-LS-S-F-R30	yes	no	French	0.4147	0.4145	0	0.2416	0.2546	0
SQUID_SEARCHERAI	yes	no	French	0.4279	0.4141	0.4177	0.2594	0.2554	0.2473
CLOSE_SBERT_BM25	yes	yes	French	0.4318	0.4128	0.4139	0.2675	0.2531	0.2432
gwca_lightstem-phrase-qexp	no	no	French	0.4294	0.4114	0.4161	0.2524	0.2475	0.2453
SQUID_W2V	yes	no	French	0.4232	0.4106	0.4174	0.2583	0.2497	0.2444
CLOSE_RERANKING	yes	yes	French	0.4166	0.4068	0.4062	0.2595	0.2508	0.2383
FADERIC_Fr-BM25-S50-LS-S-F-SC	no	no	French	0.4079	0.4034	0.4091	0.2376	0.2412	0.2384
FADERIC_Fr-BM25T-S50-LS-S-F	no	no	French	0.4044	0.4034	0.4071	0.2324	0.2414	0.235
SQUID_BasicSearcher	no	no	French	0.4149	0.3998	0.411	0.2522	0.2439	0.2425
SQUID_W2VRerank	yes	no	French	0.4154	0.3997	0.4105	0.2538	0.2442	0.242
gwca_lightstem-phrase	no	no	French	0.4052	0.3992	0.3988	0.2303	0.2375	0.2297
DARDS_BM25FRENCHBASE	no	no	French	0.3843	0.3924	0.3916	0.2083	0.2291	0.2207
semicolon_frenchAnalyzerFrStopWord	no	no	French	0.3869	0.3897		0.21	0.2273	
semicolon_frenchAnalyzerFrStopNum	no	no	French	0.3861	0.3895		0.2086	0.2277	
DARDS_BM25FRENCHBOOSTURL	no	no	French	0.3859	0.3866	0.3945	0.2151	0.2241	0.2243
gwca_lightstem-phrase-qexp-rerank3f	no	no	French	0.3872	0.3863	0.3942	0.2099	0.216	0.2168
gwca_lightstem-phrase-qexp-rerank2f	no	no	French	0.4059	0.3833	0.3905	0.2302	0.2117	0.2131
RAFJAM_BasicRuns	no	no	French	0.374	0.3804	0.3807	0.2018	0.2207	0.2123
gwca_word2vec-nostem	no	no	French	0.3843	0.3801	0.384	0.2083	0.2205	0.2176
CLOSE_QUEREXPANSION	yes	no	French	0.3725	0.3795	0.3736	0.2029	0.2213	0.2062
DARDS_BM25FRENCHRERANK100	no	no	French	0.3755	0.3756	0.3758	0.1982	0.2075	0.202
UGA_T5_French	yes	yes	French	0.3757	0.3717	0.3801	0.2223	0.2209	0.2207
SQUID_BOOST	no	no	French	0.3586	0.3693	0.3736	0.2024	0.2243	0.2172
DARDS_BM25FRENCHSPAM	no	no	French	0.3605	0.368	0.3643	0.1916	0.2126	0.2019
UGA_BM25_French	no	no	French	0.354	0.3541	0.3526	0.1904	0.2027	0.1936
seupd2223-JIHUMING-10_fr_fr_5gram	no	no	French, English	0.3413	0.3447	0.3533	0.1788	0.1926	0.192
seupd2223-JIHUMING-09_fr_fr_4gram	no	no	French, English	0.3364	0.3423	0.348	0.1763	0.1911	0.1888
seupd2223-hiball_BERT	yes	yes	English	0.3119	0.3418		0.1732	0.1991	
seupd2223-JIHUMING-08_fr_fr_3gram	no	no	French, English	0.3307	0.3384	0.3454	0.1725	0.1893	0.1881
seupd2223-JIHUMING-07_fr_fr	no	no	French, English	0.3271	0.3367	0.3443	0.1746	0.1883	0.1878
RAFJAM_PseudoRelQERuns	no	no	French	0.3516	0.3355	0.349	0.1971	0.1843	0.1872
FADERIC_En-BM25-S50-KS-S-F-SP-R30	yes	no	English	0.3031	0.3296	0.3262	0.1626	0.1931	0.1809
RAFJAM_SynQERuns	no	no	French	0.3193	0.3295	0.3231	0.1614	0.1876	0.1719
CLOSE_RERANKING_ENGLISH	yes	yes	English	0.3113	0.3285	0.3373	0.1822	0.1941	0.192
IRC_BM25+monoT5	yes	yes	English	0.3034	0.3256	0.3376	0.1642	0.19	0.1895
UGA_T5_English	yes	yes	English	0.2886	0.3202	0.3347	0.1576	0.1863	0.1936
RAFJAM_AllQERuns	no	no	French	0.3209	0.3172	0.3138	0.1652	0.1785	0.1676
IRC_BM25+colBERT	yes	yes	English	0.2883	0.3132	0.3209	0.1551	0.1769	0.1736
IRC_d2q+BM25	yes	no	English	0.2746	0.3072	0.3211	0.1347	0.168	0.1736
DARDS_BM25TRANSLATEDQUERIES	no	no	French, English	0.3072	0.304	0.3182	0.1525	0.1587	0.1644
semicolon_fusedRankAllEnglish	no	yes	English	0.2921	0.3032		0.1452	0.1608	
seupd2223-JIHUMING-12_fr_fr_4gram_ner	no	no	French, English	0.2868	0.298	0.3046	0.1369	0.1468	0.1433
IRC_E5_base	yes	no	English	0.2891	0.297	0.3131	0.1629	0.1599	0.1661
seupd2223-hiball_BASELINE	no	no	English	0.279	0.2955		0.1363	0.1576	
soup_kml	no	no	English	0.2705	0.2941	0.3042	0.1304	0.1559	0.1567
soup_kbase	no	no	English	0.2693	0.294	0.3021	0.1303	0.1551	0.1548
IRC_RRF(BM25+Bo1-XSqrA_M-PL2)	no	yes	English	0.2842	0.2939	0.3068	0.1355	0.1516	0.1557
soup_kngml	no	no	English	0.2698	0.2939	0.3039	0.1297	0.1558	0.1565
semicolon_Ngram34	no	no	English	0.2868	0.2938		0.1441	0.1557	
semicolon_porter2-1p4-eng	no	no	English	0.2739	0.2912		0.1303	0.1516	
soup_lng	no	no	English	0.2714	0.2899	0.2986	0.1338	0.1535	0.1526
HIBALL_AI-MERGED	no	no	English	0.2652	0.2887		0.1255	0.1506	
UGA_BM25_English	no	no	English	0.2689	0.2873	0.2992	0.1326	0.151	0.1536
seupd2223-hiball_RRF60	no	no	English	0.2664	0.2866		0.1247	0.1462	
soup_kmls	no	no	English	0.2739	0.2862	0.2988	0.1331	0.1492	0.152
QEVALS_LMDirichlet	no	no	French	0.2896	0.2819	0.2805	0.1572	0.1684	0.1633
QEVALS_BM25DFLT	no	no	French	0.2999	0.2806	0.285	0.1688	0.1694	0.1687
ows-bm25-10-variants-prompt-2	no	yes	English	0.256	0.2792	0.2872	0.1225	0.1432	0.1432
ows-pl2-10-variants-prompt-2	no	yes	English	0.2636	0.2776	0.2881	0.1285	0.1381	0.1393
QEVALS_BM25CSTM	no	no	French	0.2966	0.2776	0.2845	0.1653	0.1661	0.1681
ows-bm25-5-variants-prompt-2	no	yes	English	0.2556	0.2762	0.2838	0.1243	0.1401	0.1389
QEVALS_IB	no	no	French	0.3009	0.276	0.2833	0.1763	0.1634	0.1664
ows-lgd-10-variants-prompt-2	no	yes	English	0.2662	0.2759	0.2875	0.1275	0.1364	0.1384
ows-pl2-5-variants-prompt-2	no	yes	English	0.2631	0.2759	0.2876	0.1303	0.136	0.139
QEVALS_DFR	no	no	French	0.2976	0.2746	0.2824	0.1686	0.1626	0.1659
CLOSE_JSCLEANER_BM25	no	no	English	0.2647	0.2694	0.2803	0.1286	0.141	0.1419
NEON_1b	no	no	English	0.2269	0.2294	0.243	0.1338	0.139	0.1478
NEON_3b	no	no	English	0.2017	0.226	0.2387	0.1226	0.1384	0.1442
NEON_1a	no	no	English	0.2201	0.2241	0.2393	0.1287	0.1356	0.1446
NEON_2br	no	no	English	0.2177	0.2219	0.2282	0.1279	0.1319	0.1351
NEON_4b	no	no	English	0.2054	0.2187	0.2282	0.1213	0.1324	0.1351
HIBALL_AI-FIXED	no	no	English	0.0908	0.0923		0.0332	0.0319	
AVERAGE				**0.3203**	**0.3256**	**0.3234**	**0.1739**	**0.1850**	**0.1790**

For "WT-LT" the formula is:

$$\frac{\text{NDCG}_{WT}(r) - \text{NDCG}_{LT}(r)}{\text{NDCG}_{WT}(r)}$$

With such definitions, large negative values for columns "WT-ST" and "WT-LT" mean that the systems are able to generalize well on the new test collections, as the WT heldout queries are processed on the same document corpus as the training data, which is not the case of the ST and LT datasets.

What we see in Table 2 is that the systems that are the more robust to the evolution of test collection are not the top ones: for instance the NEON_3b run is almost at the bottom on Table 3 but does increase its NDCG values at ST, as well as at LT. We also see that the best systems according to NDCG at ST, FADERIC_Fr-BM25-S50-LS-S-F-SC-R20W6, FADERIC_Fr-BM25-S50-LS-S-F-R30 and SQUID_SEARCHERAI, are stable or decreasing their NDCG values at ST.

On average (last line of Table 2), the systems increase less their results on ST than on LT, which is surprising. This surprising point will need further explorations as it looks contradictory to what we were expecting. Another element worth noticing is that the NDCG changes WT-ST and WT-LT behave consistently: for most of the systems the absolute value for WT-ST is smaller than the absolute value of WT-LT.

2.6   Run Rankings

We have so far studied our first problem, which was a comparison of performance of a single system in an evolving setup. Next, we would like to study how do the submitted runs compare to each other, either in terms of the absolute NDCG scores achieved on the collections, or in terms of NDCG changes between the collections. For this, we display the ranking of runs according in all these tasks, see Table 3.

In addition, we also calculated the Pearson correlation between the rankings. The correlation between the rankings (in terms of NDCG scores) achieved on WT and ST is very high (0.95). The correlation between both WT and ST and between ST and LT rankings is slightly lower – 0.71 and 0.70, respectively. This corresponds with the high overlaps of the documents and also queries between WT and ST collections and slightly smaller overlaps of the LT collection.

The correlation between the ranking according to the NDCG score achieved on the WT dataset and the ranking of the performance change is negative. The Pearson correlation is −0.65 for the ST dataset and −0.51 on the LT dataset. This means that the better the system initially performs, harder it is to improve it. Not surprisingly, there is thus also a negative correlation between the ranking achieved on the ST dataset and the ranking of the change between the ST and WT dataset (−0.42). However, there is no such correlation (0.05) between the ranking achieved on the LT dataset and ranking of the change between the WT and LT datasets.

Table 2. Changes in the NDCG scores. Table is sorted according to the highest change between the ST and WT collection.

System	NDCG			NDCG Change	
	WT	ST	LT	WT-ST	WT-LT
NEON_3b	0.2017	0.226	0.2387	-0.1205	-0.1835
IRC_d2q+BM25	0.2746	0.3072	0.3211	-0.1188	-0.1694
UGA_T5_English	0.2886	0.3202	0.3347	-0.1095	-0.1598
seupd2223-hiball_BERT	0.3119	0.3418		-0.0959	
soup_kbase	0.2693	0.294	0.3021	-0.0918	-0.1218
ows-bm25-10-variants-prompt-2	0.256	0.2792	0.2872	-0.0907	-0.1219
soup_kngml	0.2698	0.2939	0.3039	-0.0894	-0.1264
HIBALL_AI-MERGED	0.2652	0.2887		-0.0887	
FADERIC_En-BM25-S50-KS-S-F-SP-R30	0.3031	0.3296	0.3262	-0.0875	-0.0763
soup_kml	0.2705	0.2941	0.3042	-0.0873	-0.1246
IRC_BM25+colBERT	0.2883	0.3132	0.3209	-0.0864	-0.1131
ows-bm25-5-variants-prompt-2	0.2556	0.2762	0.2838	-0.0806	-0.1104
seupd2223-hiball_RRF60	0.2664	0.2866		-0.0759	
IRC_BM25+monoT5	0.3034	0.3256	0.3376	-0.0732	-0.1128
UGA_BM25_English	0.2689	0.2873	0.2992	-0.0685	-0.1127
soup_lng	0.2714	0.2899	0.2986	-0.0682	-0.1003
NEON_4b	0.2054	0.2187	0.2282	-0.0648	-0.1111
semicolon_porter2-1p4-eng	0.2739	0.2912		-0.0632	
seupd2223-hiball_BASELINE	0.279	0.2955		-0.0592	
CLOSE_RERANKING_ENGLISH	0.3113	0.3285	0.3373	-0.0553	-0.0836
ows-pl2-10-variants-prompt-2	0.2636	0.2776	0.2881	-0.0532	-0.0930
ows-pl2-5-variants-prompt-2	0.2631	0.2759	0.2876	-0.0487	-0.0932
soup_kmls	0.2739	0.2862	0.2988	-0.0450	-0.0910
seupd2223-JIHUMING-12_fr_fr_4gram_ner	0.2868	0.298	0.3046	-0.0391	-0.0621
semicolon_fusedRankAllEnglish	0.2921	0.3032		-0.0381	
ows-lgd-10-variants-prompt-2	0.2662	0.2759	0.2875	-0.0365	-0.0801
IRC_RRF(BM25+Bo1-XSqrA_M-PL2)	0.2842	0.2939	0.3068	-0.0342	-0.0796
RAFJAM_SynQERuns	0.3193	0.3295	0.3231	-0.0320	-0.0120
SQUID_BOOST	0.3586	0.3693	0.3736	-0.0299	-0.0419
seupd2223-JIHUMING-07_fr_fr	0.3271	0.3367	0.3443	-0.0294	-0.0526
IRC_E5_base	0.2891	0.297	0.3131	-0.0274	-0.0831
semicolon_Ngram34	0.2868	0.2938		-0.0245	
seupd2223-JIHUMING-08_fr_fr_3gram	0.3307	0.3384	0.3454	-0.0233	-0.0445
DARDS_BM25FRENCHBASE	0.3843	0.3924	0.3916	-0.0211	-0.0190
DARDS_BM25FRENCHSPAM	0.3605	0.368	0.3643	-0.0209	-0.0106
NEON_2br	0.2177	0.2219	0.2282	-0.0193	-0.0483
CLOSE_QUEREXPANSION	0.3725	0.3795	0.3736	-0.0188	-0.0030
NEON_1a	0.2201	0.2241	0.2393	-0.0182	-0.0873
CLOSE_JSCLEANER_BM25	0.2647	0.2694	0.2803	-0.0178	-0.0590
seupd2223-JIHUMING-09_fr_fr_4gram	0.3364	0.3423	0.348	-0.0176	-0.0345
RAFJAM_BasicRuns	0.374	0.3804	0.3807	-0.0172	-0.0180
FADERIC_Fr-BM25-S50-LS-S-F-SC-R20W6	0.4169	0.4239		-0.0168	
HIBALL_AI-FIXED	0.0908	0.0923		-0.0166	
NEON_1b	0.2269	0.2294	0.243	-0.0111	-0.0710
seupd2223-JIHUMING-10_fr_fr_5gram	0.3413	0.3447	0.3533	-0.0100	-0.0352
semicolon_frenchAnalyzerFrStopNum	0.3861	0.3895		-0.0089	
semicolon_frenchAnalyzerFrStopWord	0.3869	0.3897		-0.0073	
DARDS_BM25FRENCHBOOSTURL	0.3859	0.3866	0.3945	-0.0019	-0.0223
DARDS_BM25FRENCHRERANK100	0.3755	0.3756	0.3758	-0.0003	-0.0008
UGA_BM25_French	0.354	0.3541	0.3526	-0.0003	0.0040
FADERIC_Fr-BM25-S50-LS-S-F-R30	0.4147	0.4145		0.0005	
gwca_lightstem-phrase-qexp-rerank3f	0.3872	0.3863	0.3942	0.0024	-0.0181
FADERIC_Fr-BM25T-S50-LS-S-F	0.4044	0.4034	0.4071	0.0025	-0.0067
DARDS_BM25TRANSLATEDQUERIES	0.3072	0.304	0.3182	0.0105	-0.0359
UGA_T5_French	0.3757	0.3717	0.3801	0.0107	-0.0118
gwca_word2vec-nostem	0.3843	0.3801	0.384	0.0110	0.0008
FADERIC_Fr-BM25-S50-LS-S-F-SC	0.4079	0.4034	0.4091	0.0111	-0.0030
RAFJAM_AllQERuns	0.3209	0.3172	0.3138	0.0116	0.0222
gwca_lightstem-phrase	0.4052	0.3992	0.3988	0.0149	0.0158
CLOSE_RERANKING	0.4166	0.4068	0.4062	0.0236	0.0250
QEVALS_LMDirichlet	0.2896	0.2819	0.2805	0.0266	0.0315
SQUID_W2V	0.4232	0.4106	0.4174	0.0298	0.0138
SQUID_SEARCHERAI	0.4279	0.4141	0.4177	0.0323	0.0239
SQUID_BasicSearcher	0.4149	0.3998	0.411	0.0364	0.0094
SQUID_W2VRerank	0.4154	0.3997	0.4105	0.0378	0.0118
gwca_lightstem-phrase-qexp	0.4294	0.4114	0.4161	0.0420	0.0310
CLOSE_SBERT_BM25	0.4318	0.4128	0.4139	0.0441	0.0415
RAFJAM_PseudoRelQERuns	0.3516	0.3355	0.349	0.0458	0.0074
gwca_lightstem-phrase-qexp-rerank2f	0.4059	0.3833	0.3905	0.0557	0.0380
QEVALS_BM25CSTM	0.2966	0.2776	0.2845	0.0641	0.0408
QEVALS_BM25DFLT	0.2999	0.2806	0.285	0.0644	0.0497
QEVALS_DFR	0.2976	0.2746	0.2824	0.0773	0.0511
QEVALS_IB	0.3009	0.276	0.2833	0.0828	0.0585
AVERAGE	**0.3226**	**0.3273**	**0.3359**	**-0.0195**	**-0.0376**

Table 3. Ranking of the submitted systems in terms of NDCG scores (columns 2–4), absolute changes in NDCG scores between WT and ST dataset (column 5), absolute changes in NDCG scores between WT and LT dataset (column 6). Column 7 shows the sum of the Borda count applied to ranking on ST dataset and Borda count of ranking change between ST and WT dataset. Column 8 shows the same value, but for the LT dataset. The darker color means better performance.

System	Ranking NDCG WT	Ranking NDCG ST	Ranking NDCG LT	Ranking NDCG Change ST-WT	Ranking NDCG Change LT-WT	Perf(ST) + Change (ST-WT)	Perf(LT) + Change (LT-WT)
seupd2223-hiball_BERT	34	29	64	4	62	113	0
UGA_T5_English	46	37	30	3	3	106	93
FADERIC_En-BM25-S50-KS-S-F-SP-R30	38	33	31	9	22	104	73
IRC_d2q+BM25	52	40	33	2	2	104	91
FADERIC_Fr-BM25-S50-LS-S-F-SC-R20W6	5	1	62	42	62	103	2
DARDS_BM25FRENCHBASE	18	13	13	34	34	99	79
IRC_BM25+colBERT	47	39	34	11	8	96	84
IRC_BM25+monoT5	37	36	28	14	9	96	89
soup_kbase	58	47	42	5	7	94	77
FADERIC_Fr-BM25-S50-LS-S-F-R30	9	2	63	51	62	93	1
SQUID_BOOST	25	24	20	29	29	93	77
CLOSE_RERANKING_ENGLISH	35	35	29	20	18	91	79
soup_kml	56	46	40	10	5	90	81
soup_kngml	57	49	41	7	4	90	81
CLOSE_QUEREXPANSION	23	21	19	37	41	88	66
DARDS_BM25FRENCHSPAM	24	25	21	35	39	86	66
RAFJAM_BasicRuns	22	19	16	41	36	86	74
FADERIC_Fr-BM25T-S50-LS-S-F	13	9	8	52	40	85	78
HIBALL_AI-MERGED	62	53	64	8	62	85	0
semicolon_frenchAnalyzerFrStopNum	16	15	64	46	62	85	0
semicolon_frenchAnalyzerFrStopWord	15	14	64	47	62	85	0
seupd2223-JIHUMING-07_fr_fr	31	31	27	30	26	85	73
RAFJAM_SynQERuns	33	34	32	28	37	84	57
seupd2223-JIHUMING-08_fr_fr_3gram	30	30	26	33	28	83	72
DARDS_BM25FRENCHBOOSTURL	17	16	11	48	33	82	82
seupd2223-hiball_BASELINE	51	45	64	19	62	82	0
FADERIC_Fr-BM25-S50-LS-S-F-SC	10	8	7	57	42	81	77
ows-bm25-10-variants-prompt-2	66	59	49	6	6	81	71
SQUID_SEARCHERAI	3	3	1	63	52	80	73
CLOSE_RERANKING	6	7	9	60	53	79	64
semicolon_fusedRankAllEnglish	43	42	64	25	62	79	0
seupd2223-JIHUMING-09_fr_fr_4gram	29	28	25	39	32	79	69
seupd2223-JIHUMING-12_fr_fr_4gram_ner	48	43	39	24	24	79	63
seupd2223-hiball_RRF60	60	55	64	13	62	78	0
soup_lng	55	52	45	16	13	78	68
SQUID_W2V	4	6	2	62	49	78	75
semicolon_porter2-1p4-eng	53	51	64	18	62	77	0
UGA_BM25_English	59	54	43	15	10	77	73
gwca_lightstem-phrase-qexp-rerank3f	14	17	12	53	35	76	79
NEON_3b	72	69	59	1	1	76	66
CLOSE_SBERT_BM25	1	4	4	67	58	75	64
gwca_lightstem-phrase	12	12	10	59	50	75	66
gwca_lightstem-phrase-qexp	2	5	3	66	54	75	69
DARDS_BM25FRENCHRERANK100	21	22	18	50	43	74	65
seupd2223-JIHUMING-10_Fr_Fr_5gram	28	27	22	45	31	74	73
ows-bm25-5-variants-prompt-2	67	62	52	12	12	72	62
SQUID_BasicSearcher	8	10	5	64	47	72	74
IRC_E5_base	45	44	37	31	19	71	70
IRC_RRF(BM25+Bo1-XSqrA_M-PL2)	50	48	38	27	21	71	67
UGA_BM25_French	26	26	23	49	45	71	58
gwca_word2vec-nostem	19	20	15	56	44	70	67
SQUID_W2VRerank	7	11	6	65	48	70	72
UGA_T5_French	20	23	17	55	38	68	71
soup_kmls	54	56	44	23	16	67	66
ows-pl2-10-variants-prompt-2	64	61	46	21	15	64	65
semicolon_Ngram34	49	50	64	32	62	64	0
gwca_lightstem-phrase-qexp-rerank2f	11	18	14	69	56	59	56
ows-pl2-5-variants-prompt-2	65	65	47	22	14	59	65
NEON_4b	71	72	61	17	11	57	54
ows-lgd-10-variants-prompt-2	61	64	48	26	20	56	58
DARDS_BM25TRANSLATEDQUERIES	36	41	35	54	30	51	61
RAFJAM_AllQERuns	32	38	36	58	51	50	39
RAFJAM_PseudoRelQERuns	27	32	24	68	46	46	56
CLOSE_JSCLEANER_BM25	63	67	56	40	25	39	45
NEON_2br	70	71	60	36	27	39	39
NEON_1a	69	70	58	38	17	38	51
NEON_1b	68	68	57	44	23	34	46
HIBALL_AI-FIXED	73	73	64	43	62	30	0
QEVALS_LMDirichlet	44	57	55	61	55	28	16
QEVALS_BM25DFLT	40	58	50	71	59	17	17
QEVALS_BM25CSTM	42	60	51	70	57	16	18
QEVALS_IB	39	63	53	73	61	10	12
QEVALS_DFR	41	66	54	72	60	8	12

We also provided the normalized results to the participants. The normalization was done according to [7] and the mean and standard deviation of the scores of all submitted runs were calculated. These scores were then used to calculate the score in normal distribution and this score was subsequently shifted using CDF into 0–1 space. However, the correlation of the original ranking and ranking according to the normalized values is highly correlated: 0.93, 0.95, and 0.88 for WT, ST and LT datasets, respectively. We thus further do not work with the normalized results.

Last, we calculated a combination of both rankings (ranking in terms of absolute values and ranking in terms of change). For this, we first calculated a Borda count of the ranking in terms of absolute values and Borda count of the ranking in terms of relative change and then we simply summed these two Borda counts: these results are displayed in two last columns in the Table 3. As the correlation between the absolute performance and performance change is negative, the best performing runs in terms of this measure are often mediocre in one measure and well performing in the another – for instance seupd2223-hiball_BERT run achieves high performance change, while it is mediocre in terms of NDCG achieved on ST dataset.

2.7   Discussion and Conclusion

This task was a first attempt at collectively investigate the impact of the evolution of the data on search system's performances. Having 14 participating teams submitting runs confirmed that this topic was of interest to the community.

The dataset released for this task consisted in a sequence of test collections corresponding to different times. The collections were composed of documents and queries coming from Qwant, and relevance judgment coming from a click model and manual assessment. While the manual assessment is ongoing at the time of the paper's publication, performances of participants' submitted runs were measured using the click logs.

The results show that the best approaches were based on query expansion techniques, and embeddings or Large Language Models. The effect of the translation of the documents and queries provided by the lab has a clear impact: the best results were obtained on the original French data.

Since each subset had substantial overlaps, the correlations between systems rankings was pretty high. As for the robustness of the systems towards dataset changes, we observed that the systems that are the more robust to the evolution of test collection were not the best performing ones.

Further evaluations will be carried out in the near future with the manual assessment of the pooled sets. A thorough analysis of the results will be necessary to study the impact of queries on the results (their nature, topic, difficulty, etc.). Further analysis work will be necessary to fully establish the robustness of the systems and the specific impact of dataset evolution on the performances.

3   Task 2 - Classification

As the meanings of words and phrases evolve over time, sentiment classifiers may struggle to accurately capture the changing linguistic landscape [4], resulting in decreased effectiveness in capturing sentiments expressed in text. Recent research shows that this is particularly the case when one is dealing with social media data [3]. Understanding the extent of this performance drop and its implications is crucial for maintaining accuracy and reliable sentiment analysis models in the face of linguistic drift. The objective of this task aimed to quantitatively measure the performance degradation of sentiment classifiers over time, providing insights into the impact of language evolution on sentiment analysis tasks and identifying strategies to mitigate the effects of temporal dynamics. Participants of this task were invited to submit classification outputs of their systems that attempted to mitigate the temporal performance drop.

The aim of Task 2 was ultimately to answer the following research questions:

- **RQ1:** *What types of models offer better short-term temporal persistence?*
- **RQ2:** *What types of models offer better long-term temporal persistence?*
- **RQ3:** *What types of models offer better overall temporal persistence?*

To assess the extent of the performance drop of models in shorter and longer temporal gaps, we provided training data pertaining to a specific year (2016), as well as test datasets pertaining to a close (2018) and a more distant (2021) year. In addition to measuring performance in each of these years separately, this setup enabled evaluating relative performance drops by comparing performance across years.

3.1   Description of the Task

In this section, we introduce the task of temporal persistence classification, as the focus of a recent shared task [1]. The goal of this task was to develop classifiers that can effectively mitigate performance drops over short and long periods of time compared to a test set from the same time frame as the training data.

The shared task was in turn divided into two sub-tasks:

Sub-Task 1: Short-Term Persistence: In this sub-task, participants were asked to develop models that demonstrated performance persistence over short periods of time. Specifically, the performance of the models was expected to be maintained within a temporal gap of two years between the training and test data.

Sub-Task 2: Long-Term Persistence: This sub-task focused on developing models that demonstrated performance persistence over a longer period of time. The classifiers were expected to mitigate performance drops over a temporal gap of five years between the training and test data.

By providing a comprehensive training dataset, two practice sets, and three testing sets, the shared competition aimed to stimulate the development of classifiers that can effectively handle temporal variations and maintain performance persistence over different time distances. Participants were expected to submit solutions for both sub-tasks, showcasing their ability to address the challenges of temporal variations in performance.

3.2   Dataset

In this section, we present the process of constructing our final annotated corpus for the task. The large-scale dataset TM-Senti was originally described in [8], from which we extract samples that we use in this shared task. TM-Senti was chosen for the task as it provided a sufficiently longitudinal dataset (covering multiple years) and for using a consistent data collection and annotation strategy, which means that only the temporal evolution of data changes with other potentially confounding factors removed.

Temporal Granularity. In the shared task, the **training** set covered a time period with a gap of 2 years, from 2014 to 2016. For the practice sets, within and distance time sets were introduced. The Practice-2016 set had a time gap of 0 years from the training data, given that it overlapped with the training period. In addition, the Practice-2018 set was also provided as a distant test set to practice with, having a temporal gap of two years from the training data.

For the test sets, the within set had a time gap of 0 years, covering the same period as the within Practice-2016 set. The Test-short set had a time gap of 2 years, coinciding with the distant Practice-2018 set. Lastly, the Test-long set had a time gap of 5 years, representing a long-term evaluation scenario.

By using these different time gaps, the shared task aimed to assess the models' performance persistence over varying temporal distances from the training data.

Un-labelled Data. The data was sampled from Twitter using the Twitter academic API. Then, duplicates and near duplicates were removed. We also enforced a diversity of users and removed tweets from most frequent users with bot-like behaviour. Finally, user mentions were replaced by '@user' for anonymization, except for verified users that remained unchanged. For all these preprocessing steps, we relied on the same pipeline and script used by [6].

Test Set Annotation. The test set was annotated using Amazon Mechanical Turk (AMT)[1]. AMT candidate workers were filtered based on them successfully passsing two *qualification tasks*. The first, built-in in the system, seeks to find workers with certain experience and located in English-speaking countries to ensure, to a certain extent, high command of the English language and high familiarity with AMT. The second qualification task consisted in presenting each candidate annotator with 5 tweets, and only workers that correctly annotated 3 or more were allowed to proceed to the actual annotation task.

---

[1] https://www.mturk.com/.

In total, we annotated 4,032 tweets, divided into 1874 for positive, 741 neutral and 1417 negative. Each tweet was annotated by 5 different workers, and the tweet's final label was decided by computing the *mode* of the array of annotations. Table 4 shows instances of the dataset, with labels and number of agreements between 5 and 3. In terms of overall statistics, 8.5% of the tweets were annotated with full agreement, 22.8% with 4 annotators agreeing, 46% with 3 agreements, and the remaining 22.5% with 2 agreements, which were mostly decided between positive and neutral, and negative and neutral.

Table 4. Tweets where 5, 4 and 3 annotators agreed. Tweets labeled as neutral tend to be factual or posing questions, whereas high agreement positive and negative tweets tend to be more emotional, occasionally backed by the use of stronger words.

#agree	Tweet	Label
5	I say this a lot But I m just so in love with Evan	pos
	Online classes r a joke	neg
	Shout out to me for living 17 min away from school	neu
4	Honestly just a Hi from you already makes my day	pos
	Been one of them weeks and I just want to burst out crying	neg
	What s your fave throwback song to jam out to on Thursdays...	neu
3	Not a good idea to mix everything but great night	pos
	just had the worst nightmare I don t want to go back to sleep	neg
	Waiting to find a man that can dance like Chris Brown	neu

Data Preprocessing we preprocess our dataset to ensure its quality with respect to the following criteria:

- Diversity: All retweets and replies are eliminated.
- Consistency: We prioritise posts written in English and impose a length restriction such that all posts contain at least 5 words and are at most 140 character long.
- Fluency: Posts containing URL links are eliminated. In addition, we select posts which contain at least one stop word as a proxy for fluency.

Before sampling, all emojis and emoticons are deleted from the body of text.

Data Sampling. In the second stage, we sample from the preprocessed data previously obtained. As we aim for a well-balanced annotated set, the sampling strategy is defined in terms of: 1) sentiment distribution, 2) time span and 3) post length. For 1), we use the distant labels provided by [8] to obtain a balanced distribution between the negative and positive classes. For 2), we sample an equal number of posts for each month within the specified temporal window in each dataset. Finally for 3), we partition the data into four bins with respect to the word length of each post (i.e., each post falls into one of the following bins: [5,10), [10,15), [15,20) and [20, 20+]) and uniformly sample from each bin.

The resulting distribution of data is shown in Table 5.

Table 5. Dataset statistics summary of training, practice and testing sets.

Dataset	Time Period	Size
Training	Feb 2014–Dec 2016	49608
Practice-2016 [within]	Jan 2016–Dec 2016	1344
Practice-2018 [distant]	Jan 2018–Dec 2018	1344
Test-within	Jan 2016–Dec 2016	908
Test-short	Jan 2018–Dec 2018	908
Test-long	Jan 2021–Aug 2021	908

3.3 Evaluation

The performance of the submissions was evaluated in two ways:

1. **Macro-averaged F1-score:** This metric measured the overall F1-score on the testing set for the sentiment classification sub-task. The F1-score combines precision and recall to provide a balanced measure of model performance. A higher F1-score indicated better performance in terms of both positive and negative sentiment classification.

$$F - macro = \frac{2 \cdot precision \cdot recall}{precision + recall} \tag{1}$$

2. **Relative Performance Drop (RPD):** This metric quantified the difference in performance between the "within-period" data and the short- or long-term distant testing sets. RPD was computed as the difference in performance scores between two sets. A negative RPD value indicated a drop in performance compared to the "within-period" data, while a positive value suggested an improvement.

$$RPD = \frac{f_{score_{t_j}} - f_{score_{t_0}}}{f_{score_{t_0}}} \tag{2}$$

where t_0 represents performance when time gap is 0; t_j represents performance when time gap is short or long as in was introduced in previous work [2].

The submissions were ranked primarily based on the macro-averaged F1-score. This ranking approach emphasized the overall performance of the sentiment classification models on the testing set. The higher the macro-averaged F1-score, the higher the ranking of the submission.

3.4   Results

Our shared task consisted of two subtasks: Short-term persistence (Sub-task A) and Long-term persistence (Sub-task B). Sub-task A focused on developing models that demonstrated performance persistence within a two-year gap from the training data, while Sub-task B required models that exhibited performance persistence over a longer period, surpassing the two-year gap. Additionally, an unlabeled corpora covering all periods of training, development, and testing was provided to teams interested in data-centric approaches. Along with the data, participating teams received python-based baseline code, and evaluation scripts[2]. The shared task progressed through two phases and results are discussed in the following paragraphs.

3.5   Practice Phase

The initial phase was the practice phase, where participants received three distantly annotated sets, training set, within time practice set and short-term practice set. The training set was used for model training, while the two labeled practice set allowed participants to refine their systems before the subsequent phase. Moreover, we limited the sharing practice sets to within-time (Practice-2016) and single distance practice sets the short-term set (Practice-2018). This decision was made because participants were requested to take part in both subtasks and reduce over-fitting. The results of this phase were not considered in final models ranking.

Table 6. Performance comparison for practice set

Team Name	F1 Score Within	F1 Score Short	Overall Drop	Overall Score
Pablojmed	0.8244 (1)	0.7976 (1)	−0.0325 (2)	0.811
saroyehun	0.8170 (2)	0.7917 (2)	−0.0310 (1)	0.8043
Baseline	*0.7879 (3)*	*0.7611 (3)*	*−0.0340 (3)*	*0.7745*

As it can be seen from Table 6, **Pablojmed** showcased outstanding performance, surpassing the **Baseline** model with the highest scores in F1 Score Within (0.8244) and F1 Score Short (0.7976), as well as the highest Overall Score (0.811). **saroyehun** also demonstrated remarkable performance achieving the lowest Overall Drop (−0.0310), as well as outperforming the **Baseline** model in F1 Score Within (0.8170) and F1 Score Short (0.7917). The results highlight the potential of both **Pablojmed** and **saroyehun**'s submissions for enhancing the baseline model's results.

---

[2] https://clef-longeval.github.io/.

3.6   Evaluation Phase

During the evaluation phase, participants were provided with three human-annotated testing sets, namely Test-within, Test-short and Test-long (See 3.2 for datasets details). The performance of participants on this phase was used to determine the overall rankings on the task.

Table 7. Performance comparison for evaluation set.

Team Name	F1 Score Within	F1 Score Short	F1 Score Long	RPD Within-Short	RPD Within-Long	Overall Drop	Overall Score
Pablojmed	0.7377 (2)	0.6739 (3)	_0.6971_ (1)	−0.0866 (5)	−0.0550 (3)	−0.0708 (4)	_0.7029_
Baseline	_0.7459_ (1)	_0.6839_ (1)	_0.6549_ (4)	_-0.0830_ (4)	_−0.1220_ (5)	_−0.1025_ (5)	_0.6949_
Cordyceps	0.7246 (3)	0.6771 (2)	0.6751 (3)	_−0.0656_ (1)	−0.0683 (4)	−0.0669 (3)	0.6923
saroyehun	0.7203 (4)	0.6674 (4)	0.6874 (2)	−0.0735 (2)	−0.0457 (2)	−0.0596 (2)	0.6917
pakapro	0.5033 (5)	0.4648 (5)	0.4910 (5)	−0.0765 (3)	_−0.0243_ (1)	_−0.0504_ (1)	0.4863

Short-term Temporal Persistence: From Table 7, we can see that still the **Baseline** model is the best for achieving the highest short-term F1 Score (0.6839) among all the teams, indicating that _RoBERTA_ architecture has a better performance in capturing short-term patterns compared to the other models. In same time, **Cordyceps** obtained the lowest short-term RPD value (-0.0656), suggesting a smaller drop in performance compared to the **Baseline** model. This indicates that **Cordyceps** may offer better short-term temporal persistence despite not having the highest Short-term F1 Score.

Long-term Temporal Persistence: In term of long-term persistence, **Pablojmed** achieved the highest f score (0.6971), indicating better performance in capturing long-term patterns compared to the other models. However, when considering the long-term RPD measure, **pakapro** obtained the lowest value (−0.0243), suggesting a smaller drop in performance compared to the other models. This suggests that pessimistic models as in **pakapro** may provide a relatively stable long-term temporal persistence despite not having the highest long-term F1 Score. Although **Pablojmed** obtained the highest F1 Score Long (0.6971), the model that offers better long-term temporal persistence, considering RPD, is pakapro. Despite its lower F1 Score Long (0.4910), **pakapro** achieved the smallest long-term RPD (−0.0243) compared to the other models. This suggests that **pakapro** maintains its performance more consistently over a longer period, indicating better long-term temporal persistence.

Overall Temporal Persistence: Considering the overall scores, **Pablojmed** achieved the highest overall score (0.7029) with (−0.0708) overall RPD, indicating better overall temporal persistence compared to the other models. However, **pakapro** offers better overall temporal persistence based on the Overall Drop metric. Indicating that **pakapro**'s approach may be more persistent over time in our case despite its low F1 Scores. Overall, the best model is **Pablojmed** demonstrating better overall F score and higher temporal persistence than **Baseline model**. Additionally, the **Baseline model** performed best in short-term temporal persistence, and **pakapro** shows promise for long-term temporal persistence despite not having the highest long-term F1 Score.

Systems Temporal Ranking: The **Baseline model**, ranks first in within-time and short-term F1 Score but drops to fourth place in long-term F1 Score. **Pablojmed** and **Cordyceps** interchange the second and third positions in both the within-time F1 Score and short-term F1 Score categories. This suggests a relatively consistent ranking between these two models within these specific categories. **saroyehun** consistently ranks fourth in both within-time F1 Score and short-term F1 Score. **pakapro** shows worst performance among all and ranks fifth in all three F scores demonstrate consistent performance across different timeframes compared to the other models.

It is important to note that ranking consistency varies across the different measures. We can see that low RPD does not indicate better performance rather stable metric over different sets. For example, if we look at the RPD metric, we see that **pakapro** achieves the best ranking in long-term and Overall Drop. This indicates a lower drop in performance over longer time-frames. However, when considering the F1 Score, **pakapro** ranks fifth in all three categories: F1 Score Within, F1 Score Short, and F1 Score Long. This demonstrates that a low RPD does not necessarily indicate better performance in terms of F1 Score.

In all cases, submitted systems demonstrated their highest performance when evaluated using the within-time held-out set. Moreover, the overall performance of participating teams seems to have dropped between the practice phase and the final evaluation phase. Given that participants are likely to have submitted their best models from the practice phase, it might be the case that this drop is a result of participants having overoptimism on the practice set.

3.7 Discussion

Only two out of the four teams have submitted technical reports for their used models. In the following, we delve into the discussion and interpretation of the findings concerning the three research questions we raised in relation to our classification task. These interpretations are solely based on the evaluation matrix, which is further explained in Sect. 3.3.

- Regarding **RQ1**, which aimed to identify the types of models offering better short-term temporal persistence, we observed that the **Baseline model** achieved the highest short-term F1 Score among all the teams. This indicates its strong performance in maintaining consistency over a shorter time frame compared to its initial performance using within-time set. Additionally, when examining the short-term RPD values, we found that **Cordyceps** exhibited the smallest drop in performance compared to the **Baseline model**.
- Regarding **RQ2**, which investigated the models offering better long-term temporal persistence, we observed that **Pablojmed** achieved the highest F1 Score

for the long-term. This indicates its superior ability to maintain performance over an extended period. Notably, **pakapro** demonstrated a smaller long-term RPD compared to the other models, suggesting its potential for maintaining performance stability over time.

– Regarding **RQ3**, this research question aimed to identify the models offering better overall temporal persistence. In this regard, **Pablojmed** ranked as the top performing system, achieving the highest overall score. Its relatively low overall RPD further supports its consistency across different time frames. Interestingly, **pakapro** demonstrated promising results for long-term temporal persistence, despite not achieving the highest long-term F1 Score.

By delving into the evaluation matrix results, we provided insights into the performance trends observed among the participating systems. However, it is essential to acknowledge that the absence of the submission from a certain number of systems may have influenced the overall interpretation of the findings. To address this limitation, we made our leaderbored available for future submissions in Codalab[3]. This should ensure more robust and unbiased assessment for the temporal persistence of text classifiers within the research community.

3.8 Conclusion

Overall findings highlight the importance of evaluating temporal persistence in model performance. The identified models showcase varying levels of persistence in both short-term and long-term persistence. These insights provide valuable guidance for future research and development efforts aimed at improving temporal consistency in machine learning models. In future shared tasks, we aim to incorporate evolving training sets as well as expanding out temporal persistence investigation to more tasks including stance detection and topic categorization.

Acknowledgements. This work is supported by the ANR Kodicare bi-lateral project, grant ANR-19-CE23-0029 of the French Agence Nationale de la Recherche, and by the Austrian Science Fund (FWF, grant I4471-N). This work is also supported by a UKRI/EPSRC Turing AI Fellowship to Maria Liakata (grant no. EP/V030302/1) and The Alan Turing Institute (grant no. EP/N510129/1) through project funding and its Enrichment PhD Scheme for Iman Bilal. This work has been using services provided by the LINDAT/CLARIAH-CZ Research Infrastructure (https://lindat.cz), supported by the Ministry of Education, Youth and Sports of the Czech Republic (Project No. LM2018101) and has been also supported by the Ministry of Education, Youth and Sports of the Czech Republic, Project No. LM2018101 LINDAT/CLARIAH-CZ.

References

1. Alkhalifa, R., et al.: Longeval: longitudinal evaluation of model performance at CLEF 2023. In: Kamps, J., et al. (eds.) ECIR 2023. Lecture Notes in Computer Science, vol. 13982, pp. 499–505. Springer, Cham (2023). https://doi.org/10.1007/978-3-031-28241-6_58

---

[3] https://codalab.lisn.upsaclay.fr/competitions/12762.

2. Alkhalifa, R., Kochkina, E., Zubiaga, A.: Opinions are made to be changed: temporally adaptive stance classification. In: Proceedings of the 2021 Workshop on Open Challenges in Online Social Networks, pp. 27–32 (2021)
3. Alkhalifa, R., Kochkina, E., Zubiaga, A.: Building for tomorrow: Assessing the temporal persistence of text classifiers. arXiv preprint arXiv:2205.05435 (2022)
4. Alkhalifa, R., Zubiaga, A.: Capturing stance dynamics in social media: open challenges and research directions. Int. J. Digit. Hum. **3**, 1–21 (2022)
5. Galuščáková, P., et al.: Longeval-retrieval: French-English dynamic test collection for continuous web search evaluation (2023)
6. Loureiro, D., Barbieri, F., Neves, L., Espinosa Anke, L., Camacho-collados, J.: TimeLMs: diachronic language models from Twitter. In: Proceedings of the 60th Annual Meeting of the Association for Computational Linguistics: System Demonstrations, pp. 251–260. Association for Computational Linguistics, Dublin, Ireland (2022). https://doi.org/10.18653/v1/2022.acl-demo.25, https://aclanthology.org/2022.acl-demo.25
7. Urbano, J., Lima, H., Hanjalic, A.: A new perspective on score standardization. In: International ACM SIGIR Conference on Research and Development in Information Retrieval, pp. 1061–1064 (2019)
8. Yin, W., Alkhalifa, R., Zubiaga, A.: The emojification of sentiment on social media: Collection and analysis of a longitudinal Twitter sentiment dataset. arXiv preprint arXiv:2108.13898 (2021)

Overview of PAN 2023: Authorship Verification, Multi-Author Writing Style Analysis, Profiling Cryptocurrency Influencers, and Trigger Detection
Condensed Lab Overview

Janek Bevendorff[1], Ian Borrego-Obrador[2], Mara Chinea-Ríos[2],
Marc Franco-Salvador[2], Maik Fröbe[3], Annina Heini[4], Krzysztof Kredens[4],
Maximilian Mayerl[5], Piotr Pęzik[4], Martin Potthast[6], Francisco Rangel[2],
Paolo Rosso[7], Efstathios Stamatatos[8], Benno Stein[1], Matti Wiegmann[1(✉)],
Magdalena Wolska[1], and Eva Zangerle[5]

[1] Bauhaus-Universität Weimar, Weimar, Germany
pan@webis.de
[2] Symanto Research, Valencia, Spain
[3] Friedrich Schiller University Jena, Jena, Germany
[4] Aston University, Birmingham, UK
[5] University of Innsbruck, Innsbruck, Austria
[6] Leipzig University and ScaDS.AI, Leipzig, Germany
[7] Universitat Politècnica de València, Valencia, Spain
[8] University of the Aegean, Samos, Greece
https://pan.webis.de

Abstract. The paper gives a brief overview of three shared tasks which have been organized at the PAN 2023 lab on digital text forensics and stylometry hosted at the CLEF 2023 conference. The tasks include authorship verification across discourse types, multi-author writing style analysis, profiling cryptocurrency influencers with few-shot learning, and trigger detection. Authorship verification and multi-author analysis continue and advance from past editions of PAN and influencer profiling and trigger detection are new tasks with novel research questions and evaluation resources. All four tasks alilgn with the goals of all shared tasks at PAN: to advance the state of the art in text forensics and stylometry while ensuring objective evaluation on newly developed benchmark datasets.

1 Introduction

PAN is a workshop series and a networking initiative for stylometry and digital text forensics. The workshop's goal is to bring together scientists and practitioners studying technology to analyze texts regarding their originality, authorship, trust, and ethicality. Since its inception in 2009, PAN has been the venue for 69 shared tasks on computational challenges related to authorship analysis, computational ethics, and determining the originality of a piece of writing.

A. Arampatzis et al. (Eds.): CLEF 2023, LNCS 14163, pp. 459–481, 2023.
https://doi.org/10.1007/978-3-031-42448-9_29

Over the years, the respective organizing committees have assembled and studied 60 datasets evaluation resources,[1] nine of which are community contributions.

The 2023 edition of PAN at CLEF continues in the same spirit and presents four new shared tasks. First, cross-discourse type authorship verification asks if two given documents are written by the same or by different authors, where one document is in a written (essays, emails) and one in a spoken (interviews, speech transcriptions) register. The task iterates on the previous edition by defining a much more difficult setting based on the resources established last year. 10 participants submitted solutions. Second, multi-author writing style analysis asks at which position in the document the authorship changes. The task iterates on the previous edition by presenting a completely new dataset of Reddit comments while relying on the established problem definition. 6 participants submitted solutions. Third, profiling cryptocurrency influencers with few-shot learning requests participants to profile the influence, interest, and intent of Twitter users given at most 10 tweets from their timelines. The task proposes a completely new challenge, including a new evaluation resource for author profiling in a new, and difficult, few-shot setting, i.e., only little data is available to make a decision. 27 participants submitted solutions. Fourth, trigger detection asks to assign a warning label to a given fan fiction document if it contains potentially harmful content. The task presents a completely new problem, including a new evaluation resource for computational ethics. 6 participants submitted solutions.

PAN is committed to reproducible research in IR and NLP, hence all participants are asked to submit their software (instead of just their predictions) through the submission software TIRA. With the recent updates to the TIRA platform [11], all submissions to PAN were made as publicly available docker containers. In the following sections, we briefly outline the 2023 tasks and their results.

2 Cross-Discourse Type Authorship Verification

Authorship verification is the task of deciding whether a document has been written by a certain author. In general, a number of documents of known authorship by the author in question are available and the task aims at identifying stylistic similarities/differences between the known document and the disputed text. In its simplest form, only one document of known authorship is given and, in that case, authorship verification can be seen as determining whether two texts have been written by the same author [23]. Any authorship attribution case can be decomposed into a series of authorship verification tasks, therefore focusing on authorship verification is fundamental in testing the ability of computational approaches to recognize the writing style characteristics of authors.

One factor that may affect the difficulty of the authorship verification task is the length of the considered texts. In addition, it is critical to examine whether there are thematic similarities among the involved documents since the topic factor may be misleading (e.g., two documents may appear to be similar due to

---

[1] https://pan.webis.de/data.html.

a common theme rather than the writing style). It is even more challenging in cases the documents belong to different genres or discourse types (e.g., essay vs. email) that considerably affect the stylistic properties of documents.

Several previous editions of PAN included authorship verification tasks [1, 2,41,41,62,64]. There were also attempts to focus on *cross-domain authorship attribution* where the documents of known and unknown authorship belong to different domains (e.g., thematic areas or genres) [1,2,64]. Recent PAN editions focused on fan-fiction texts (i.e., non-professional fiction published online by fans of well-known works) where the documents of known and unknown authorship come from different fandoms (e.g., Harry Potter, Sherlock Holmes) allowing us to build large-scale datasets. The obtained results indicate that this task can be handled with relatively high accuracy [1,2]. In the last edition of PAN, a more challenging scenario was considered, focusing on *cross-discourse type authorship verification* where the documents of known and unknown authorship belong to different discourse types (i.e., essays, emails, text messages, and business memos) [62]. The discourse type also affects the text length (e.g., essays are much longer than text messages). The obtained results indicate that it is extremely difficult to recognize the writing style characteristics related to the personal style of authors across discourse types.

In the current edition of PAN, we continue to focus on cross-discourse type authorship verification of document pairs. In contrast to previous versions of the task where only discourse types of written language were used, we also consider oral language. This provides the opportunity to study the ability of authorship verification methods to handle the different forms of expression in written and oral language.

Dataset

A new dataset has been created based on the recent Aston 100 Idiolects Corpus in English[2] including a rich set of discourse types written by around 100 individuals. All individuals have similar ages (18–22) and are native English speakers. The topic of text samples is not restricted. Part of this corpus was also used to build the datasets of the PAN-2022 edition of the task [62]. In more detail, we consider four discourse types: two from written language (i.e., emails and essays) and two from oral language (i.e., interviews and speech transcriptions). All possible six combinations of document pairs are examined.

Since the length of emails can be very short, we concatenate consecutive messages (ordered by date) so that at least text samples of at least 2,000 characters are obtained. In addition, since separate interview utterances are included in the corpus, we also concatenate consecutive utterances to obtain text samples of at least 2,000 characters. All text samples in the corpus have been pre-processed to replace named entities with general tags. This helps to reduce the topic factor.

In order to provide training and test datasets, we first split the available individuals into two non-overlapping sets of equal size. In more detail, the text

---

[2] https://fold.aston.ac.uk/handle/123456789/17.

Table 1. Statistics of the PAN'23 datasets used in the cross-discourse type authorship verification task.

	Training	Test
Text pairs		
Positive	4,418 (50.0%)	4,828 (50.0%)
Negative	4,418 (50.0%)	4,828 (50.0%)
Email - Speech transcription	1,036 (11.7%)	1,074 (11.1%)
Essay - Email	1,454 (16.5%)	1,618 (16.8%)
Essay - Interview	884 (10.0%)	938 (9.7%)
Essay - Speech transcription	256 (2.9%)	206 (2.1%)
Interview - Email	4,564 (51.7%)	5,214 (54.0%)
Speech transcription - Interview	642 (7.3%)	606 (6.3%)
Text length (avg. chars)		
Email	2,308	2,346
Essay	9,894	10,770
Interview	2,503	2,501
Speech transcription	2,395	2,537

samples of 56 individuals are used for the training dataset and the test dataset is obtained from another set of 56 individuals. Both sets of authors have similar gender distribution. Each dataset comprises a set of document pairs and in each pair, the documents belong to different discourse types. Given that the distribution of text samples over the discourse types is not balanced, the distribution of document pairs over the six possible combinations of discourse types is not homogeneous as can be seen in Table 1. However, it is similar between training and test datasets. In addition, both datasets are balanced regarding same-author and different-author pairs. This is also true when each specific combination of discourse types is considered separately.

Evaluation Setup and Results

The evaluation framework is similar to the one used in recent shared tasks at PAN [1,2,62]. Formally, one has to approximate the target function ϕ : $(d_k, d_u) \rightarrow \{T, F\}$, d_k being a text of known authorship and d_u being a text of unknown or disputed authorship. If $\phi(d_k, d_u) = T$, then the author of d_k is also the author of d_u and if $\phi(d_k, d_u) = F$, then the author of d_k is not the same as the author of d_u. In the current edition of the task, d_k and d_u belong to different discourse types of written or oral language.

For each text pair of the test dataset, participants have to produce a scalar score a_i (in the $[0, 1]$ range) indicating the probability both texts are written by the same author. It is possible for participants to leave text pairs unanswered by submitting a score of precisely $a_i = 0.5$. As concerns the set of evaluation

Table 2. Final results for the cross-discourse type authorship verification task at PAN'23. Submitted systems are ranked by their mean performance across five evaluation metrics. The best result per column is shown in bold.

Systems	AUROC	$c@1$	F_1	$F_{0.5u}$	Brier	Overall
Ibrahim, et al. (reduced-graph) [19]	**0.616**	**0.572**	0.617	0.562	0.746	**0.623**
Ibrahim, et al. (resolving-globe) [19]	**0.616**	**0.572**	0.617	0.562	0.746	**0.623**
Guo, et al. (irregular-strategist) [14]	0.581	0.557	0.621	**0.571**	0.742	0.614
Ibrahim, et al. (golden-ottoman) [19]	0.598	0.546	0.622	0.550	0.744	0.612
BASELINE (cngdist)	0.516	0.499	0.666	0.555	0.741	0.595
Petropoulos (graceful-chianti) [40]	0.526	0.514	0.624	0.549	0.743	0.591
Petropoulos (clever-daemon) [40]	0.525	0.516	0.622	0.550	0.743	0.591
BASELINE (galicia22)	0.504	0.502	0.650	0.552	0.740	0.589
Valdez Valenzuela, et al. (GNN-SHORT) [70]	0.511	0.508	0.655	0.555	0.705	0.587
Sun, et al. (SDML epoch 8) [66]	0.504	0.502	0.632	0.546	0.747	0.586
Sun, et al. (SDML epoch 24) [66]	0.505	0.501	0.601	0.536	0.749	0.578
Guo, et al. (uniform-reward) [14]	0.595	0.555	0.460	0.527	0.723	0.572
Valdez Valenzuela, et al. (GNN-FULL) [70]	0.517	0.512	0.628	0.549	0.644	0.570
Sun, et al. (SDML epoch 35) [66]	0.511	0.508	0.558	0.526	0.749	0.570
Valdez Valenzuela, et al. (GNN-MED) [70]	0.503	0.502	0.602	0.534	0.709	0.570
BASELINE (najafi22)	0.601	0.569	0.466	0.543	0.595	0.555
Huang, et al. (isochoric-paint) [18]	0.563	0.563	0.511	0.550	0.563	0.550
Liu, et al. (coincident-sound) [30]	0.548	0.548	0.544	0.547	0.548	0.547
Lv (radioactive-copyright) [33]	0.553	0.553	0.504	0.540	0.553	0.541
Huang, et al. (steel-coriander) [18]	0.500	0.500	0.651	0.551	0.500	0.540
Li, et al. (wan-ocean) [28]	0.500	0.500	0.646	0.550	0.500	0.539
Lv, et al. (tender-bugle) [33]	0.551	0.551	0.501	0.537	0.551	0.538
Lv, et al. (cold-rotor) [33]	0.550	0.550	0.465	0.524	0.550	0.528
Qiu, et al. (corn-mall) [42]	0.540	0.540	0.421	0.499	0.540	0.508
Qiu, et al. (poky-deck) [42]	0.540	0.540	0.421	0.499	0.540	0.508
Liu, et al. (perpendicular-field) [30]	0.534	0.534	0.421	0.493	0.534	0.503
Liu, et al. (foggy-raster) [30]	0.533	0.533	0.424	0.493	0.533	0.503
BASELINE (compressor)	0.506	0.051	0.626	0.076	0.750	0.402
Sanjesh, et al. (calm-lyrics) [58]	0.525	0.500	0.030	0.068	0.729	0.370
Sanjesh, et al. (null-midpoint) [58]	0.523	0.499	0.031	0.066	0.730	0.370
Sanjesh, et al. (Multi-Feature Classifier) [58]	0.501	0.01	0.000	0.000	0.750	0.252

measures, the set of measures used in the last edition of PAN is also adopted. These include the area under ROC (AUROC), $c@1$ that rewards unanswered cases over wrong predictions, F_1, $F_{0.5u}$, and the complement of Brier score (so that higher scores correspond to better performance) [62]. The average of these diverse measures is used as the final score to rank participants.

Two simple approaches are used as baselines: a compression-based approach based on Prediction by Partial Matching (PPM) [67] and a naive distance-based character n-gram model [21]. In addition, two submissions from the previous edition of the task at PAN-2022 are also used as baselines [62]. One of them is based on a pre-trained language model (T5) combined with a convolutional neural network [39] while the other uses a graph-based Siamese network [34]. We received submissions from 11 research teams and a total number of 27 runs (i.e., at most three runs per participant were allowed). The performance of each run was evaluated using the TIRA experimentation framework. The evaluation results on the test dataset of all submitted software and the baselines can be seen in Table 2.

The difficulty of the task and the specific dataset including discourse types from both written and oral language is reflected in the obtained results. In general, the performance of most submitted systems is quite low, nearly surpassing a random guess baseline. The most successful approaches are based on pre-trained language models (e.g., BERT) enhanced by contrastive learning. However, a naive baseline based on character n-grams is quite competitive. A more detailed analysis of the evaluation results and the submissions is available in the task overview paper [63].

3   Multi-Author Writing Style Analysis

Authorship identification tasks are based on the intrinsic analysis of writing styles. Multi-author writing style analysis of multi-author documents aims to identify text positions at which the authorship changes based on an intrinsic style analysis. With advancing task definitions, data sets, and evaluation procedures, this PAN task has evolved steadily since 2016. The task in 2016 was to identify individual authors within a document and group these fragments [56]. In 2017, participants were asked to assess whether a given document is multi-authored. We asked participants to identify the positions of style changes if the document was indeed multi-authored [69]. For the challenges between 2018 and 2021, we asked participants to predict whether a given document is single- or multi-authored [22]. Additionally, we asked for the number of authors of multi-author documents [81]. In 2020 and 2021, we asked participants to detect paragraph-level style changes for multi-author documents [80]. In 2021, participants had to assign all paragraphs of the text uniquely to some author [77]. In 2022, participants were asked to identify all positions of writing style changes both on the paragraph- and the sentence-level [78].

Multi-Author Writing Style Analysis at PAN'23

Methods for multi-author writing style analysis are the key enabling technology for author identification tasks. The analysis of writing styles allows for performing intrinsic plagiarism detection (i.e., detecting plagiarism without the use of a reference corpus). As part of PAN@CLEF, we continue to develop benchmarks and challenges to advance research in this important field.

The multi-author writing style analysis task at PAN'23 asks participants to identify all positions of writing style change on the paragraph level for a given text. For each pair of consecutive paragraphs, the goal is to assess whether there was a style change between those paragraphs. In previous years, we employed different tasks of different complexity, that were carried out on the same data sets. However, the previously used data sets exhibited substantial topic diversity, which allowed the participants to leverage topic information as a style change

Table 3. Overall results for the style change detection task. The best result for each data set is given in bold.

Systems	Easy F_1	Medium F_1	Hard F_1
Ye et al. [76]	0.983	0.830	**0.821**
Hashemi et al. [15]	**0.984**	**0.843**	0.812
Kucukkaya et al. [24]	0.982	0.810	0.772
Huang et al. [17]	0.968	0.806	0.769
Chen et al. [6]	0.914	0.820	0.676
Jacobo et al. [20]	0.793	0.591	0.498

signal. Therefore, at PAN'23, we provide three data sets of increasing difficulty w.r.t. the multi-author writing style analysis task: *Easy*: The paragraphs of a document cover a variety of topics, allowing approaches to make use of topic information to detect authorship changes. *Medium:* The topical variety in a document is small (though still present), forcing the approaches to focus more on style to effectively solve the detection task. *Hard:* All paragraphs in a document are on the same topic.

Data Set and Evaluation

As a departure from the data sets of previous years, the data sets for this year's edition of the Multi-Author Writing Style Analysis task are based on user posts on Reddit[3]. In an effort to generate both realistic and diverse texts for the data sets, we chose parts of Reddit (so-called *subreddits*) that tend to generate longer and more meaningful discussions by users to extract our data from. The following subreddits were chosen: *r/worldnews, r/politics, r/askhistorians*, and *r/legaladvice*.

Like in previous years, we performed various cleaning steps to ensure the documents generated for the task consisted of well-formed texts. Quotes, all forms of markdown, multiple line breaks or whitespaces, frequently used emojis, hyperlinks as well as trailing and leading whitespaces were removed.

Following this, the collected user posts were split into paragraphs, and then documents for the data sets were generated from the paragraphs of a single given Reddit post. This was done to ensure at least a basic level of topical coherence for all the paragraphs in the final document. To generate style changes, a random set of authors for the given post was chosen, and paragraphs written by those authors were concatenated to form the final document. For the first time this year, this mixing of paragraphs into documents was not done fully randomly, but instead uses a newly developed procedure that allows us to (1) generate more topically and stylistically coherent documents and (2) tweak the difficulty of the produced data set. For this, both semantic as well as stylistic properties of

---

[3] https://www.reddit.com/.

the paragraphs were extracted into a feature vector, and paragraphs were then mixed based on the similarity of those vectors, where those similarities were configured to be (1) relatively large for the *easy* data set, (2) moderate for the *medium* data set, and (3) small for the *hard* data set.

All generated documents were written by between two and four authors, with an even distribution of the number of authors over the documents. Overall, each data set consists of 6,000 documents. Like in previous years, training, test, and validation splits are provided for all three data sets, with the test sets being withheld until the evaluation phase of the competition. The training sets contain 70% of the documents in each data set, while the test and validation sets contain 15% each.

The effectiveness of the models is evaluated independently on the three datasets using macro-averaged F1-score value across all documents.

Results

The Multi-Author Writing Style Analysis task received six software and note-book paper submissions. The individual results achieved by the participants are presented in Table 3. For both the *easy* and *medium* data set, the submission by Hashemi et al. achieved the highest performance, while the approach by Ye et al. performed best on the *hard* data set. Further details on the approaches taken can be found in the overview paper [79].

4   Author Profiling

Author profiling is the problem of distinguishing between classes of authors by studying how language is shared by people. This helps in identifying authors' individual characteristics, such as age, gender, or language variety, among others. During the years 2013–2022, we addressed several of these aspects in the shared tasks organized at PAN.[4] In 2013 the aim was to identify gender and age in social media texts for English and Spanish [50]. In 2014 we addressed age identification from a continuous perspective (without gaps between age classes) in the context of several genres, such as blogs, Twitter, and reviews (in Trip Advisor), both in English and Spanish [47]. In 2015, apart from age and gender identification, we addressed also personality recognition on Twitter in English, Spanish, Dutch, and Italian [52]. In 2016, we addressed the problem of cross-genre gender and age identification (training on Twitter data and testing on blogs and social media data) in English, Spanish, and Dutch [53]. In 2017, we addressed gender and language variety identification in Twitter in English, Spanish, Portuguese, and Arabic [51]. In 2018, we investigated gender identification on Twitter from a multimodal perspective, considering also the images linked within tweets; the dataset was composed of English, Spanish, and Arabic tweets [49]. In 2019 our

---

[4] To generate the datasets, we have followed a methodology that complies with the EU General Data Protection Regulation [45].

focus was on profiling and discriminating bots from humans on the basis of textual data only [46] and targeting both English and Spanish tweets. In 2020, we focused on profiling fake news spreaders [44], in two languages, English and Spanish. The ease of publishing content on social media has also increased the amount of disinformation that is published and shared. The goal of this shared task was to profile those authors who have shared some fake news in the past. In 2021 the focus was on profiling hate speech spreaders in social media [43]. The goal was to identify Twitter users who can be considered haters, depending on the number of tweets with hateful content that they had spread. The task was set in English and Spanish. Finally, in 2022, we focused on profiling irony and stereotype spreaders on English tweets [55]. The shared task goal was to profile highly ironic authors and those that employ irony to convey stereotypical messages, e.g. towards women or the LGTB community.

Profiling Cryptocurrency Influencers with Few-shot Learning

Cryptocurrencies have massively increased their popularity in recent years [59]. The promise of independence from central authorities, the possibilities offered by the different projects, and the new influencer-driven gold rush make cryptocurrencies a trendy topic in social media. Additionally, we believe that due to the early stage and complexity of the crypto ecosystem, many users trust social media influencers to bridge the gap in their lack of knowledge to later take investment decisions. As a consequence, profiling those influential actors becomes relevant.

Producing a sufficient number of high-quality annotations for author profiling is challenging. Profiling influencers, in particular, has high requirements in the economic and temporal cost, the psychological and linguistic expertise needed by the annotator, and the congenital subjectivity involved in the annotation task [3,68]. Additionally, in a real environment, i.e. when traders want to leverage social media signals to forecast the market, profiling needs to be done in real-time in a few milliseconds. This difficult, expensive, and high-speed data collection process implies data scarcity: models need to work with as little data as possible and still perform.

In this shared task, we aim to profile cryptocurrency influencers in social media from a low-resource perspective, that is, using little data. Moreover, we proposed to profile types of influencers also using a low-resource setting. Specifically, we focus on English Twitter posts for three different sub-tasks: (1) *SubTask1-Low-resource influencer profiling*: profile authors according to their degree of influence (non-influencer, nano, micro, macro, mega); (2) *SubTask2-Low-resource influencer interest profiling*: profile authors according to their main interests or areas of influence (technical information, price update, trading matters, gaming, other); and (3) *SubTask3-Low-resource influencer intent profiling*: profile authors according to the intent of their messages (subjective opinion, financial information, advertising, announcement).

Table 4. Datasets statistics including the per-class numbers of users, where the tasks are the following. SubTask1: Low-resource influencer profiling; SubTask2: Low-resource influencer interest profiling; and SubTask3: Low-resource influencer intent profiling.

Task	Partition	Total number of users per class
1	train	macro:32, mega:32, micro:30, nano:32, non-influencer:32
	test	macro:42, mega:45, micro:46, nano:45, non-influencer:42
2	train	technical information:64; trading matters:64; price update:64; gaming:64; other:64
	test	technical information:42; trading matters:112; price update:108; gaming:40; other:100
3	train	announcement:64; subjective opinion:64; financial information:64; advertising:64
	test	announcement:37; subjective opinion:160; financial information:43; advertising:52

Dataset and annotation

As in previous years, a new dataset has been created from English tweets posted by users on Twitter. We built the datasets as follows: first, we identified those who are crypto influencers, and next, we classified their interest and intent.

We identify crypto influencer candidates with two conditions: (1) user with tweets that contain the *ticker* hashtag for different crypto projects e.g. *$ETH, $BTC, $UNI etc.* ; and (2) tweets with mentions in the name of the crypto projects e.g. *Ethereum, Bitcoin, Uniswap.* Next, we extract the number of followers for those users. Finally, we use a follower scale to determine their influence grade. This scale adjusted as much as possible to the most commonly accepted definition of influencer tiers:[5]

- Non-influencer: Individuals with a minimal social media following; typically ranging from 0 to 1,000 followers. Lacks the ability to sway opinions or impact decisions through their online presence.
- Nano-influencers: Individuals with a small, dedicated social media following; typically ranging from 1,000 to 10,000 followers.
- Micro-influencers: Individuals with a moderately sized social media following ranging from 10,000 to 100,000 followers. They often have a more focused and engaged audience.
- Macro-influencers: Individuals with a substantial social media following; ranging from 100,000 to 1 million followers. They have a wide reach and may cover a broader range of topics or industries.
- Mega-influencers: Individuals with an extensive social media following; more than 1 million followers. They often have a significant impact on popular culture and possess considerable influence across multiple platforms.

For the interest and intent datasets, we applied the following criteria after the influencer identification. For each influencer, three human annotators classified

[5] https://zerogravitymarketing.com/the-different-tiers-of-influencers-and-when-to-use-each/.
https://twitter.com/latermedia/status/1385337617340829701.
https://izea.com/resources/influencer-tiers/.

Table 5. Participant and baseline results of the profiling cryptocurrency influencers shared task. Results in terms of macro F_1 for all three sub-tasks (ST), ordered by weighted average. Bold indicates the leading approach for each task.

Systems	Macro F_1		
	ST1 (Influence)	ST2 (Interest)	ST3 (Intent)
Cano-Caravaca (terra-classic)	61.14	63.15	**67.46**
Villa-Cueva et al. (stellar) [72]	58.44	**67.12**	64.46
(MRL-LLP)	57.44	62.00	65.74
Balanzá García (holo)	**62.32**	57.50	61.81
Giglou et al.(symbol) [12]	52.31	61.21	65.83
Cardona-Lorenzo (vechain)	55.51	60.16	60.28
Carbonell Granados (shiba-inu)	50.38	58.47	66.15
Ferri-Molla et al. (magic) [35]	57.14	55.68	61.62
Li et al.(neo) [29]	55.10	61.63	57.62
Iranzo Sánchez (iota)	54.43	64.55	50.62
t5-large (label tuning) - FS	49.34	56.48	59.91
Huallpa (hive)	52.94	51.48	59.08
Llanes Lacomba (api3)	49.18	46.07	63.12
Labadie et al.(dogecoin) [26]	50.80	51.72	52.59
Casamayor Segarra (tron)	50.13	49.77	53.43
user-char-lr	35.25	52.95	60.21
de Castro Isasi (terra)	48.74	44.60	54.83
Rodríguez Ferrero (harmony)	47.93	54.41	45.83
LDSE	50.20	44.92	51.96
Jaramillo-Hernández (waves)	55.06	42.35	49.21
Girish et al. [13]	37.92	46.66	50.42
Espinosa et al. (core) [9]	34.76	43.47	55.34
Coto et al. (ethereum)	46.68	–	55.94
García Bohigues (sushiswap)	46.64	19.23	22.58
t5-large (bi-encoders) - ZS	12.76	33.34	32.71
random	15.92	20.81	18.41
Kumar et al. [25]	50.21	–	–
Siino et al. (alchemy-pay) [60]	38.51	–	–
Siino et al. (nexo) [61]	38.34	–	–
Lomonaco et al. (wax) [31]	37.62	–	-
Valles Silva (solana)	15.92	–	–
Muslihuddeen et al. (icon) [38]	12.90	–	–

the interest and intent for a random tweet sample. We used majority voting to select the final class.

Table 4 presents the statistics of the datasets and the number of users for each class. Due to the low resources task nature, the number of tweets shared with our participants is small. For SubTask1 the maximum number of tweets is 10; for SubTask2 and SubTask3, the number of tweets per user is limited to 1.

On average more than 20 teams participated in each subtask. Most of the participants addressed our few-shot scenario using neural Transformers [71], including the best-performing system, which used DeBERTaV3 [16]. We compare the participants' results with different baselines covering diverse concepts such as transfer [73] and few-shot learning [8, 36, 37]:

- *random*: labels are randomly selected with equal probability.
- *t5-large (bi-encoders) - ZS* : Zero shot (ZS) text classification employing a t5-large model with bi-encoders [36].
- *t5-large (label tuning) - FS* : Few shot (FS) text classification employing a t5-large model with a label-tuning training strategy [36]
- *Character n-grams with logistic regression* (user-char-lr): We use [1..5] character *n*-grams with a TF-IDF weighting calculated using all texts.
- *Low-Dimensionality Statistical Embedding* (LDSE): This method [48] represents documents on the basis of the probability distribution of occurrence of their tokens in the different classes. The distribution of weights for a given document is expected to be closer to the weights of its corresponding class.

Results

Table 5 shows the participants' scores and baseline results. Our result analysis shows that around 46% of the final submissions outperformed our best baseline for each subtask. In addition, only one submission performed worse than the *random* baseline. Finally, the best systems could achieve an improvement of up to 10% absolute macro F_1 score compared to our best baselines.

Further details on the participants' approaches and results can be found in the task overview paper [7].

5 Trigger Detection

A trigger in psychology is a stimulus that elicits negative emotions or feelings of distress. In general, triggers include a broad range of stimuli—such as smells, tastes, sounds, textures, or sights—which may relate to possibly distressing acts or events of whatever type, for instance, violence, trauma, death, eating disorders, or obscenity. In order to proactively apprise the audience that a piece of media (writing, audio, video, etc.) contains potentially distressing material, the use of "trigger warnings"—labels indicating the type of potentially triggering content present— has become common not only in online communities but also in institutionalized education, making it possible for a sensitive audience to prepare themselves for the content and better manage their reactions. We cast this

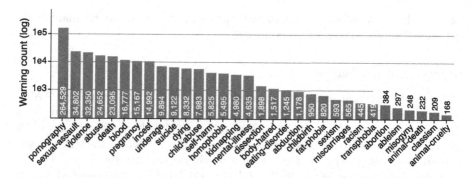

Fig. 1. Distribution of the 32 classes in the PAN23-trigger-detection dataset.

Table 6. Descriptive statistics of the training, validation, and test split of the dataset.

Training Dataset		Validation Dataset		Test Dataset	
Total Works	307,102	Total Works	17,104	Total Works	17,040
< 512 words	15,233	< 512 words	861	< 512 words	813
< 4,096 words	261,156	< 4,096 words	14,571	< 4,096 words	14,555
Mean no. words	2,400	Mean no. words	2,386	Mean no. words	2,388
Median no. words	2,126	Median no. words	2,115	Median no. words	2,101
90pct no. words	4,579	90pct no. words	4,550	90pct no. words	4,558

setting as a computational problem of identifying whether or not a given document contains triggering content, and if so, of what kind. In the present edition of the shared task, we asked participants to work with a corpus in which documents have been pre-tagged with content descriptors by the author (see below). Specifically, we modeled trigger detection as a multi-label document classification challenge of assigning each document all appropriate trigger warnings, but not more.

In this pilot edition of the Trigger Detection task at PAN 2023, our aim was to establish the computational problem of identifying whether or not a given document contains triggering content, and if so, of what type. As data, we created PAN23-trigger-detection, a new evaluation resource of fan fiction from Archive of our Own (Ao3) in which trigger warnings have been assigned by the authors, hence we rely on user-generated labels and follow the authors' understanding of triggers and which documents require a warning. The warnings are assigned via AO3's freeform content descriptor system ("tags"), which are custom, high-dimensional, and mostly contain non-warning descriptors, so we developed a distant-supervision strategy to detect if a freeform tag corresponds to one of 32 predefined warnings which we compiled from institutional content warning guidelines.

We formalize trigger detection as a multi-label document classification task as follows: Given a fan fiction document, assign all appropriate trigger warnings from the given set. The task is primarily evaluated with the standard measures for multi-label classification, micro and macro F_1. In total, 6 participants submitted software to Trigger Detection 2023.

Dataset and Evaluation

For trigger detection 2023, we created a new evaluation resource, PAN23-trigger-detection, consisting of 341,246 fan fiction works downloaded by us from Archive of our Own (Ao3) and annotated in a multi-label setting with any of 32 warning labels. Figure 1 shows the distribution of the labels over the test dataset; Table 6 shows the statistics of the standard splits of our dataset.

Since there was no authoritative (closed-set) label set, we complied the 32 labels for our dataset from two institutionally-prescribed trigger lists: the University of Reading list of "themes that require trigger warnings" [54] and the University of Michigan list of content warnings [32]. The two largely overlapping guidelines list 21 categories of triggering concepts, including health-related (*eating disorders, mental illness*), sexually-oriented (*sexual assault, pornography*) as well as verbal (*hate speech, racial slurs*), and physical abuse (*animal cruelty, blood, suicide*). The lists were preprocessed to unfold compound categories into individual elements (e.g. "Animal cruelty or animal death" $\rightarrow$ "animal cruelty", "animal death") and lower-cased. The final set of trigger warnings comprises 35 categories, although we removed the rarest three labels since there were too few annotated documents with those labels in the final dataset.

We initially downloaded all ca. 10 million works released between August 13, 2008 (the platform launch) and August 09, 2021, from `archiveofourown.org` and extracted the document text and metadata (i.e. the freeform tags) from the scraped HTML. To download the HTML page of each work, we scraped the output of the search function to get the work ID and then constructed a direct URL to that works page. Since the search function was limited to 10,000 works per page, we constructed queries to search for all works released on one particular day, for each day in the release window.

We annotated all works in our corpus with appropriate trigger warnings by replacing each freeform tag assigned to the work with all corresponding warnings or removing the freeform tag if there is no corresponding warning. The underlying replacement table, which maps from freeform tag to trigger warning, was created by (1) manually annotating 2,000 most common tags, (2) efficiently identifying sub-structures of the tag graph that indicate a trigger warning, annotating each node in the structure with that warning, and (3) merging both results, giving priority to the manual annotations. This method is presented in more detail by Wiegmann et al. [75].

From the resulting corpus of annotated fan fiction works, we sampled pan23-trigger-detection by discarding all works that had no warning assigned, were originally published pre-2009 (as opposed to posted since AO3 also archives works from older fan fiction sites), had freeform tags that could not decidedly be mapped, was not in English (ca 8% of the works), had less than 50 or more than 6,000 words (outliers; ease of computation), less than 2 or more than 66 freeform tags (confidence threshold), less than 1,000 hits (views), less than 10

Table 7. Participant scores of the Trigger Detection task at PAN 2023. Shown are only the core metrics. The table is sorted by macro F_1, the primary metric. Bold indicates the leading approach for each metric.

Systems	Macro			Micro			Acc
	Prec	Rec	F_1	Prec	Rec	F_1	
Sahin et al. [57]	0.37	0.42	**0.352**	0.73	0.74	0.74	0.59
Su et al. [65]	**0.54**	0.30	**0.350**	0.80	0.71	**0.75**	**0.62**
XGBoost baseline	0.52	0.25	0.301	**0.88**	0.57	0.69	0.53
Cao, H. et al. [5]	0.24	0.29	0.228	0.43	0.79	0.56	0.18
Cao, G. et al. [4]	0.28	0.22	0.225	0.58	0.66	0.62	0.32
Felser et al. [10]	0.11	**0.63**	0.161	0.27	**0.82**	0.40	0.27
Shashirekha et al. [27]	0.10	0.04	0.048	0.82	0.50	0.63	0.52

kudos (likes; popularity threshold). We also removed all (near) duplicates. The resulting dataset (cf. Table 6) had 341,246 fan fiction works remaining, from which we stratified sampled 90:5:5 intro training, validation, and test datasets, i.e. we kept the label distribution equal across the standard splits.

We evaluate the submitted approaches through precision, recall, and F_1 at both micro- and macro average, as well as with subset accuracy, which measures accuracy on a per-example level (i.e. if all labels of one example are set correctly). We slightly favor the macro over the micro F_1 scores due to the label imbalance. We also favor recall over precision, since trigger warning assignment is a high-recall task where false negatives cause more harm than false positives. As a baseline approach, we supplied an XGBoost approach trained on TF·IDF document vectors of a max. 1,000 examples per-class random down-sampling of the training data.

Submissions and Results

The 6 submissions to trigger detection 2023 utilized a broad set of techniques, from hierarchical transformer structures to strategic feature engineering via semi-supervised topic modeling. Table 7 shows the final results, ordered by macro F_1. Most submissions focussed on improving the long document aspect of the task (most documents are longer than the input size of the SotA classification models) by using chunking and hierarchical neural networks and coping with the label imbalance (the most common label (*pornography*) is an order of magnitude more common than the other labels) by using adapted class balancing or custom loss functions. The best-performing approaches used hierarchical transformers to use the strong contextualization of the architecture while overcoming its input size limitation.

Sahin et al. submitted a hierarchical transformer architecture that achieved the top macro F_1 score (by a slim margin of 0.002) and second in micro F_1 and accuracy while having a relatively high recall within the top approaches. The approach first segments the document into chunks (200 words with 50 words overlap) and then pre-trains a RoBERTa transformer on the chunks to learn the genre. The architecture then embeds all chunks of a document using the pre-trained transformer, followed by an LSTM for each label (in a one-vs-all setting) which predicts the class from a sequence of chunk-embeddings (RoBERTa's [CLS] token). To cope with the label imbalance, the approach up-scales the class weight in the loss function for the rare half of the classes.

Su et al. also submitted a siamese transformer that achieved the second-best macro F_1 score (by a slim margin of 0.002) and the top scores in micro F_1 and accuracy while notably favoring precision over recall. The approach first segments the documents into 505-word chunks encodes the first and last chunks using a pre-trained RoBERTa, mean-pools the contextual embeddings (ignoring the [CLS] token), and classifying based on the pooled embeddings using a 1D convolutional neural network.

Cao, H. et al. submitted a voting-based transformer that favors recall over precision. The approach segments the training documents into chunks, assigns each chunk the labels from its source document, and trains a single RoBERTa-based classifier on each chunk. To make predictions, the documents are again chunked, the labels for each chunk are predicted, and a label is assigned to the document if it is assigned to more than half of the chunks. The training data was dynamically over- and under-sampled: pornography was under-sampled to 5,000 and other labels to 2,000. Examples with rare labels were replicated 8–10 times.

Cao, G. et al. also submitted a voting-based transformer that achieved very balanced results, neither favoring macro over micro or precision over recall. The approach chunks and votes similarly to Cao, H. et al. but builds two different models to overcome the data imbalance, one for pornography and one for the other 31 classes. The pornography model was trained on a random selection of 40,000 works with and 40,000 works without the pornography warning. The model for the other labels removes works with only the pornography warning, under-samples frequent classes to 3,000 examples, and over-samples the rare labels by replicating works 4–6 times.

Felser et al. submitted a multi-layer perceptron classifier based on fasttext-based document embeddings and coarse-grained label priors determined through a combination of semi-supervised topic modeling and supervised learning. This approach achieved the top micro and macro recall, although at the cost of precision on the test dataset. The document embeddings were created by training a fasttext model from the training data, generating the embeddings for each unique word in a document, scaling those by term frequency, and adding and norming those scaled word vectors over the document. The topic modeling features were created by, first, grouping the 32 labels semantically into 6 groups, second, bootstrapping a seeded LDA with the 50 most relevant bi-grams of each group (determined through a TF·IDF-like approach for n-gram weighting which also down-grades pornographic terms), and third, training a classifier to predict the label from the topic model, where the classifiers label confidence serves as the final feature for the MLP.

Lastly, Shashirekha et al. present an LSTM-based approach using GloVE-embeddings and which is third in micro F_1 with very high precision but rather weak in macro averages.

A more extensive evaluation and comparison of the approaches and the insights they give us into trigger detection can be found in the extended overview paper [74].

Acknowledgments. The work from Symanto has been partially funded by the Pro^{2}Haters - Proactive Profiling of Hate Speech Spreaders (CDTi IDI-20210776), the XAI-DisInfodemics: eXplainable AI for disinformation and conspiracy detection during infodemics (MICIN PLEC2021-007681), the OBULEX - *OBservatorio del Uso de Lenguage sEXista en la red* (IVACE IMINOD/2022/106), and the ANDHI - ANomalous Diffusion of Harmful Information (CPP2021-008994) R&D grants.

The work of Paolo Rosso was in the framework of the FairTransNLP research project (PID2021-124361OB-C31), funded by MCIN/AEI/10.13039/501100011033 and by ERDF, EU A way of making Europe.

This work has been partially supported by the OpenWebSearch.eu project (funded by the EU; GA 101070014).

References

1. Bevendorff, J., et al.: Overview of PAN 2021: authorship verification, profiling hate speech spreaders on twitter, and style change detection. In: Candan, K.S., et al. (eds.) CLEF 2021. LNCS, vol. 12880, pp. 419–431. Springer, Cham (2021). https://doi.org/10.1007/978-3-030-85251-1_26

2. Bevendorff, J., et al.: Overview of PAN 2020: authorship verification, celebrity profiling, profiling fake news spreaders on Twitter, and style change detection. In: Arampatzis, A., et al. (eds.) CLEF 2020. LNCS, vol. 12260, pp. 372–383. Springer, Cham (2020). https://doi.org/10.1007/978-3-030-58219-7_25

3. Bobicev, V., Sokolova, M.: Inter-annotator agreement in sentiment analysis: machine learning perspective. In: Proceedings of the International Conference Recent Advances in Natural Language Processing (2017)

4. Cao, G., et al.: A dual-model classification method based on RoBERTa for trigger detection. In: Aliannejadi, M., Faggioli, G., Ferro, N., Vlachos, M. (eds.) Working Notes of CLEF 2023 - Conference and Labs of the Evaluation Forum, CEUR-WS.org (2023)

5. Cao, H., et al.: Trigger warning labeling with RoBERTa and resampling for distressing content detection. In: Aliannejadi, M., Faggioli, G., Ferro, N., Vlachos, M. (eds.) Working Notes of CLEF 2023 - Conference and Labs of the Evaluation Forum, CEUR-WS.org (2023)

6. Chen, H., Han, Z., Li, Z., Han, Y.: A writing style embedding based on contrastive learning for multi-author writing style analysis. In: Aliannejadi, M., Faggioli, G., Ferro, N., Vlachos, M. (eds.) Working Notes of CLEF 2023 - Conference and Labs of the Evaluation Forum, CEUR-WS.org (2023)

7. Chinea-Rios, M., Borrego-Obrador, I., Franco-Salvador, M., Rangel, F., Rosso, P.: Profiling cryptocurrency influencers with few-shot learning at PAN 2023. In: CLEF 2022 Labs and Workshops, Notebook Papers, CEUR-WS.org (2023)

8. Chinea-Rios, M., Müller, T., Sarracén, G.L.D.l.P., Rangel, F., Franco-Salvador, M.: Zero and few-shot learning for author profiling. arXiv preprint arXiv:2204.10543 (2022)

9. Espinosa, D., Sidorov, G.: Using BERT to profiling cryptocurrency influencers. In: Aliannejadi, M., Faggioli, G., Ferro, N., Vlachos, M. (eds.) Working Notes of CLEF 2023 - Conference and Labs of the Evaluation Forum, CEUR-WS.org (2023)

10. Felser, J., Demus, C., Labudde, D., Spranger, M.: FoSIL at PAN?23: trigger detection with a two stage topic classifier. In: Aliannejadi, M., Faggioli, G., Ferro, N., Vlachos, M. (eds.) Working Notes of CLEF 2023 - Conference and Labs of the Evaluation Forum, CEUR-WS.org (2023)

11. Fröbe, M., et al.: Continuous integration for reproducible shared tasks with TIRA.io. In: Kamps, J., et al. (eds.) ECIR 2023. Lecture Notes in Computer Science, vol. 13982, pp. 236–241. Springer, Berlin (2023). https://doi.org/10.1007/978-3-031-28241-6_20

12. Giglou, H.B., Rahgouy, M., Oskuee, J.D.M.M., Tekanlou, H.B.A., Seals, C.D.: Leveraging large language models with multiple loss learners for few-shot author profiling. In: Aliannejadi, M., Faggioli, G., Ferro, N., Vlachos, M. (eds.) Working Notes of CLEF 2023 - Conference and Labs of the Evaluation Forum, CEUR-WS.org (2023)

13. Girish, K., Hegdev, A., Balouchzahi, F., Lakshmaiah, S.H.: Profiling cryptocurrency influencers with sentence transformers. In: Aliannejadi, M., Faggioli, G., Ferro, N., Vlachos, M. (eds.) Working Notes of CLEF 2023 - Conference and Labs of the Evaluation Forum, CEUR-WS.org (2023)

14. Guo, M., Han, Z., Chen, H., Qi, H.: A contrastive learning of sample pairs for authorship verification. In: Aliannejadi, M., Faggioli, G., Ferro, N., Vlachos, M. (eds.) Working Notes of CLEF 2023 - Conference and Labs of the Evaluation Forum, CEUR-WS.org (2023)

15. Hashemi, A., Shi, W.: Enhancing writing style change detection using transformer-based models and data augmentation. In: Aliannejadi, M., Faggioli, G., Ferro, N., Vlachos, M. (eds.) Working Notes of CLEF 2023 - Conference and Labs of the Evaluation Forum, CEUR-WS.org (2023)

16. He, P., Liu, X., Gao, J., Chen, W.: Deberta: Decoding-enhanced BERT with disentangled attention. In: International Conference on Learning Representations (2021). https://openreview.net/forum?id=XPZIaotutsD

17. Huang, M., Huang, Z., Kong, L.: Encoded classifier using knowledge distillation for multi-author writing style analysis. In: Aliannejadi, M., Faggioli, G., Ferro, N., Vlachos, M. (eds.) Working Notes of CLEF 2023 - Conference and Labs of the Evaluation Forum, CEUR-WS.org (2023)

18. Huang, Z., Kong, L., Huang, M.: Authorship verification based on CoSENT. In: Aliannejadi, M., Faggioli, G., Ferro, N., Vlachos, M. (eds.) Working Notes of CLEF 2023 - Conference and Labs of the Evaluation Forum, CEUR-WS.org (2023)

19. Ibrahim, M., et al.: Enhancing authorship verification using sentence-transformers. In: Aliannejadi, M., Faggioli, G., Ferro, N., Vlachos, M. (eds.) Working Notes of CLEF 2023 - Conference and Labs of the Evaluation Forum, CEUR-WS.org (2023)

20. Jacobo, G., Dehesa, V., Rojas, D., Gómez-Adorno, H.: Authorship verification machine learning methods for style change detection in texts. In: Aliannejadi, M., Faggioli, G., Ferro, N., Vlachos, M. (eds.) Working Notes of CLEF 2023 - Conference and Labs of the Evaluation Forum, CEUR-WS.org (2023)

21. Kestemont, M., Stover, J., Koppel, M., Karsdorp, F., Daelemans, W.: Authenticating the writings of Julius Caesar. Expert Syst. Appl. **63**, 86–96 (2016)

22. Kestemont, M., et al.: Overview of the author identification task at PAN 2018: cross-domain authorship attribution and style change detection. In: CLEF 2018 Labs and Workshops, Notebook Papers (2018)
23. Koppel, M., Winter, Y.: Determining if two documents are written by the same author. J. Assoc. Inf. Sci. Technol. **65**(1), 178–187 (2014)
24. Kucukkaya, I.E., Sahin, U., Toraman, C.: ARC-NLP at PAN 23: transition-focused natural language inference for writing style detection. In: Aliannejadi, M., Faggioli, G., Ferro, N., Vlachos, M. (eds.) Working Notes of CLEF 2023 - Conference and Labs of the Evaluation Forum, CEUR-WS.org (2023)
25. Kumar, A., Saeed, A.A., Trinh, L.H.M.: Profiling cryptocurrency influencers with few shot learning. In: Aliannejadi, M., Faggioli, G., Ferro, N., Vlachos, M. (eds.) Working Notes of CLEF 2023 - Conference and Labs of the Evaluation Forum, CEUR-WS.org (2023)
26. Labadie, R., Sarvazyan, A.M.: Reshape or update? metric learning and fine-tuning for low-resource influencer profiling. In: Aliannejadi, M., Faggioli, G., Ferro, N., Vlachos, M. (eds.) Working Notes of CLEF 2023 - Conference and Labs of the Evaluation Forum, CEUR-WS.org (2023)
27. Lakshmaiah, S.H., Hegde, A., Balouchzahi, F.: Trigger detection in social media text. In: Aliannejadi, M., Faggioli, G., Ferro, N., Vlachos, M. (eds.) Working Notes of CLEF 2023 - Conference and Labs of the Evaluation Forum, CEUR-WS.org (2023)
28. Li, J., Zhang, Q., Huang, M.: Author verification of text fragments based on the Bert model. In: Aliannejadi, M., Faggioli, G., Ferro, N., Vlachos, M. (eds.) Working Notes of CLEF 2023 - Conference and Labs of the Evaluation Forum, CEUR-WS.org (2023)
29. Li, Z., Han, Z., Cai, J., Huang, Z., Huang, S., Kong, L.: CLEM4PCI: profiling cryptocurrency influencers with few-shot learning via contrastive learning and ensemble model. In: Aliannejadi, M., Faggioli, G., Ferro, N., Vlachos, M. (eds.) Working Notes of CLEF 2023 - Conference and Labs of the Evaluation Forum, CEUR-WS.org (2023)
30. Liu, X., Kong, L., Huang, M.: Text-segment interaction for authorship verification using BERT-based classification. In: Aliannejadi, M., Faggioli, G., Ferro, N., Vlachos, M. (eds.) Working Notes of CLEF 2023 - Conference and Labs of the Evaluation Forum, CEUR-WS.org (2023)
31. Lomonaco, F., Siino, M., Tesconi, M.: Text enrichment with Japanese language to profile cryptocurrency influencers. In: Aliannejadi, M., Faggioli, G., Ferro, N., Vlachos, M. (eds.) Working Notes of CLEF 2023 - Conference and Labs of the Evaluation Forum, CEUR-WS.org (2023)
32. LSA, U.: An Introduction to Content Warnings and Trigger Warnings (2023). https://sites.lsa.umich.edu/inclusive-teaching-sandbox/wp-content/uploads/sites/853/2021/02/An-Introduction-to-Content-Warnings-and-Trigger-Warnings-Draft.pdf. Accessed 10 May 2023
33. Lv, J., Han, Y., Dong, Q.: Application of R-drop in author authorship verification. In: Aliannejadi, M., Faggioli, G., Ferro, N., Vlachos, M. (eds.) Working Notes of CLEF 2023 - Conference and Labs of the Evaluation Forum, CEUR-WS.org (2023)
34. Martinez-Galicia, J.A., Embarcadero-Ruiz, D., Rios-Orduna, A., Gómez-Adorno, H.: Graph-based siamese network for authorship verification. In: CLEF 2022 Labs and Workshops, Notebook Papers, CEUR-WS.org (2022)
35. Mollá, I.F., Jordà, J.S.: Profiling cryptocurrency influencers with few-shot learning. Overview for PAN at CLEF 2023. In: Aliannejadi, M., Faggioli, G., Ferro, N.,

Vlachos, M. (eds.) Working Notes of CLEF 2023 - Conference and Labs of the Evaluation Forum, CEUR-WS.org (2023)

36. Mueller, T., Pérez-Torró, G., Franco-Salvador, M.: Few-shot learning with Siamese networks and label tuning. In: Proceedings of the Annual Meeting of the Association for Computational Linguistics, pp. 8532–8545 (2022)

37. Müller, T., Pérez-Torró, G., Basile, A., Franco-Salvador, M.: Active few-shot learning with fasl. arXiv preprint arXiv:2204.09347 (2022)

38. Muslihuddeen, H., Sathvika, P., Sankar, S., Ostwal, S., Kumar, D.A.: Profiling cryptocurrency influencers using few-shot learning. In: Aliannejadi, M., Faggioli, G., Ferro, N., Vlachos, M. (eds.) Working Notes of CLEF 2023 - Conference and Labs of the Evaluation Forum, CEUR-WS.org (2023)

39. Najafi, M., Tavan, E.: Text-to-text transformer in authorship verification via stylistic and semantical analysis. In: CLEF 2022 Labs and Workshops, Notebook Papers, CEUR-WS.org (2022)

40. Petropoulos, P.: Contrastive learning for authorship verification using BERT and bi-LSTM in a Siamese architecture. In: Aliannejadi, M., Faggioli, G., Ferro, N., Vlachos, M. (eds.) Working Notes of CLEF 2023 - Conference and Labs of the Evaluation Forum, CEUR-WS.org (2023)

41. Potthast, M., Gollub, T., Rangel, F., Rosso, P., Stamatatos, E., Stein, B.: Improving the Reproducibility of PAN's shared tasks: In: Kanoulas, E., et al. (eds.) CLEF 2014. LNCS, vol. 8685, pp. 268–299. Springer, Cham (2014). https://doi.org/10.1007/978-3-319-11382-1_22

42. Qiu, Y., Qi, H., Han, Y., Huang, K.: Authorship verification based on SimCSE. In: Aliannejadi, M., Faggioli, G., Ferro, N., Vlachos, M. (eds.) Working Notes of CLEF 2023 - Conference and Labs of the Evaluation Forum, CEUR-WS.org (2023)

43. Rangel, F., De-La-Peña-Sarracén, G.L., Chulvi, B., Fersini, E., Rosso, P.: Profiling hate speech spreaders on Twitter task at PAN 2021. In: Faggioli, G., Ferro, N., Joly, A., Maistro, M., Piroi, F. (eds.) CLEF 2021 Labs and Workshops, Notebook Papers, CEUR-WS.org (2021)

44. Rangel, F., Giachanou, A., Ghanem, B., Rosso, P.: Overview of the 8th author profiling task at PAN 2019: profiling fake news spreaders on Twitter. In: CLEF 2020 Labs and Workshops, Notebook Papers. CEUR Workshop Proceedings (2020)

45. Rangel, F., Rosso, P.: On the implications of the general data protection regulation on the organisation of evaluation tasks. Lang. Law/Linguagem Direito 5(2), 95–117 (2019)

46. Rangel, F., Rosso, P.: Overview of the 7th author profiling task at pan 2019: bots and gender profiling. In: CLEF 2019 Labs and Workshops, Notebook Papers (2019)

47. Rangel, F., et al.: Overview of the 2nd author profiling task at PAN 2014. In: CLEF 2014 Labs and Workshops, Notebook Papers (2014)

48. Rangel, F., Rosso, P., Franco-Salvador, M.: A low dimensionality representation for language variety identification. In: Proceedings of the CICLING, pp. 156–169 (2016)

49. Rangel, F., Rosso, P., Montes-y-Gómez, M., Potthast, M., Stein, B.: Overview of the 6th author profiling task at PAN 2018: multimodal gender identification in Twitter. In: CLEF 2019 Labs and Workshops, Notebook Papers (2018)

50. Rangel, F., Rosso, P., Moshe Koppel, M., Stamatatos, E., Inches, G.: Overview of the author profiling task at PAN 2013. In: CLEF 2013 Labs and Workshops, Notebook Papers (2013)

51. Rangel, F., Rosso, P., Potthast, M., Stein, B.: Overview of the 5th author profiling task at PAN 2017: gender and language variety identification in Twitter. Working Notes Papers of the CLEF (2017)

52. Rangel, F., Rosso, P., Potthast, M., Stein, B., Daelemans, W.: Overview of the 3rd author profiling task at PAN 2015. In: CLEF 2015 Labs and Workshops, Notebook Papers (2015)

53. Rangel, F., Rosso, P., Verhoeven, B., Daelemans, W., Potthast, M., Stein, B.: Overview of the 4th author profiling task at PAN 2016: cross-genre evaluations. In: CLEF 2016 Labs and Workshops, Notebook Papers (2016). ISSN 1613–0073

54. Read., U.: Guide to policy and procedures for teaching and learning; Guidance on content warnings on course content ('trigger' warnings) (2023). https://www.reading.ac.uk/cqsd/-/media/project/functions/cqsd/documents/qap/trigger-warnings.pdf. Accessed 10 May 2023

55. Reynier, O.B., Berta, C., Francisco, R., Paolo, R., Elisabetta, F.: Profiling irony and stereotype spreaders on Twitter (irostereo) at PAN 2022. In: CLEF 2021 Labs and Workshops, Notebook Papers (2022)

56. Rosso, P., Rangel, F., Potthast, M., Stamatatos, E., Tschuggnall, M., Stein, B.: Overview of PAN'16–new challenges for authorship analysis: cross-genre profiling, clustering, diarization, and obfuscation. In: Experimental IR Meets Multilinguality, Multimodality, and Interaction. 7th International Conference of the CLEF Initiative (CLEF 16) (2016)

57. Sahin, U., Kucukkaya, I.E., Toraman, C.: ARC-NLP at PAN 2023: hierarchical long text classification for trigger detection. In: Aliannejadi, M., Faggioli, G., Ferro, N., Vlachos, M. (eds.) Working Notes of CLEF 2023 - Conference and Labs of the Evaluation Forum, CEUR-WS.org (2023)

58. Sanjesh, R., Mangai, A.: A Multi-feature custom classification approach to authorship verification. In: Aliannejadi, M., Faggioli, G., Ferro, N., Vlachos, M. (eds.) Working Notes of CLEF 2023 - Conference and Labs of the Evaluation Forum, CEUR-WS.org (2023)

59. Sawhney, R., Agarwal, S., Mittal, V., Rosso, P., Nanda, V., Chava, S.: Cryptocurrency bubble detection: a new stock market dataset, financial task & hyperbolic models. In: Proceedings of the 2022 Conference of the North American Chapter of the Association for Computational Linguistics: Human Language Technologies, pp. 5531–5545 (2022)

60. Siino, M., Tesconi, M., Tinnirello, I.: Profiling cryptocurrency influencers with few-shot learning using data augmentation and ELECTRA. In: Aliannejadi, M., Faggioli, G., Ferro, N., Vlachos, M. (eds.) Working Notes of CLEF 2023 - Conference and Labs of the Evaluation Forum, CEUR-WS.org (2023)

61. Siino, M., Tinnirello, I.: XLNet with data augmentation to profile cryptocurrency influencers. In: Aliannejadi, M., Faggioli, G., Ferro, N., Vlachos, M. (eds.) Working Notes of CLEF 2023 - Conference and Labs of the Evaluation Forum, CEUR-WS.org (2023)

62. Stamatatos, E., et al.: Overview of the authorship verification task at PAN 2022. In: CLEF 2022 Labs and Workshops, Notebook Papers, CEUR-WS.org (2022)

63. Stamatatos, E., et al.: Overview of the authorship verification task at PAN 2023. In: CLEF 2023 Labs and Workshops, Notebook Papers, CEUR-WS.org (2023)

64. Stamatatos, E., Potthast, M., Rangel, F., Rosso, P., Stein, B.: Overview of the PAN/CLEF 2015 evaluation lab. In: Mothe, J., et al. (eds.) CLEF 2015. LNCS, vol. 9283, pp. 518–538. Springer, Cham (2015). https://doi.org/10.1007/978-3-319-24027-5_49

65. Su, Y., Han, Y., Qi, H.: Siamese networks in trigger detection task. In: Aliannejadi, M., Faggioli, G., Ferro, N., Vlachos, M. (eds.) Working Notes of CLEF 2023 - Conference and Labs of the Evaluation Forum, CEUR-WS.org (2023)

66. Sun, Y., Afanaseva, S., Patil, K.: Stylometric and neural features combined deep Bayesian classifier for authorship verification. In: Aliannejadi, M., Faggioli, G., Ferro, N., Vlachos, M. (eds.) Working Notes of CLEF 2023 - Conference and Labs of the Evaluation Forum, CEUR-WS.org (2023)

67. Teahan, W.J., Harper, D.J.: Using compression-based language models for text categorization. In: Croft, W.B., Lafferty, J. (eds.) Language Modeling for Information Retrieval. The Springer International Series on Information Retrieval, vol. 13, pp. 141–165. Springer, Netherlands (2003). https://doi.org/10.1007/978-94-017-0171-6_7

68. Troiano, E., Padó, S., Klinger, R.: Emotion ratings: How intensity, annotation confidence and agreements are entangled. arXiv preprint arXiv:2103.01667 (2021)

69. Tschuggnall, M., et al.: Overview of the author identification task at PAN 2017: style breach detection and author clustering. In: CLEF 2017 Labs and Workshops, Notebook Papers (2017)

70. Valenzuela, A.V., Adorno, H.G., Galicia, J.A.M.: Heterogeneous-graph convolutional network for authorship verification. In: Aliannejadi, M., Faggioli, G., Ferro, N., Vlachos, M. (eds.) Working Notes of CLEF 2023 - Conference and Labs of the Evaluation Forum, CEUR-WS.org (2023)

71. Vaswani, A., et al.: Attention is all you need. In: Advances in Neural Information Processing Systems, vol. 30 (2017)

72. Villa-Cueva, E., Valles-Silva, J.M., Lopez-Monroy, A.P., Sanchez-Vega, F., Lopez-Santillan, J.R.: Integrating fine-tuned language models and entailment-based approaches for low-resource tweet classification. In: Aliannejadi, M., Faggioli, G., Ferro, N., Vlachos, M. (eds.) Working Notes of CLEF 2023 - Conference and Labs of the Evaluation Forum, CEUR-WS.org (2023)

73. Weiss, K., Khoshgoftaar, T.M., Wang, D.D.: A survey of transfer learning. J. Big Data **3**(1), 1–40 (2016). https://doi.org/10.1186/s40537-016-0043-6

74. Wiegmann, M., Wolska, M., Potthast, M., Stein, B.: Overview of the trigger detection task at PAN 2023. In: CLEF 2023 Labs and Workshops, Notebook Papers, CEUR-WS.org (2023)

75. Wiegmann, M., Wolska, M., Schröder, C., Borchardt, O., Stein, B., Potthast, M.: Trigger warning assignment as a multi-label document classification problem. In: Proceedings of the 61th Annual Meeting of the Association for Computational Linguistics (Volume 1: Long Papers), Association for Computational Linguistics, Toronto, Canada (2023)

76. Ye, Z., Zhong, C., Qi, H., Han, Y.: Supervised contrastive learning for multi-author writing style analysis. In: Aliannejadi, M., Faggioli, G., Ferro, N., Vlachos, M. (eds.) Working Notes of CLEF 2023 - Conference and Labs of the Evaluation Forum, CEUR-WS.org (2023)

77. Zangerle, E., Mayerl, M., Potthast, M., Stein, B.: Overview of the style change detection task at PAN 2021. In: Faggioli, G., Ferro, N., Joly, A., Maistro, M., Piroi, F. (eds.) CLEF 2021 Labs and Workshops, Notebook Papers, CEUR-WS.org (2021)

78. Zangerle, E., Mayerl, M., Potthast, M., Stein, B.: Overview of the style change detection task at PAN 2022. In: CLEF 2022 Labs and Workshops, Notebook Papers, CEUR-WS.org (2022)

79. Zangerle, E., Mayerl, M., Potthast, M., Stein, B.: Overview of the style change detection task at PAN 2023. In: CLEF 2023 Labs and Workshops, Notebook Papers, CEUR-WS.org (2023)

80. Zangerle, E., Mayerl, M., Specht, G., Potthast, M., Stein, B.: Overview of the style change detection task at PAN 2020. In: CLEF 2020 Labs and Workshops, Notebook Papers (2020)
81. Zangerle, E., Tschuggnall, M., Specht, G., Stein, B., Potthast, M.: Overview of the style change detection task at PAN 2019. In: CLEF 2019 Labs and Workshops, Notebook Papers (2019)

Overview of the CLEF 2023 SimpleText Lab: Automatic Simplification of Scientific Texts

Liana Ermakova[1(✉)], Eric SanJuan[2], Stéphane Huet[2], Hosein Azarbonyad[3], Olivier Augereau[4], and Jaap Kamps[5]

[1] Université de Bretagne Occidentale, HCTI, Brest, France
`liana.ermakova@univ-brest.fr`
[2] Avignon Université, LIA, Avignon, France
[3] Elsevier, Amsterdam, The Netherlands
[4] ENIB, Lab-STICC UMR CNRS 6285, Brest, France
[5] University of Amsterdam, Amsterdam, The Netherlands

Abstract. There is universal consensus on the importance of objective scientific information, yet the general public tends to avoid scientific literature due to access restrictions, its complex language or their lack of prior background knowledge. Academic text simplification promises to remove some of these barriers, by improving the accessibility of scientific text and promoting science literacy. This paper presents an overview of the CLEF 2023 SimpleText track addressing the challenges of text simplification approaches in the context of promoting scientific information access, by providing appropriate data and benchmarks, and creating a community of IR and NLP researchers working together to resolve one of the greatest challenges of today. The track provides a corpus of scientific literature abstracts and popular science requests. It features three tasks. First, *content selection* (what is in, or out?) challenges systems to select passages to include in a simplified summary in response to a query. Second, *complexity spotting* (what is unclear?) given a passage and a query, aims to rank terms/concepts that are required to be explained for understanding this passage (definitions, context, applications). Third, *text simplification* (rewrite this!) given a query, asks to simplify passages from scientific abstracts while preserving the main content.

Keywords: Scientific text simplification · (Multi-document) summarization · Contextualization · Background knowledge · Comprehensibility · Scientific information distortion

1 Introduction

Scientific literacy is an important ability for people. It is one of the keys to critical thinking, objective decision-making, and judgment of the validity and significance of findings and arguments, which allows discerning facts from fiction. Thus, having basic scientific knowledge may also help maintain one's health, both physiological and mental. The COVID-19 pandemic provides a good example of such a matter. Understanding the issue itself, choosing to use or avoid a particular treatment or prevention procedure can become crucial. However, the recent pandemic has also shown that simplification can be modulated by political needs and the scientific information can be distorted [13]. Thus, the evaluation of the alteration of scientific information during the simplification process is crucial but underrepresented in the state-of-the-art.

Digitization and open access have made scientific literature available to every citizen. While this is an important first step, there are several remaining barriers preventing laypersons to access objective scientific knowledge in the literature. In particular, scientific texts are often hard to understand as they require solid background knowledge and use tricky terminology. Although there were some recent efforts on text simplification (e.g. [18]), removing such understanding barriers between scientific texts and the general public in an automatic manner is still an open challenge. The CLEF 2023 SimpleText track[1] brings together researchers and practitioners working on the generation of simplified summaries of scientific texts. It is an evaluation lab that follows up on the CLEF 2021 SimpleText Workshop [10] and CLEF 2022 SimpleText Track [12]. All perspectives on automatic science popularisation are welcome, including but not limited to: Natural Language Processing (NLP), Information Retrieval (IR), Linguistics, Scientific Journalism, etc.

SimpleText provides data and benchmarks for discussion of challenges of automatic text simplification by bringing in the following interconnected tasks [11]:

Task 1: What is in (or out)? Select passages to include in a simplified summary, given a query.

Task 2: What is unclear? Given a passage and a query, rank terms/concepts that are required to be explained for understanding this passage (definitions, context, applications, ..).

Task 3: Rewrite this! Given a query, simplify passages from scientific abstracts.

A total of 74 teams registered for our SimpleText track at CLEF 2023. A total of 20 teams submitted 139 runs in total. The statistics for these runs submitted are presented in Table 1.

The bulk of this paper presents the tasks with the datasets and evaluation metrics used, as well as the results of the participants, in three self-contained sections: Sect. 2 on the first task about content selection, Sect. 3 on the second task about complexity spotting, and Sect. 4 on the third task about text simplification proper. We end with Sect. 5 discussing the results and findings, and lessons for the future.

2   Task 1: What is in (or Out)?

Given a popular science article targeted to a general audience, this task aims at retrieving passages that can help to understand this article, from a large corpus of academic abstracts and bibliographic metadata. Relevant abstracts should relate to any of the topics in the source article. These passages can be complex and require further simplification to be carried out in tasks 2 and 3. Task 1 focuses on content retrieval.

---

[1] https://simpletext-project.com.

Table 1. CLEF 2023 Simpletext official run submission statistics

Team	Task 1	Task 2.1	Task 2.2	Task 3	Total runs
Elsevier	10				10
Maine (Aiirlab)	10	3	3	2	18
uninib_DoSSIER	2				2
UAms	10	1		2	13
LIA	7				7
MiCroGerk		4	4	3	11
Croland		2	2		
NLPalma		1	1	1	3
Pandas				6	6
QH				3	3
SINAI		4	2		
irgc				4	4
CYUT				4	4
UOL-SRIS		1			1
Smroltra		10	10	1	21
TeamCAU		3	3	1	7
TheLangVerse		1	1	1	3
ThePunDetectives		2	2	2	6
UBO		7	1	1	9
RT				1	1
Total runs	39	39	29	32	139

2.1   Data

Corpus: DBLP Abstracts.   We use the Citation Network Dataset: DBLP+Citation, ACM Citation network (12th version released in 2020).[2] This contains a total of 4,894,063 scientific articles. A JSON dump of the corpus is made available for participants. In addition, an ElasticSearch index is provided to participants with access through an API.

Topics: Press Articles.   Topics are a selection of press articles from the technology section of *The Guardian*[3] newspaper (topics G01 to G20) and the *Tech Xplore*[4] website (topics T01 to T20). URLs to original articles and textual content of each topic are provided to participants. All passages retrieved from DBLP by participants are expected to have some overlap (lexical or semantic) with the article content.

---

[2] https://www.aminer.org/citation.
[3] https://www.theguardian.com/uk/technology.
[4] https://techxplore.com/.

Queries as Facets. For each popular news article, multiple keyword queries are provided, leading to a grand total of 114 requests. It has been manually checked that each query allows retrieving relevant articles related to the topic of the press article.

Qrels. Quality relevance of abstracts w.r.t. topics are given in both the train qrels (released prior to submissions) and the test qrels.

Train Qrels Relevance annotations are provided on a 0–2 scale (the higher the more relevant) for 29 queries associated with the first 15 articles from The Guardian (G01–G15). Specifically, it extends the 2022 qrels released with a significant increase in the depth of judgments of abstracts per query.

Test Qrels Relevance annotations are provided on a 0–2 scale (the higher the more relevant) for 34 queries associated with the 5 articles from The Guardian (G16–G20, 17 queries) and 5 articles from Tech Xplore (T01–T05, 17 queries). These qrels were based on pooling the submissions of 2023 participants.

2.2 Attended Results

Ad Hoc Passage Retrieval. Participants should retrieve, for each topic and each query, all passages from DBLP abstracts, related to the query and potentially relevant to be inserted as a citation in the paper associated with the topic. Some abstracts could be very complex for non-experts. We encourage participants to take into account passage complexity as well as its credibility/influentialness.

Output Format. Results should be provided in a TREC-style JSON or TSV format with the following fields:

run_id Run ID starting with: team_id_task_id_method_used, e.g. *UBO_task_1_TFIDF*
manual Whether the run is manual {0,1}
topic_id Topic ID
query_id Query ID used to retrieve the document (if one of the queries provided for the topic was used; 0 otherwise)
doc_id ID of the retrieved document (to be extracted from the JSON output)
rel_score Relevance score of the passage (higher is better)
comb_score General score that may combine relevance and other aspects: readability, credibility or authoritativeness
passage Text of the selected abstract.

For each query, the maximum number of distinct DBLP references (doc_id field) must be 100 and the total length of passages should not exceed 1,000 tokens. The idea of taking into account complexity is to have passages easier to understand for non-experts, while the credibility score aims at guiding them on the expertise of authors and the value of publication w.r.t. the article topic. For example, complexity scores can be evaluated using readability and credibility scores using bibliometrics.

An example of the output is shown in Table 2. For each topic, the maximum number of distinct DBLP references (_id json field) was 100 and the total length of passages was not to exceed 1,000 tokens.

Table 2. CLEF 2023 SimpleText Task 1 on content selection: example of output

Run	M/A	Topic	Query	Doc	Rel	Comb	Passage
ST1_task1_1	0	G01	G01.1	1564531496	0.97	0.85	A CDA is a mobile user device, similar to a Personal Digital Assistant (PDA). It supports the citizen when dealing with public authorities and proves his rights - if desired, even without revealing his identity ...
ST1_task1_1	0	G01	G01.1	3000234933	0.9	0.9	People are becoming increasingly comfortable using Digital Assistants (DAs) to interact with services or connected objects ...
ST1_task1_1	0	G01	G01.2	1448624402	0.6	0.3	As extensive experimental research has shown individuals suffer from diverse biases in decision-making ...

2.3 Evaluation Metrics

Passage relevance is assessed based on:

- manual relevance assessment of a pool of passages (relevance scores provided by participants is used to measure ranking quality)

In addition to topical relevance, additional aspects such as the text complexity and the credibility or importance of the retrieved results are key in the use-case of the track. Hence we provide additional analysis in terms of:

- readability level analysis of the retrieved results, providing an indication of the accessibility of the retrieved abstracts.
- manual assessment by non-expert users of credibility and complexity.

2.4 Participants' Approaches

Elsevier (Elsevier in the Table 3) [6] submitted a total of 10 runs to Task 1, exploring the effectiveness of a stream of neural rankers, both applied in a zero-shot way as well as with unsupervised fine-tuning on scientific documents.

University of Amsterdam (UAms_*) [16] submitted a total of 10 runs for Task 1. First, they submitted 3 baseline rankers to improve the pool of judgments: an elastic run using keyword (non-phrase) queries, and a cross-encoder reranking of the top 100 and top 1k results from Elastic. Second, they submitted 4 runs aiming to address the credibility of the retrieved results, taking into account the recency and number of citations of each paper. Third, they submitted 3 runs aiming to address the readability of the retrieved results.

University of Avignon (LIA_*) submitted a total of 7 runs to Task 1, using a range of lexical and neural ranking models. These runs were used to analyze pool diversity and reusability of the resulting test collection and to investigate the aggregation of several queries to their associated article.

University of Maine (AIIR Lab, maine_*) [19] submitted a total of 5 runs to Task 1, experimenting with several cross-encoders and bi-encoder models, in comparison to lexical models.

University of Milano Bicocca (unimibDoSSIER_*) [20] submitted a total of 2 runs to Task 1, with a range of domain-specific approaches for scientific documents, including probabilistic lexical ranking, hierarchical document classification, and pseudo-relevance feedback (PRF).

2.5   Results

Retrieval Effectiveness. Table 3 shows the results of the CLEF 2023 Simpletext Task 1, based on the 34 test queries. The main measure of the task is NDCG@10, and the table is sorted on this measure for convenience.

A number of observations stand out. First and foremost, we see in general that the top of the Table is dominated by neural rankers, in particular, cross-encoders trained on MSMarco applied in a zero-shot way (or variants thereof), perform well for ranking scientific abstracts on NDCG@10 and other early precision measures. Traditional lexical retrieval models perform reasonably but at some distance from the top-scoring runs, with the neural runs typically re-ranking such a lexical baseline run.

Second, looking at more recall-oriented measures, such as MAP and bpref, the picture is more mixed. This is indicating some approaches privilege precision over recall, whereas other approaches seem to promote all recall levels.

Third, some submissions aimed to balance the topical relevance with the readability or credibility of the results. We observe that these runs still achieve competitive retrieval effectiveness, despite removing or down-ranking highly relevant abstracts that have for example a high text complexity or are dated with low numbers of citations.

Analysis of Readability. Table 4 shows several statistics over to the top 10 results retrieved for the entire topic set for Task 1:

- citation analysis (impact factor based on ACM records and average number of references per document)
- textual analysis (document length and FKGL scores)

We make a number of observations.

First, it appears that the most effective ranking models tend to retrieve abstracts that are not only longer, but also exhibit greater length variability. These retrieved abstracts often have higher impact factors and extensive bibliographies. There also seems to be a discernible difference between the lengths of abstracts retrieved by lexical-based systems compared to those retrieved by neural-based systems.

Table 3. Evaluation of SimpleText Task 1 (Test qrels).

Run	MRR	Precision		NDCG		Bpref	MAP
		10	20	10	20		
ElsevierSimpleText_run8	0.8082	0.5618	0.3515	0.5881	0.4422	0.2371	0.1633
ElsevierSimpleText_run7	0.7136	0.5618	0.4103	0.5704	0.4627	0.2626	0.1915
maine_CrossEncoder1	0.7309	0.5265	0.4500	0.5455	0.4841	0.3337	0.2754
maine_CrossEncoderFinetuned1	0.7338	0.4971	0.4000	0.4859	0.4295	0.3443	0.2385
ElsevierSimpleText_run5	0.6600	0.4765	0.3838	0.4826	0.4186	0.2542	0.1828
ElsevierSimpleText_run2	0.7010	0.4676	0.4059	0.4791	0.4282	0.2528	0.1942
ElsevierSimpleText_run6	0.6402	0.4676	0.3853	0.4723	0.4185	0.2557	0.1809
ElsevierSimpleText_run4	0.6774	0.4529	0.3794	0.4721	0.4116	0.2485	0.1898
ElsevierSimpleText_run9	0.5933	0.4735	0.3176	0.4655	0.3595	0.1758	0.1238
ElsevierSimpleText_run1	0.6821	0.4588	0.3824	0.4626	0.4071	0.2573	0.1823
maine_CrossEncoderFinetuned2	0.7082	0.4706	0.3926	0.4617	0.4089	0.3259	0.2253
UAms_CE1k_Filter	0.6403	0.4765	0.3559	0.4533	0.3743	0.2727	0.1936
ElsevierSimpleText_run3	0.6502	0.4471	0.3779	0.4460	0.3994	0.2558	0.1785
UAms_ElF_Cred44	0.6888	0.4324	0.3338	0.4103	0.3499	0.2395	0.1719
UAms_CE100	0.6779	0.3971	0.3456	0.4016	0.3642	0.2658	0.1792
maine_Pl2TFIDF	0.5626	0.4176	0.2809	0.4014	0.3218	0.2155	0.1364
UAms_Elastic	0.6424	0.4059	0.3456	0.3910	0.3541	0.2501	0.1895
UAms_ElF_Cred53	0.6429	0.4088	0.3382	0.3883	0.3468	0.2454	0.1833
UAms_ElF_Cred44Read	0.6625	0.3971	0.3147	0.3723	0.3282	0.2123	0.1403
UAms_CE1k_Combine	0.5880	0.4147	0.3515	0.3706	0.3398	0.2700	0.1865
UAms_CE1k	0.5880	0.4147	0.3515	0.3706	0.3398	0.2700	0.1865
UAms_ElF_Read25	0.6076	0.3735	0.3074	0.3539	0.3190	0.2194	0.1522
UAms_ElF_Cred53Read	0.6088	0.3676	0.3059	0.3469	0.3153	0.2133	0.1456
maine_tripletloss	0.5502	0.3382	0.2176	0.3353	0.2561	0.1335	0.0696
unimib_DoSSIER_2	0.5201	0.2853	0.2515	0.2980	0.2683	0.1898	0.1141
unimib_DoSSIER_4	0.5202	0.2853	0.2441	0.2972	0.2632	0.1873	0.1111
run-LIA.bm25	0.4536	0.1912	0.1338	0.2192	0.1700	0.1384	0.0515
run-LIA.all-MiniLM-L6-v2.query	0.3505	0.2000	0.1662	0.2019	0.1767	0.1956	0.0667
run-LIA.all-MiniLM-L6-v2.query-topic	0.3655	0.1765	0.1485	0.1912	0.1647	0.2043	0.0591
run-LIA.all-mpnet-base-v2.query-topic	0.3506	0.1647	0.1294	0.1835	0.1517	0.2073	0.0523
run-LIA.all-mpnet-base-v2.query	0.3302	0.1647	0.1529	0.1802	0.1644	0.1956	0.0602
run-LIA.lda	0.3138	0.1824	0.1456	0.1666	0.1488	0.1402	0.0521
run-LIA.es	0.3056	0.1118	0.0912	0.1277	0.1080	0.1935	0.0342

Second, in terms of readability levels, the overwhelming majority of systems retrieve abstracts with an FKGL of around 14 – corresponding to university-level texts. This is entirely as expected since the corpus is based on scientific text, known to be written for experts with higher text complexity than for example newspaper articles.

Third, two systems retrieve abstracts with an FKGL of 11–12 – corresponding to the exit level of compulsory education, and the reading level of the average newspaper reader targeted by the use case of the track. These runs still achieved very reasonable retrieval effectiveness (NDCG@10 0.37–0.45 in Table 3) while only retrieving abstracts with the desirable readability level.

Table 4. Text Analysis of SimpleText Task 1 output.

Run	Impact	#Refs	Length Mean	Length Median	FKGL Mean	FKGL Median
ElsevierSimpleText_run1	1.88	0.95	965.02	921.00	13.80	13.80
ElsevierSimpleText_run2	2.24	1.36	1017.57	981.00	13.98	13.90
ElsevierSimpleText_run3	1.80	0.94	951.64	912.00	13.71	13.75
ElsevierSimpleText_run4	2.10	1.21	1011.10	994.00	13.95	13.90
ElsevierSimpleText_run5	1.78	0.71	993.14	972.50	13.76	13.80
ElsevierSimpleText_run6	1.59	0.65	995.65	975.50	13.75	13.90
ElsevierSimpleText_run7	2.37	0.94	1101.23	1075.50	13.87	13.80
ElsevierSimpleText_run8	0.60	0.50	1089.90	1045.00	14.09	14.00
ElsevierSimpleText_run9	0.71	0.54	1016.96	991.00	13.66	13.70
UAms_CE100	3.20	1.64	1028.78	975.00	14.59	14.50
UAms_CE1k	2.41	1.24	1071.67	985.50	14.70	14.60
UAms_CE1k_Combine	0.84	0.49	924.38	839.00	10.84	11.20
UAms_CE1k_Filter	1.09	0.62	988.00	913.50	12.40	12.70
UAms_ElF_Cred44	3.32	1.62	973.03	970.50	13.60	14.50
UAms_ElF_Cred44Read	1.85	1.34	799.29	851.00	13.18	14.20
UAms_ElF_Cred53	2.89	1.49	938.41	932.00	13.73	14.40
UAms_ElF_Cred53Read	1.70	1.28	774.76	823.00	13.29	14.30
UAms_ElF_Read25	1.60	1.25	767.70	819.00	13.09	14.20
UAms_Elastic	2.84	1.45	922.36	917.00	13.49	14.30
maine_CrossEncoder1	4.22	2.86	961.17	923.00	14.64	14.60
maine_CrossEncoderFinetuned1	4.41	3.37	1003.75	988.00	15.01	14.80
maine_CrossEncoderFinetuned2	3.49	3.04	988.86	951.50	14.95	14.80
maine_Pl2TFIDF	3.35	2.58	893.29	894.00	14.03	14.00
maine_tripletloss	4.76	3.29	969.09	973.50	14.69	14.60
unimib_DoSSIER_2	1.44	1.33	1024.48	994.00	14.77	14.60
unimib_DoSSIER_4	1.44	1.33	238.63	212.00	15.11	15.00

3 Task 2: What is Unclear?

The goal of this task is to identify key concepts that need to be contextualized with a definition, example, and/or use-case and provide useful and understandable explanations for them. Thus, there are two subtasks:

- to retrieve up to 5 difficult terms in a given passage from a scientific abstract
- to provide an explanation (one/two sentences) of these difficult terms (e.g. definition, abbreviation deciphering, example, etc.)

For each passage, participants should provide a ranked list of difficult terms with corresponding difficulty scores on a scale of 0–2 (2 to be the most difficult terms, while the meaning of terms scored 0 can be derived or guessed) and definitions (optional). Passages (sentences) are considered to be independent, i.e. difficult term repetition is allowed. Detected concept spans and term and term difficulty are evaluated.

3.1 Data

Datasets for Task 2.1. To build the **test set** for Task 2.1, 116,763 sentences from the DBLP abstracts were extracted. Then, a set of 1262 distinct sentences were manually evaluated to measure the performance of different models in terms of their ability in detecting difficult terms and their difficulty scores. A pooling mechanism is used to further annotate 5,142 distinct pairs sentence-term manually in which each evaluated source sentence contained the results of all participants.

Datasets for Task 2.2. A set of 203 difficult terms (within sentences from Task 1) with their ground truth annotations are provided in the training set for Task 2.2 for the definition generation part. For the evaluation of runs for this task, we use ∼800 terms with ground truth definitions. From this set, ∼300 terms are annotated using a pooling mechanism (based on the submitted runs) to make sure that the majority of runs have enough annotated samples in the test set. There are a total of 15,056 sentences containing at least one of these terms in our test set. For the abbreviation expansion evaluation, we manually annotate a set of ∼1K manually abbreviations. We additionally expand this dataset by mining 4,374 extra abbreviations from the sentences from Task 1. We use the Schwartz and Hearst [26] algorithm to extract these extra abbreviations and their expansion from the test set. There are 38,416 sentences in the test set containing at least one of the ∼5K abbreviations. We use this set of sentences for the final evaluation of this subtask.

Input Format. The train and the test data are provided in JSON and TSV formats with the following fields:

snt_id a unique passage (sentence) identifier
doc_id a unique source document identifier
query_id a query ID
query_text difficult terms should be extracted from sentences with regard to this query
source_snt passage text

Input example:

```
[{"query_id":"G14.2",
  "query_text":"end to end encryption",
  "doc_id":"2884788726",
  "snt_id":"G14.2_2884788726_2",
```

```
"source_snt": "However, in information-centric networking (ICN)
    ↳  the end-to-end encryption makes the content caching
    ↳  ineffective since encrypted content stored in a cache is
    ↳  useless for any consumer except those who know the
    ↳  encryption key."},

{"snt_id": "G06.2_2548923997_3",
 "doc_id":2548923997,
 "query_id": "G06.2",
 "query_text": "self driving",
 "source_snt": "These communication systems render self-driving
    ↳  vehicles vulnerable to many types of malicious attacks,
    ↳  such as Sybil attacks, Denial of Service (DoS), black
    ↳  hole, grey hole and wormhole attacks."}]
```

Output Format. Results should be provided in a TREC-style JSON or TSV format with the following fields:

run_id Run ID starting with (team_id)_(task_id)_(method_used), e.g. UBO_task_2.1_TFIDF

manual Whether the run is manual {0, 1}.

snt_id a unique passage (sentence) identifier from the input file.

term Term or another phrase to be explained.

term_rank_snt term difficulty rank within the given sentence.

difficulty difficulty scores of the retrieved term on the scale 0–2 (2 to be the most difficult terms, while the meaning of terms scored 0 can be derived or guessed)

definition (only used for Task 2.2) short (one/two sentence) explanations/definitions for the terms. For the abbreviations, the definition would be the extended abbreviation.

Output example Task 2.1:

```
[{"snt_id": "G14.2_2884788726_2",
 "term": "content caching",
 "difficulty":1.0,
 "term_rank_snt":1,
 "run_id": "team1_task_2.1_TFIDF",
 "manual":0}]
```

Output example Task 2.2:

```
[{"snt_id": "G14.2_2884788726_2",
 "term": "content caching",
 "difficulty":1.0,
 "term_rank_snt":1,
 "definition": "Content caching is a performance optimization
    ↳  mechanism in which data is delivered from the closest
    ↳  servers for optimal application performance.",
 "run_id": "team1_task_2.2_TFIDF_BLOOM",
 "manual":0}]
```

3.2   Evaluation Metrics

In this section, we describe different evaluation metrics used to evaluate the performance of submissions for Task 2.1 and Task 2.2.

Task 2.1. We have evaluated the performance of different submissions for Task 2.1 based on:

- correctness of detected term limits: this metric reflects whether the retrieved difficult terms are well limited or not. This is a binary label assigned to each retrieved term.
- difficulty scores: we used a three-scale terms difficulty score which reflects how difficult the term is in the context for an average user and how necessary it is to provide more context about the term: 0 score corresponds to an easy term (explanation might be given but not required); 1 corresponds to somewhat difficult (explanation could help); 2 corresponds to difficult (explanation is necessary).

Task 2.2. For this task, we use the following evaluation metrics:

- **BLEU** score [24] between the reference (ground truth definition) and the predicted definitions.
- **ROUGE L F-measure** [17] which measures the ROUGE F-measure based on the Longest Common Subsequence between the reference and the predicted definitions.
- **Semantic match** between the reference and predicted definitions measured using the *all-mpnet-base-v2*[5] sentence transformer model which is an advanced model for sentence similarity. This measure is the average semantic similarity between reference and predicted definitions for all detected terms.
- **Exact match** which is only used for the task of **abbreviation extension** in which we ask the participants to provide only extensions for the detected difficult abbreviations. This metric measures the number of exact matches between the reference and predicted extensions for abbreviations.
- **Partial match** which measures the number of non-identical abbreviation extensions (between reference and predicted extensions) which have a Levenshtein distance lower than 4 characters. This corresponds to slight variations (such as plural/non-plural) between reference and predicted abbreviation extensions.

3.3   Participants' Approaches

National Polytechnic Institute of Mexico (NLPalma) [23] submitted a total of 2 runs for Task 2, a single run for each of Task 2.1 and Task 2.2. They experimented with BLOOMZ to produce description-style prompts given by text input on a task and a binary classifier based on BERT-multilingual for term difficulty.

University of Amsterdam (UAms) [16] submitted a single run for Task 2 focusing on complexity spotting. Their approach aimed to demonstrate the relative effectiveness of simple and straightforward approaches, and made use of standard TF-IDF based term-weighting using the large test set as a source for within-domain term statistics.

---

[5] https://huggingface.co/sentence-transformers/all-mpnet-base-v2.

University of Cadiz/Split (Smroltra) [25] submitted a total of 20 runs for Task 2, with both 10 runs for Task 2.1 and 10 runs for Task 2.2. They experimented with a range of keyword extraction approaches (KeyBERT, RAKE, YAKE!, BLOOM, T5, TextRank) for the first task, and a Wikipedia extraction approach, BERT, and BLOOMZ for the second task.

University of Guayaquil/Jaén (SINAI) [22] submitted a total of 6 runs for Task 2, with 4 runs for Task 2.1 and 2 runs for Task 2.2. They investigated zero-shot and few-shot learning strategies over the auto-regressive model GPT-3, and in particular effective prompt engineering.

University of Kiel (TeamCAU) [4] submitted 6 runs for Task 2, based on three different large pre-trained language models (SimpleT5, AI21, and BLOOM). They made three and corresponding submissions to both Task 2.1 and 2.2, and also note the complexities of adapting models with limited train data.

University of Kiel/Split/Malta (MicroGerk) [7] submitted a total of 8 runs for Task 2, with 4 runs for Task 2.1 and 4 runs for Task 2.2. They experimented with a range of models (YAKE!, TextRank, BLOOM, GPT-3) for the first task, and a range of models (Wikipedia, SimpleT5, BLOOMZ, GPT-3) for the second task.

University of Southern Maine (Aiirlab) [19] submitted a total of 6 runs for Task 2, consisting 3 runs for Task 2.1 and 3 runs for Task 2.2. They experimented with key-word extraction approaches (YAKE!, KBIR) and IDF weighting for the first task, and definition detection in top-ranked documents based on a trained classifier.

University of Western Brittany (UBO) [8] submitted a total of 8 runs for Task 2, no less than 7 runs for Task 2.1 and a single run for Task 2.2. They experimented with a range of keyword extraction approaches (FirstPhrase, TF-IDF, YAKE!, TextRank, SingleRank, TopicRank, PositionRank) for the first task and a Wikipedia extraction approach for the second task.

University of Split (Croland) submitted a total of 4 runs for Task 2, specifically 2 runs for Task 2.1 and 2 runs for Task 2.2. They applied GPT-3 and TF-IDF for difficult term detection. They extracted definitions from Wikipedia and applied GPT-3 to generate explanations.

University of Liverpool (UOL-SRIS) submitted a single run for Task 2, specifically for Task 2.1 by applying KeyBERT.

University of Kiel/Cadiz/Gdansk (TheLangVerse) submitted a total of 2 runs for Task 2, a single run for both Task 2.1 and Task 2.2 using GPT-3.

3.4 Results

We evaluate the performance of the submissions separately for the difficult terms spotting (Task 2.1) and definition extraction/generation (Task 2.2) using separate test sets created per task. In this section, we describe the main results of different submissions per task.

Table 5. SimpleText Task 2.1: Results for the official runs

	Total	Evaluated	+Limits	Score	+Limits
SINAI_task_2.1_PRM_ZS_TASK2_1_V1	11081	1322	1185	556	507
UAms_Task_2_RareIDF	675090	1293	1145	309	241
SINAI_task_2.1_PRM_FS_TASK2_1_V1	10768	1235	1122	440	405
Smroltra_task_2.1_keyBERT_FKgrade	11099	1215	1061	379	341
Smroltra_task_2.1_keyBERT_F	11099	1215	1061	223	171
UOL-SRIS_2.1_KeyBERT	23757	1215	1061	0	0
MiCroGerk_task_2.1_TextRank	21516	1275	1002	482	391
Smroltra_task_2.1_TextRank_FKgrade	10056	1275	1002	456	363
SINAI_task_2.1_PRM_ZS_TASK2_1_V2	10952	1075	965	366	330
SINAI_task_2.1_PRM_FS_TASK2_1_V2	8836	1004	915	346	316
Smroltra_task_2.1_YAKE_D	11112	1576	905	627	422
MiCroGerk_task_2.1_YAKE	23790	1576	905	582	362
Smroltra_task_2.1_YAKE_Fscore	11112	1576	905	409	209
MiCroGerk_task_2.1_GPT-3	15892	968	889	487	459
UBO_task_2.1_FirstPhrases	14088	1032	831	210	161
UBO_task_2.1_PositionRank	13881	1071	825	237	181
UBO_task_2.1_SingleRank	14088	981	748	200	151
UBO_task_2.1_TfIdf	14340	1206	740	263	187
UBO_task_2.1_TextRank	14088	960	722	189	139
Smroltra_task_2.1_RAKE_AUI	10660	1016	713	378	288
Smroltra_task_2.1_RAKE_F	10660	1016	713	255	170
UBO_task_2.1_TopicRank	13912	824	663	174	144
UBO_task_2.1_YAKE	14337	1118	576	265	116
MiCroGerk_task_2.1_BLOOM	9600	608	535	235	218
Aiirlab_task_2.2_KBIR	4797	498	429	158	135
TeamCAU_task_2.1_ST5	2234	484	418	222	201
Smroltra_task_2.1_SimpleT5	2234	460	406	259	239
Smroltra_task_2.1_SimpleT5_COLEMAN_LIEAU	2234	460	406	168	152
TheLangVerse_task_2.2_openai-curie-finetuned	2234	445	391	0	0
ThePunDetectives_task_2.1_SimpleT5	152072	428	371	110	91
Aiirlab_task_2.2_YAKEIDF	4790	465	241	154	75
Aiirlab_task_2.2_YAKE	4790	486	234	169	78
TeamCAU_task_2.1_AI21	100	10	6	3	2
Smroltra_task_2.1_Bloom	100	4	2	1	1
TeamCAU_task_2.1_BLOOM	100	1	1	0	0

Task 2.1: Difficult Term Spotting. In this section, we describe the results of the submissions on Task 2.1. A total of 12 teams submitted runs for Task 2.1. There were in total 39 runs. Table 5 shows the results for different runs. We show the total

Table 6. SimpleText Task 2.2: Results for the official runs

Run	Evaluated	BLEU	ROUGE	Semantic
UBO_task_2.1_FirstPhrases_Wikipedia	393	29.73	0.41	0.80
Croland_task_2_PKE_Wiki	43	33.68	0.46	0.70
MiCroGerk_task_2.2_GPT-3_Wikipedia	932	26.38	0.41	0.75
Smroltra_task_2.2_Text_Wiki	547	17.59	0.33	0.75
Smroltra_task_2.2_RAKE_Wiki	337	16.95	0.32	0.74
Smroltra_task_2.2_YAKE_Wiki	436	16.94	0.32	0.73
TeamCAU_task_2.1_BLOOM	10	10.46	0.27	0.48
MiCroGerk_task_2.2_GPT-3_BLOOMZ	1,108	9.07	0.40	0.83
Smroltra_task_2.2_keyBERT_Wiki	302	8.60	0.23	0.69
MiCroGerk_task_2.2_GPT-3_GPT-3	1,108	7.73	0.38	0.83
NLPalma_task_2.2_BERT_BLOOMZ	537	7.22	0.39	0.76
Smroltra_task_2.2_Bloomz	23	7.15	0.30	0.69
TeamCAU_task_2.1_AI21	22	6.38	0.31	0.78
TheLangVerse_task_2.2_openai-curie-finetuned	444	5.03	0.25	0.74
Croland_task_2_GPT3	69	4.83	0.27	0.77
SINAI_task_2.1_PRM_FS_TASK2_2_V1	649	4.23	0.21	0.78
MiCroGerk_task_2.2_GPT-3_simpleT5	1,108	4.22	0.28	0.77
TeamCAU_task_2.1_ST5	379	3.33	0.20	0.60
Smroltra_task_2.2_SimpleT5	392	3.09	0.22	0.72
SINAI_task_2.1_PRM_ZS_TASK2_2_V1	649	3.08	0.19	0.69
Smroltra_task_2.2_keyBERT_dict	120	2.07	0.14	0.51
Smroltra_task_2.2_YAKE_WN	48	1.88	0.15	0.44
Aiirlab_task_2.2_KBIR	556	1.62	0.15	0.50
Smroltra_task_2.2_keyBERT_WN	328	1.33	0.14	0.45
Aiirlab_task_2.2_YAKEIDF	179	1.13	0.14	0.41
Aiirlab_task_2.2_YAKE	165	1.10	0.15	0.43
Smroltra_task_2.2_RAKE_WN	70	0.00	0.14	0.46

number of evaluated terms and the number of terms with correct term limits. We present results for correctly attributed scores regardless of the correctness of term limits and the number of correctly limited terms with correctly attributed scores (+Limits). The SINAI_task_2.1_PRM_ZS_TASK2_1_V1 run has the highest number of correctly detected terms and scores among all the runs for this task.

Participants used Large Language Models (LLMs) as well as unsupervised methods. We received many partial runs due to token constraints of LLMs or their execution time. We also observe that the results of the same methods depend heavily on implementation, fine-tuning, and/or used prompts. Results of difficult term detection by LLMs are comparable to RareIDF, TextRank and YAKE! Term difficulty scores assigned by models are quite different from the lay annotations.

Table 7. SimpleText Task 2.2: Results for the official runs on the abbreviation expansion task

Run	Evaluated	BLEU	ROUGE	Semantic	Exact	Partial
MiCroGerk_task_2.2_GPT-3_BLOOMZ	854	13.87	0.68	0.76	326	185
MiCroGerk_task_2.2_GPT-3_GPT-3	855	11.86	0.64	0.73	294	166
MiCroGerk_task_2.2_GPT-3_Wikipedia	855	4.68	0.43	0.60	205	109
MiCroGerk_task_2.2_GPT-3_Wikipedia	618	5.01	0.56	0.64	198	109
NLPalma_task_2.2_BERT_BLOOMZ	345	6.83	0.39	0.52	50	47
Smroltra_task_2.2_SimpleT5	185	0.00	0.12	0.39	8	7
TeamCAU_task_2.1_ST5	141	1.48	0.14	0.40	6	3
TheLangVerse_task_2.2_openai-curie-finetuned	204	1.60	0.14	0.42	1	2
SINAI_task_2.1_PRM_ZS_TASK2_2_V1	228	1.61	0.13	0.55	1	0
TeamCAU_task_2.1_AI21	10	1.87	0.14	0.38	0	0
SINAI_task_2.1_PRM_FS_TASK2_2_V1	228	1.35	0.10	0.53	0	0
UBO_task_2.1_FirstPhrases_Wikipedia	116	5.09	0.19	0.47	0	0
Aiirlab_task_2.2_KBIR	202	1.17	0.07	0.44	0	0
Smroltra_task_2.2_RAKE_Wiki	27	0.54	0.04	0.14	0	0
Smroltra_task_2.2_Bloomz	4	0	0.22	0.61	0	0
Aiirlab_task_2.2_YAKEIDF	19	0	0.10	0.40	0	0
Smroltra_task_2.2_keyBERT_WN	188	0	0.04	0.27	0	0
Smroltra_task_2.2_keyBERT_Wiki	163	0.21	0.02	0.13	0	0
Smroltra_task_2.2_keyBERT_dict	46	0	0.04	0.34	0	0
Smroltra_task_2.2_RAKE_WN	21	0	0.04	0.24	0	0
Smroltra_task_2.2_YAKE_WN	32	0	0.02	0.21	0	0
Smroltra_task_2.2_YAKE_Wiki	31	0	0.03	0.11	0	0
Smroltra_task_2.2_Text_Wiki	50	0	0.02	0.10	0	0
Aiirlab_task_2.2_YAKE	9	0	0.13	0.36	0	0
TeamCAU_task_2.1_BLOOM	3	0	0	0.14	0	0

Task 2.2: Difficult Term Explanation. For this task, 10 teams submitted 29 runs in total. The main results for Task 2.2 are shown in Table 6. The low number of evaluated sentences for most runs is due to the fact that most runs are done on a small set of sentences from the test set. The rest of the runs also achieved strong performance in terms of the semantic similarity of their provided definitions with the ground truth definitions. As the results show, UBO_task_2.1_-FirstPhrases_Wikipedia, Croland_task_2_PKE_Wiki, and MiCroGerk_task_2.2_GPT-3_-Wikipedia runs achieved a strong performance in terms of the BLEU score. This result shows that although these runs do not use the same set of words as the ground truth definitions to define difficult terms, they still provide an explanation for the terms that are semantically similar to the ground truth ones. The Wikipedia-based runs have the highest similarity with the ground truth definitions.

Table 7 shows the performance of the runs on the abbreviation expansion task. *MiCroGerk* run has the highest performance on this task. This best-performing model is

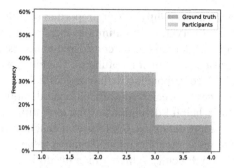

Fig. 1. Histogram of the difficulties of the definitions on a scale of 1–3 (1 - easy; 2 - difficult; 3 - very difficult)

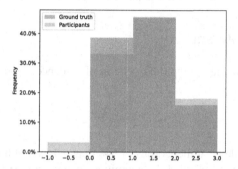

Fig. 2. Difference between term difficulty and definition difficulty on a scale of 1–3 (1 - easy; 2 - difficult; 3 - very difficult). Positive values on X axis show helpful definitions. 0 refers to unhelpful definitions. Negative values increase the difficulty.

able to provide an expansion for 326 identical expansions the true expansions and 185 partially correct expansions. In general, LLMs (BLOOMz, GPT-3) have the best performance for abbreviation expansion. Note, that the provided scores are averaged over the number of evaluated instances favoring small runs. Many partial runs are due to token/-time constraints of LLMs. Besides, evaluation results depend on the terms extracted in Task 2.1.

Analysis of Definitions' Difficulty. In order to analyze the helpfulness of the provided definitions, a master's student in translation and technical writing manually assigned scores of difficulty on a scale of 1–3 (1 - easy; 2 - difficult; 3 - very difficult) to 353 definitions for 82 distinct terms. The analyzed definitions are taken from participants' runs as well as from the ground truth.

Figure 1 shows the relative distribution of easy, difficult, and very difficult definitions in the participants' runs as well as in our ground truth. The figure provides evidence that in the majority of cases (more than 50% both in the runs and the ground truth), the definitions are considered by a non-expert in computer science to be easy.

In our ground truth, there are a slightly higher proportion of difficult definitions and a slightly lower proportion of very difficult definitions than in the participants' runs.

Although the majority of definitions are considered to be easy, this evidence is not enough to make a conclusion about their helpfulness. Therefore, we decided to compare the term difficulty and the corresponding definitions' difficulty. Figure 2 presents the histogram of the differences between term difficulty and definition difficulty. Positive values of the X axis show helpful definitions as the term difficulty is higher than the difficulty of the corresponding definition. 0 refers to an unhelpful definition as it has the same difficulty as the terms it should explain. Negative values on the X axis increase the difficulty, i.e. definition difficulty is higher than the difficulty of the corresponding term. The results suggest that 30%–40% of definitions are either unhelpful or even more difficult than the corresponding terms. Our ground truth does not have harmful definitions in contrast to the runs of the participants.

4   Task 3: Rewrite This!

This task aims to provide a simplified version of sentences extracted from scientific abstracts.

4.1   Data

As in 2022, we provide a parallel corpus of 648 manually simplified sentences as train data [12]. This year, we evaluated the submitted runs by comparing them against the new 245 manually simplified sentences extracted from relevant passages for Task 1.

Input Format. The train and the test data are provided in JSON and TSV formats with the following fields:

snt_id a unique passage (sentence) identifier
doc_id a unique source document identifier
query_id a query ID
query_text difficult terms should be extracted from sentences with regard to this query
source_snt passage text

Input example:

```
{"snt_id":"G11.1_2892036907_2",
 "source_snt":"With the ever increasing number of unmanned
    ⌐ aerial vehicles getting involved in activities in the
    ⌐ civilian and commercial domain, there is an increased need
    ⌐ for autonomy in these systems too.",
 "doc_id":2892036907,
 "query_id":"G11.1",
 "query_text":"drones"}
```

Output Format. Results should be provided in a TREC-style JSON or TSV format with the following fields:

run_id Run ID starting with (team_id)_(task_3)_(method_used), e.g. UBO_BLOOM
manual Whether the run is manual $\{0, 1\}$.
snt_id a unique passage (sentence) identifier from the input file.
simplified_snt simplified passage .

Output example (JSON format):

```
{"run_id": "BTU_task_3_run1",
"manual":1,
"snt_id": "G11.1_2892036907_2",
"simplified_snt": "Drones are increasingly used in the civilian
⌐  and commercial domain and need to be autonomous."}
```

4.2   Evaluation Metrics

To evaluate the simplification results, we used the EASSE implementation [2] of the following metrics:

- **FKGL:** Flesch-Kincaid Grade Level is a readability metric that relies on average sentence lengths and number of syllables per word [14];
- **SARI** metric compares the system's output to multiple simplification references and the original sentence based on the words added, deleted, and kept by a system [28];
- **BLEU** is a precision-oriented metric that relies on the proportion of shared n-gram in a system's output and references [24];
- **Compression ratio**;
- **Sentence splits**;
- **Levenshtein similarity** measures the number of edits (insertions, deletions, or substitutions) needed to transform one sentence into another;
- **Exact copies**;
- **Additions proportion**;
- **Deletions proportion**;
- **Lexical complexity score** computed by taking the log-ranks of each word in the frequency table [2].

4.3   Participants' Approaches

Chaoyang University of Technology (CYUT) [27] submitted four runs for Task 3, experimenting with the GPT-4 API provided by OpenAI. They experiment with three different prompts, even using GPT-4 to suggest better prompts for the task.

National Polytechnic Institute of Mexico (NLPalma) [23] submitted a single run for Task 3. They experimented with BLOOMZ with different prompts for generate text simplifications.

University of Amsterdam [16] submitted two runs (*UAms_**) for Task 3, using the zero-shot application of GPT-2 based text simplification model. Their approach aimed to address one of the main issues in text generation approaches, which are prone to 'hallucinate' and generate spurious content unwarranted by the input. Specifically, by post-processing the generated output to ensure grounding on input sentences, spurious generated output was identified and removed.

University of Applied Sciences, Cologne [9] submitted four runs (*irgc_**) for Task 3, with two runs using T5, one run using PEGASUS, and the final run exploiting ChatGPT. They perform detailed analysis

University of Cadiz/Split (Smroltra) [25] submitted a single run for Task 3. They experimented with a SimpleT5 model for text simplification.

University of Kiel [4] submitted a single run (*TeamCAU_**) for Task 3, based on the SimpleT5 pre-trained language model.

University of Kiel/Cadiz/Gdansk [21] submitted two runs for Task 3 (as *Pun Detective*). They used SimpleT5 and GPT-3 models under resource constrained conditions such as the limited task specific train data, and showed the SimpleT5 model outperforming GPT-3 in key metrics.

University of Kiel/Split/Malta (MicroGerk) [7] submitted a total of 3 runs for Task 3. They experimented with BLOOMZ, GPT-3, and SimpleT5 models for text simplification.

University of Southern Maine (AIIR Lab) [19] submitted a total of 2 runs for Task 3. They experimented with two models, a GPT-2 based model and an OpenAI DaVinci model for generating text simplifications.

University of Zurich (Andermatt) [3] submitted 6 runs (*Pandas_**) for Task 3, experimenting with four large pretrained language models: T5, Alpaca 5B, and Alpaca LoRA. They exploit Task 2 data as additional train data, and experiment with prompt engineering.

University of Zurich (Hou) [15] submitted three runs (*QH_**) for Task 3, adapting the Multilingual Unsupervised Sentence Simplification (MUSS) model to HuggingFace's BART, and using a T5-Large model. They experiment with a template consisting of 5 control tokens and also add the original request.

University of Kiel/Gdansk/Cadiz (TheLangVerse) submitted a single run for Task 3. They experimented with a finetuned OpenAI Curie model for text simplification.

University of Western Brittany (UBO) [8] submitted a single run for Task 3. They experimented with a SimpleT5 model (and with BLOOM) to generate simplifications.
    Another team from the

University of Western Brittany (not in the Table) [5] experimented with ChatGPT for scientific text simplification, conducting a qualitative experiment with various analysis of the prompts and generated output.

Table 8. Results for task 3 (task number removed from the run_id)

run_id	count	FKGL	SARI	BLEU	Compression ratio	Sentence splits	Levenshtein similarity	Exact copies	Additions proportion	Deletions proportion	Lexical complexity score
Identity_baseline	245	13.64	15.09	26.22	1.00	1.00	1.00	1.00	0.00	0.00	8.64
AiirLab_task3_davinci	243	11.17	47.10	18.68	0.75	1.00	0.68	0.0	0.20	0.45	8.59
AiirLab_task3_run1	245	9.86	30.07	15.93	1.26	1.67	0.80	0.0	0.30	0.17	8.47
CYUT_run1	245	9.63	47.98	14.81	0.87	1.14	0.56	0.0	0.47	0.55	8.35
CYUT_run2	245	8.43	44.93	12.09	0.76	1.06	0.56	0.0	0.46	0.62	8.31
CYUT_run3	245	10.00	46.81	14.70	0.81	1.02	0.59	0.0	0.44	0.57	8.36
CYUT_run4	245	9.24	47.69	15.41	0.78	1.03	0.58	0.0	0.41	0.58	8.32
MiCroGerk_BLOOMZ	245	12.54	32.01	22.24	0.92	0.99	0.89	0.0	0.13	0.21	8.54
MiCroGerk_GPT-3	245	10.74	46.90	16.98	0.72	1.01	0.67	0.0	0.19	0.47	8.67
MiCroGerk_simpleT5	245	12.96	25.43	21.26	0.91	0.99	0.92	0.0	0.09	0.18	8.52
NLPalma_BLOOMZ	245	9.61	35.66	5.76	0.68	1.00	0.51	0.0	0.35	0.66	8.26
Pandas_alpaca-lora-alpaca-simplifier-alpaca-simplifier	245	10.96	38.31	17.88	0.74	1.00	0.77	0.0	0.10	0.36	8.51
Pandas_alpaca-lora-both-alpaca-normal-tripple	245	12.02	36.10	20.89	0.89	1.05	0.82	0.0	0.16	0.29	8.57
Pandas_alpaca-lora-both-alpaca-simplifier-tripple_10	244	11.71	36.38	19.62	0.89	1.07	0.78	0.0	0.16	0.31	8.55
Pandas_alpaca-lora-simplifier-alpaca-short	245	12.90	31.88	24.08	0.93	1.02	0.89	0.0	0.13	0.20	8.58
Pandas_clean-alpaca-lora-simplifier-alpaca-short	245	12.90	31.88	24.08	0.93	1.02	0.89	0.0	0.13	0.20	8.58
Pandas_submission_ensemble	245	10.51	40.25	17.40	0.77	1.09	0.73	0.0	0.15	0.40	8.52
QH_run1	245	12.45	26.46	21.23	0.94	1.07	0.92	0.0	0.11	0.17	8.50
QH_run2	245	13.05	24.40	21.33	0.96	1.03	0.92	0.0	0.12	0.15	8.48
QH_run3	245	12.74	27.56	20.24	0.90	1.01	0.91	0.0	0.09	0.19	8.50
Smroltra_SimpleT5	245	12.88	26.25	21.43	0.90	1.00	0.91	0.0	0.09	0.19	8.54
TeamCAU_ST5	245	12.77	27.19	21.06	0.90	1.00	0.91	0.0	0.10	0.20	8.52
TheLangVerse_openai-curie-finetuned	245	12.21	30.78	18.92	0.86	1.00	0.86	0.0	0.11	0.24	8.49
ThePunDetectives_GPT-3	245	7.52	41.56	6.10	0.46	0.97	0.50	0.0	0.16	0.68	8.46
ThePunDetectives_SimpleT5	245	12.92	25.87	21.79	0.91	0.99	0.92	0.0	0.09	0.18	8.53
UAms_Large_KIS150	245	10.50	33.02	14.59	1.26	1.48	0.76	0.0	0.34	0.20	8.45
UAms_Large_KIS150_Clip	245	11.12	33.47	16.59	1.01	1.23	0.82	0.0	0.24	0.23	8.48
UBO_SimpleT5	245	12.33	30.89	21.08	0.88	1.05	0.89	0.0	0.10	0.22	8.51
irgc_ChatGPT_2stepTurbo	245	12.31	46.98	16.86	0.94	1.04	0.63	0.0	0.37	0.46	8.46
irgc_pegasusTuner007plus_plus	245	12.74	23.28	17.42	1.23	1.28	0.83	0.0	0.22	0.15	8.55
irgc_t5	245	9.56	37.83	15.85	0.76	1.35	0.73	0.0	0.15	0.38	8.49
irgc_t5_noaron	245	9.55	37.84	15.84	0.76	1.35	0.73	0.0	0.15	0.38	8.49

4.4 Results

A total of 14 teams submitted 32 runs for Task 3, mainly LLMs. Table 8 presents the results of participants' runs according to the automatic evaluation listed in Section 4.2. Surprisingly, all systems modified the original sentences (Exact copies = 0). While many participants applied the same LLMs, such as GPT-3 and T5, their results differ a lot.

All runs improved the FKGL readability and lexical complexity scores with regard to the identity baseline (i.e. source sentences) suggesting that systems produced shorter sentences with simpler and shorter words on average. Note, that shorter words are not necessarily simpler as in the case of numerous abbreviations. Original sentences have an FKGL score of around 14 – corresponding to university-level texts. The majority of the submitted runs are scored lower than 11–12 according to FKGL – corresponding to the exit level of compulsory education.

All runs largely improved the SARI score compared to the original sentences. However, the source sentences have the highest vocabulary overlap with reference sentences.

Information Distortion. In order to analyze information distortion [12], a master student in translation and technical writing manually annotated 249 pairs of source sentences and simplifications submitted by the participants corresponding to 13 distinct source sentences. Sentences were assigned with binary labels corresponding to the

Table 9. Information distortion type statistics

Information distortion type	Instances	
	#	%
Incorrect syntax	9	3.61
Unresolved anaphora due to simplification	32	12.85
Unnecessary repetition/iteration	9	3.61
Spelling, typographic or punctuational errors	115	46.18
Contresens	18	7.22
Topic shift	3	1.20
Omission of essential details with regard to a query	45	18.07
Oversimplification	31	12.44
Insertion of false or unsupported information	8	3.21
Insertion of unnecessary details with regard to a query	3	1.20
Redundancy	3	1.20
Style	3	1.20
Non-sense	2	0.80

Table 10. Statistics on the levels of the difficulty of simplified sentences and information distortion severity on the scale of 1–7

	1	2	3	4	5	6	7
syntax complexity	230	19					
lexical complexity	54	114	62	19			
information loss severity	151	40	24	10	13	2	7
information loss severity %	60.64	16.06	9.63	4.01	5.22	0.80	2.81

occurrence of the information distortion types. Table 9 provides statistics on the information distortion identified in the participants' runs. The most common errors (46%) are spelling, typographic, and punctuational ones. It is followed by information loss (18%), unresolved anaphora due to simplification (13%), and oversimplification (12%). In 60% of cases, information loss was judged to be low (see Table 10).

4.5 Difficult Terms and Simplification

A master student in translation and technical writing manually assigned difficulty scores on a scale of 1–7 to the syntax and vocabulary of 249 simplified sentences from the

Table 11. Comparison of manually simplified and source sentences in Task 3

Metric (Avg)	Source snt	Simplified snt
FKGL	15.16	12.12
# Abbreviations	0.24	0.13
# Difficult terms	0.41	0.28

participants' runs corresponding to 13 distinct source sentences. Table 10 provides evidence that automatic simplification is effective in terms of reducing syntax difficulty. However, lexical difficulty, i.e. the presence of difficult scientific terms, is much higher, remaining the main barrier to understanding a scientific text.

In order to evaluate the quality of our train data (648 manually simplified sentences), we compared simplified and source sentences according to the following metrics:

- FKGL readability score that relies on average sentence lengths and number of syllables per word [14];
- Average number of abbreviations per sentence. The list of abbreviations was taken from Task 2.1.
- Average number of difficult terms per sentence. The list of difficult terms was constructed from the data used for the evaluation of Task 2.1.

Table 11 reports the scores of manually simplified and source sentences used in Task 3 according to these three metrics. The table provides evidence that our manual simplifications reduce text difficulty not only in terms of readability score, but our simplified sentences have more than 50% less difficult terms and abbreviations. These results also provide evidence that our tasks are closely interconnected.

5 Discussion and Conclusions

We introduced the CLEF 2023 SimpleText track, containing three interconnected shared tasks on scientific text simplification. Conceptually, we envisage a system pipeline retrieving relevant abstracts or passages for Task 1 (Content Selection); in order to detect difficult terms to be explained for Task 2 (Complexity Spotting); and simplify the ultimate selected sentences for Task 3 (Text Simplification). We evaluated the term difficulty, their explanations, and simplifications with regard to the queries from Task 1.

For Task 1, we created a large corpus of scientific abstracts, a set of popular science requests with detailed relevance judgments on the level of relevance of scientific abstracts to the request and the broader context of a newspaper article on this topic. The abstracts of scientific papers retrieved for these requests were used in the follow-up tasks. In 2023, we dramatically extended the qrels and introduced a additional evaluation measures that takes into account the complexity or credibility of the retrieved abstracts.

For Task 2 and 3, we created a corpus of sentences extracted from the abstracts of scientific publications, with manual annotations of term complexity and their definitions (Task 2). Our manual simplifications (Task 3) reduce text difficulty not only in terms of readability score but also have 50% less difficult terms and abbreviations than the source sentences. These results confirm the interconnection of the SimpleText tasks, and the value of researching their key dependencies.

We refer to the preceding sections for details of the different approaches to the tasks, and their effectiveness. A few general observations stand out. First, even when deploying similar models, the results of the same methods depend heavily on implementation, fine-tuning, and/or used prompts. Second, efficiency is of key importance in addition to effectiveness. We have received many partial runs due to token/time constraints of LLMs. Results of difficult term detection by LLMs are comparable to those of unsupervised methods. Third, robustness of the approaches remains challenging. Specifically, no less than 30%–40% of definitions are either unhelpful or even more difficult than the corresponding terms. Fourth, automatic simplification is effective in terms of reducing syntax difficulty and optimizing the FKGL score. However, lexical difficulty, i.e. the presence of difficult scientific terms, is much higher, remaining the main barrier to understanding a scientific text. The most common errors introduced in simplifications are spelling, typographic, and punctuational ones (46%), followed by information loss (18%), unresolved anaphora (13%), and oversimplification (12%).

So the general upshot of the CLEF 2023 SimpleText track is both that we observed great progress, but at the same time that there is also still a lot of room for improvements. In the future, we plan to classify difficult term explanations (definitions, examples, abbreviation deciphering etc.) and evaluate systems according to the usefulness and complexity of the provided explanations of scientific terms. We will further explore information distortion introduced by simplification.

Acknowledgments. *This research was funded, in whole or in part, by the French National Research Agency (ANR) under the project ANR-22-CE23-0019-01. We would like to thank Sarah Bertin, Radia Hannachi, Silvia Araújo, Pierre De Loor, Olga Popova, Diana Nurbakova, Quentin Dubreuil, Helen McCombie, Aurianne Damoy, Angelique Robert, and all other colleagues and participants who helped run this track.*

References

1. Aliannejadi, M., Faggioli, G., Ferro, N., Vlachos, M. (eds.): Working Notes of CLEF 2023: Conference and Labs of the Evaluation Forum. CEUR Workshop Proceedings. CEUR-WS.org (2023)
2. Alva-Manchego, F., Martin, L., Scarton, C., Specia, L.: EASSE: easier automatic sentence simplification evaluation. In: Proceedings of the 2019 Conference on Empirical Methods in Natural Language Processing and the 9th International Joint Conference on Natural Language Processing (EMNLP-IJCNLP): System Demonstrations, pp. 49–54. Association for Computational Linguistics, Hong Kong (2019). https://doi.org/10.18653/v1/D19-3009, https://aclanthology.org/D19-3009
3. Andermatt, P.S., Fankhauser, T.: UZH_Pandas at SimpleTextCLEF-2023: alpaca LoRA 7B and LENS model selection for scientific literature simplification. In: [1] (2023)

4. Anjum, A., Lieberum, N.: Automatic simplification of scientific texts using pre-trained language models: a comparative study at CLEF symposium 2023. In: [1] (2023)
5. Bertin, S.: Scientific simplification, the limits of ChatGPT. In: [1] (2023)
6. Capari, A., Azarbonyad, H., Tsatsaronis, G., Afzal, Z.: Elsevier at simpletext: passage retrieval by fine-tuning GPL on scientific documents. In: [1] (2023)
7. Davari, D.R., Prnjak, A., Schmitt, K.: CLEF 2023 SimpleText task 2, 3: identification and simplification of difficult terms. In: [1] (2023)
8. Dubreuil, Q.: UBO team @ CLEF SimpleText 2023 track for task 2 and 3 - using IA models to simplify scientific texts. In: [1] (2023)
9. Engelmann, B., Haak, F., Kreutz, C.K., Nikzad-Khasmakhi, N., Schaer, P.: Text simplification of scientific texts for non-expert readers. In: [1] (2023)
10. Ermakova, L., et al.: Overview of SimpleText 2021 - CLEF workshop on text simplification for scientific information access. In: Candan, K.S., et al. (eds.) CLEF 2021. LNCS, vol. 12880, pp. 432–449. Springer, Cham (2021). https://doi.org/10.1007/978-3-030-85251-1_27
11. Ermakova, L., SanJuan, E., Huet, S., Augereau, O., Azarbonyad, H., Kamps, J.: CLEF 2023 simpletext track - what happens if general users search scientific texts? In: Kamps, J., et al. (eds.) ECIR 2023. LNCS, vol. 13982, pp. 536–545. Springer, Cham (2023). https://doi.org/10.1007/978-3-031-28241-6_62
12. Ermakova, L., et al.: Overview of the CLEF 2022 simpletext lab: automatic simplification of scientific texts. In: Barrón-Cedeño, A., et al. (eds.) CLEF 2022. LNCS, vol. 13390, pp. 470–494. Springer, Cham (2022). https://doi.org/10.1007/978-3-031-13643-6_28
13. Ermakova, L.N., Nurbakova, D., Ovchinnikova, I.: COVID or not COVID? Topic shift in information cascades on Twitter. In: Linguistics, A.F.C. (ed.) 3rd International Workshop on Rumours and Deception in Social Media (RDSM) Collocated with COLING 2020. Proceedings of the 3rd International Workshop on Rumours and Deception in Social Media (RDSM), Barcelona (On line), Spain, pp. 32–37 (2020). https://hal.archives-ouvertes.fr/hal-03066857
14. Flesch, R.: A new readability yardstick. J. Appl. Psychol. **32**(3), 221–233 (1948). ISSN 0021-9010
15. Hou, R., Qin, X.: An evaluation of MUSS and T5 models in scientific sentence simplification: a comparative study. In: [1] (2023)
16. Hutter, R., Sutmuller, J., Adib, M., Rau, D., Kamps, J.: University of Amsterdam at the CLEF 2023 SimpleText track. In: [1] (2023)
17. Lin, C.Y., Hovy, E.: Automatic evaluation of summaries using N-gram co-occurrence statistics. In: Proceedings of the 2003 Conference of the North American Chapter of the ACL on Human Language Technology, vol. 1, pp. 71–78. ACL (2003)
18. Maddela, M., Alva-Manchego, F., Xu, W.: Controllable text simplification with explicit paraphrasing (2021). http://arxiv.org/abs/2010.11004
19. Mansouri, B., Durgin, S., Franklin, S., Fletcher, S., Campos, R.: AIIR and LIAAD labs systems for CLEF 2023 SimpleText. In: [1] (2023)
20. Mendoza, O.E., Pasi, G.: Domain context-centered retrieval for the content selection task in the simplification of scientific literature. In: [1] (2023)
21. Ohnesorge, F., Gutierrez, M.A., Plichta, J.: Scientific text simplification and general audience. In: [1] (2023)
22. Ortiz-Zambrano, J.A., Espin-Riofrio, C., Montejo-Ráez, A.: SINAI participation in SimpleText task 2 at CLEF 2023: GPT-3 in lexical complexity prediction for general audience. In: [1] (2023)
23. Palma, V.M., Preciado, C.P., Sidorov, G.: NLPalma @ CLEF 2023 SimpleText: BLOOMZ and BERT for complexity and simplification task. In: [1] (2023)
24. Papineni, K., Roukos, S., Ward, T., Zhu, W.J.: BLEU: a method for automatic evaluation of machine translation. In: Proceedings of the 40th Annual Meeting on ACL, pp. 311–318. ACL (2002)

25. Dadić, P., Popova,O.: CLEF 2023 SimpleText tasks 2 and 3: enhancing language compre-
    hension: addressing difficult concepts and simplifying ccientific texts using GPT, BLOOM,
    KeyBert, simple T5 and more. In: [1] (2023)
26. Schwartz, A.S., Hearst, M.A.: A simple algorithm for identifying abbreviation definitions in
    biomedical text. In: Biocomputing 2003, pp. 451–462 (2002)
27. Wu, S.H., Huang, H.Y.: A prompt engineering approach to scientific text simplification:
    CYUT at SimpleText2023 task3. In: [1] (2023)
28. Xu, W., Napoles, C., Pavlick, E., Chen, Q., Callison-Burch, C.: Optimizing statistical
    machine translation for text simplification. Trans. ACL **4**, 401–415 (2016)

Overview of Touché 2023: Argument and Causal Retrieval

Alexander Bondarenko[1], Maik Fröbe[1], Johannes Kiesel[2], Ferdinand Schlatt[1],
Valentin Barriere[3], Brian Ravenet[4], Léo Hemamou[5], Simon Luck[6],
Jan Heinrich Reimer[1(✉)], Benno Stein[2], Martin Potthast[7],
and Matthias Hagen[1]

[1] Friedrich-Schiller-Universität Jena, Jena, Germany
touche@webis.de, {alexander.bondarenko,heinrich.reimer}@uni-jena.de
[2] Bauhaus-Universität Weimar, Weimar, Germany
[3] Centro Nacional de Inteligencia Artificial (CENIA), Santiago, Chile
[4] CNRS-LISN, Université Paris-Saclay, Gif-sur-Yvette, France
[5] Sanofi R&D France, Chilly-Mazarin, France
[6] Alma Mater Studiorum – Università di Bologna, Bologna, Italy
[7] Leipzig University and ScaDS.AI, Leipzig, Germany
https://touche.webis.de/

Abstract. This paper is a condensed overview of Touché: the fourth edition of the lab on argument and causal retrieval that was held at CLEF 2023. With the goal to create a collaborative platform for research on computational argumentation and causality, we organized four shared tasks: (a) argument retrieval for controversial topics, where participants retrieve web documents that contain high-quality argumentation and detect the argument stance, (b) causal retrieval, where participants retrieve documents that contain causal statements from a generic web crawl and detect the causal stance, (c) image retrieval for arguments, where participants retrieve from a focused web crawl images showing support or opposition to some stance, and (d) multilingual multi-target stance classification, where participants detect the stance of comments on proposals from an online multilingual participatory democracy platform.

Keywords: Argument retrieval · Causal retrieval · Image retrieval · Stance classification · Argument quality · Causality

1 Introduction

Making informed decisions and forming opinions on a matter often involves not only weighing pro and con arguments but also considering cause–effect relationships for one's actions [1]. Nowadays, everybody has the chance to acquire knowledge and find any kind of information on the Web (facts, opinions, arguments, etc.) on almost any topic, which can help to make decisions or get an

---

L. Hemamou—Independent view, not influenced by Sanofi R&D France.

A. Arampatzis et al. (Eds.): CLEF 2023, LNCS 14163, pp. 507–530, 2023.
https://doi.org/10.1007/978-3-031-42448-9_31

overview of different standpoints. However, conventional search engines are primarily optimized for returning *relevant* results and hardly address the deeper analysis of arguments (e.g., argument quality and stance) or analysis of causal relationships. To close this gap, with the Touché lab's four shared tasks,[1] we intended to solicit the research community to develop respective approaches. In 2023, we organized the four following shared tasks:

1. Argumentative document retrieval from a generic web crawl to provide an overview of arguments and opinions on controversial topics.
2. Retrieval of web documents from a generic web crawl to find evidence if a causal relationship between two events exists (*new task*).
3. Image retrieval to corroborate and strengthen textual arguments and to provide a quick overview of public opinions on controversial topics.
4. Stance classification of comments on proposals from the multilingual participatory democracy platform CoFE,[2] written in different languages to support opinion formation on socially important topics (*new task*).

Touché follows the traditional TREC[3] methodology: documents and topics are provided to participants, who then submit their results (up to five runs) for each topic to be assessed by human assessors.

All teams that participated in the fourth Touché lab used BM25(F) [36,37] for first-stage retrieval (except Task 4). The final ranked lists (runs) were often created based on argument quality estimation and predicted stance (Task 1), based on the presence of causal relationships in documents (Task 2), and exploiting the contextual similarity between images and queries and using the predicted stance for images (Task 3). The participants trained their own feature-based and neural classifiers to predict argument quality and stance. Also, many often used ChatGPT with various prompt-engineering methods. To predict the stance for multilingual texts (Task 4), the participants used transformer-based models exploiting a few-step fine-tuning, data augmentation, and label propagation techniques. A more comprehensive overview of all submitted approaches is provided in the extended overview paper [8].

The corpora, topics, and judgments created at Touché are freely available to the research community and can be found on the lab's website.[4] Parts of the data are also already available via the BEIR [42] and ir_datasets [28] resources.

2   Lab Overview and Statistics

In the fourth edition of the Touché lab, we received 41 registrations from 21 countries (vs. 58 registrations in 2022). The majority of the lab registrations came

---

[1] The term 'touché' is commonly "used to acknowledge a hit in fencing or the success or appropriateness of an argument, an accusation, or a witty point." [https://merriam-webster.com/dictionary/touche]

[2] https://futureu.europa.eu

[3] https://trec.nist.gov/

[4] https://touche.webis.de/

from Germany (10 registered teams), followed by China and India (4 teams each), France (3 teams), Italy, Malaysia, and Sweden (2 teams each), Bangladesh, Botswana, Bulgaria, Canada, Guinea, Ireland, Netherlands, Nigeria, Mexico, Romania, Spain, Syria, Thailand, and United Kingdom (1 team each). Out of the 41 registered teams, 7 actively participated (1 team submitted results for Task 1 and Task 2 each, 3 teams participated in Task 3, and 2 teams in Task 4) by making valid result submissions (previous editions had more active participants, with 23 active teams in 2022, 27 participating teams in 2021 and 17 teams in 2020).

We used TIRA [19] as the submission platform for Touché 2023 through which the participants could either submit software or upload run files.[5] Software submissions increase reproducibility, as the software can later be executed on different data of the same format. Overall, 5 out of the 7 active teams made software submissions. To submit software, a team implemented their approach in a Docker image that they then uploaded to their dedicated Docker registry in TIRA. Software submissions in TIRA are immutable, and after the docker image had been submitted, the teams specified the to-be-executed command— the same Docker image can thus be used for multiple software submissions (e.g., by changing some parameters). A team could upload as many Docker images or software submissions as needed (the images were not public while the shared tasks were ongoing). To improve reproducibility, TIRA executes software in a sandbox by blocking the internet connection. This ensures that the software is fully installed in the Docker image, which eases running the software later. For the execution, the participants could select the resources that their software had available for execution out of four options: (1) 1 CPU core with 10 GB RAM, (2) 2 cores with 20 GB RAM, (3) 4 cores with 40 GB RAM, or (4) 1 CPU core with 10 GB RAM and 1 Nvidia GeForce GTX 1080 GPU with 7 GB RAM. Also, the participants were able to run their software multiple times using different resources to investigate the scalability and reproducibility (e.g., whether the software executed on a GPU yields the same results as on a CPU). TIRA used a Kubernetes cluster with 1,620 CPU cores, 25.4 TB RAM, and 24 GeForce GTX 1080 GPUs to schedule and execute the software submissions, allocating the resources that the participants selected for their submissions.

3   Task 1: Argument Retrieval for Controversial Questions

The goal of the first task was to support individuals who search for opinions and arguments on socially important controversial topics like "Are social networking sites good for our society?". The previous three task iterations explored different granularities of argument retrieval and analysis: debates on various topics crawled from several online debating portals and their concise gist. For the fourth edition of the task, our focus shifted towards retrieving argumentative web documents from the web crawl corpus ClueWeb22 [31]. The topics and manual judgments from the previous task iterations were provided to the participants to enable approaches that leverage training and parameter tuning.

---
[5] https://tira.io

Table 1. Example topic for Task 1: Argument Retrieval for Controversial Questions.

Number	34
Title	Are social networking sites good for our society?
Description	Democracy may be in the process of being disrupted by social media, with the potential creation of individual filter bubbles. So a user wonders if social networking sites should be allowed, regulated, or even banned
Narrative	Highly relevant arguments discuss social networking in general or particular networking sites, and its/their positive or negative effects on society. Relevant arguments discuss how social networking affects people, without explicit reference to society

3.1 Task Definition

Given a controversial topic and a collection of web documents, the task was to retrieve and rank documents by relevance to the topic, by argument quality, and to (optionally) detect the document's stance. Participants of Task 1 needed to retrieve documents from the ClueWeb22-B crawl for 50 search topics.

To lower the entry barrier for participants who could not index the whole ClueWeb22-B corpus on their side, we provided a first-stage retrieval possibility via the API of the BM25F-based search engine ChatNoir [7] and a smaller version of the corpus that contained one million documents per topic. For the identification of arguments (claims and premises) in documents, participants could use any existing argument tagging tool such as the TARGER API [12] hosted on our own servers or develop their own tools if necessary.

3.2 Data Description

Topics. For the task on argument retrieval for controversial questions (Task 1), we provided 50 search topics that represent various debated societal issues. These issues were chosen from the online debate portals (debatewise.org, idebate.org, debatepedia.org, and debate.org), with the largest number of user-generated comments and thus representing the highest societal interest. For each such case, we formulated a topic's *title* (i.e., a question on a controversial issue), a *description* specifying the particular search scenario, and a *narrative* that served as a guideline for the human assessors (see Table 1).

Document Collection. The retrieval collection was the ClueWeb22 (Category B) corpus [31] that contains 200 million multilingual most frequently visited web pages like Wikipedia articles or news websites. The indexed corpus was available via the ChatNoir API[6] and its module[7] integrated in PyTerrier [29].

---

[6] https://github.com/chatnoir-eu/chatnoir-api
[7] https://github.com/chatnoir-eu/chatnoir-pyterrier

3.3   Evaluation Setup

Our human assessors labeled the ranked results by the task participants both for their general topical relevance and for the rhetorical argument quality [44], i.e., "well-writtenness": (1) whether the document contains arguments and whether the argument text has a good style of speech, (2) whether the argument text has a proper sentence structure and is easy to follow, and (3) whether it includes profanity, has typos, etc. Also, the documents' stance towards the argumentative search topics was labeled as 'pro', 'con', 'neutral', or 'no stance'.

Analogously to the previous Touché editions, our volunteer assessors annotated the document's topical relevance with three levels: 0 (not relevant), 1 (relevant), and 2 (highly relevant). The argument quality was also labeled with three labels: 0 (low quality or no arguments in the document), 1 (average quality), and 2 (high quality). We provided the annotators with detailed annotation guidelines, including examples. In the training phase, we asked 4 annotators to label the same 20 randomly selected documents (initial Fleiss' kappa values: relevance $\kappa = 0.39$ (fair agreement), argument quality $\kappa = 0.34$ (fair agreement), and $\kappa = 0.51$ (moderate agreement) for labeling the stance) and in the follow-up discussion clarified potential misinterpretations. Afterward, each annotator independently judged the results for disjoint subsets of the topics (i.e., each topic was judged by one annotator only). We used this annotation policy due to a high annotation workload. Our human assessors labeled in total 747 documents pooled from 8 runs using a top-10 pooling strategy implemented in the TrecTools library [32].

3.4   Submitted Approaches and Evaluation Results

In 2023, only one team participated in Task 1 and submitted seven runs. We, thus, decided to evaluate all the participant's runs and an additional baseline. Table 2 shows the results of all submitted runs with respect to relevance, argument quality, and stance detection (more detailed results for each submitted run, including the 95% confidence intervals, are in Tables 9 and 10 in Appendix B). Below, we briefly describe the submitted approaches to the task.

The task's baseline run, *Puss in Boots*, used the results that ChatNoir [7] returned for the topics' titles without any pre-processing used as queries. ChatNoir is an Elasticsearch-based search engine for the ClueWeb and Common Crawl web corpora that employs BM25F ranking (fields: document title, keywords, main content, and the whole document) and SpamRank scores [15]. The document stance for the baseline run was predicted after summarizing the document's main content with BART [25],[8] by zero-shot prompting the Flan-T5 model [13].[9]

Team *Renji Abarai* [33] submitted in total seven runs. Their baseline run simply used the top-10 results returned by ChatNoir for the pre-processed topics' titles used as queries. During pre-processing, stop words were first removed

---

[8] Pre-trained model: https://huggingface.co/facebook/bart-large-cnn; minimum length: 64; maximum length: 256.

[9] Pre-trained model: https://huggingface.co/google/flan-t5-base; maximum generated tokens: 3; the prompt is given in Appendix A.

Table 2. Results of all runs submitted to Task 1: Argument Retrieval for Controversial Questions. Reported are the mean nDCG@10 for relevance and argument quality and macro-averaged F_1 for stance detection. Since Renji Abarai re-ranked the same set of documents for all the runs, this yields identical stance detection results. The task baseline run by Puss in Boots is shown in bold.

Team	Run Tag	nDCG@10		F_1 macro
		Rel.	Qua.	Stance
Puss in Boots	**ChatNoir** [7]	**0.834**	**0.831**	**0.203**
Renji Abarai	stance_ChatGPT	0.747	0.815	0.599
Renji Abarai	stance-certainNO_ChatGPT	0.746	0.811	0.599
Renji Abarai	ChatGPT_mmGhl	0.718	0.789	0.599
Renji Abarai	ChatGPT_mmEQhl	0.718	0.789	0.599
Renji Abarai	meta_qual_score	0.712	0.771	0.599
Renji Abarai	baseline	0.708	0.766	0.599
Renji Abarai	meta_qual_prob	0.697	0.774	0.599

using their own handcrafted list of terms; the remaining query terms were then lowercased and lemmatized with the Stanza NLP library [34]. For the other six runs, the results of the baseline run were re-ranked based on the predicted argument quality and predicted document stance. Argument quality was predicted using either a meta-classifier (random forests) trained on the class predictions and class probabilities of six base classifiers or by prompting ChatGPT. Each base classifier (feedforward neural network, LightGBM [22], logistic regression, naïve Bayes, SVM, and random forests) was trained in two variants: (1) using a set of 32 handcrafted features (e.g., sentiment, spelling errors, the ratio of arguments in documents, etc.) and (2) using documents represented with the instruction-based fine-tuned embedding model INSTRUCTOR [40]. All the classifiers were trained on the manual argument quality labels from the Touché 2021 Task 1 [9], which was also used to select examples for a few-shot prompting ChatGPT. The resulting ranked lists submitted by Renji Abarai differed in the type of argument quality classifiers used for re-ranking, whether predicted classes or probabilities were used, or if the predicted document stance was considered. The document stance for all the runs was predicted using ChatGPT.

4   Task 2: Evidence Retrieval for Causal Questions

The goal of the Touché 2023 lab's second task was to support users who search for answers to causal yes-no questions like "Do microwave ovens cause cancer?", supported by relevant evidence instances. In general, such causal questions ask if something causes or does not cause something else.

Table 3. Example topic for Task 2: Evidence Retrieval for Causal Questions.

Number	39
Title	Do microwave ovens cause cancer?
Cause	microwave ovens
Effect	cancer
Description	A user has recently learned that radiation waves can cause cancer. They are wondering if their microwave oven produces radiation waves and if these are dangerous enough to cause cancer
Narrative	Highly relevant documents will provide information on a potential causal connection between microwave ovens and cancer. This includes documents stating or giving evidence that the first is (or is not) a cause of the other. Documents stating that there is not enough evidence to decide either way are also highly relevant. Relevant documents may contain implicit information on whether the causal relationship exists or does not exist. Documents are not relevant if they either mention one or both concepts, but do not provide any information about their causal relation

4.1 Task Definition

Given a causality-related topic and a collection of web documents, the task was to retrieve and rank documents by relevance to the topic. For 50 search topics, participants of Task 2 needed to retrieve documents from the ClueWeb22-B crawl that contain relevant causal evidence. An optional task was to detect the document's *causal stance*. A document can provide supportive evidence (a causal relationship between the cause and effect from the topic holds), refutative (a causal relationship does not hold), or neutral (in some cases holds and in some does not). Like in Task 1, ChatNoir [7] could be used for first-stage retrieval.

4.2 Data Description

Topics. The 50 search topics for Task 2 described scenarios where users search for confirmation of whether some causal relationship holds. For example, a user may want to know the possible reason for a current physical condition. Each of these topics had a *title* (i.e., a causal question), *cause* and *effect* entities, a *description* specifying the particular search scenario, and a *narrative* serving as a guideline for the assessors (see Table 3). The topics were manually selected from a corpus of causal questions [10] and a graph of causal statements [21] such that they spanned a diverse set of domains.

Document Collection. The same document collection as in Task 1 was used.

4.3 Evaluation Setup

Relevance assessments were gathered with volunteer human assessors. The assessors were instructed to label documents as *not relevant* (0), *relevant* (1), or *highly*

Table 4. Relevance results of all runs submitted to Task 2: Evidence Retrieval for Causal Questions. Reported are the mean nDCG@5 for relevance and macro-averaged F_1 for stance detection; Puss in Boots baseline is in bold. The dagger[†] indicates a statistically significant improvement ($p < 0.05$, Bonferroni corrected) over the Puss in Boots baseline. Team He-Man did not detect the stance.

Team	Run Tag	Relevance nDCG@5	Stance F_1 macro
He-Man	no_expansion_rerank	0.657[†]	–
Puss in Boots	**ChatNoir [7]**	**0.585**	**0.256**
He-Man	gpt_expansion_rerank	0.374	–
He-Man	causenet_expansion_rerank	0.268	–

relevant (2). The direction of causality was considered, i.e., a document stating that B causes A was considered as off-topic (not relevant) for the topic "Does A cause B?". The document's stance was also labeled to evaluate the optional stance detection task. The labeling procedure was analogous to Task 1, where volunteer assessors participated in training and a discussion. Agreement on the same 20 randomly selected documents across 4 annotators was measured with Fleiss' kappa. Before the discussion, the agreement was $\kappa = 0.58$ for relevance and $\kappa = 0.55$ for stance assessment (both indicate a moderate agreement). After discussing discrepancies, similar to Task 1, each annotator labeled a disjoint set of topics. We pooled the top-5 documents from each submitted run (plus additional baseline) and labeled 718 documents in total.

4.4 Submitted Approaches and Evaluation Results

One team *He-Man* [20] participated in Task 2 and submitted three runs. Like the baseline run *Puss in Boots*, all three participant runs used ChatNoir [7] for first-stage retrieval. For two runs, first, the cause and effect events were extracted from the topic title field using dependency tree parsing. Next, query expansion and query reformulation approaches were applied. In the query expansion approach, the topic title was expanded with semantically related concepts from the CauseNet, a graph of causal relations [21]. For this, all relations in the CauseNet-Precision variant were embedded using BERT [16]. Next, the embedding's cosine similarity was compared with the embedding of the topic's relation. The top-5 terms from the documents linked to the matched CauseNet relation were then used to expand the query. The second approach, the query reformulation technique, fed the deconstructed topic title in a semi-structured JSON format to ChatGPT. The chatbot was then prompted to generate new variants of the query, exchanging causes, effects, and causal phrases. All three query variants (original topic title, expanded query, and reformulated query) were then submitted to ChatNoir. Finally, all approaches re-ranked the results using a position bias. Documents containing the causal relationship from the topic earlier in the document were ranked higher. To detect the position of the relation, the same dependency tree parsing developed for the query deconstruction was used.

The task's baseline run of *Puss in Boots* additionally predicted the document stance by first summarizing a document's main content with BART [25],[10] and then zero-shot prompting the Flan-T5 model [13].[11]

Table 4 shows the results of Task 2 (more detailed results for each submitted run, including the 95% confidence intervals, is in Table 11 in Appendix B). We report nDCG@5 for relevance and macro-averaged F_1 for stance detection. The Puss in Boots baseline was more effective in terms of relevance than two of the three participant runs. The final participant run was, however, able to statistically outperform the baseline. The participant team opted not to detect the stance. Therefore, only the baseline run could be evaluated for stance detection, achieving an F_1 score of 0.256.

5   Task 3: Image Retrieval for Arguments

The goal of the third task was to provide argumentation support through image search. The retrieval of relevant images should provide both a quick visual overview of frequent arguments on some topic and for compelling images to support one's argumentation. To this end, the second edition of this task continued with the retrieval of images that indicate an agreement or disagreement to some stance on a given topic as two separate lists, similar to textual argument search.

5.1   Task Definition

Given a controversial topic and a collection of web documents with images, the task was to retrieve for each stance (pro and con) images that show support for that stance. Participants of Task 3 should retrieve and rank images, possibly utilizing the corresponding web documents, from a focused crawl of 55,691 images and for a given set of 50 search topics (which were used by other tasks in previous years) [24]. Like in the last edition of this task, the focus is on providing users with an overview of public opinions on controversial topics, for which we envision a system that provides not only textual but also visual support for each stance in the form of images. Participants were able to use the approximately 6,000 relevance judgments from the last edition of the task for training supervised approaches [23].[12] Similar to the other tasks, participants were free to use any additional existing tools and datasets or develop their own.

5.2   Data Description

Topics. Task 3 employs 50 controversial topics from earlier Touché editions (e.g., used in 2021), but which were not used in the first edition of this task. Like for

---

[10] Pre-trained model: https://huggingface.co/facebook/bart-large-cnn; minimum length: 64; maximum length: 256.

[11] Pre-trained model: https://huggingface.co/google/flan-t5-base; maximum generated tokens: 3; the prompt is given in Appendix A.

[12] https://webis.de/data.html#touche-corpora

Task 1 (cf. Sect. 3), we provided for each topic a title, description, and narrative. Description and narrative were adapted as needed to fit to the image retrieval setting.

Document Collection. This task's document collection stems from a focused crawl of 55,691 images and associated web pages from late 2022. We downloaded the top 100 images and associated web pages from Google's image search for a total of 2,209 queries. Nearly half of the queries (namely 1,050) were created like in the first edition of this task, by appending filter words like "good", "meme", "stats", "reasons", or "effects" to a manually created query for each topic. The remaining 1,159 queries were collected from participants in an open call, which allowed anyone to submit queries until the end of December 2022. Of these queries, 557 were created manually (57 by team Neville Longbottom, 250 by team Hikaru Sulu, and 250 by us) and the remaining were created using ChatGPT by team Neville Longbottom: they asked ChatGPT for a list of pro and con arguments for each topic, then for an image description illustrating the respective arguments, and then for a search query to match the description. From the search results we attempted to download 147,264 images, but discarded 5,666 for which we could not download the image, 6,619 for which the image was more than 2,000 pixels wide or high,[13] 20,696 for which an initial text recognition using Tesseract[14] yielded more than 20 words,[15] 8,538 for which the web page could not be downloaded, 484 for which the web page contained no text, and 45,254 for which we could not find the image URL in the web page DOM. After a duplicate detection using pHash[16] as in the previous year, the final dataset contains 55,691 images. The dataset contains various resources for each image, including the associated page for which it was retrieved as HTML page and as detailed web archive,[17] information on how the image was ranked by Google, and information from Google's Cloud Vision API,[18] e.g., detected text and objects.

5.3   Evaluation Setup

Our two volunteer human assessors labeled the ranked results by the task participants (i.e., the images) for their relevance to the topic's narrative. First, assessors decided whether an image is on topic (yes or no). If so, they also decided whether an image is relevant according to the pro-side of the narrative or its con-side, or both: 0 (not relevant), 1 (relevant), and 2 (highly relevant), though we did not distinguish between levels 1 and 2 in our final evaluation. However, assessors were instructed that an image could not be highly relevant for both pro and con,

---

[13] As one of our suggested use case for image retrieval for arguments is getting a quick overview, we excluded overly large images.

[14] https://github.com/tesseract-ocr/tesseract

[15] To sharpen our focus on images, this year we tried to exclude images that are actually screenshots of text documents.

[16] https://www.phash.org/

[17] Archived using https://github.com/webis-de/scriptor

[18] https://cloud.google.com/vision

which could be useful for a stricter evaluation in the future. We provided the assessors with guidelines, discussed several examples, and also were in close contact during the assessment phase to discuss further edge cases. Achieved Fleiss' κ values (measured on three topics for which all images were assessed by both assessors) were for on-topic 0.38 (fair), for pro 0.34 (fair), and for con 0.31 (fair). Without the distinction of levels 1 and 2, the agreement increases to 0.45 for pro (moderate) and 0.52 for con (moderate). Our human assessors labeled in total 6692 images.

Although rank-based metrics for single image grids exist [45], none have been proposed so far for a 'pro-con' layout. Therefore, like the last year, participants' submitted results were evaluated by the simple ratio of relevant images among 20 retrieved images, namely 10 images for each stance (precision@10). We again used three increasingly strict definitions of relevance, corresponding to three precision@10 evaluation measures: being on-topic, being in support of some stance (i.e., an image is "argumentative"), and being in support of the stance for which the image was retrieved.

5.4 Submitted Approaches and Evaluation Results

In total, three teams participated in Task 3 and submitted 12 runs total, not counting the submitted queries described above. Table 5 shows the results of all submitted runs (more detailed results for each submitted run, including the 95% confidence intervals, are in Tables 12, 13, and 14 in Appendix B). Overall, scores are considerably lower than last year, where precision@10 for stance relevance was as high as 0.425. We attribute this to the new set of topics, which contained a much more questions that were hard to picture.

As a baseline (team *Minsc*), we used the model of [11], which was developed by a collaboration of two teams that participated in last year's task: Aramis and Boromir.[19] The approach employed standard retrieval together with a set of handcrafted features for argumentativeness detection. For retrieval, the approach used Elasticsearch's BM25 (default settings: $k_1 = 1.2$ and $b = 0.75$) with each image (document) represented by the text from the web page around the image and text recognized in the image using Tesseract.[14] For argumentativeness detection, the approach used a neural network classifier based on thirteen different features (color properties, image type, and textual features), and trained on the ground-truth annotations from last year. The features are calculated from, amongst others, the query, the image text, the HTML text around the image, the interrelation and sentiments of the mentioned texts, and the colors in the image. The approach used random stance assignment. Since this baseline performed much worse than anticipated, we expect a bug in the implementation.

Team *Hikaru Sulu* submitted two valid runs. Their approach used CLIP [35] to calculate the similarity between keywords and images and retrieved, per topic, the images most similar to one of the keywords. For the first run, they used the

---

[19] Since no stance model convincingly outperformed naive baselines in their evaluation, we use the simple both-sides baseline that assigns each image to both stances.

Table 5. Relevance results of all runs submitted to Task 3: Image Retrieval for Argumentation. Reported are the mean precision@10 for all three definitions of relevance; Minsc baseline is in bold. The dagger† indicates a statistically significant improvement ($p < 0.05$, Bonferroni corrected) over the baseline.

Team	Run Tag	On-topic	Arg.	Stance
		\multicolumn{3}{c}{Precision@10}		
Neville Longbottom	clip_chatgpt_args.raw	$0.785^\dagger$	$0.338^\dagger$	$0.222^\dagger$
Neville Longbottom	clip_chatgpt_args.debater	$0.684^\dagger$	$0.341^\dagger$	$0.216^\dagger$
Hikau Sulu	Keywords	$0.664^\dagger$	$0.350^\dagger$	$0.185^\dagger$
Hikaru Sulu	Topic-title	$0.770^\dagger$	$0.335^\dagger$	$0.179^\dagger$
Neville Longbottom	bm25_chatgpt_args.raw	0.572	0.274	$0.166^\dagger$
Jean-Luc Picard	No stance detection	$0.523^\dagger$	$0.292^\dagger$	0.162
Neville Longbottom	bm25_chatgpt_args.diff	0.442	0.240	0.150
Jean-Luc Picard	Text+image text stance detection	$0.502^\dagger$	0.272	0.144
Jean-Luc Picard	BM25 Baseline	$0.536^\dagger$	$0.268^\dagger$	0.141
Jean-Luc Picard	Text stance detection	$0.498^\dagger$	$0.262^\dagger$	0.136
Neville Longbottom	bm25_chatgpt_args.debater	0.416	0.201	0.128
Minsc	**Aramis**	**0.376**	**0.194**	**0.102**
Jean-Luc Picard	Image text stance detection	0.369	0.196	0.098

topic title as a keyword, but for the second run, they extracted all nouns and verbs from the topic title and extended that list with synonyms and antonyms from WordNet [18]. The stance was determined randomly, which performed in their internal evaluation better than using different keywords for pro and con. As Table 5 shows, the extended list lead to retrieving more on-topic images, but less argumentative ones.

Team *Jean-Luc Picard* [30] submitted five valid runs. Their first run used the web page text indexed by PyTerrier's BM25 [29] (default settings: $k_1 = 1.2$ and $b = 0.75$). For the other runs, they used a pipeline of query preprocessing, the same BM25-based retrieval as their first run, stance detection, and re-ranking. For query preprocessing, they created a parse tree of the topic and filtered out frequent words to create a short query. The runs correspond to four different stance detection approaches: (1) random or (2) using zero-shot classification based on the pre-trained BART MultiNLI model[20] that assigns the image to pro, contra, or neutral (i.e., will be discarded) based on the (a) web page text, (b) the image text, or (c) both texts. After that, images were re-ranked: for each topic, images were generated with Stable Diffusion [38] using the preprocessed query as prompt, then SIFT keypoints were identified [27] and matched between generated and retrieved images, and then the result list was re-ranked per the number of matched keypoints in descending order. Similar to the internal evaluation of team Hikaru Sulu, a random stance assignment performed best.

---

[20] https://huggingface.co/facebook/bart-large-mnli

Team *Neville Longbottom* [17] submitted five valid runs. They first employed ChatGPT[21] to generate image descriptions for each topic and stance (neither description nor narrative were used). Then, they retrieved images with these descriptions, either (1) using the web page text close to the image indexed via PyTerrier's BM25 [29] (default settings: $k_1 = 1.2$ and $b = 0.75$) or (2) using CLIP [35] for ranking images by their similarity to the description. For runs 3–5, the approach continued by re-ranking the result list, either (a) by penalizing the BM25-score of an image with the BM25-score of the image for the respective other stance's description (re-ranking the results of run (1)) or (b) by using IBM's debater pro-con score [3] between the topic title and the text close to the image on the web page (2 runs; re-ranking results of run (1) or (2)). The CLIP method without re-ranking performed best.

6   Task 4: Multilingual Multi-target Stance Classification

The goal of the fourth task was to build technologies to analyze peoples' opinions on a wide range of socially important subjects and to facilitate building useful tools for analyzing society or helping decision-makers. Large-scale deployment of such tools faces challenges like multilingualism or high variability of the topics of interest, and hence of the target of the stances. In this edition of the Touché lab, we proposed a new task on multilingual multi-target stance classification of comments to proposals from the Conference on the Future of Europe (CoFE), an online participatory democracy platform.

6.1   Task Definition

Given a proposal on a socially important issue, its title, and its topic, the task is to classify whether a comment on the proposal is 'in favor', 'against, or 'neutral' towards the commented proposal. The participants needed to classify multilingual comments from 6 different languages[22] into the 3 stance classes. We also provided an automatic English translation of all proposals and titles initially written in any of the 24 official EU languages (plus Catalan and Esperanto). Comments to the proposals can be written in a different language than the proposal itself.

Within the task, we organized two subtasks: (1) *Cross-debate Classification*, where the participants were not allowed to use comments from debates included in the test set, and (2) *All-data-available Classification*, where the participants could use all the available data. Also, the participants could use any additional existing tools or datasets like Debating Europe [5] or X-Stance [43].

---

[21] https://chat.openai.com/chat
[22] German, English, Greek, French, Italian, and Hungarian.

Table 6. A data instance for Task 4: Multilingual Multi-Target Stance Classification.

Number	34
Title	Set up a program for returnable food packaging made from recyclable materials
Proposal	The European Union could set up a program for returnable food packaging made from recyclable materials (e.g. stainless steel, glass). These packaging would be produced on the basis of open standards and cleaned according to [...]
Comment	Ja, wir müssen den Verpackungsmül reduzieren
Label	In favor

Table 7. Characteristics of the datasets used in Task 4.

Dataset	Stance classes	Languages	Proposals	Comments
CF_S	2	25	2,731	7,002
CF_{E-D}	3	21	936	1,414
CF_U	–	25	2,892	13,213
CF_{E-T}	3	17	771	1,228

6.2   Data Description

The proposals and comments used in Task 4 stem from the Conference on the Future of Europe (CoFE)[23], an online debating platform where users can write proposals and comment on the suggested ideas. The dataset contains about 4,000 proposals and 20,000 comments written in 26 languages [4,6] with all languages. English, German, and French were the most commonly used languages to write proposals and comments on the platform. An example of a proposal,[24] a corresponding comment, and stance of the comment towards the proposal is shown in Table 6.

For training stance classifiers, participants were provided with three datasets: (1) CF_{E-D}, a small set of comment–proposal pairs that were manually annotated with three stance labels, (2) CF_S, a larger set of comment–proposal pairs that are self-annotated by the comment authors as either 'in favor' or 'against' (no 'neutral' comments included because users could not select a 'neutral' label), and (3) CF_U, a large set of unlabeled comment–proposal pairs. The fourth dataset, CF_{E-T}, was built in the same way as the CF_{E-D} dataset, but was held out for testing the submitted approaches (see Table 7).

---

[23] https://futureu.europa.eu
[24] From https://futureu.europa.eu/en/processes/GreenDeal/f/1/proposals/83

6.3 Submitted Approaches and Evaluation Results

Two teams participated in Task 4 and submitted 8 runs in total. Below, we briefly describe the participants' approaches plus additional baseline runs.

Team *Cavalier* was our baseline that implemented three stance classifiers. For Subtask 1 (cross-debate classification) we implemented two baseline classifiers: The first one ('Cavalier Simple') always predicts the majority class ('in favor'). The second baseline ('Cavalier Subtask 1') is based on the transformer-based multilingual masked language model XLM-R [4, 14]. This model was first fine-tuned on the X-Stance dataset [43] and the CF_S dataset to classify just two stance classes ('in favor' or 'against') and subsequently fine-tuned again on the Debating Europe dataset [5] to classify all three stance classes ('in favor', 'against', or 'neutral'). All comments or proposals appearing in the test set, CF_{E-T}, were removed before fine-tuning. The baseline classifier for Subtask 2 (all-data-available classification) used the same model and training scheme as 'Cavalier Subtask 1', but comments or proposals that appeared in the test set were not removed from the training data.

Team *Silver Surfer* [2] submitted six valid runs to Subtask 2. Their approaches were based on fine-tuning pre-trained English and multilingual transformers: a RoBERTa model[25] [26], an XLM-R model[26] [14] and an English BERT model[27] [16]. To increase the size of the training data, the team applied data augmentation using back-translation [41] (i.e., translating texts to other languages and then back to the original language) and used label spreading [46] to transfer labels from the CF_{E-D} dataset to the CF_U dataset. Run 1 used an XGBoost classifier trained on comment metadata (e.g., number of upvotes/downvotes, endorsements) and the output probabilities from four fine-tuned transformer models: (1) RoBERTa fine-tuned on the CF_S dataset, (2) RoBERTa fine-tuned on the CF_{E-D} dataset, both translated to English, (3) XLM-R fine-tuned on the CF_S dataset, and (4) XLM-R fine-tuned on the CF_{E-D} dataset. Neither data augmentation nor label spreading was used in this run. Runs 2 and 3 used a RoBERTa or XLM-R model, respectively, fine-tuned on the CF_{E-D} dataset (English comments only for RoBERTa, all comments for XLM-R) after applying label spreading using CF_S dataset. For Run 4, they fine-tuned an XLM-R model on the CF_{E-D} dataset with back-translation, and for Run 5, they fine-tuned a RoBERTa model on all non-English comments from the CF_{E-D} dataset after translating the comments to English. For the team's last run, Run 6, they translated non-English comments from two datasets to English and fine-tuned a BERT model, first on translated comments from the CF_S dataset (binary stance) and subsequently on translated comments from the CF_{E-D} dataset (all three stance classes).

Team *Queen of Swords* [39] submitted two valid runs to Subtask 1. Their runs used English and multilingual BERT models [16] that were fine-tuned on a combination of the labeled (CF_S and CF_{E-D}) and unlabeled datasets (CF_U).

---

[25] `roberta-base.`

[26] `xlm-roberta-large.`

[27] `bert-base-uncased.`

Table 8. Results of Task 4 (Multilingual Multi-Target Stance Classification) for two subtasks (Sub.) evaluated using macro-averaged F_1 (per language and overall) and overall accuracy (Acc.). Sorted by overall F_1. The Cavalier baseline is shown in bold.

Team	Sub.	F_1 macro							Acc.
		En	Fr	De	It	Hu	El	All	
Subtask 1: Cross-Debate Classification									
Cavalier Subtask 1	**1**	**57.2**	**54.6**	**58.8**	**68.5**	**50.9**	**56.6**	**59.3**	**67.3**
Queen of swords (Run 1)	1	44.8	41.3	34.5	37.7	40.5	38.9	41.7	60.5
Queen of swords (Run 2)	1	35.1	31.5	26.2	40.9	43.0	35.7	32.4	61.6
Cavalier Simple	**1**	**24.4**	**24.2**	**20.3**	**25.1**	**29.3**	**17.1**	**23.7**	**55.2**
Subtask 2: All-Data-Available Classification									
Cavalier Subtask 2	**2**	**59.4**	**54.9**	**54.6**	**54.9**	**52.8**	**54.2**	**57.7**	**63.0**
Silver Surfer (Run 3)	2	36.7	33.9	30.2	37.8	38.0	33.3	35.0	55.1
Silver Surfer (Run 5)	2	35.3	30.4	26.1	35.3	34.8	27.8	32.9	53.7
Silver Surfer (Run 4)	2	35.0	30.3	20.0	37.5	41.7	25.0	32.3	52.4
Silver Surfer (Run 2)	2	28.5	25.6	24.3	32.9	21.5	22.8	27.0	46.3
Silver Surfer (Run 1)	2	26.3	21.1	18.9	19.1	30.0	23.3	23.9	46.1
Silver Surfer (Run 6)	2	41.4	23.2	21.2	14.1	22.8	32.8	17.7	21.6

Labels for the CF_S dataset were derived from a BERT model fine-tuned on the labeled datasets. For training the English BERT model, all non-English texts were translated to English using the `deep-translator` Python package.[28]

The submitted approaches were evaluated using the macro-averaged F_1-score and the accuracy of classifying the correct stance for each comment-proposal pair. Table 8 shows the system's classification effectiveness for the most common languages and across all languages. None of the submitted runs outperformed the baseline (Cavalier) in the two subtasks. For cross-debate classification (Subtask 1), the best participant approach (Queen of Swords, Run 1) achieved a macro-averaged F_1-score of 41.7 slightly worse than the baseline (Cavalier Subtask 1, macro-avg. F_1: 59.3). With all data available for training (Subtask 2), the best submitted run (Silver Surfer, Run 3) achieved an F_1-score of 35.0 compared to the baseline (Cavalier Subtask 2, macro-avg. F_1: 57.7).

7   Conclusion

The fourth edition of the Touché lab on argument and causal retrieval featured four tasks: (1) argument retrieval for controversial topics, (2) causal retrieval, (3) image retrieval for arguments, and (4) multilingual multi-target stance classification. In contrast to the prior iterations of the Touché lab, the main challenge for the participant was to apply argument analysis methodology on long web

---

[28] https://pypi.org/project/deep-translator/#google-translate-1

documents from the ClueWeb22-B dataset. Furthermore, we expanded the lab's scope by introducing novel tasks that aimed to retrieve evidence for causal relationships and predict the stance of multilingual texts.

Out of the 41 registered teams, 7 participated in the tasks and submitted a total of 29 runs. The participants often used well-performing techniques from previous editions, such as sparse retrieval and fine-tuning transformer-based models for argument quality estimation and stance prediction, but now increasingly used generative language models like ChatGPT with various prompt-engineering techniques. All teams that participated in Tasks 1 and 2 used the search engine Chat-Noir as their first-stage retrieval model, and then re-ranked documents based on the predicted argument quality and stance (Task 1), and based on the presence of causal relationships (Task 2). For Task 3, each of the top 4 runs employed CLIP embeddings [35] to find similar images to some text, which means dense retrieval approaches outperformed traditional approaches this year. However, still, no approach was able to predict an image's stance better than random. To predict the stance of multilingual texts (Task 4), participants used BERT-based models exploiting a few-step fine-tuning, label propagation, data augmentation, and translation.

We plan to continue Touché as a collaborative platform for researchers in argument retrieval. All Touché resources are freely available, including topics, manual relevance, argument quality, and stance judgments, and submitted runs from participating teams. These resources and other events such as workshops will help to further foster the community working on argument retrieval.

Acknowledgment. This work has been partially supported by the Deutsche Forschungsgemeinschaft (DFG) in the project "ACQuA 2.0: Answering Comparative Questions with Arguments" (project 376430233) as part of the priority program "RATIO: Robust Argumentation Machines" (SPP 1999). V. Barriere's work was funded by the National Center for Artificial Intelligence CENIA FB210017, Basal ANID. This work has been partially supported by the OpenWebSearch.eu project (funded by the EU; GA 101070014).

References

1. Ajzen, I.: The social psychology of decision making. In: Social psychology: Handbook of basic principles, pp. 297–325. Guilford Press (1996)
2. Avila, J.P., Rodrigo, A., Centeno, R.: Silver surfer team at Touché task 4: Testing data augmentation and label propagation for multilingual stance detection. In: Working Notes of CLEF 2023 - Conference and Labs of the Evaluation Forum. CEUR Workshop Proceedings. CEUR-WS.org (2023)
3. Bar-Haim, R., Kantor, Y., Venezian, E., Katz, Y., Slonim, N.: Project debater APIs: Decomposing the AI grand challenge. In: Proceedings of EMNLP 2021, pp. 267–274. ACL (2021). https://doi.org/10.18653/v1/2021.emnlp-demo.31
4. Barriere, V., Balahur, A.: Multilingual multi-target stance recognition in online public consultations. Mathematics **11**(9), 2161 (2023)

5. Barriere, V., Balahur, A., Ravenet, B.: Debating Europe: A multilingual multi-target stance classification dataset of online debates. In: Proceedings of PoliticalNLP 2022, pp. 16–21. ELRA (2022). https://aclanthology.org/2022.politicalnlp-1.3

6. Barriere, V., Jacquet, G., Hemamou, L.: CoFE: A new dataset of intra-multilingual multi-target stance classification from an online European participatory democracy platform. In: Proceedings of AACL-IJCNLP 2022 (2022)

7. Bevendorff, J., Stein, B., Hagen, M., Potthast, M.: Elastic ChatNoir: Search engine for the ClueWeb and the common crawl. In: Pasi, G., Piwowarski, B., Azzopardi, L., Hanbury, A. (eds.) ECIR 2018. LNCS, vol. 10772, pp. 820–824. Springer, Cham (2018). https://doi.org/10.1007/978-3-319-76941-7_83

8. Bondarenko, A., et al.: Overview of Touché 2023: Argument and causal retrieval. In: Working Notes of CLEF 2023 - Conference and Labs of the Evaluation Forum. CEUR Workshop Proceedings. CEUR-WS.org (2023)

9. Bondarenko, A., et al.: Overview of Touché 2021: Argument retrieval. In: Proceedings of CLEF 2021. CEUR Workshop Proceedings, vol. 2936, pp. 2258–2284. CEUR-WS.org (2021). https://ceur-ws.org/Vol-2936/paper-205.pdf

10. Bondarenko, A., et al.: CausalQA: A benchmark for causal question answering. In: Proceedings of COLING 2022, pp. 3296–3308. ICCL (2022). https://aclanthology.org/2022.coling-1.291

11. Carnot, M.L., et al.: On stance detection in image retrieval for argumentation. In: Proceedings of SIGIR 2023. ACM (2023). https://doi.org/10.1145/3539618.3591917

12. Chernodub, A., et al.: TARGER: Neural argument mining at your fingertips. In: Proceedings of ACL 2019, pp. 195–200. ACL (2019). https://doi.org/10.18653/v1/p19-3031

13. Chung, H.W., et al.: Scaling instruction-finetuned language models. arXiv (2022). https://doi.org/10.48550/arXiv.2210.11416

14. Conneau, A., et al.: Unsupervised cross-lingual representation learning at scale. In: Proceedings of ACL 2020, pp. 8440–8451. ACL (2020). https://doi.org/10.18653/v1/2020.acl-main.747

15. Cormack, G.V., Smucker, M.D., Clarke, C.L.A.: Efficient and effective spam filtering and re-ranking for large web datasets. Inf. Retrieval J. **14**(5), 441–465 (2011). https://doi.org/10.1007/s10791-011-9162-z

16. Devlin, J., Chang, M., Lee, K., Toutanova, K.: BERT: Pre-training of deep bidirectional transformers for language understanding. In: Proceedings of NAACL-HLT 2019, pp. 4171–4186. ACL (2019). https://doi.org/10.18653/v1/n19-1423

17. Elagina, D., Heizmann, B.A., Koch, M., Lahmann, G., Ortlepp, C.: Neville longbottom at Touché 2023: Image retrieval for arguments using ChatGPT, CLIP and IBM debater. In: Working Notes of CLEF 2023 - Conference and Labs of the Evaluation Forum. CEUR Workshop Proceedings. CEUR-WS.org (2023)

18. Fellbaum, C.: WordNet: An Electronic Lexical Database. MIT Press, Cambridge (1998)

19. Fröbe, M., et al.: Continuous integration for reproducible shared tasks with TIRA.io. In: Kamps, J., et al. (eds.) ECIR 2023. LNCS, vol. 13982, pp. 236–241. Springer, Cham (2023). https://doi.org/10.1007/978-3-031-28241-6_20

20. Gaden, A., Reinhold, B., Zeit-Altpeter, L., Rausch, N.: Evidence retrieval for causal questions using query expansion and reranking. In: Working Notes of CLEF 2023 - Conference and Labs of the Evaluation Forum. CEUR Workshop Proceedings. CEUR-WS.org (2023)

21. Heindorf, S., Scholten, Y., Wachsmuth, H., Ngonga Ngomo, A.C., Potthast, M.: CauseNet: Towards a causality graph extracted from the web. In: 29th ACM International Conference on Information and Knowledge Management (CIKM 2020), pp. 3023–3030. ACM (2020). https://doi.org/10.1145/3340531.3412763

22. Ke, G., et al.: LightGBM: a highly efficient gradient boosting decision tree. In: Proceedings of NeurIPS 2017, pp. 3146–3154. NeurIPS (2017). https://proceedings.neurips.cc/paper/2017/file/6449f44a102fde848669bdd9eb6b76fa-Paper.pdf

23. Kiesel, J., Potthast, M., Stein, B.: Dataset Touché22-image-retrieval-for-arguments (2022). https://doi.org/10.5281/zenodo.6786948

24. Kiesel, J., Potthast, M., Stein, B.: Dataset Touché23-image-retrieval-for-arguments (2023). https://doi.org/10.5281/zenodo.7497994

25. Lewis, M., et al.: BART: Denoising sequence-to-sequence pre-training for natural language generation, translation, and comprehension. In: Proceedings of ACL 2020, pp. 7871–7880. ACL (2020). https://doi.org/10.18653/v1/2020.acl-main.703

26. Liu, Y., et al.: RoBERTa: A robustly optimized BERT pretraining approach. arXiv preprint arXiv:1907.11692 (2019)

27. Lowe, D.G.: Distinctive image features from scale-invariant keypoints. Int. J. Comput. Vision **60**(2), 91–110 (2004). https://doi.org/10.1023/B:VISI.0000029664.99615.94

28. MacAvaney, S., Yates, A., Feldman, S., Downey, D., Cohan, A., Goharian, N.: Simplified data wrangling with ir_datasets. In: Proceedings of the 44th International ACM SIGIR Conference on Research and Development in Information Retrieval, SIGIR 2021, pp. 2429–2436. ACM (2021). https://doi.org/10.1145/3404835.3463254

29. Macdonald, C., Tonellotto, N., MacAvaney, S., Ounis, I.: PyTerrier: Declarative experimentation in Python from BM25 to dense retrieval. In: Proceedings of CIKM 2021, pp. 4526–4533. ACM (2021). https://doi.org/10.1145/3459637.3482013

30. Möbius, M., Enderling, M., Bachinger, S.: Jean-Luc Picard at Touché 2023: Comparing image generation, stance detection and feature matching for image retrieval for arguments. In: Working Notes of CLEF 2023 - Conference and Labs of the Evaluation Forum. CEUR Workshop Proceedings. CEUR-WS.org (2023)

31. Overwijk, A., Xiong, C., Callan, J.: ClueWeb22: 10 billion web documents with rich information. In: Proceedings of the 45th International ACM SIGIR Conference on Research and Development in Information Retrieval (SIGIR 2022), pp. 3360–3362. ACM (2022). https://doi.org/10.1145/3477495.3536321

32. Palotti, J.R.M., Scells, H., Zuccon, G.: TrecTools: An open-source Python library for information retrieval practitioners involved in TREC-like campaigns. In: Proceedings of SIGIR 2019, pp. 1325–1328. ACM (2019). https://doi.org/10.1145/3331184.3331399

33. Plenz, M., Buchmüller, R., Bondarenko, A.: Argument quality prediction for ranking documents. In: Working Notes of CLEF 2023 - Conference and Labs of the Evaluation Forum. CEUR Workshop Proceedings. CEUR-WS.org (2023)

34. Qi, P., Zhang, Y., Zhang, Y., Bolton, J., Manning, C.D.: Stanza: A python natural language processing toolkit for many human languages. In: Proceedings of ACL 2020, pp. 101–108. ACL (2020). https://doi.org/10.18653/v1/2020.acl-demos.14

35. Radford, A., et al.: Learning transferable visual models from natural language supervision. In: Proceedings of ICML 2021. Proceedings of Machine Learning Research, vol. 139, pp. 8748–8763. PMLR (2021). https://proceedings.mlr.press/v139/radford21a.html

36. Robertson, S.E., Walker, S., Jones, S., Hancock-Beaulieu, M., Gatford, M.: Okapi at TREC-3. In: Proceedings of TREC 1994. NIST Special Publication, vol. 500–225, pp. 109–126. NIST (1994)
37. Robertson, S.E., Zaragoza, H., Taylor, M.J.: Simple BM25 extension to multiple weighted fields. In: Proceedings of CIKM 2004, pp. 42–49. ACM (2004). https://doi.org/10.1145/1031171.1031181
38. Rombach, R., Blattmann, A., Lorenz, D., Esser, P., Ommer, B.: High-resolution image synthesis with latent diffusion models. In: Proceedings of CVPR 2022, pp. 10674–10685. IEEE (2022). https://doi.org/10.1109/CVPR52688.2022.01042
39. Schaefer, K.: Queen of swords at Touché 2023: Intra-multilingual multi-target stance classification using BERT. In: Working Notes of CLEF 2023 - Conference and Labs of the Evaluation Forum. CEUR Workshop Proceedings. CEUR-WS.org (2023)
40. Su, H., et al.: One embedder, any task: Instruction-finetuned text embeddings. arXiv (2022). 10.48550/arXiv.2212.09741
41. Sugiyama, A., Yoshinaga, N.: Data augmentation using back-translation for context-aware neural machine translation. In: Proceedings of DiscoMT@EMNLP 2019, pp. 35–44. ACL (2019). https://doi.org/10.18653/v1/D19-6504
42. Thakur, N., Reimers, N., Rücklé, A., Srivastava, A., Gurevych, I.: BEIR: A heterogeneous benchmark for zero-shot evaluation of information retrieval models. In: Proceedings of NeurIPS 2021. NeurIPS (2021). https://datasets-benchmarks-proceedings.neurips.cc/paper/2021/hash/65b9eea6e1cc6bb9f0cd2a47751a186f-Abstract-round2.html
43. Vamvas, J., Sennrich, R.: X-stance: A multilingual multi-target dataset for stance detection. In: Proceedings of SwissText/KONVENS 2020. CEUR-WS.org (2020). https://ceur-ws.org/Vol-2624/paper9.pdf
44. Wachsmuth, H., et al.: Computational argumentation quality assessment in natural language. In: Proceedings of EACL 2017, pp. 176–187. ACL (2017). https://doi.org/10.18653/v1/e17-1017
45. Xie, X., et al.: Grid-based evaluation metrics for web image search. In: Proceedings of WWW 2019, pp. 2103–2114. ACM (2019). https://doi.org/10.1145/3308558.3313514
46. Zhou, D., Bousquet, O., Lal, T.N., Weston, J., Schölkopf, B.: Learning with local and global consistency. In: Proceedings of NIPS 2003, pp. 321–328. MIT Press (2003). https://proceedings.neurips.cc/paper/2003/hash/87682805257e619d49b8e0dfdc14affa-Abstract.html

A   Zero-Shot Prompts

The zero-shot prompts used for the stance prediction baselines are given in Listing 1 (for Task 1, see Sect. 3) and in Listing 2 (for Task 2, see Sect. 4).

```
Given a query, predict the stance of a given text. The stance should be one
    of the following four labels:
PRO: The text contains opinions or arguments in favor of the query "<query>".
CON: The text contains opinions or arguments against the query "<query>".
NEU: The text contains as many arguments in favor of as it contains against
    the query "<query>".
UNK: The text is not relevant to the query "<query>", or it only contains
    factual information.

Text: <summary>
```

Listing 1: Zero-shot prompt to predict the stance of a document towards a query (Task 1). The placeholder `<query>` is replaced by the topic titles, and `<summary>` for a short summary of the retrieved document's text. The `UNK` label is mapped to `NO`.

```
Given a query, predict the stance of a given text. The stance should be one
    of the following four labels:
SUP: According to the text, <cause> causes <effect>.
REF: According to the text, <cause> does not cause <effect>.
UNK: The text is not relevant to <cause> and <effect>.

Text: <summary>
```

Listing 2: Zero-shot prompt to predict the causal stance of a document towards a query (Task 2). The placeholders `<cause>` and `<effect>` are replaced with the query's cause and effect entities, and `<summary>` with a short summary of the retrieved document's text. The `UNK` label is mapped to `NO`. The `NEU` label is not considered in the prompt.

Table 9. Relevance results of all runs submitted to Task 1: Argument Retrieval for Controversial Questions. Reported are the mean nDCG@10 and the 95% confidence intervals. The baseline Puss in Boots is shown in bold.

Team	Run Tag	nDCG@10		
		Mean	Low	High
Puss in Boots	**ChatNoir** [7]	**0.834**	**0.791**	**0.875**
Renji Abarai	stance_ChatGPT	0.747	0.687	0.812
Renji Abarai	stance-certainNO_ChatGPT	0.746	0.678	0.810
Renji Abarai	ChatGPT_mmGhl	0.718	0.653	0.775
Renji Abarai	ChatGPT_mmEQhl	0.718	0.650	0.779
Renji Abarai	meta_qual_score	0.712	0.641	0.782
Renji Abarai	baseline	0.708	0.632	0.775
Renji Abarai	meta_qual_prob	0.697	0.622	0.765

Table 10. Quality results of all runs submitted to Task 1: Argument Retrieval for Controversial Questions. Reported are the mean nDCG@10 and the 95% confidence intervals. The baseline Puss in Boots is shown in bold.

Team	Run Tag	nDCG@10		
		Mean	Low	High
Puss in Boots	**ChatNoir** [7]	**0.831**	**0.786**	**0.873**
Renji Abarai	stance_ChatGPT	0.815	0.764	0.862
Renji Abarai	stance-certainNO_ChatGPT	0.811	0.754	0.863
Renji Abarai	ChatGPT_mmEQhl	0.789	0.730	0.846
Renji Abarai	ChatGPT_mmGhl	0.789	0.731	0.842
Renji Abarai	meta_qual_prob	0.774	0.712	0.830
Renji Abarai	meta_qual_score	0.771	0.710	0.832
Renji Abarai	baseline	0.766	0.698	0.823

B   Full Evaluation Results of Touché 2023: Argument and Causal Retrieval

Table 11. Relevance results of all runs submitted to Task 2: Evidence Retrieval for Causal Questions. Reported are the mean nDCG@5 and the 95% confidence intervals. The baseline Puss in Boots is shown in bold.

Team	Run Tag	nDCG@5		
		Mean	Low	High
He-Man	no_expansion_rerank	0.657	0.564	0.740
Puss In Boots	**ChatNoir** [7]	**0.585**	**0.503**	**0.673**
He-Man	gpt_expansion_rerank	0.374	0.284	0.469
He-Man	causenet_expansion_rerank	0.268	0.172	0.368

Table 12. On-topic relevance results of all runs submitted to Task 3: Image Retrieval for Argumentation. Reported are the mean precision@10 and the 95% confidence intervals. The baseline Minsc is shown in bold.

Team	Run Tag	Precision@10		
		Mean	Low	High
Neville Longbottom	clip_chatgpt_args.raw	0.785	0.714	0.852
Hikaru Sulu	Keywords	0.770	0.704	0.831
Neville Longbottom	clip_chatgpt_args.debater	0.684	0.601	0.764
Hikaru Sulu	Topic-title	0.664	0.581	0.746
Neville Longbottom	bm25_chatgpt_args.raw	0.572	0.510	0.636
Jean-Luc Picard	BM25 Baseline	0.536	0.458	0.608
Jean-Luc Picard	No stance detection	0.523	0.442	0.598
Jean-Luc Picard	Text+image text stance detection	0.502	0.429	0.573
Jean-Luc Picard	Text stance detection	0.498	0.419	0.567
Neville Longbottom	bm25_chatgpt_args.diff	0.442	0.378	0.507
Neville Longbottom	bm25_chatgpt_args.debater	0.416	0.350	0.481
Minsc	**Aramis**	**0.376**	**0.310**	**0.442**
Jean-Luc Picard	Image text stance detection	0.369	0.301	0.433

Table 13. Argumentativeness results of all runs submitted to Task 3: Image Retrieval for Argumentation. Reported are the mean precision@10 and the 95% confidence intervals. The baseline Minsc is shown in bold.

Team	Run Tag	Precision@10		
		Mean	Low	High
Hikaru Sulu	Topic-title	0.350	0.291	0.415
Neville Longbottom	clip_chatgpt_args.debater	0.341	0.271	0.410
Neville Longbottom	clip_chatgpt_args.raw	0.338	0.273	0.404
Hikaru Sulu	Keywords	0.335	0.275	0.395
Jean-Luc Picard	No stance detection	0.292	0.220	0.367
Neville Longbottom	bm25_chatgpt_args.raw	0.274	0.211	0.338
Jean-Luc Picard	Text+image text stance detection	0.272	0.208	0.339
Jean-Luc Picard	BM25 Baseline	0.268	0.198	0.334
Jean-Luc Picard	Text stance detection	0.262	0.198	0.325
Neville Longbottom	bm25_chatgpt_args.diff	0.240	0.176	0.309
Neville Longbottom	bm25_chatgpt_args.debater	0.201	0.146	0.263
Jean-Luc Picard	Image text stance detection	0.196	0.149	0.247
Minsc	**Aramis**	**0.194**	**0.144**	**0.248**

Table 14. Stance relevance results of all runs submitted to Task 3: Image Retrieval for Argumentation. Reported are the mean precision@10 and the 95% confidence intervals. The baseline Minsc is shown in bold.

Team	Run Tag	Precision@10		
		Mean	Low	High
Neville Longbottom	clip_chatgpt_args.raw	0.222	0.174	0.268
Neville Longbottom	clip_chatgpt_args.debater	0.216	0.155	0.281
Hikaru Sulu	Topic-title	0.185	0.149	0.221
Hikaru Sulu	Keywords	0.179	0.140	0.219
Neville Longbottom	bm25_chatgpt_args.raw	0.166	0.127	0.208
Jean-Luc Picard	No stance detection	0.162	0.118	0.206
Neville Longbottom	bm25_chatgpt_args.diff	0.150	0.108	0.196
Jean-Luc Picard	Text+image text stance detection	0.144	0.108	0.185
Jean-Luc Picard	BM25 Baseline	0.141	0.105	0.183
Jean-Luc Picard	Text stance detection	0.136	0.101	0.177
Neville Longbottom	bm25_chatgpt_args.debater	0.128	0.091	0.170
Minsc	**Aramis**	**0.102**	**0.076**	**0.129**
Jean-Luc Picard	Image text stance detection	0.098	0.067	0.132

Author Index

A

Adams, Griffin 370
Aidos, Helena 343
Alam, Firoj 251
Alkhalifa, Rabab 440
Altun, Bahadir 161
Amigó, Enrique 316
Andrei, Alexandra-Georgiana 370
Araujo, Lourdes 174
Arslan, Mustafa Bora 161
Augereau, Olivier 482
Azarbonyad, Hosein 482
Azizov, Dilshod 251

B

Barriere, Valentin 507
Barrón-Cedeño, Alberto 251
Ben Abacha, Asma 370
Bergamaschi, Roberto 343
Bevendorff, Janek 459
Bilal, Iman 440
Birolo, Giovanni 343
Bloch, Louise 370
Bondarenko, Alexander 507
Bonnet, Pierre 416
Borchert, Florian 135
Borkakoty, Hsuvas 440
Borrego-Obrador, Ian 459
Bosser, Anne-Gwenn 397
Botella, Christophe 416
Braker, Jan 186
Brüngel, Raphael 370

C

Camacho, David 60
Camacho-Collados, Jose 440
Carrillo-de-Albornoz, Jorge 316
Caselli, Tommaso 251
Cavalla, Paola 343

D

Da San Martino, Giovanni 251
Dagliati, Arianna 343
de Carvalho, Mamede 343
Del Moro, Mirko 3
Deneu, Benjamin 416
Denton, Tom 416
Deshayes, Jérôme 370
Deveaud, Romain 440
Dhanani, Farhan 148
Di Camillo, Barbara 85, 343
Di Nunzio, Giorgio Maria 15, 343
Dogariu, Mihai 370
Doucet, Antoine 276
Drăgulinescu, Ana-Maria 370
Duque, Andres 174

E

Eggel, Ivan 416
El-Ebshihy, Alaa 21, 440
Elsayed, Tamer 251
Ermakova, Liana 397, 482
Espinosa-Anke, Luis 440
Estopinan, Joaquim 416
Eyuboglu, Ahmet Bahadir 161

C (right column, continued)

Chamidullin, Rail 416
Chatzakou, Despoina 72
Cheema, Gullal S. 251
Chinea-Ríos, Mara 459
Chiò, Adriano 343
Clifton, Ann 48
Coman, Ioan 370
Constantin, Mihai-Gabriel 370
Correia, Joana 48
Coustaty, Mickaël 276
Cremonesi, Paolo 97
Crestani, Fabio 294

A. Arampatzis et al. (Eds.): CLEF 2023, LNCS 14163, pp. 531–534, 2023.
https://doi.org/10.1007/978-3-031-42448-9

F
Fabregat, Hermenegildo   174
Faggioli, Guglielmo   343
Fariselli, Piero   343
Farré-Maduell, Eulália   227
Ferrari Dacrema, Maurizio   97
Ferro, Nicola   85, 97, 343
Fink, Tobias   21
Franco-Salvador, Marc   121, 459
Frenda, Simona   34
Friedrich, Christoph M.   370
Fröbe, Maik   459, 507

G
Galassi, Andrea   3, 251
Galuščáková, Petra   21, 440
García Dominguez, Jose Manuel   343
García Seco de Herrera, Alba   370
Garmash, Ekaterina   48
Gasco, Luis   227
Glotin, Hervé   416
Goëau, Hervé   416
Goeuriot, Lorraine   21, 440
González, José Ángel   121
Gonzalez-Saez, Gabriela   21, 440
Gonzalo, Julio   316
Gromicho, Marta   343
Guazzo, Alessandro   85, 343

H
Hagen, Matthias   507
Halvorsen, Pål   370
Hamdi, Ahmed   276
Haouari, Fatima   251
Hasanain, Maram   251
Heini, Annina   459
Hemamou, Léo   507
Hicks, Steven A.   370
Hrúz, Marek   416
Huang, Hong-Yi   198
Huertas-Tato, Javier   60
Huet, Stéphane   482

I
Idrissi-Yaghir, Ahmad   370
Iommi, David   21
Ionescu, Bogdan   370

J
Jat, Sharmistha   48
Jatowt, Adam   397
Jha, Debesh   370
Joly, Alexis   416
Jones, Rosie   48

K
Kadurin, Artur   109
Kahl, Stefan   416
Kamps, Jaap   482
Karatzas, Dimosthenis   276
Karlgren, Jussi   48
Katsimpras, Georgios   227
Kiesel, Johannes   507
Klinck, Holger   416
Kochkina, Elena   440
Kocián, Matěj   276
Kompatsiaris, Ioannis   72
Konstantinou, Apostolos   72
Kovalev, Vassili   370
Krallinger, Martin   227
Kredens, Krzysztof   459
Krithara, Anastasia   227
Kutlu, Mucahid   161, 251

L
Larcher, Théo   416
Leblanc, Cesar   416
Li, Chengkai   251
Liakata, Maria   440
Lima López, Salvador   227
Lin, Jimmy   209
Llorca, Ignacio   135
Longato, Enrico   85, 343
Longhin, Francesca   85
Losada, David E.   294
Loureiro, Daniel   440
Luck, Simon   507

M
Madeira, Sara C.   343
Manera, Umberto   343
Marchesin, Stefano   343
Marcos, Diego   416
Martín, Alejandro   60
Martinez-Romo, Juan   174
Martín-Rodilla, Patricia   294
Matas, Jiří   276

Mayerl, Maximilian   459
Menotti, Laura   343
Miller, Tristan   397
Morante, Roser   316
Mulhem, Philippe   21, 440
Müller, Henning   370, 416

N
Nakov, Preslav   251
Nentidis, Anastasios   227

P
Paliouras, Georgios   227
Palma Preciado, Victor Manuel   397
Papachrysos, Nikolaos   370
Parapar, Javier   294
Pasin, Andrea   97
Patel, Yash   276
Patti, Viviana   34
Pęzik, Piotr   459
Picek, Lukáš   416
Piroi, Florina   21, 440
Planqué, Robert   416
Plaza, Laura   316
Popel, Martin   440
Popescu, Adrian   370
Potthast, Martin   459, 507
Prokopchuk, Yuri   370

R
Radzhabov, Ahmedkhan   370
Rafi, Muhammad   148
Rangel, Francisco   459
Ravenet, Brian   507
Reimer, Jan Heinrich   507
Riegler, Michael A.   370
Rosso, Paolo   34, 121, 316, 459
Rückert, Johannes   370
Ruggeri, Federico   3, 251

S
Sakhovskiy, Andrey   109
SanJuan, Eric   482

Sarvazyan, Areg Mikael   121
Schäfer, Henning   370
Schapranow, Matthieu-P.   135
Schlatt, Ferdinand   507
Schöler, Johanna   370
Schreieder, Tobias   186
Semenova, Natalia   109
Servajean, Maximilien   416
Servan, Christophe   440
Sidorov, Grigori   397
Silvello, Gianmaria   343
Šimsa, Štěpán   276
Skalický, Matyáš   276
Snider, Neal   370
Sonmezer, Ekrem   161
Spina, Damiano   316
Stamatatos, Efstathios   459
Ştefan, Liviu-Daniel   370
Stein, Benno   459, 507
Storås, Andrea M.   370
Struß, Julia Maria   251
Šulc, Milan   276, 416

T
Tahir, Muhammad Atif   148
Tanaka, Edgar   48
Tavazzi, Eleonora   343
Tavazzi, Erica   343
Tayyar Madabushi, Harish   440
Thambawita, Vajira   370
Theodosiadou, Ourania   72
Trescato, Isotta   343
Tsikrika, Theodora   72
Tudosie, Serban Cristian   3
Tutubalina, Elena   109

U
Uřičář, Michal   276

V
Vannoni, Francesco   3
Vellinga, Willem-Pier   416

Vettoretti, Martina   343
Vezzani, Federica   15
Vrochidis, Stefanos   72

W
Wiegmann, Matti   459
Wolska, Magdalena   459
Wu, Shih-Hung   198

X
Xie, Yuqing   209

Y
Yetisgen, Meliha   370
Yim, Wen-Wai   370

Z
Zaghouani, Wajdi   251
Zangerle, Eva   459
Zhong, Wei   209
Zhu, Winstead   48
Zubiaga, Arkaitz   440

Printed in the United States
by Baker & Taylor Publisher Services